Geotechnical, Geological and Earthquake Engineering

Volume 51

More information about this series at http://www.springer.com/series/6011

George A. Papagiannopoulos
George D. Hatzigeorgiou
Dimitri E. Beskos

Seismic Design Methods for Steel Building Structures

Springer

George A. Papagiannopoulos
School of Sciences and Technology,
Structural Technology and Applied
Mechanics Laboratory
Hellenic Open University
Patras, Greece

George D. Hatzigeorgiou
School of Sciences and Technology,
Structural Technology and Applied
Mechanics Laboratory
Hellenic Open University
Patras, Greece

Dimitri E. Beskos
Department of Disaster Mitigation
for Structures
College of Civil Engineering
Tongji University
Shanghai, China

Department of Civil Engineering
University of Patras
Patras, Greece

ISSN 1573-6059 ISSN 1872-4671 (electronic)
Geotechnical, Geological and Earthquake Engineering
ISBN 978-3-030-80686-6 ISBN 978-3-030-80687-3 (eBook)
https://doi.org/10.1007/978-3-030-80687-3

This Springer imprint is published by the registered company Springer Nature Switzerland AG.
The registered company address is: Gewerbestrasse 11, 6330 Cham, Switzerland

To our families

Foreword

Over the last 30 years, a variety of well-engineered steel structures, which were designed according to building codes that were valid during their construction time, experienced severe damage worldwide from strong earthquakes in Northridge, California, 1994; Kobe, Japan, 1995; and Chi-Chi, Taiwan, 1999, among other places. Most of this damage was the result of excessive seismic displacements and rotations of member connections, and many research programs were launched during the last three decades to improve the seismic design of steel structures. Remarkable improvements have been achieved in both analysis and design methodologies by engineers and scholars of experience and insight, and this book by G. A. Papagiannopoulos, G. D. Hatzigeorgiou, and D. E. Beskos offers a comprehensive educational panorama of these recent technical developments.

The book, following an introduction to the seismic design of structures in association with the framework of performance-based engineering, offers an in-depth description of the two prevailing methods of seismic design—that is, the force-based and the displacement-based design methods as applied to steel structures with the former method paving the basis of Eurocode 8 (EC8). The book is written with a reader-centric philosophy where, while the intellectual framework of ductility-based and energy-based plastic designs is treated with rigor, educational, real-world examples are offered at the end of every chapter. While the thrust of the book is on *Seismic Design Methods*, as its title indicates, the book also covers advanced inelastic analysis techniques which account for geometric nonlinearities and damage. The book offers a balanced presentation of the basic framework of seismic design methods together with a comprehensive review of the computation techniques for steel structures under earthquake loads. This enables the reader to appreciate the way in which a computer software for nonlinear inelastic analysis delivers its solution, without necessarily investigating every detail of the computation.

The book comes in a timely manner since the majority of structural steel produced today comes from recycled materials. This recycling usage renders structural steel cost-effective, which is in-line with the current trends for sustainable engineering.

Addy Family Centennial Professor in
Civil Engineering, Southern Methodist
University, Dallas, TX, USA

Nicos Makris

Preface

This book presents 11 seismic design methods for steel building structures that have been developed during the last 25 years or so. The emphasis is on the description and application of these design methods up to the phase of the optimum section selection of the members (beams, columns, and braces) of the building that ensure safety and economy. Design of connections, floor slabs, foundations, and other building design details are not considered here. Conceptual design is also not discussed in this book. Thus, for a given type of structure, which is usually the product of conceptual design, the methods presented here start with the load determination, especially the seismic one, proceed with the structural response to those loads, and conclude with the selection of member sizes in an iterative manner capable of satisfying through strength and deformation checking established safety and economy requirements. All the design methods satisfy capacity design rules and follow performance-based design principles.

The structural material used plays an important role in seismic design. Steel is the ideal material for seismically resistant structures because of its high strength and ductility (high dissipation of energy during inelastic deformation). However, those two material properties are not enough to ensure strength, stability, and ductility at the structural level. The structural members should be appropriately connected and sized in order for the structure to develop the desirable strength, stability, and dissipative properties. A properly seismically designed steel building structure has definitive advantages over a reinforced concrete one, as the reader can find out in many books on steel structures.

The book consists of 13 chapters with the first two dealing with fundamental concepts, principles, and methods for structural seismic design and structural seismic analysis, respectively, with the emphasis being placed on steel building structures. These two chapters serve the purpose of providing the necessary background information for the remaining 11 chapters of the book while constituting a brief up-to-date review on the subject of structural earthquake engineering emphasizing steel building structures. Chapters 3 and 4 describe the two most important and well-known methods of seismic design, the force-based and the displacement-based design methods, respectively, as applied to steel structures with the former one presented on the basis of Eurocode 8 (EC8). Chapter 5 describes the hybrid force-displacement seismic design method, which comes as a result of combining the advantages of the two previous design methods. Chapters 6 and 7 present the ductility-based and energy-based plastic design methods, respectively, which succeed in achieving global collapse mechanisms in steel structures. Chapters 8 and 9 develop force-based seismic design methods for steel structures that employ equivalent modal damping ratios and modal behavior factors, respectively, instead of a single value for the behavior factor for all modes as it is the case of the force-based design of EC8. Chapters 10 and 11 employ advanced nonlinear time history analysis methods for steel structural design with the latter utilizing in addition the concept of damage. Finally, Chaps. 12 and 13 present two special design methods for steel structures, which introduce external energy dissipation mechanisms to reduce seismic forces on the structure itself, namely, seismic base isolation and supplemental dampers.

For all these 11 design methods, two to three numerical examples are presented in detail to illustrate every method and demonstrate its advantages as applied to the seismic design of steel building structures.

This book has a unique characteristic in comparison with all the existing books on seismic design of steel structures: it presents not just one seismic method but 11 methods with many numerical examples worked in detail. Thus, the reader can easily understand in depth the basic principles behind every method, the similarities and differences between the methods, and their advantages and disadvantages.

The book is a valuable source of information for both practicing engineers and researchers working in the area of seismic design of steel building structures. It can be also used as a text for graduate-level courses of MS and PhD programs. The authors hope that this book will be a useful tool for seismic analysis and design of steel building structures.

Closing this preface, the authors would like to express their thanks and appreciation to all those colleagues and friends who have helped them in various ways in this scientific endeavor: Professor N. Makris for reading the whole book and writing the foreword just before this preface; Professors S. A. Anagnostopoulos, T. L. Karavasilis, G. S. Kamaris, and K. A. Skalomenos as well as Drs A. S. Tzimas, A. A. Vasilopoulos, E. V. Muho, and N. A. Kalapodis for reading parts of the book and offering valuable suggestions; and Professors E. M. Marino and D. R. Sahoo for their help in connection with two examples. Finally, many thanks are due to the publishing team of Springer for their help and cooperation during all the phases of the production of this book.

Patras, Greece
March 2021

George A. Papagiannopoulos
George D. Hatzigeorgiou
Dimitri E. Beskos

Acknowledgments

The authors would like to acknowledge with thanks the following organizations which have granted to them permission for reprinting figures and tables from their publications in this book.

AISC = American Institute of Steel Construction
CAEE = Canadian Association for Earthquake Engineering
ECCS = European Convention for Constructional Steelwork
ICC = International Code Council
IITDCE = Indian Institute of Technology Delhi, Civil Engineering, India
MSP = Mathematical Sciences Publishers
PAEE = Portuguese Association for Earthquake Engineering
UMCEE = University of Michigan, Civil and Environmental Engineering, USA
UPCE = University of Patras, Civil Engineering, Greece

Contents

Chapter 1
Fundamentals of Seismic Structural Design

Abstract This chapter presents fundamental aspects of seismic structural design and together with the next chapter devoted to seismic structural analysis serve as an introduction and basic reference to all subsequent chapters dealing with specific seismic design methods. After some historical notes on the development of seismic design methods, the basic characteristics of earthquakes pertaining to their generation and propagation, magnitude and intensity, duration, attenuation, frequency content and their distance from the source are briefly described. Basic concepts in seismic design, such as inelastic deformation, energy of dissipation, ductility, damage, behavior factor and overstrength, torsional effects and response combination rules are presented and discussed. Capacity design rules and their importance in design are explained. Performance-based design in its two forms, deterministic and probabilistic is briefly discussed and its importance in design is stressed. Performance levels defined by pairs of seismic hazard and performance objectives are stated. Finally, related to performance-based design subjects, such as seismic intensity measures, fragility curves and incremental dynamic analysis are also briefly presented.

Keywords Seismic structural design · Characteristics of earthquakes · Ductility · Energy of dissipation · Behavior factor · Damage · Capacity design · Performance-based design

1.1 Introduction

Seismic design of structures aims at designing structures capable of resisting seismic forces and limiting displacements. This design is done in accordance with seismic codes, which are revised and improved from time to time on the basis of research involving experimental and analytical/numerical work. This research is conducted in order to satisfy needs arising from observations and lessons learned after major catastrophic earthquakes. This book deals with the seismic design of building structures in general and building structures made of steel in particular.

G. A. Papagiannopoulos et al., *Seismic Design Methods for Steel Building Structures*, Geotechnical, Geological and Earthquake Engineering 51,
https://doi.org/10.1007/978-3-030-80687-3_1

Seismic design of structures started at the beginning of the twentieth century and was based on the assumption that seismic forces were horizontal forces equal to a small percentage of the weight of the structure that could be applied statically on the elastically behaving structure. With the works of Housner (1959), Veletsos and Newmark (1960) and Newmark and Hall (1982), the concepts of elastic and inelastic response spectra, ductility and energy of dissipation capacity were developed on the basis of a dynamic rather than a static framework and formed the basis of the first generation of seismic design codes. The advent of high speed computers and their capability to produce more realistic dynamic response results, led to the development of the second generation of seismic design codes by the end of the 1970s. The third generation of seismic design codes came as a result of the lessons learned from the Northridge and Kobe catastrophic earthquakes in 1994 and 1995, respectively. Because of the great economic losses during those earthquakes, more emphasis was placed on controlling deformation and damage of the structure. In addition, a new design philosophy, the performance-based seismic design (PBSD) one was introduced, which requires the satisfaction of multi-level design objectives associated with multi-level seismic actions. This design philosophy, which aims at achieving a good balance between safety and economy, will be discussed in more detail in Sect. 1.6.

The publication of FEMA 445 (2006) marks the beginning of the development of the fourth generation of seismic design codes, more descriptively presented in FEMA 695 (2009) and FEMA 750 (2009). The new concepts developed here have to do with the estimation of losses in terms of casualties, repair costs and functionality interruption as well as with the assessment of structural performance in terms of response and uncertainty at the level of earthquake hazard. Thus, emphasis was placed on designing not only safe but resilient structures as well.

In addition to improvements in analysis and performance criteria for a better prediction and control of the structural response during the aforementioned conventional design methods, other design approaches of an unconventional character have been also developed. These approaches include: (1) the various structural isolation ones, which through rubber or friction bearings reduce the seismic forces transmitted to a structure, (2) the various passive energy dissipation ones, which through frictional or viscous/viscoelastic damping devices reduce the seismic structural response and (3) the various active control ones, which control seismic response by modifying structural characteristics during the seismic excitation. The present book discusses a number of conventional seismic design methods, such as the force-based design (FBD) and the displacement-based design (DBD) methods, where the seismic forces and displacements are the main design parameters, respectively, as well as some of the unconventional seismic design methods of the structural isolation and the passive energy dissipation type ones mentioned above.

Closing this introduction, a number of review type of articles on seismic design of building structures in general and steel structures in particular can be referred. Thus, the works of Anagnostopoulos (1997), Roesset (1997), Kappos (2002) and Fajfar (2018) from the first and those of Roeder and MacRae (1997), Hamburger (2009),

Elghazouli (2010) and Landolfo (2018) from the second category of structures can be mentioned.

1.2 Basic Characteristics of Earthquakes

Most of the material in this section is based on the articles of Kappos (2002) and Hamburger (2009).

The crust of the earth consists of a number of small and large tectonic plates. Zones of weakness of the crust near the plate boundaries are called faults. Because of the movement of the plates, rupture and slip along the fault can suddenly occur resulting in stress wave propagation in all directions and ground motion (earthquake) in the area around the fault. This phenomenon lasts until all the accumulated stress energy has been released and a new stable state has been achieved.

Faults that exhibit a lateral horizontal displacement along their axis are called strike slip faults. Faults that exhibit a vertical slip along their axis are called normal faults. The propagating seismic waves are either body or surface waves. Body waves are either longitudinal (P-waves) or shear (S-waves) with particle motion along or perpendicular to the wave propagation direction, respectively. Surface waves propagate near the surface of the earth and are of the Rayleigh or Love type. The ground motions are usually recorded by accelerographs, instruments that record ground accelerations as functions of time.

The magnitude of an earthquake is based either on the fault strength or the released energy. The Richter magnitude M_L is defined as the logarithm of the maximum seismic amplitude corrected to a distance of 100 km or as a function of the logarithm of the released energy. A value of $M_L \geq 5$ usually signifies earthquakes that may cause structural damage. Another magnitude is the moment magnitude M_w based on the seismic moment expressing the deformation at the fault. This is the most frequently used index for earthquake magnitude. The intensity of an earthquake is empirically measured by its effect on humans, buildings and the environment in general. The most well-known earthquake intensity measure is the modified Mercalli intensity index I_{MM}.

As previously mentioned, strong ground motions are recorded as accelerations versus time for all three motion components (two horizontal and one vertical) and are available from databanks mostly in the USA, Japan and Europe. Figure 1.1 shows such three components of the Nishi-Akashi accelerogram from the 1995 Kobe earthquake. The most widely used such seismic databanks are those of PEER (2009) and COSMOS (2013). The maximum acceleration value of an accelerogram (ground acceleration versus time) is called peak ground acceleration (*PGA*) and is frequently used in hazard estimation. By numerically integrating with respect to time once and twice an accelerogram, one can determine the ground velocity and displacement versus time, respectively. The peak ground velocity (*PGV*) and peak ground displacement (*PGD*) are better indicators of structural damage than the *PGA*.

Other earthquake aspects of importance for assessing seismic hazard are:

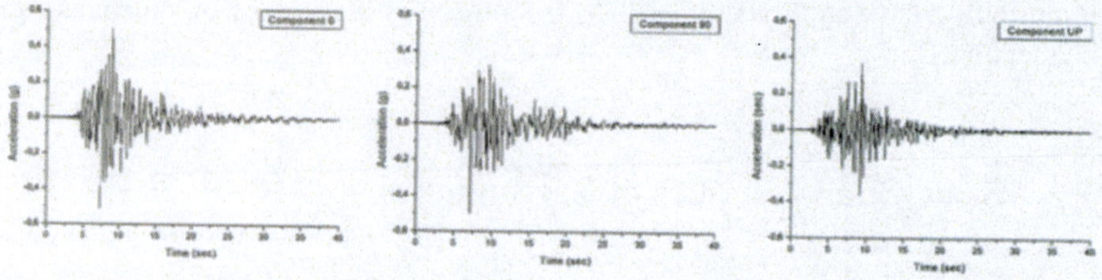

Fig. 1.1 The three components of the Nishi-Akashi accelerogram (1995 Kobe earthquake)

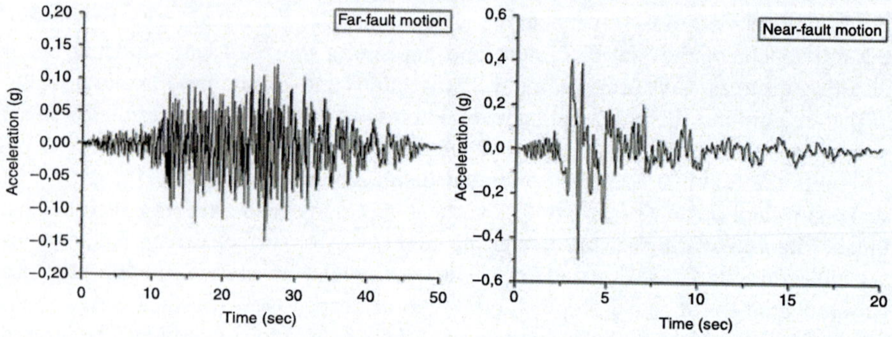

Fig. 1.2 Far-fault (left) and neat-fault (right) seismic motions

1. The proximity of a seismic record to the source, which categorizes seismic motions as far- and near-fault ones. Far-fault seismic motions are recorded at distances larger than 20 km and up to about 100 km from the fault. Most of the seismic codes assume far-fault seismic motions to describe the seismic loads and for this reason they are also called ordinary motions. Near-fault seismic motions are recorded at distances less than 15 km from the fault and are qualitatively quite different than the far-fault ones. Near-fault motions are characterized by strong pulses usually at the start of the motion and this inflicts considerable damage to the structure. Figure 1.2 depicts typical far-fault and near-fault seismic motions clearly showing their differences.

2. The attenuation of an earthquake, which is defined as the decrease of its amplitude with distance during its propagation away from the seismic source. This does not imply that as the distance from the source increases, the destructiveness of an earthquake always decreases because site effects may amplify the seismic amplitude. Attenuation relationships are empirical expressions providing the magnitude, intensity, or *PGA* in terms of the epicentral distance and focal depth of the earthquake.
3. The duration of an earthquake, which affects the seismic response of a structure. Short duration motions usually have a duration less than 20 s, while long duration motions a duration more than 20 s and up to 150 s. It has been found that longer duration earthquakes lead to higher maximum values of displacements and lower

structural collapse capacity than shorter duration earthquakes (Tirca et al. 2015; Chandramohan et al. 2016; Bravo-Haro and Elghazouli 2018).

4. The directivity of an earthquake, i.e., the direction of the fault rupture, which strongly affects the generated seismic motion. Thus, the motion recorded at a site in the rupture propagation direction, has a higher amplitude and shorter duration than the motion recorded at an equidistant site in the opposite direction.
5. The frequency content of an earthquake, which is usually characterized by a single parameter, the predominant period T_p corresponding to the predominant frequency f_p in the Fourier transformed earthquake signal. However, work by Rathje et al. (1998) has revealed that the traditional parameter T_p has a large uncertainty in its prediction expression in terms of earthquake magnitude, distance and site dependence and the mean period T_m appears to be a better (more reliable) frequency content parameter.
6. The site of an earthquake, which can significantly affect earthquake ground motions. Site effects have to do with the material properties of the soil layers (elastic moduli, density, damping, nonlinearities) and the topography of the site (hills and valleys). Thus, material and geometric aspects of the site can modify seismic waves during their propagation and amplify or deamplify those waves. More on the subject of site effects can be found in the book of Kramer (1996). Recently, Hashash et al. (2018), through extensive parametric studies and large scale simulation involving linear and nonlinear analyses have developed linear and nonlinear site amplification functions for the response spectrum and smooth Fourier amplitude spectrum for two regions in the USA. Concerning the soil material behavior in site response analyses, studies by Kim et al. (2016) have shown that equivalent linear analyses provide very good results only for small strains and that for large strain cases nonlinear analyses should be used.
7. The sequence of earthquakes, consisting of the main-shock and its after-shocks. Thus, a structure that has been damaged by the main earthquake, is weaker when subjected to subsequent earthquakes because the time between seismic events is usually not enough for repairing that structure. That means that the structure has to be designed for a sequence of earthquakes rather than a single earthquake. One can mention here the works of Fragiacomo et al. (2004) and Loulelis et al. (2012) dealing with the seismic response of steel frames to repeated earthquakes.

All the above aspects related to earthquakes are essential in assessing seismic hazard at a particular site and hence in defining seismic design loads, especially in connection with the construction of design spectra, discussed in the next chapter. Design seismic loads for a structure are based on ground motions that have a certain probability of exceedance during the lifetime of the structure (usually 50 years for buildings) or a certain return period in years. One can consult Sect. 1.6 for more details.

1.3 Basic Concepts in Seismic Design

Structural design for loads other than seismic ones, such as dead and live loads, or wind and snow loads, which are or considered to be static, assumes elastic structural behavior until the loads reach their maximum values. Thus, no damage is considered. On the contrary, seismic structural design assumes that damage can occur under seismic forces due to inelastic deformation. In other words, it is accepted that only a part of the inputted seismic energy will create structural vibration and that the remaining part will be dissipated by inelastic deformation and damage. The amount of this energy of dissipation and the resulting damage depend on the intensity of the earthquake. More specifically, it is accepted in seismic design that (1) no damage occurs in structural and non-structural elements for the frequent minor earthquake, (2) no damage occurs in structural elements but non-structural damage can occur for occasional moderate earthquakes, (3) damage can occur in both structural and non-structural elements but this damage is not very serious and structural collapse is avoided for rare major earthquakes.

One should note that structural and non-structural elements refer to skeletal members and partitions or cladding or various facilities, respectively. Furthermore, the situation described by the third state above, is called ultimate limit state (ULS). Thus, the acceptance of damage in seismic design results in seismic forces smaller than the ones that would act on the structure if that were to remain elastic even during the ULS. Smaller seismic design forces lead to lighter and thus more economic structures. For structures where any damage cannot be tolerated, even for major earthquakes, as it is, e.g., the case of nuclear power plants, the design is elastic because full safety is much more important than economy.

It is thus apparent that in seismic design the most important performance objectives are the amount of damage and the associated repair costs. Unfortunately, damage control cannot be easily achieved in seismic design because it has to do with acceptable risk as a result of the unpredictable nature of seismic forces. Thus, the goal on seismic design is to design a structure with a capacity larger than or equal to its seismic demand. Seismic demand denotes the effect the earthquake imposes on the structure, i.e., the seismic response of the structure to the earthquake, while seismic capacity denotes the ability of the structure to resist seismic effects by its strength, stiffness and ductility.

Coming back to the energy of dissipation concept, one can categorize structures to be designed seismically, as dissipative and low or non-dissipative ones, depending on their capacity to dissipate energy. Figure 1.3 (Landolfo et al. 2017) shows the lateral load F versus lateral displacement δ of two steel plane moment resisting frames (MRFs) with different plastic mechanisms of collapse. The load F increases monotonically until the yield strength or buckling value F_y of the frame corresponding to the yield displacement δ_y and the maximum or ultimate displacement δ_u signifying the limit beyond which collapse suddenly occurs. One can observe that even though the two frames have the same yield strength and yield displacement, frame (a) has a much larger δ_u than frame (b), which implies a much

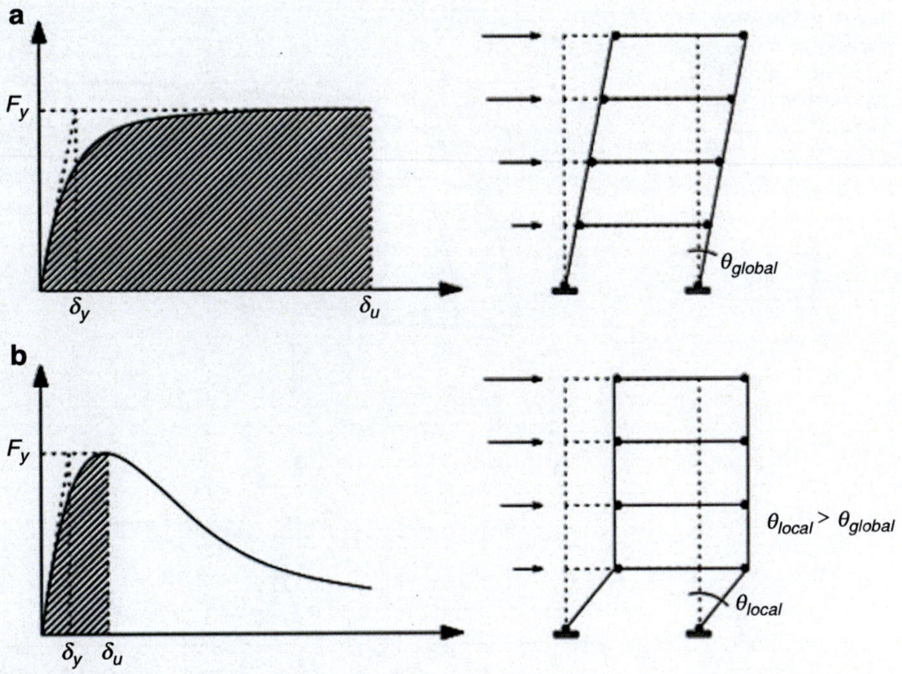

Fig. 1.3 Ductility of frames (after Landolfo et al. 2017, reprinted with permission from ECCS)

larger energy of dissipation (shown in Fig. 1.3 by the shaded areas) for frame (a) than for frame (b). By defining ductility μ as the ability of the frame to deform inelastically without losing its bearing capacity, i.e., as

$$\mu = \delta_u / \delta_y \tag{1.1}$$

one can easily see that frame (a) has a much larger ductility than frame (b). Thus, frame (a) is a dissipative structure, while frame (b) is a low-dissipative structure.

Two plane steel MRFs under lateral cyclic loading, as the one due to an earthquake, exhibit a series of hysteretic loops in a force-displacement, $F - \delta$, diagram, as shown in Fig. 1.4 (Landolfo et al. 2017), where F_y, δ_y and δ_u have the same meaning as in the previous case of Fig. 1.3. One can observe from Fig. 1.4 that even though both frames have the same F_y, δ_y and δ_u, the frame with the $F - \delta$ diagram of case (a) dissipates more energy than the one of case (b) before failure, as it is evident from its larger area within its hysteretic loops than that of the frame of case (b). Thus, the frame corresponding to Fig. 1.4a, is a dissipative structure, while the one corresponding to Fig. 1.4b is a low-dissipative structure.

Figure 1.5 shows typical ideal elastoplastic curves in a force F-displacement δ diagram corresponding to three different plane steel MRFs with yield strengths $F_{y,1}$, $F_{y,2}$ and $F_{y,3}$ and corresponding yield displacements $\delta_{y,1}$, $\delta_{y,2}$ and $\delta_{y,3}$. One can

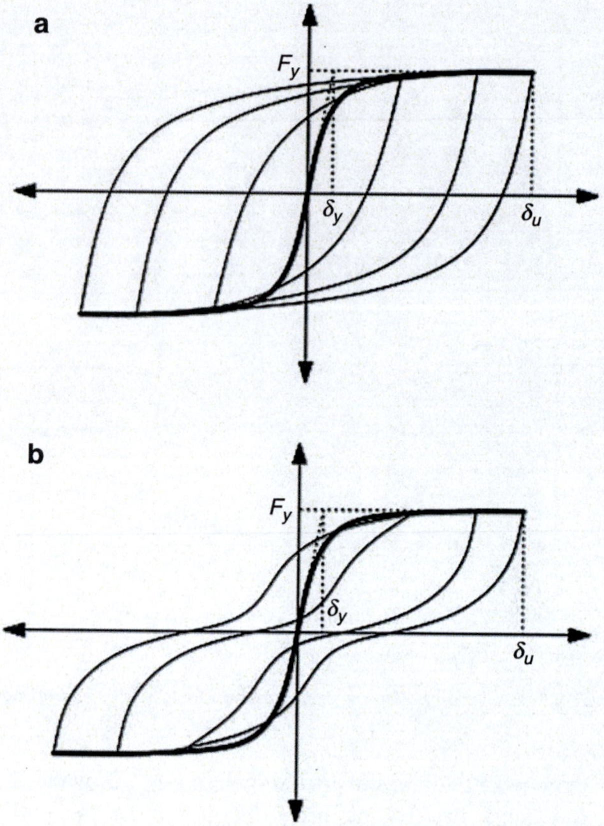

Fig. 1.4 Dissipative capacity of frames (after Landolfo et al. 2017, reprinted with permission from ECCS)

observe from Fig. 1.5 that for the same displacement capacity δ_c, as the yield strength increases, the ductility demand μ_d, in view of Eq. (1.1), decreases.

As it was mentioned at the beginning of this section, acceptance of inelastic deformation and damage in seismic design results in seismic forces smaller than those that would act on the structure if it were to remain elastic. Figure 1.6 depicts the variation of the seismic base shear V of a framed structure with a reference lateral displacement δ of that structure (e.g., the one at the roof) under monotonic lateral load representing seismic load.

Two V-δ curves are shown there: one on the assumption of linear elastic behavior and one on the assumption of ideal elastoplastic behavior with corresponding base shear forces V_{el} and V_{in} and displacements δ_{el} and $\delta_{in,}$ respectively. By accepting damage due to inelastic deformation, a part of the inputted seismic energy on the structure is dissipated and hence the elastic seismic base shear V_{el} (i.e., the one without energy dissipation) is reduced to the value of V_{in} or design base shear V_d, which is actually the yield strength of the structure V_y. Thus, one can define the behavior (or strength reduction) factor q as

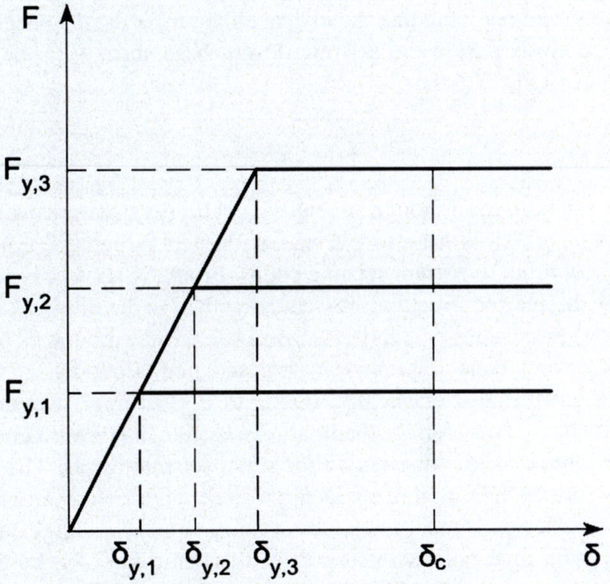

Fig. 1.5 Force *F*-displacement δ diagram for ideal elastoplastic systems

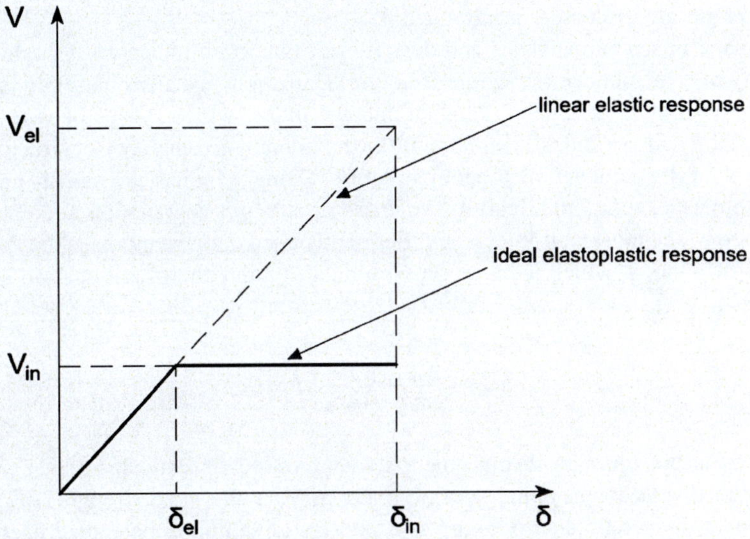

Fig. 1.6 Base shear *V*—displacement δ relation for linear elastic and ideal elastoplastic systems

$$q = V_{el}/V_y \tag{1.2}$$

Then, by assuming the validity of the equal displacement rule (Veletsos and Newmark 1960) stating that the maximum inelastic displacement δ_{in} is equal to the maximum elastic one (which is really valid only for structures with first natural periods ≥0.50 s) and realizing that the displacement δ_{el} is the displacement obtained by an elastic analysis due to the seismic design base shear V_{in}, one has from the similar triangles in Fig. 1.6 that

$$\delta_{in} = q\,\delta_{el} \tag{1.3}$$

The above relation, even though in reality is valid for structures with first natural periods ≥0.50 s, which is usually the case with steel structures, enjoys a general applicability according to current seismic codes. From the above discussion, it turns out that highly dissipative structures associated with high ductility can be seismically designed with higher values of q and thus lower seismic design forces provided some detailing and design code requirements are satisfied. Low-dissipative structures should be designed almost elastically, i.e., with q close to 1. In general, seismic design of structures considering them as dissipative is recommended for large seismic forces, while as low-dissipative for small seismic forces. The magnitude of seismic forces, due to their inertial character, also depends on the structural mass and hence dissipative design is recommended for multi-storey buildings because of their large masses. As a final note with respect to the definition of q by Eq. (1.2), one should stress the fact that this definition takes into account overstrength of the system because of its redundancy as well as material overstrength. In Sect. 1.5.1 as well as in Chap. 3 dealing with EC8 (2004), these two concepts of behavior factor and overstrength are discussed in more detail.

Damage in seismic analysis and design is an important indicator of the lost load capacity of a member or the whole structure. Damage is usually quantified in terms of a dimensionless parameter D called damage index, which ranges between 0 and 1. Thus, the damage index is equal to 0 for the undamaged member or structure and to 1 for the fully damaged member or structure. Damage indices are usually based on displacement or rotational ductility, stiffness or strength degradation and energy of dissipation. The most well-known and frequently used damage index, is the Park and Ang (1985) index, which is based on both ductility and energy dissipation and has the form

$$\mathrm{D_{PA}} = \frac{\delta_m}{\delta_u} + \frac{\beta}{V_y \delta_u} \int dE_h \tag{1.4}$$

where δ_u is the ultimate displacement under monotonic static loading, δ_m is the maximum displacement under seismic loading, V_y is the yield strength, dE_h is the incremental hysteretic energy and β is a non-negative non-dimensional coefficient determined from experimental calibration. This index, which was originally developed for reinforced concrete members, can also be used for steel members with $\beta = 0.25$ (Castiglioni and Pucinotti 2009). From the many modifications of D_{PA} one can mention the one due to Ghosh et al. (2011) with δ_m/δ_u in Eq. (1.4) being replaced by $(\delta_m\text{-}\delta_y)/(\delta_u\text{-}\delta_y)$, where δ_y is the yield displacement and the one due to Kunnath et al. (1992) of the form

$$D_{PA} = \frac{\theta_m - \theta_r}{\theta_u - \theta_r} + \frac{\beta}{M_y \theta_u} \int dE_h \quad (1.5)$$

where θ is the member end rotation, with θ_m being its maximum value and θ_r the recoverable rotation during unloading and M_y is the yield moment capacity of the member. In closing, one should mention that in many cases damage is also quantified by deformation measures, such as the interstorey drift ratio (*IDR*) or the member plastic rotation θ_p (usually given as a multiple of the member yield rotation θ_y in FEMA 356 (2000)). More information about damage can be found in Chap. 11.

1.4 Capacity Seismic Design

As it has been stated in the previous section, the goal of the seismic design of a building structure is to produce a structure with an acceptable level of safety at a logical cost. In contrast to the conventional or direct method of design for the main loads (dead, live, wind, snow), seismic loads are taken into account by accepting that the structure will deform inelastically and hence it will experience a certain level of damage. Thus, a controlled dissipation of energy in a state of damage, is essential for the survival of the structure under strong earthquake excitations. Consequently, control of its mode of failure is the most important requirement for the safety of the structure. The capacity design philosophy has been developed exactly in order to ensure that this control of the mode of the failure or collapse of the structure will be effective during its inelastic deformation.

In accordance with Elnashai (1995), the difference between the philosophies of the direct and the capacity methods of design can be described with the aid of the following simple example: Consider the members B_1, B_2 and B_3 framing at the rigid joint K of a plane steel frame under their corresponding seismic design actions M (bending moment), V (shear force) and N (axial force), as shown in Fig. 1.7. According to the direct design method, every member is dimensioned in order to

Fig. 1.7 Seismic design actions in members B_1, B_2 and B_3 framing at a joint K of a plane steel frame

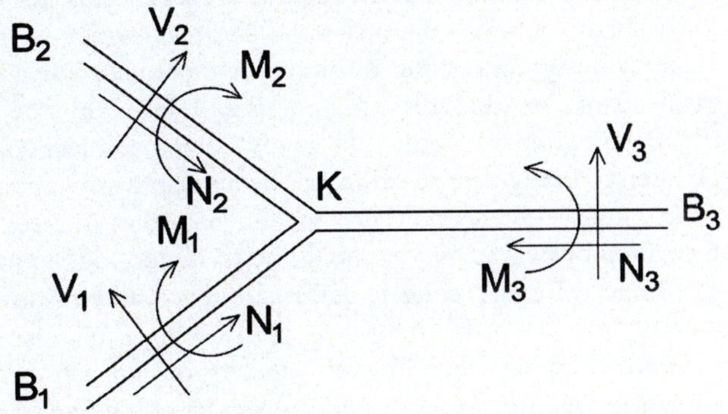

safely carry its own actions as stipulated by the code (EC3 2009 for steel structures). However, according to the capacity design method, the designer decides first which members will yield first and if the other members will yield afterwards or will remain elastic.

Here it is assumed that member B_3 will contribute to the development of the desired mode of failure and thus will act as a zone of energy dissipation, while members B_1 and B_2 will remain elastic. The capacity design procedure starts with dimensioning of the member B_3 for the applied actions M_3, V_3 and N_3 and with a strength limit equal to the plastic resistance of the member. Then the overstrength of member B_3 is estimated by taking into account the possible upper yield limit strength of the steel material, the possible section area increase in comparison with its theoretical value and the work hardening of the steel material. After that, the action forces in members B_1 and B_2 are recalculated elastically so as to achieve equilibrium at the joint K among the forces in members B_1 and B_2 and the new strength forces of member B_3. Thus, one calculates the new (higher) action forces M_1', V_1', N_1' and M_2', V_2', N_2', which are used for the dimensioning of the two members B_1 and B_2 with limit strength equal to their elastic strength, since these members are elastic and do not dissipate energy.

In conclusion, capacity design of a steel framed structure aims at (1) achieving as high as possible strength and ductility through inelastic deformation in those members of the structure which can dissipate a large amount of seismic energy and (2) securing adequate yield strength in the remaining members of the structure, which will remain elastic. Thus, one does first ultimate strength checks for the former members on the basis of actions coming from the static analysis of the structure under design loads and then elastic behavior checks for the latter members on the basis of the maximum actions, which are transferred from the neighboring former members.

In order to ensure the possibility of energy dissipation in the framed structure during its response to the seismic design action without the occurrence of partial or full collapse prematurely (before the structure has utilized all its strength), the inelastic response of the structure should be dissipative and distributed to the possible larger number of structural members in areas with limited length (plastic hinges). This presupposes that the avoidance of all possible brittle modes of failure, which could possibly precede, can be secured. Thus, for framed building structures, the rule of "strong columns-weak beams", which is intimately associated with capacity design, has the goal of creating a structure that can fail by global collapse mechanism as shown, e.g., in Fig. 1.3a. On the contrary, the frame of Fig. 1.3b exhibits a "soft first storey" collapse mechanism and fails prematurely, i.e., before being able to utilize all its available strength.

In closing this section, one should note that in steel moment resisting frames, formation of plastic hinges should be restricted to the ends of their beams, while in steel braced frames inelastic deformation should be restricted to their tensile diagonals or to shear or bending type plastic hinges in seismic links. Plastic hinges in columns designed according to capacity design can only appear at their ends, as shown in Fig. 1.3. More on this subject can be found in Chap. 3. A review type of article on seismic capacity design of structures has been recently published by Fardis (2018).

1.5 Some Additional Design Aspects

This section discusses some additional seismic design aspects pertaining to framed building structures in general and those made of steel in particular. These aspects have to do with the behavior (strength reduction) factor and overstrength, torsional effects in space framed buildings and combination rules for multi-component seismic analysis of building structures.

1.5.1 Behavior (Strength Reduction) Factor and Overstrength

In Sect. 1.3 the definition of the behavior (strength reduction) factor was given by Eq. (1.2). This factor, even though is given in seismic codes in terms of constant values for various types of structures and materials, as explicitly described in Chap. 3 for the case of EC8 (2004), in reality depends on ductility μ, fundamental structural period T and site (soil) conditions (Miranda and Bertero 1994; Santa-Ana and Miranda 2000). In addition, the behavior factor also depends on the strength reserves existing in the structure due to overstrength. These strength reserves enable structures to sustain without damage larger seismic forces than the design ones. The definition of Eq. (1.2) includes overstrength in the behavior factor q.

According to Rahgozar and Humar (1998) overstrength for steel framed structures can come from the following sources: (1) sources of uncertainty, like the difference between the actual material strength and its nominal value used in design, which is lower than or equal to the former one, or the use of discrete member sizes, which leads to steel sections with a strength larger than or equal to the design one; (2) sources than cannot be accounted for due to lack of knowledge, like use of conservative models, or the effect of non-structural or structural elements (e.g., infill walls or reinforced concrete slabs); (3) sources that are not usually accounted for seismic capacity determination, like minimum code requirements for conservative design, or architectural additions that will increase strength, or control design by other loading cases (e.g., wind loads); (4) sources pertinent to design simplification, like use of single-degree-of-freedom (SDOF) spectra for capacity estimation of multi-degree-of-freedom (MDOF) systems, or redistribution of internal forces in the inelastic range with the former case leading to overstrength or understrength.

On the basis of the above discussion, one can conclude that in a rational design, only the reliable sources of extra strength should be taken into account for reducing

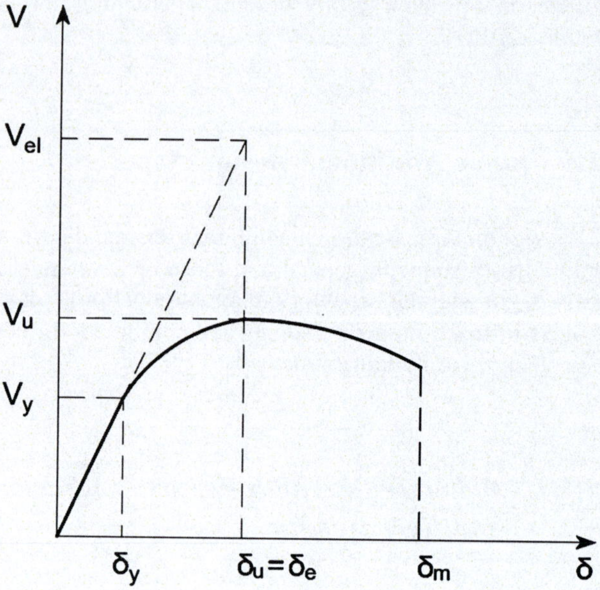

Fig. 1.8 Force V-displacement δ of an inelastic structure (subscripts y, el, u and m of δ stand for first yield, elastic, ultimate and maximum, respectively)

the seismic design load. The other sources of overstrength can be used only for estimating capacity. Thus, only the source of overstrength due to the redistribution of internal forces should be accounted for in design. This redistribution is the result of the difference between ultimate strength and the strength at first yield. With reference Fig. 1.8, which represents the base shear force V-displacement δ relation of an inelastic structure with $V_u > V_y$ the definition of q by Eq. (1.2) remains the same, while one has

$$q = q_\mu q_o \tag{1.6}$$

with

$$q_\mu = V_{el}/V_u, \;\; q_o = V_u/V_y \tag{1.7}$$

where q_μ is the part of q corresponding to ductility and q_o the one corresponding to overstrength due to redistribution of internal forces. Rahgozar and Humar (1998) have studied by nonlinear static (pushover) analysis the response of moment resisting steel frames and concentrically braced steel frames with 2 to 30 storeys and found out that q_o varies between the values of 1.5 to 3.5. Additional information on the subject of overstrength can be found in Bertero (1989) and Kappos (1999, 2002) for reinforced concrete structures and in Hamburger (2009) for steel structures.

1.5.2 Torsional Effects in Space Framed Buildings

Three-dimensional building frames can experience torsional motion in addition to lateral translational motions during strong earthquake excitations. Torsion increases the structural response and damage and hence has to be taken into account in design. Torsional motion can be induced by various sources, such as stiffness and/or mass eccentricities, eccentric placement of non-structural elements, incoherent seismic motions at their foundation and asymmetric yielding. Thus, torsional effects are not restricted only to irregular and asymmetric buildings but can also affect symmetric ones as well. Because some of the causes of torsional motions are not apparent, seismic codes require the consideration of accidental eccentricities in otherwise regular and symmetric buildings.

Extensive work on torsional effects on reinforced concrete and steel buildings by Anagnostopoulos et al. (2015b) during the last 15 years or so has revealed a number of problems associated with code provisions, which were based on simplified one-storey inelastic buildings. Restricting the present discussion to steel structures, Kyrkos and Anagnostopoulos (2011a) have performed extensive parametric studies on EC8 (2004) code designed 1, 3 and 5-storey non-symmetric concentrically braced steel building frames under a set of 10, two-component, semi-artificial spectrum compatible motions and found that the flexible edge frames exhibit higher ductility demands and interstorey drifts than the stiff edge frames of the building. These results, found on the basis of comprehensive inelastic models, are opposite to those reported in other works based on the simplified, one-storey, shear-beam type of models. The significant differences in demands between the two building sides indicate that there is a need for reconsideration of the related code provisions. Kyrkos and Anagnostopoulos (2011b, 2013) proposed a design modification that leads to a more uniform distribution of ductility demands on the elements of all building edges.

The related problem of accidental eccentricity in symmetric framed buildings has been also considered by Stathopoulos and Anagnostopoulos (2010) and Anagnostopoulos et al. (2015a) in conjunction with reinforced concrete and steel buildings, respectively. The results of their investigation indicate that accidental design eccentricities lead to small reduction in response demands and the effort expended in taking them into account is not practically justified for the benefit gained. Similar conclusions have been reached by De la Llera and Chopra (1994, 1995) and Basu et al. (2014), who have also proposed alternative simple methods for computing effects of accidental eccentricities.

1.5.3 Combination Rules for Multicomponent Seismic Analysis

In seismic design (IBC 2018; EC8 2004), the maximum structural response to the combined action of the three earthquake components (two horizontal and one vertical) is obtained by first determining separately the maximum responses due to those three components and then combining them by using one combination rule, like that of the square-root-of-sum-of-squares (SRSS), or the 30% rule of Rosenblueth and Contreras (1977). An alternative to the 30% rule is the 40% rule due to Newmark (1975). In most of the cases, only the two horizontal seismic components are considered and applied along the two axes of the structure alternately. However, the combined response depends on the incident angle, i.e., the angle θ between the two seismic components 1 and 2 and the structural axes x and y, as shown in Fig. 1.9. At this point one should stress that while seismic components are uncorrelated along the axes 1 and 2 they are not so along the axes x and y.

The complete quadratic combination-3 (CQC3) rule due to Menun and Der Kiureghian (1998) provides the response in terms of the incident angle and also an expression for the critical angle, i.e., the incident angle θ resulting in the maximum response. An expression for the maximum response directly in terms of the critical angle is provided. All these expressions are exact for any elastic structure and any seismic design spectrum.

Lopez et al. (2001) have presented a comparison between the SRSS, 30% and 40% combination rules on the basis of the more rational and accurate results of the CQC3 rule and found out that (1) for a 9-storey asymmetric-plan reinforced concrete building the critical response is overestimated by at most 17% according to the 40%-rule, 9% by the 30%-rule and 16% by the SRSS-rule and (2) for a 20-storey symmetrical-plan steel building the response results for all three rules are within 10% of the critical response.

Reyes-Salazar et al. (2012) in their work studied also the accuracy of the 30% and the SRSS rules on two multistorey steel buildings assuming material and geometric nonlinearities and using dynamic nonlinear analyses. They used two components of seismic horizontal motion (correlated along the x and y axes and uncorrelated along the 1 and 2 axes of Fig. 1.9) applied simultaneously (exact solution) and separately (approximate solution according to the 30% and SRSS rules) in alternate ways. Twenty seismic motions were used for the simulation. They also considered the effect of the incident angle on the response. Their conclusions were that for the inelastic behavior, the above two rules underestimate the axial load by about 10%

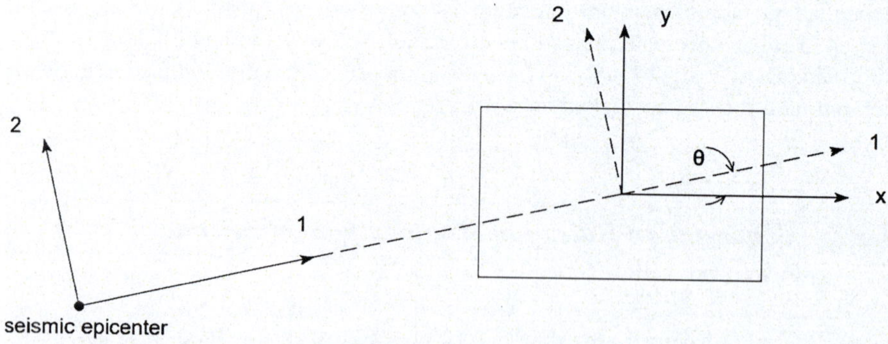

Fig. 1.9 Seismic components 1 and 2 and structural axes x and y

and overestimate the base shear by about 10%. However, the uncertainty in estimating axial loads was found to be larger than that in estimating base shears. In general, they found that the level of approximation depends on the degree of correlation of the components, the type of structure, the response parameter, the location of the member and the level of structural deformation. Finally, they proposed replacement of the 30% rule by a 45% rule.

1.5.4 Seismic Design Codes

Specific provisions for the seismic design of building structures have been established in the various seismic codes. Among the most well-known such codes, one can currently mention the EC8 (2004) of the European Union countries, the IBC (2018) and the ASCE/SEI 7-16 (2017) of the USA, the BSLJ (2013) of Japan, the NZS 1170.5 (2004) of New Zealand and the GB 50011-2010 (2010) of the P.R. China. Specific provisions for the seismic design of steel building structures can be found in the AISC (2005) code of the USA. A critical comparison of major seismic building codes, namely those of USA, Japan, New Zealand, Europe, Canada, Chile and Mexico has been recently published by Pinto (2013).

1.6 Performance-Based Seismic Design

Performance-based seismic design (PBSD) or in more general terms Performance-based earthquake engineering (PBEE) is a general design methodology or better a general design philosophy or framework. It started with the publication of the SEAOC (1995) and FEMA 273 (1997) documents, which established the first generation of PBSD framework characterized by its deterministic nature. This design framework introduces a number of structure performance levels associated with corresponding levels of seismic excitation (hazard) to use them as design criteria. At those levels of seismic intensity, deformation and damage should not exceed pre-defined limits in order to achieve a design that balances safety and economy.

Table 1.1 shows the performance levels and corresponding seismic intensity (hazard) levels for a structure in accordance with SEAOC (1995) and FEMA 273 (1997) as well as the articles of Ghobarah (2001, 2004). In that table, the four performance levels and corresponding seismic intensities are as follows: Fully operational (FO), Operational (O), Life safety (LS), Near collapse (NC) performance levels and frequent (43-year return period or 50% in 30 years probability of exceedance), occasional (72-year return period or 50% in 50 years probability of exceedance), rare (475-year return period or 10% in 50 years probability of exceedance) and very rare (970-year return period or 5% in 50 years probability of exceedance) earthquakes. The case of an extremely rare earthquake associated with collapse is also shown in this table. Table 1.1 also shows the damage levels

Table 1.1 Seismic performance and hazard levels with associated damage and *IDR* levels for moment resisting building structures

Performance levels	FO/IO	O	LS	NC/CP	C
Seismic hazard levels	Frequent earthquake FOE 43 years/50% in 30 years	Occasional earthquake 72 years/ 50% in 50 years	Rare earthquake DBE 475 years/ 10% in 50 years	Very rare earthquake MCE 970 years/5% in 50 years	Extremely rare earthquake 2475 years/2% in 50 years
Damage levels	No damage	Repairable damage	Irreparable damage	Severe damage	Collapse
IDR levels	≤0.7%	≤1.5%	≤2.5%	≤5.0%	>5.0%

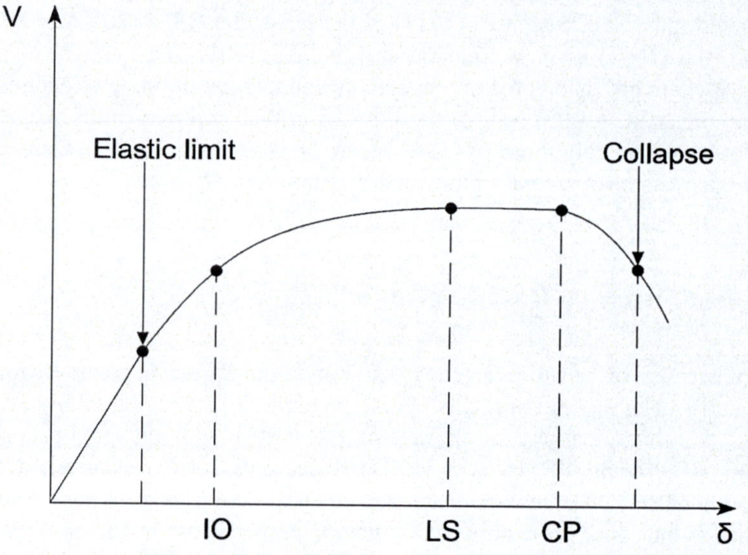

Fig. 1.10 Three performance levels (IO, LS, CP) and their relative position in a *V*-*δ* curve

and interstorey drift ratios (*IDR*) associated with the above performance levels. One can also consider three performance levels and corresponding seismic intensities reading as follows (FEMA 273 1997): Immediate occupancy (IO), Life safety (LS) and Collapse prevention (CP) with corresponding earthquakes the frequent occurred earthquake (FOE) with a 43-year return period, the design basis earthquake (DBE) with a 475-year return period and the maximum considered earthquake (MCE) with a 949-year return period. These performance levels are also shown in Table 1.1. Figure 1.10 depicts the positioning of those three performance levels in relation with the base shear force-displacement or *V*-*δ* curve of a building structure.

Implementation of the above PBSD framework in seismic design can be done by conducting first a seismic design on the basis of two performance levels, such as the

IO and LS levels, usually employed in current seismic codes (EC8 2004) and then check the structure for all the other levels (the CP level for the three level case).

Because of some limitations of the first generation PBSD or PBEE framework and its deterministic nature for an inherently probabilistic problem due to the uncertain character of seismic motions, the second generation of PBEE was developed and appeared in final form in the FEMA P-58 (2012) document based on the work of the Pacific Earthquake Engineering Research (PEER) Center (Moehle and Deierlein 2004; Hamburger et al. 2012; Gunay and Mosalam 2013). This new version of PBEE is probabilistic in nature and expresses the probable performance value of an earthquake loss measure as

$$\lambda(DV) = \iiint G(DV/DM)|dG(DM/EDP)|dG(EDP/IM)|d\lambda(IM) \tag{1.8}$$

where DV is the value of a decision variable or a performance measure, e.g., repair cost, for a given damage measure DM, EDP is an engineering demand parameter, e.g., a response quantity like element plastic rotation demand, for a given earthquake intensity measure IM and the integration is over a range of seismic hazards, considering uncertainty in hazard response, damage and its consequences. The loss measure includes repair costs, repair time and deaths. The FEMA P-58 (2012) methodology for performance assessment involves (1) the building performance model and the earthquake hazards definition in order to simulate the building response and (2) the collapse fragility development in order to combine it with the building response and calculate the desired performance.

During the last few years, in the framework of continuous research for improving PBEE, the concept of earthquake resilience in seismic building design has received considerable attention. Resilience requires seismic design of buildings that takes into account structural functionality after the earthquake. This concept of resilience has been recently extended from the performance of individual buildings and facilities to that of the whole built environment and the city or the community (Krawinkler and Deierlein 2014). In this book, performance-based seismic design methods are usually refer to those of the first generation of PBEE.

1.7 Some Key Aspects of Probabilistic PBSD

This section briefly describes some key aspects of probabilistic performance-based seismic design with emphasis being given to applications in steel structures. More specifically, earthquake intensity measures, fragility curves and the incremental dynamic analysis are briefly presented and discussed.

1.7.1 Earthquake Intensity Measures

As Eq. (1.8) clearly indicates, the earthquake intensity measure *IM* plays a very important role in structural performance assessment. Selection of an appropriate *IM* is still an object of research. The *PGA* was used as an *IM* in the past, while the spectral acceleration at the first natural period of the structure $S_a(T_1)$, is the most widely used *IM* nowadays. The former *IM* accounts only for the motion, while the latter for both the motion and its effect on the structure. Modifications and improvements of the $S_a(T_1)$ intensity measure have also been reported (Katsanos et al. 2010). For example, Shome and Cornell (1999) found that consideration of $S_a(T_2)$ and $S_a(T_3)$ in addition to $S_a(T_1)$ significantly improves the efficiency of $S_a(T_1)$ for tall buildings. Also, Baker and Cornell (2005) introduced a vector-valued *IM*, which consists of two values instead of one value characterizing the so-called scalar-valued *IM*, such as *PGA* or $S_a(T_1)$. This vector-valued *IM* can be found in the works of Vamvatsikos and Cornell (2005a), Luco and Cornell (2007) and Tothong and Luco (2007).

1.7.2 Fragility Curves

In order to be compatible with the probabilistic PBEE framework, the damage measures (*DM*) appearing in Eq. (1.8) should be defined in terms of fragility relations or curves. Typical fragility curves for a component or structure describe the probability of exceedance of the limiting value of a damage measure (*DM*), e.g., the maximum *IDR* or the Park and Ang (1985) damage index, in terms of a seismic intensity measure, e.g. the *PGA* or the $S_a(T_1)$, as shown in Fig. 1.11. It is obvious that the *p*(*DM*) to exceed the limit damage value, increases for increasing values of the *IM*. The mathematical form for such a fragility function or curve is (Porter et al. 2007)

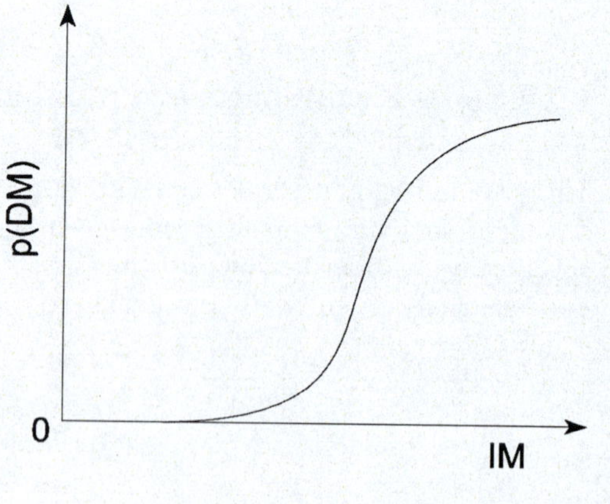

Fig. 1.11 Typical structural fragility curve

$$f_i(IM) = \Phi[\, ln\,(IM/\theta_i)/\beta_i] \tag{1.9}$$

where $f_i(IM)$ is the conditional probability that a structure will be damaged at a damage state i or higher, IM is an intensity measure, Φ is the standard normal (Gaussian) cumulative distribution function, θ_i is the median value of the probability distribution and β_i is the logarithmic standard deviation. Among the plethora of works dealing with fragility curves for steel frames, one can mention those of Noh et al. (2012), Skalomenos et al. (2015) and Diaz et al. (2018).

1.7.3 Incremental Dynamic Analysis

The incremental dynamic analysis (IDA), developed by Vamvatsikos and Cornell (2002, 2004, 2005b), determines by nonlinear dynamic analyses the response of a structure to a set of earthquake motions scaled to a number of intensity levels so as to drive the structure all the way from elastic behavior to collapse. A typical IDA curve for one structure under one earthquake motion is a curve relating an intensity measure (IM) for the earthquake to a damage measure (DM) for the structural response. Figure 1.12 depicts such a typical IDA curve for a steel frame under one earthquake with $S_a(T_1)$ and IDR_{max} as the IM and DM, respectively. In that figure, three limit-states, the IO, CP and the global instability (GI) ones, as defined in Vamvatsikos and Cornell (2004), are also shown. If one co-plots many such IDA curves constructed for the same structure under many earthquakes, he can see pictorially the dispersion of the results. Vamvatsikos and Cornell (2005a) have constructed IDA curves for a structure under many earthquakes for two different IMs, the PGA and the $S_a(T_1)$ and observed how the dispersion due to motion-to-motion variability decreases as one goes from the PGA to $S_a(T_1)$ intensity measure proving that the latter is a better IM than the former one. This demonstrates the importance of the IM on the IDA computational effort, since an IM resulting in low

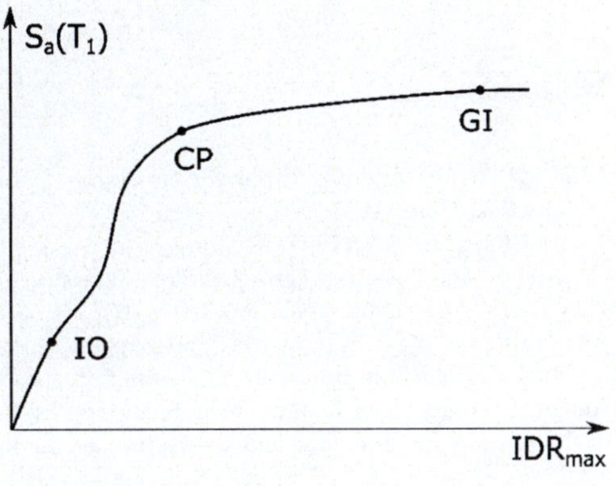

Fig. 1.12 Typical IDA curve for a steel frame under one earthquake

dispersion implies that one needs to consider a fewer number of ground motions for acceptable results.

The IDA is a powerful analysis tool in the probabilistic environment of PBEE, since it can provide comprehensive information needed for both demand and capacity estimation (Vamvatsikos and Cornell 2002, 2004).

1.8 Conclusions

A summary of the most important conclusions coming out from the preceding sections of this chapter is as follows:

1. The development of seismic structural design can be placed at the beginning of the twentieth century but only after the early 1960s started gaining momentum. At the beginning the emphasis was on human safety. Nowadays the emphasis is on reduction of losses not only of humans but also of repair costs and the time required for re-functionality of structures.
2. The present knowledge on the basic characteristics of earthquakes, such as generation and propagation, magnitude, intensity, duration, attenuation, frequency content, their different variation with time in near-fault and far-fault regions, their possible sequential character and their possible site amplification as well as on the effects of these characteristics on the structural response, has resulted in design improvements.
3. Well established design concepts, such as inelastic deformation, energy of dissipation, ductility, behavior (strength reduction) factor, overstrength, damage assessment and capacity design have been placed in the comprehensive design framework of performance-based design involving various design levels characterized by pairs of seismic hazard and design objectives in a deterministic or probabilistic framework. Even though full performance-based design has not been implemented as yet in seismic design codes, this implementation is expected to take place soon.

References

AISC (2005) Seismic provisions for structural steel buildings. American Institute of Steel Construction, Chicago, IL

Anagnostopoulos SA (1997) Buildings. In: Beskos DE, Anagnostopoulos SA (eds) Computer analysis and design of earthquake resistant structures: a handbook. Computational Mechanics Publications, Southampton, pp 369–440

Anagnostopoulos SA, Kyrkos MT, Papalymperi A, Plevri E (2015a) Should accidental eccentricity be eliminated from Eurocode 8? Earthq Struct 8:463–484

Anagnostopoulos SA, Kyrkos MT, Stathopoulos KG (2015b) Earthquake induced torsion in buildings: critical review and state of the art. Earthq Struct 8:305–377

ASCE/SEI 7-16 (2017) Minimum design loads and associated criteria for buildings and other structures. American Society of Civil Engineers, Reston, VI

Baker JW, Cornell AC (2005) A vector-valued ground motion intensity measure consisting of spectral acceleration and epsilon. Earthq Eng Struct Dyn 34:1193–1217

Basu D, Constantinou MC, Whittaker AS (2014) An equivalent accidental eccentricity to account for the effects of torsional ground motion on structures. Eng Struct 69:1–11

Bertero VV (1989) State-of-the-art report: ductility based structural design. In: Proceedings of 9th World Conference on Earthquake Engineering, Tokyo-Kyoto, Japan 1988, vol 3. Maruzen Press, Tokyo, pp 673–686

Bravo-Haro MA, Elghazouli AY (2018) Influence of earthquake duration on the response of steel moment frames. Soil Dyn Earthq Eng 115:634–651

BSLJ (2013) Building standard law of Japan. Building Center of Japan, The Ministry of Land, Infrastructure, Transport and Tourism of Japan, Tokyo, Japan

Castiglioni CA, Pucinotti R (2009) Failure criteria and cumulative damage models for steel components under cyclic loading. J Constr Steel Res 65:751–765

Chandramohan R, Baker JW, Deierlein GG (2016) Quantifying the influence of ground motion duration on structural collapse capacity using spectrally equivalent records. Earthquake Spectra 32:927–950

COSMOS (2013) Consortium of organizations for strong motion observation systems, San Francisco. http://www.cosmos-eq.org/

De la Llera JC, Chopra AK (1994) Evaluation of code accidental-torsion provisions from building records. J Struct Eng ASCE 120:597–616

De la Llera JC, Chopra AK (1995) Estimation of accidental torsion effects for seismic design of buildings. J Struct Eng ASCE 121:102–114

Diaz SA, Pujades LG, Barbat AH, Hidalgo-Leiva DA, Vargas-Alzate YF (2018) Capacity, damage and fragility models for steel buildings: a probabilistic approach. Bull Earthq Eng 16:1209–1243

EC8 (2004) Eurocode 8, Design of structures for earthquake resistance, Part 1: general rules, seismic actions and rules for buildings, EN 1998-1-1. European Committee for Standardization (CEN), Brussels

Elghazouli AY (2010) Assessment of European seismic design procedures for steel framed structures. Bull Earthq Eng 8:65–89

Elnashai AS (1995) Local ductility in steel structures subjected to earthquake loading. In: Mazzolani FM, Gioncu V (eds) Behavior of steel structures in seismic areas. E & FN Spon – Chapman & Hall, London, pp 133–148

Fajfar P (2018) Analysis in seismic provisions for buildings: past, present and future, the 5th Prof. Nicholas Ambraseys lecture. Bull Earthq Eng 16:2567–2608

Fardis MN (2018) Capacity design: early history. Earthq Eng Struct Dyn 47:2887–2896

FEMA 273 (1997) NEHRP guidelines for the seismic rehabilitation of buildings. Federal Emergency Management Agency, Washington, DC

FEMA 356 (2000) Prestandard and commentary for the seismic rehabilitation of buildings. Federal Emergency Management Agency, Washington, DC

FEMA 445 (2006) Next-generation performance-based seismic design guidelines. Federal Emergency Management Agency, Washington, DC

FEMA 695 (2009) Quantification of building seismic performance factors. Federal Emergency Management Agency, Washington, DC

FEMA 750 (2009) Recommended seismic provisions for new buildings and other structures. Federal Emergency Management Agency, Washington, DC

FEMA P-58 (2012) Guidelines for performance-based seismic design of buildings. Federal Emergency Management Agency, Washington, DC

Fragiacomo M, Amadio C, Macorini L (2004) Seismic response of steel frames under repeated earthquake ground motions. Eng Struct 26:2021–2035

GB 50011-2010 (2010) Code for seismic design of buildings. Ministry of Housing and Urban-Rural Development of the People's Republic of China, China Architecture and Building Press, Beijing. (in Chinese)
Ghobarah A (2001) Performance-based design in earthquake engineering: state of development. Eng Struct 23:878–884
Ghobarah A (2004) On drift limits associated with different damage levels. In: Fajfar P, Krawinkler H (eds) Performance-based seismic design concepts and implementation, PEER report 2004/05. University of California, Berkeley, CA, pp 321–332
Ghosh S, Datta D, Katakdhond AA (2011) Estimation of the Park-Ang damage index for planar multi-storey frames using equivalent single-degree systems. Eng Struct 33:2509–2524
Gunay S, Mosalam KM (2013) Peer performance-based earthquake engineering methodology, revisited. J Earthq Eng 17:829–858
Hamburger RO (2009) Earthquakes and seismic design, facts for steel buildings number 3. American Institute of Steel Construction, Chicago, IL
Hamburger RO, Rojahn C, Heintz JA, Mahoney MG (2012) FEMA P58: next-generation building seismic performance assessment methodology. In: Proceedings of 15th World Conference on Earthquake Engineering, Lisboa, Portugal
Hashash YMA, Harmon J, Ilhan O, Stewart JP, Rathje EM, Campbell KW, Silva WJ, Goulet CA (2018) Modeling of site amplification via large scale nonlinear simulations with application to North America. Geotech Spec Publ GSP 291:523–537
Housner GW (1959) Behavior of structures during earthquakes. J Eng Mech Div ASCE 85:109–130
IBC (2018) International building code. International Code Council, Washington, DC
Kappos AJ (1999) Evaluation of behavior factors on the basis of ductility and overstrength studies. Eng Struct 21:823–835
Kappos AJ (2002) Earthquake loading. In: Kappos AJ (ed) Dynamic loading and design of structures. Spon Press, London, pp 109–173
Katsanos EI, Sextos AG, Manolis GD (2010) Selection of earthquake ground motion records: a state-of-the-art review from a structural engineering perspective. Soil Dyn Earthq Eng 30:157–169
Kim B, Hashash YMA, Stewart JP, Rathje EM, Harmon JA, Musgrove MI, Campbell KW, Silva WJ (2016) Relative differences between nonlinear and equivalent linear 1-D site response analyses. Earthquake Spectra 32:1845–1865
Kramer SL (1996) Geotechnical earthquake engineering. Prentice Hall, Englewood Cliffs, NJ
Krawinkler H, Deierlein GG (2014) Challenges towards achieving earthquake resilience through performance-based earthquake engineering. In: Fischinger M (ed) Performance-based seismic engineering: vision for an earthquake resilient society. Springer, Dordrecht, pp 3–23
Kunnath SK, Reinhorn AM, Lobo RF (1992) IDARC Version 3.0: a program for the inelastic damage analysis of reinforced concrete structures, Report no NCEER-92-0022. National Center for Earthquake Engineering Research, State University of New York, Buffalo, NY
Kyrkos MT, Anagnostopoulos SA (2011a) An assessment of code designed, torsionally stiff, asymmetric steel buildings under strong earthquakes excitations. Earthq Struct 2:109–126
Kyrkos MT, Anagnostopoulos SA (2011b) Improved earthquake resistant design of torsionally stiff asymmetric steel buildings. Earthq Struct 2:127–147
Kyrkos MT, Anagnostopoulos SA (2013) Improved earthquake resistant design of steel buildings. Soil Dyn Earthq Eng 47:144–156
Landolfo R (2018) Seismic design of steel structures: new trends of research and updates of Eurocode 8. In: Pitilakis K (ed) Recent advances in earthquake engineering in Europe. Springer International Publishing, Cham, pp 413–438
Landolfo R, Mazzolani F, Dubina D, da Silva LS, D' Aniello M (2017) Design of steel structures for buildings in seismic areas. ECCS/Wiley/Ernst & Sohn, Berlin
Lopez OA, Chopra AK, Hernandez JJ (2001) Evaluation of combination rules for maximum response calculation in multicomponent seismic analysis. Earthq Eng Struct Dyn 30:1379–1398

Loulelis DG, Hatzigeorgiou GD, Beskos DE (2012) Moment resisting steel frames under repeated earthquakes. Earthq Struct 3:231–248

Luco N, Cornell CA (2007) Structure-specific scalar intensity measures for near-source and ordinary earthquake ground motions. Earthquake Spectra 23:357–392

Menun C, Der Kiureghian A (1998) A replacement for the 30%, 40%, and SRSS rules for multicomponent seismic analysis. Earthquake Spectra 14:153–163

Miranda E, Bertero VV (1994) Evaluation of strength reduction factors for earthquake resistant design. Earthquake Spectra 10:357–379

Moehle J, Deierlein GG (2004) A framework methodology for performance-based earthquake engineering. In: Proceedings of 13th World Conference on Earthquake Engineering, Vancouver, B.C. Canada, Paper No 679

Newmark NM (1975) Seismic design criteria for structures and facilities: Trans-Alaska pipeline system. In: Proceedings of the US National Conference on Earthquake Engineering, Ann Arbor, MI. Earthquake Engineering Research Institute, Oakland, CA, pp 94–103

Newmark NM, Hall WJ (1982) Earthquake spectra and design, engineering monographs on earthquake criteria, structural design and strong motion records, vol 3. University of California, Berkeley, CA, Earthquake Engineering Research Institute

Noh HY, Lignos DG, Nair KK, Kiremidjian AS (2012) Development of fragility functions as a damage classification/prediction method for steel moment-resisting frames using a wavelet-based damage sensitive feature. Earthq Eng Struct Dyn 41:681–696

NZS 1170.5 (2004) New Zealand standard, structural design actions, part 5: earthquake actions. Standards Association of New Zealand, Wellington

Park YJ, Ang AHS (1985) Mechanistic seismic damage model for reinforced concrete. J Struct Eng ASCE 111:722–739

PEER (2009) Pacific Earthquake Engineering Research Center, Strong Ground Motion Database, Berkeley, CA. http://peer.berkeley.edu/

Pinto P (2013) Convener, Critical comparison of major seismic design codes for buildings. FIB Bulletin No 69. International Federation for Structural Concrete, Lausanne

Porter K, Kennedy R, Bachman R (2007) Creating fragility functions for performance-based earthquake engineering. Earthquake Spectra 23:471–489

Rahgozar MA, Humar JL (1998) Accounting for overstrength in seismic design of steel structures. Can J Civ Eng 25:1–15

Rathje EM, Abrahamson NA, Bray JD (1998) Simplified frequency content estimates of earthquake ground motions. J Geotech Geoenviron Eng ASCE 124:150–158

Reyes-Salazar A, Valenzuela-Beltran F, De Leon-Escobedo D, Bojorquez E, Lopez-Barraza A (2012) Accuracy of combination rules and individual effect correlation: MDOF vs SDOF systems. Steel Compos Struct 12:351–377

Roeder CW, MacRae GA (1997) Steel structures. In: Beskos DE, Anagnostopoulos SA (eds) Computer analysis and design of earthquake resistant structures: a handbook. Computational Mechanics Publications, Southampton, pp 533–561

Roesset JM (1997) Principles of earthquake resistant design. In: Beskos DE, Anagnostopoulos SA (eds) Computer analysis and design of earthquake resistant structures: a handbook. Computational Mechanics Publications, Southampton, pp 333–367

Rosenblueth E, Contreras H (1977) Approximate design for multicomponent earthquakes. J Eng Mech Div ASCE 103:895–911

Santa-Ana PR, Miranda E (2000) Strength reduction factors for multi-degree-of-freedom systems. In: Proceedings of 12th World Conference on Earthquake Engineering, Auckland, New Zealand, Paper No. 1446

SEAOC (1995) A framework for performance-based design. Structural Engineers Association of California, Vision 2000 Committee, Sacramento, CA

Shome N, Cornell CA (1999) Probabilistic seismic demand analyses of nonlinear structures, Report No. RMS-35, Reliability of Marine Structures Program. Department of Civil and Environmental Engineering, Stanford University, Stanford, CA

Skalomenos KA, Hatzigeorgiou GD, Beskos DE (2015) Modeling level selection for seismic analysis of CFT/MRFs by using fragility curves. Earthq Eng Struct Dyn 44:199–220
Stathopoulos KG, Anagnostopoulos SA (2010) Accidental eccentricity: is it important for the inelastic response of buildings to strong earthquakes? Soil Dyn Earthq Eng 30:782–797
Tirca L, Chen L, Tremblay R (2015) Assessing collapse safety on CBF buildings subjected to crustal and subduction earthquakes. J Constr Steel Res 115:634–651
Tothong P, Luco N (2007) Probabilistic seismic demand analysis using advanced ground motions intensity measures. Earthq Eng Struct Dyn 36:1837–1860
Vamvatsikos D, Cornell CA (2002) Incremental dynamic analysis. Earthq Eng Struct Dyn 31:491–514
Vamvatsikos D, Cornell CA (2004) Applied incremental dynamic analysis. Earthquake Spectra 20:523–553
Vamvatsikos D, Cornell CA (2005a) Developing efficient scalar and vector intensity measures for IDA capacity estimation by incorporating elastic spectral shape estimation. Earthq Eng Struct Dyn 34:1573–1600
Vamvatsikos D, Cornell CA (2005b) Direct estimation of seismic demand and capacity of multi-degree-of-freedom systems through incremental dynamic analysis of single-degree-of-freedom approximation. J Struct Eng ASCE 131:589–599
Veletsos AS, Newmark NM (1960) Effect of inelastic behavior on the response of simple systems to earthquake motions. In: Proceedings of 2nd World Conference on Earthquake engineering, Tokyo, Japan, vol 2, pp 895–912

Chapter 2
Fundamentals of Seismic Structural Analysis

Abstract This chapter briefly describes fundamental aspects of the analysis of building structures under seismic loads for reasons of completeness and easy reference. Emphasis is placed on steel building structures. Modeling procedures in the framework of the finite element method are presented and discussed. Linear elastic global analysis methods involving modal superposition and stepwise time integration are discussed. The special case of determining the maximum response for design purposes by combining modal superposition and elastic spectra is also presented. Nonlinear global analysis methods by stepwise time integration are described. Both material and geometric nonlinearities are considered. The special case of determining the maximum response for design purposes with the aid of inelastic spectra is also presented. The static nonlinear (pushover) method of analysis where seismic loads are applied as gradually increasing lateral forces is also briefly discussed. Some special topics, such as hysteretic material modeling under cyclic loading including deterioration and selection and scaling of earthquake records for nonlinear time history analyses are also briefly presented. Finally, special and general purpose computer programs for seismic analysis of steel building structures are presented.

Keywords Buildings · Seismic structural analysis · Seismic motions · Material and geometric nonlinearities · Elastic and inelastic spectra · Pushover analysis

2.1 Introduction

The calculation of member forces and moments as well as storey lateral displacements and member rotations of framed steel buildings under seismic design loads is necessary for the strength checking and the deformation checking, respectively, of those buildings. This calculation of the response of the building in the form of internal forces and deformation to the seismic design loads is accomplished by a global structural analysis of the mathematical model of that building.

G. A. Papagiannopoulos et al., *Seismic Design Methods for Steel Building Structures*, Geotechnical, Geological and Earthquake Engineering 51,
https://doi.org/10.1007/978-3-030-80687-3_2

When the seismic design loads are considered to be dynamic, i.e., they vary with time, the problems of the structural modeling and analysis become more difficult than the corresponding ones for the case the seismic loads are considered to be static. Thus, a global dynamic analysis of a linear elastic structure, which is usually done with the finite element method (FEM), requires in addition to the formation of the stiffness matrix, the formation of the mass and damping matrices and applies more complex methods of analysis than those used for static analysis. Furthermore, because modern seismic design codes and other design methods allow inelastic deformations of the structure under strong ground motions, the analysis problem becomes even more difficult when an inelastic dynamic analysis has to be employed. In that case, the modeling of the cyclic inelastic structural behavior and the analysis methods are more complex than those of the elastic dynamic case. One should also notice that, steel structures, because of their flexibility, can experience phenomena of elastic or inelastic instability in conjunction with large displacements resulting in additional difficulties in both modeling and analysis. In an effort to simplify the above structural modeling and analysis problems, seismic design codes also permit the use of simplified structural models and approximate elastic static analysis methods.

In general, a global analysis static or dynamic can be elastic or inelastic. An elastic analysis can be linear (first order theory with small displacements) or nonlinear (second order theory with large displacements or geometrical nonlinearities). An inelastic analysis, which is characterized by a nonlinear material behavior, can be associated with small displacements or large displacements. Obviously, a dynamic analysis with inelastic material behavior and large displacements, i.e., with material and geometric nonlinearities, can provide a more realistic structural response than any other analysis method at the expense of higher computational effort.

At this point one should stress that the type of analysis to be used is intimately connected with the class of the cross-sections of the structural members. According to EC3 (2009), commercial steel cross-sectional profiles are placed in four classes or categories on the basis of their ability to sustain local buckling and develop plastic hinge rotation. Class 1 cross-sections have the ability to develop plastic hinges with high rotation potential, while class 2 cross-sections can also develop plastic hinges but with restricted rotation potential. Class 3 cross-sections can develop yielding at their section edges but not full plastic moment of resistance due to local buckling at compression zones, while class 4 cross-sections cannot develop yielding at all in their compression zones due to local buckling. Thus, the use of inelastic global analysis is restricted to structures with class 1 cross-sections and is combined with inelastic strength checking. Structures with class 2 member cross-sections are analyzed by elastic global analysis combined with inelastic strength checking. Finally, structures with class 3 or 4 member cross-sections are analyzed by elastic global analysis combined with appropriate elastic allowable stress checking.

2.2 Elastic Global Analysis

2.2.1 Structural Modeling

All the methods of global elastic analysis are applied to a mathematical model of the steel framed building under gravity and lateral (seismic loading). The modeling of the building refers to its beams, columns, floors, infills, walls, connections etc. and the subsequent construction of its mass, stiffness and damping matrices in the framework of the finite element method (FEM). Nowadays, because of the availability of powerful personal computers, the seismic analysis and design of building structures can be easily done in practice, not only with the use of simple plane (two-dimensional or 2D) models but also with the use of detailed space (three-dimensional or 3D) models for static or dynamic elastic analysis (Anagnostopoulos 1997).

Simplified models can also be used. For example, for most of the regular buildings with floors practically rigid at their own horizontal plane (diaphragm action), the most frequently used model in its spatial form is a shear type of building with three degrees of freedom per floor (two horizontal displacements and one rotation with respect to a vertical axis) and building masses concentrated at every floor level, as shown in Fig. 2.1. The model is assumed to be subjected to gravity loads as well as to seismic motion along the two horizontal directions at its fixed base. The mass moments of inertia due to the rotational components of the motion of every node are, in general, negligible. For this reason, the corresponding degrees of freedom can be eliminated from the equations of motion through static condensation (Anagnostopoulos 1997).

Usually, a space (3D) frame like the one of Fig. 2.1 can be decomposed into plane (2D) frames so that the analysis can be done separately for each plane frame. This approximation is acceptable for regular in plan and elevation buildings, but it can lead to unacceptable results for irregular buildings. More details can be found in Chap. 3.

A typical plane elastic framed building, which can be thought of as the 2D counterpart of the 3D building of Fig. 2.1, also of the shear building type, is shown in Fig. 2.2a. This frame is under gravity loads and seismic horizontal ground motion at its fixed base. The model of this frame, as shown in Fig. 2.2b, has n degrees of freedom, i.e., one horizontal displacement at every floor level corresponding to the concentrated floor mass.

2.2.2 Dynamic Elastic Analysis

The equation of motion of an elastic 2D building under seismic excitation in the framework of the FEM has the form (Chopra 2007)

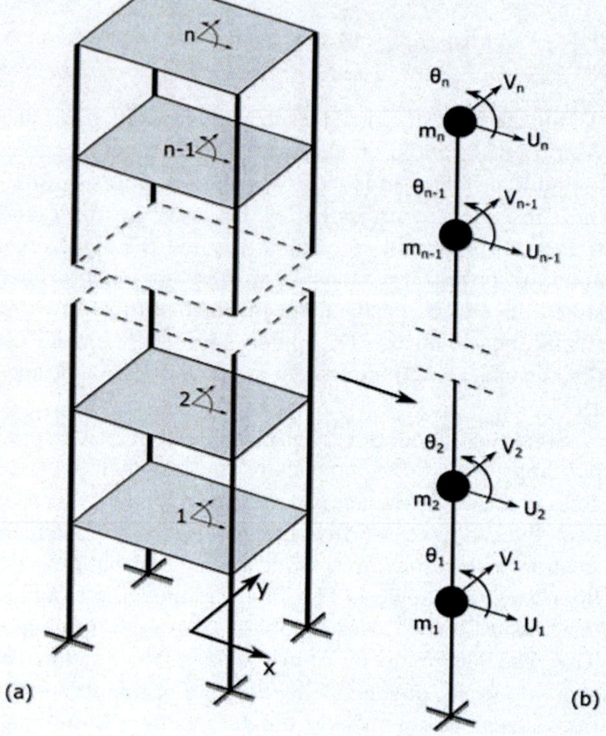

Fig. 2.1 Modeling of a space building under seismic excitation: (**a**) the building; (**b**) its model

$$M\ddot{u} + C\dot{u} + Ku = R = -MI\ddot{x}_g \tag{2.1}$$

where M, C and K are the mass, viscous damping and stiffness matrices, respectively, u is the vector of the lateral displacement of the structure relative to the ground, $\ddot{x}_g$ is the ground seismic acceleration, I is the unit vector and overdots denote differentiation with respect to time t.

For the shear building model of Fig. 2.2b, the mass of the building is concentrated at every floor and thus matrix M is diagonal. Matrix K is symmetric and consists of elements which are appropriate combinations of stiffnesses k_i obtained by adding the column stiffnesses of storey i. For all the models, matrix C is computed differently for each of the two methods of solution of Eq. (2.1). These two methods of solution are modal superposition and numerical integration, which are described in the following:

1. *Modal superposition*

This method of solution of Eq. (2.1), determines the seismic response u by an appropriate superposition (synthesis) of the structural modal shapes taking also into

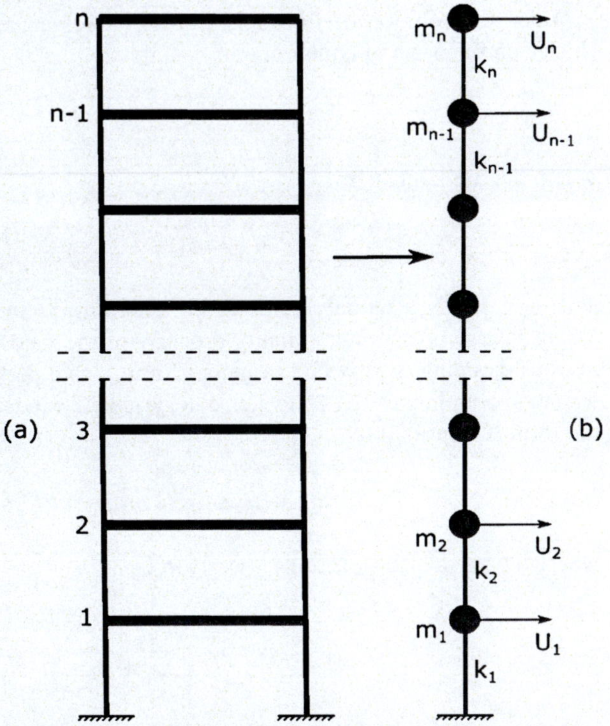

Fig. 2.2 Modeling of a plane shear building under seismic excitation: (**a**) the building; (**b**) its model

account the corresponding natural frequencies of the structure. For this reason, this method requires first the solution of the free vibration problem

$$M\ddot{u} + Ku = 0 \tag{2.2}$$

which is obtained from Eq. (2.1) by putting there $C = 0$ and $\ddot{x}_g = 0$. Its solution provides modal shapes and natural frequencies. Assuming a solution of the form

$$u = \varphi \sin \omega t \tag{2.3}$$

one receives from Eq. (2.2) the relation

$$K\varphi = \omega^2 M\varphi \tag{2.4}$$

where φ is the vector of the modal shape and ω the corresponding natural frequency. The eigenvalue problem of Eq. (2.4) is usually solved iteratively (Chopra 2007) resulting in the natural frequencies ω_i and modal shapes φ_i of the structure, where $i = 1,2,\ldots,n$. Index n denotes the total number of degrees of freedom of the structure, which, for the case of the 2D shear building of Fig. 2.2, is equal to the number of its floors.

Thus, the solution of Eq. (2.1) can be expressed as a superposition of modal shapes on the basis of equation

$$u = \Phi q \tag{2.5}$$

where the modal matrix

$$\Phi = [\varphi_1, \varphi_2, \ldots, \varphi_n] \tag{2.6}$$

and $q = q(t)$ is a modal vector to be determined. Substitution of Eq. (2.5) into Eq. (2.1), use of the orthogonality property of the modal shapes and assuming that viscous damping is provided in modal form ξ_k as a percentage of its critical value, lead to uncoupling of Eq. (2.1), i.e., to a system of n independent modal equations of the form (Chopra 2007)

$$\ddot{q}_k + 2\xi_k\omega_k\dot{q}_k + \omega_k^2 q_k = -\Gamma_k\ddot{u}_g \tag{2.7}$$

where the participation factor Γ_k is given by

$$\Gamma_k = \frac{\varphi_k^T MI}{\varphi_k^T M\varphi_k} \tag{2.8}$$

Solution of Eq. (2.7) can be accomplished by either of the following two ways:

(a) Numerically by a stepwise time integration using an algorithm, such as those employed for the numerical integration of Eq. (2.1) to be described next (e.g., Newmark's algorithm). This numerical integration can be applied either directly to Eq. (2.7) or to its closed form solution in terms of Duhamel's integral of the form

$$q_k(t) = -(\Gamma_k/\omega_k)\int_0^t \ddot{u}_g(\tau)\exp[-\xi_k\omega_{dk}(t-\tau)\sin\omega_k(t-\tau)d\tau] \tag{2.9}$$

where

$$\omega_{dk} = \omega_k\left(1-\xi_k^2\right)^{1/2} \tag{2.10}$$

Once $q_k(t)$ has been computed, one can determine $u(t)$ from Eq. (2.5) and then u_{max} for design purposes.

(b) With the aid of elastic response spectra when only the maximum value $(q_k)_{max}$ of the solution of Eq. (2.7) is required for design purposes. Elastic response spectra are curves representing maximum response values of either displacement S_d, or velocity S_v, or acceleration S_a versus the natural period T of single-degree-of-freedom (SDOF) systems to a specific seismic ground motion. They are constructed by solving the equation of motion (2.1) for the case of a SDOF system, which reads as

$$\ddot{u} + 2\xi\omega\dot{u} + \omega^2 u = -\ddot{u}_g \tag{2.11}$$

where $\omega^2 = k/m$ and $\xi = c/2m\omega$ with c being the viscous damping coefficient. Thus, after solving Eq. (2.11) by numerical stepwise time integration, one can obtain $S_d = \max u$, $S_v = \max \dot{u}$ and $S_a = \max \ddot{u}$. The response spectrum curves are given in a S versus $T = 2\pi/\omega$ diagram for various values of damping ξ. One can prove that for zero damping

$$S_v = \omega S_d \ , \quad S_a = \omega^2 S_d \tag{2.12}$$

The above relations can also be used for small amounts of damping ($\xi < 10\%$). In the latter case S_v and S_a as obtained in terms of S_d are called pseudo-velocity and pseudo-acceleration, respectively. Figure 2.3 shows the elastic response spectra S_d, S_v and S_a for the NS component of the 1940 El Centro earthquake for two values of damping ($\xi = 5\%, 10\%$).

Thus, in view of Eq. (2.5) one has

$$(u_k)_{max} = \varphi_k (q_k)_{max} = (1/\omega_k^2)\ \varphi_k \Gamma_k (S_a)_k \tag{2.13}$$

where $(S_a)_k$ is the value of spectral acceleration for the natural period $T_k = 2\pi/\omega_k$ of mode k. The maximum value of the response u_{max} is determined by combining the modal values $(u_k)_{max}$ in accordance with some combination rule, like the square root of the sum of the squares (SRSS) rule and reads

$$u_{max} = \left(u_{1\ max}^2 + u_{2\ max}^2 + \ldots + u_{n\ max}^2\right)^{0.5} \tag{2.14}$$

On the basis of the above, one can easily determine the maximum modal seismic (inertial) forces of the structure from the relation

$$(P_k)_{max} = \sum_{i=1}^{n} m_i \varphi_{ik} \omega_k^2 (q_k)_{max} = \sum_{i=1}^{n} m_i \varphi_{ik} \Gamma_k (S_a)_k \tag{2.15}$$

and hence the maximum modal seismic base shear force by

$$(V_k)_{max} = I^T (P_k)_{max} \tag{2.16}$$

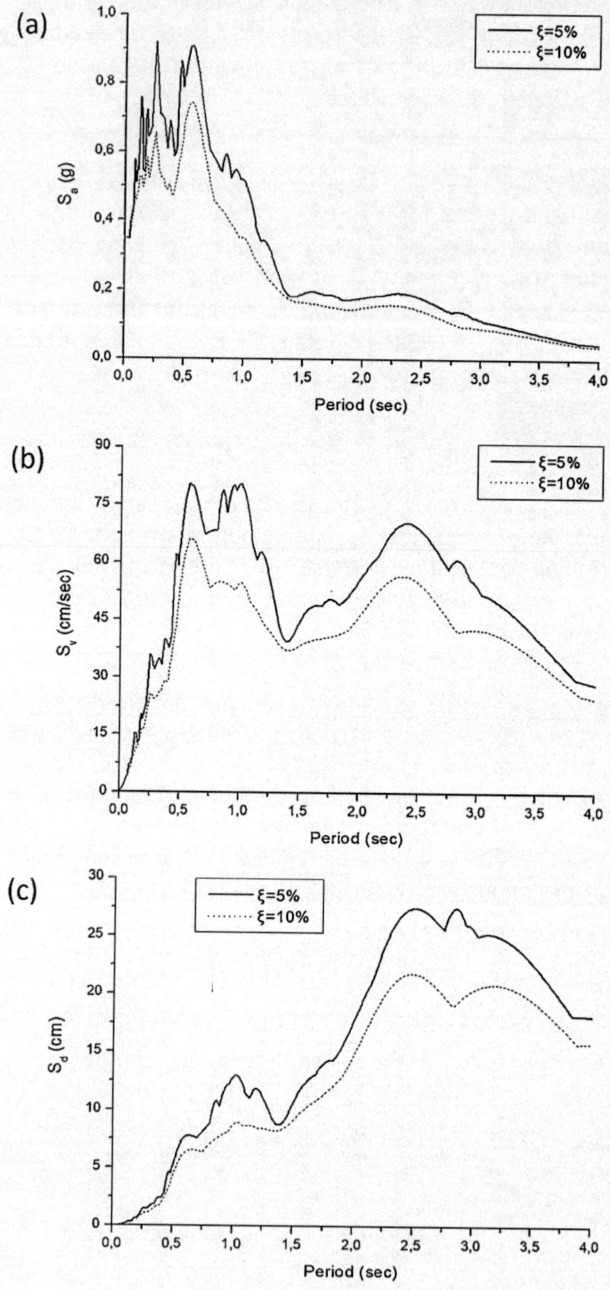

Fig. 2.3 Elastic response spectra for the NS component of the 1940 El Centro earthquake: (**a**) displacement S_d; (**b**) velocity S_v; (**c**) acceleration S_a

where I is the unit vector and T denotes transposition. The total seismic base shear force V can be finally determined with the aid of the SRSS modal combination rule in the form

$$V = \left(V_{1\ max}^2 + V_{2\ max}^2 + \ldots + V_{n\ max}^2\right)^{0.5} \tag{2.17}$$

2. *Numerical time integration*

The various existing algorithms of numerical time integration of Eq. (2.1) assume the solution u to be known at the time instant t and they determine it at the next time instant $t + \Delta t$, where Δt is the time step, which is selected appropriately. For building structures the step-by-step time integration algorithms are usually implicit ones, like those of Newmark, Houbolt and Wilson-θ, which are described in the books of Bathe (1996) and Chopra (2007). Implicit time integration algorithms, due to their inherent numerical damping, eliminate the contribution of higher modes to the response, in agreement with the basic property of a framed structure experiencing a dynamic response depending mainly on the first few modes of vibration.

The successive steps of computation according to the algorithm of Newmark for the solution of Eq. (2.1) have as follows:

1. With known initial conditions u_0 and $\dot{u}_0$, one can write Eq. (2.1) as

$$M\ddot{u}_0 = R - C\dot{u}_0 - Ku_0 \tag{2.18}$$

and solve it for $\ddot{u}_0$

2. After selecting an appropriate time step Δt, one computes

$$\begin{aligned} \overline{K} &= K + (\gamma/\beta\Delta t)C + \left(1/\beta\Delta t^2\right)M \\ \mathrm{A} &= (1/\beta\Delta t)M + (\gamma/\beta)C \\ B &= (1/2\beta)M + \Delta t[(\gamma/2\beta) - 1]C \end{aligned} \tag{2.19}$$

where β and γ are the two parameters of the algorithm

3. Assuming that at every time instant t, u_t, $\dot{u}_t$ and $\ddot{u}_t$ are known, one computes

$$\overline{\Delta R} = \Delta R + A\dot{u}_t + B\ddot{u}_t \tag{2.20}$$

and solves the equation

$$\overline{K}\Delta u = \overline{\Delta R} \tag{2.21}$$

for the displacement increment Δu, where Δ expresses a very small but finite increment and subscripts denote time instants.

4. With known Δu, one determines

$$\begin{aligned}\Delta \dot{u} &= (\gamma/\beta\Delta t)\Delta u - (\gamma/\beta)\dot{u}_t + \Delta t[1 - (\gamma/2\beta)]\ddot{u}_t \\ \Delta \ddot{u} &= (1/\beta\Delta t^2)\Delta u - (1/\beta\Delta t)\dot{u}_t - (1/2\beta)\ddot{u}_t\end{aligned} \tag{2.22}$$

and obtains the response at the next time instant $t + \Delta t$ as

$$\begin{aligned}u_{t+\Delta t} &= u_t + \Delta u \\ \dot{u}_{t+\Delta t} &= \dot{u}_t + \Delta \dot{u} \\ \ddot{u}_{t+\Delta t} &= \ddot{u}_t + \Delta \ddot{u}\end{aligned} \tag{2.23}$$

5. With known response at the time instant $t + \Delta t$, one proceeds with its computation at the next time instant $(t + \Delta t) + \Delta t$ by repeating the computations indicated by Eqs. (2.20)–(2.23).

This algorithm for $\beta = 1/4$ and $\gamma = 1/2$ (constant acceleration) is unconditionally stable, i.e., its stability does not depend on the selected time step Δt. However, under stable conditions, its accuracy depends on the time step Δt. In general, the selection of the appropriate time step Δt is of decisive importance for securing stability and accuracy of the solution. When an algorithm is unconditionally stable, like the above algorithm of Newmark, a time step Δt, e.g., less than or equal to 1/10 of the natural period T_j of the building, where j is the number of the first significant modes, is usually enough for obtaining response results of acceptable accuracy. More details about the selection of Δt for acceptable response results can be found in the book of Bathe (1996).

The viscous damping matrix C is assumed to be of the Rayleigh type and is of the form

$$C = \alpha_1 M + \alpha_2 K \tag{2.24}$$

where the damping constants α_1 and α_2 are determined in terms of two values of modal damping ratios and corresponding natural frequencies. The selected modes include the first one and one of the higher modes. It is suggested to select frequencies corresponding to periods equal to $0.02T_1$ and $1.5T_1$, where T_1 is the fundamental period (Deierlein et al. 2010). On the assumption that the two selected modes i and j have the same damping ratios ($\xi_i = \xi_j = \xi$), one has (Bathe 1996)

$$\alpha_1 = \left(\frac{2\omega_i\omega_j}{\omega_i + \omega_j}\right)\xi\ , \quad \alpha_2 = \left(\frac{2}{\omega_i + \omega_j}\right)\xi \tag{2.25}$$

2.2.3 Additional Remarks on Dynamic Elastic Analysis

This section provides a number of additional remarks pertaining to the two above mentioned methods of elastic dynamic analysis of framed buildings. The elastic force vector f of the building under the seismic motion is of the form

$$f = Ku \tag{2.26}$$

Substituting u by its expression in Eq. (2.5) and $K\varphi$ by its expression in Eq. (2.4), Eq. (2.26) takes the form

$$f = \sum_{i=1}^{n} \omega_i^2 M \varphi_i q_i(t) \tag{2.27}$$

Thus, the base shear force due to the seismic motion takes the form

$$V(t) = I^T f = \sum_{i=1}^{n} \omega_i^2 q_i(t) I^T M \varphi_i = \sum_{i=1}^{n} m_i^* \omega_i^2 q_i(t)/\Gamma_i \tag{2.28}$$

where

$$m_i^* = L_i^2/m_i \tag{2.29}$$

with

$$L_i = I^T M \varphi_i = \varphi_i^T M I \ , \ \ m_i = \varphi_i^T M \varphi_i \tag{2.30}$$

The m_i^* in Eq. (2.29) is called effective modal mass of the building for mode i and represents that part of the total building mass m responding in mode i during the seismic motion. The ratio $m_i^*/m_i < 1$ represents the relative contribution of mode i to the total response of the building. The ratio of the sum of m_i^* which are taken into account in the analysis over the total building mass m, i.e., $\sum_{i=1}^{r<n} m_i^*/m$, is used by current codes as a criterion for determining the minimum number of modes r required for the computation of the seismic response with satisfactory accuracy and small computational effort. According to the EC8 (2004), this ratio should not be lower than 0.90 for every major direction of the building which is analyzed.

Lopez and Cruz (1996) proposed practical rules for the determination of the required number of modes, so as the relative response error not to exceed 5%. According to them, the required number r of modal shapes is given in terms of the number of stories N and the ratio T/T_2, where T is the fundamental period of the building frame and T_2 is the characteristic period of the response spectrum at the intersection of the horizontal and the curved part of the spectrum line. More specifically

$$\begin{aligned}
&For\ N \leq 20\\
&r = 3\ for\ T/T_2 \leq 1.5\\
&r = 0.5\ (T/T_2 - 1.5) + 3.0\ for\ T/T_2 > 1.5\\
&For\ 20 < N \leq 50\\
&r = 4\ for\ T/T_2 \leq 1.5\\
&r = (2/3)\ (T/T_2 - 1.5) + 4.0\ for\ 1.5 < T/T_2 \leq 6.0\\
&r = 0.5\ (T/T_2 - 6.0) + 7.0\ for\ T/T_2 > 6.0
\end{aligned} \tag{2.31}$$

Thus, the method of modal superposition for elastic dynamic building analysis, not only succeeds in uncoupling the equations of motions, but also in using the first few modes significantly contributing to the response. This implies that the method of modal superposition is more efficient than the one of numerical time integration, especially in cases of seismic motions with a long duration, which require a large number of time steps. Of course, mode superposition requires the solution of the problem of free vibrations for the computation of natural frequencies and modal shapes (at least a small number of them). However, the knowledge of the first few modal shapes and natural frequencies is also very useful even for the case of using numerical time integration, because it provides a better understanding of the dynamic behavior of the building and helps in selecting an appropriate value of the time step. Needless to say, when the method of mode superposition is used for design purposes in conjunction with response spectrum analysis, its efficiency over that of numerical time integration increases considerably. In conclusion, for elastic dynamic analysis of buildings, the method of modal superposition is a better choice than the method of numerical time integration. However, for nonlinear dynamic analysis of buildings, which is described in the next section, even though several researchers have attempted the application of modal superposition, e.g., Villaverde and Hanna (1992), the numerical time integration is the best choice.

Another aspect that deserves further discussion is the way of combining modal responses for design purposes, The usual way of combining modal responses is by using the SRSS rule, e.g. in Eq. (2.12). The accuracy of this rule has been checked many times in the past and it has been found to be acceptable, provided that the building has modal shapes/natural frequencies that are not very close. In cases where there are indeed modal shapes/natural frequencies very close, the rule of the complete quadratic combination (CQC) is a better choice. For a response quantity X with modal values X_i, the CQC rule reads (Chopra 2007)

$$X = \left(\sum_{i=1}^{k} \sum_{j=1}^{k} P_{ij} X_i X_j \right)^{1/2} \tag{2.32}$$

where P_{ij} is a correlation coefficient between modes i and j given by the relation

$$P_{ij} = \frac{8\sqrt{\xi_i \xi_j}(\xi_i + \lambda_{ij}\xi_j)\lambda_{ij}^{3/2}}{\left(1 - \lambda_{ij}^2\right)^2 + 4\xi_i\xi_j\lambda_{ij}\left(1 + \lambda_{ij}^2\right) + 4\left(\xi_i^2 + \xi_j^2\right)\lambda_{ij}} \tag{2.33}$$

with $\lambda_{ij} = \omega_i/\omega_j$, ω_i and ω_j being the natural frequencies of modes i and j and ξ_i and ξ_j the viscous damping ratios of modes i and j, respectively. If all modes are apart enough among them, then $P_{ij} \approx 0$ for $i \neq j$ and since $P_{ij} = 1$ for $i = j$, expression (2.32) becomes the same with that of the SRSS rule.

2.3 Nonlinear Global Analysis

2.3.1 *Structural Modeling*

Steel building frames under strong ground motions exhibit large inelastic deformations and thus their modeling and analysis should take into account material and geometric nonlinearities.

In general, there are basically three levels of complexity for dynamic inelastic modeling of steel building frames (Anagnostopoulos 1997). At the lower complexity level, one represents the whole building or its stories by simple nonlinear model or models, respectively. Thus, a whole building can be modeled by a single-degree-of-freedom (SDOF) system, whose behavior is described by a multi-linear force-displacement relation that can be determined through a nonlinear static analysis involving gradual increase of lateral loads (pushover analysis). A whole building can be also modeled by a number of nonlinear elements equal to the number of its stories, if this building exhibits a shear building behavior. In that case, e.g., the stiffnesses k_i ($i = 1...n$) in Fig. 2.2b will not be constants as in the linear elastic case, but functions of deformation. The models of this first level of complexity cannot describe the individual inelastic behavior of the members of the framed structure as well as the stress-redistribution among the members.

According to the second level of modeling complexity, the members of frame buildings (axial elements, beams and columns) are modeled as finite elements with elastic behavior and concentration of all inelastic behavior at their two ends. These models, which restrict all their inelastic behavior at their two ends, are called models with plastic hinges or models of concentrated plasticity. The inelastic behavior at those plastic hinges is describe by an ideal elastoplastic model or a multi-linear elastoplastic model with hysteresis, i.e., with the capability to exhibit cyclic behavior without or with degradation (deterioration). Well known computer programs employing concentrated plasticity models are, e.g., the DRAIN-2DX & 3DX (Prakash et al. 1993, 1994) and Ruaumoko-2D & 3D (Carr 2005).

In the third and more advanced level of nonlinear models one has models of distributed plasticity, which are capable of simulating the propagation of yielding and plasticity along the length of beam-columns and along the height of their cross-section. The most well-known model of this category is the fiber model, where the beam-column element is discretized into subelements with behavior referred to the central cross-section of every element, which is divided to a finite number of fibers with a nonlinear stress-strain relation along the fiber length. This model despite its high accuracy, is very complicated and requires very high computational resources, especially for large space framed buildings. Among the many works on the subject of distributed plasticity modeling in conjunction with nonlinear dynamic analysis of framed buildings, one can mention those of Elnashai and Izzuddin (1993), Challa and Hall (1994) and Chi et al. (1998). As a compromise between accuracy and simplicity, a concentrated plasticity model with fibers of finite length at the two ends of a beam-column element has also been developed and included at DRAIN-3DX (Prakash et al. 1994).

The abovementioned nonlinear models of intermediate and advanced level of complexity also take into account the flexibility of the beam to column joints, either by simple rotational springs or by a combination of springs in order to simulate the panel zone effect at those joints (Challa and Hall 1994; Schneider and Amidi 1998; Chi et al. 1998). The most well-known models for panel zone effects in steel frames are the Krawinkler (1978) and the Scissors (FEMA 356 2000) models. The Krawinkler model resists panel shear by an elastoplastic plane stress element and flange shear by four rotational springs, as shown in Fig. 2.4a. It consists of 12 nodes (with two nodes at every corner of the panel) and has 28 degrees-of-freedom with only four of them being truly independent. The Scissors model consists of two rigid links (one perpendicular to the other) forming a cross, two nodes and four degrees-of-freedom, as shown in Fig. 2.4b. The accuracy of the Krawinkler model is higher than that of the Scissors model (Charney and Marshall 2006) at the expense of increased complexity.

Detailed information regarding modeling issues relevant to concentrically-braced, eccentrically-braced and buckling-restrained braced frames can be found in Bruneau et al. (2011).

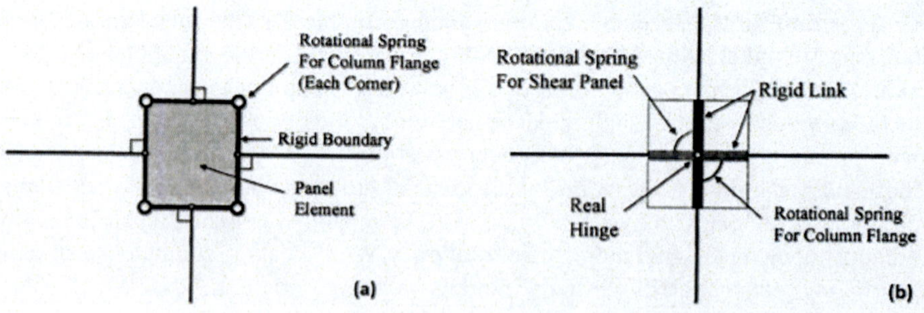

Fig. 2.4 Models for beam-column joints including panel zone effect: (**a**) Krawinkler model; (**b**) Scissors model (after Charney and Marshall 2006, reprinted with permission from AISC)

2.3.2 Dynamic Nonlinear Analysis

A dynamic nonlinear analysis of a steel framed building considers both material and geometric nonlinearities and works in a time domain FEM environment. The resulting nonlinear matrix equation of motion of a plane steel framed structure under a seismic excitation has the form

$$M\ddot{u} + C\dot{u} + F = R = -MI\ddot{u}_g \tag{2.34}$$

where F is the vector of the nodal internal forces which depend nonlinearly on the deformation and the rest of the matrices and vectors are the same as in the case of linear elastic behavior described by Eq. (2.1).

The above system of Eq. (2.34) is solved by a stepwise time integration as in the case of linear elastic behavior. However, here iterations are required at every time step due to the nonlinearity of the problem (Bathe 1996; Chopra 2007). Using the stepwise time integration of Newmark (with $\beta = 1/4$, $\gamma = 1/2$), the algorithm for solving Eq. (2.34) proceeds as follows:

1. With known initial conditions u_0 and $\dot{u}_0$, one can compute $F_0 = F(u_0, \dot{u}_0)$ and write Eq. (2.34) as

$$M\ddot{u}_0 = R - C\dot{u}_0 - F_0 \tag{2.35}$$

and solve it for $\ddot{u}_0$

2. After selecting an appropriate time step Δt, one computes

$$\begin{aligned} \mathrm{A} &= (4/\Delta t)M + 2C \\ B &= 2M \end{aligned} \tag{2.36}$$

3. Assuming that at every time instant t, u_t, $\dot{u}_t$ and $\ddot{u}_t$ are known, one computes

$$\overline{\Delta R_t} = \Delta R_t + A\dot{u}_t + B\ddot{u}_t \tag{2.37}$$

and then the tangent stiffness matrix K_t on the basis of the inelastic hysteretic material behavior

4. Utilizing vector $\overline{\Delta R_t}$ and matrix $\overline{K}_t$ of the form

$$\overline{K}_t = K_t + (2/\Delta t)C + \left(4/\Delta t^2\right)M \tag{2.38}$$

one can solve for Δu_t, as in the modified Newton-Raphson algorithm described separately below.

5. With known Δu_t, one determines

$$\begin{aligned} \Delta \dot{u}_t &= (2/\Delta t)\Delta u - 2\dot{u}_t \\ \Delta \ddot{u}_t &= (4/\Delta t^2)\Delta u - (4/\Delta t)\dot{u}_t - 2\ddot{u}_t \end{aligned} \tag{2.39}$$

and obtains the response at the next time instant $t + \Delta t$ as

$$\begin{aligned} u_{t+\Delta t} &= u_t + \Delta u_t \\ \dot{u}_{t+\Delta t} &= \dot{u}_t + \Delta \dot{u}_t \\ \ddot{u}_{t+\Delta t} &= \ddot{u}_t + \Delta \ddot{u}_t \end{aligned} \tag{2.40}$$

6. With known response at the time instant $t + \Delta t$, one proceeds with its computation at the next time instant $(t + \Delta t) + \Delta t$ by repeating the computations indicated by Eqs. (2.37)–(2.40).

The above algorithm is restricted to the materially nonlinear case. Geometric nonlinearities can be included as described in Sect. 2.3.5 (3).

As in the case of linear elastic analysis, time integration algorithms should be stable and accurate. Implicit algorithms are usually unconditionally stable, like the Newmark (with $\beta = 1/4$, $\gamma = 1/2$) algorithm, and require for accuracy an appropriately small time step Δt, which is usually selected on the basis of practical rules established only for linear problems (Chopra 2007).

Returning to step 4 above, iterations at every time instant t by the modified Newton-Raphson algorithm (Bathe 1996; Chopra 2007) have to be performed in order to determine Δu at that instant. The sequence of required computations with superscripts in parenthesis to stand for iteration cycles is as follows:

1. Initially one has that

$$u_{t+\Delta t}^{(0)} = u_t, F^{(0)} = F_t, \Delta Q^{(1)} = \Delta R_t, \overline{K}_T = \overline{K}_t \tag{2.41}$$

2. Then one can proceed to the equation

$$\overline{K}_T \Delta u^{(k)} = \Delta Q^{(k)} \tag{2.42}$$

to solve it for $\Delta u^{(k)}$

3. Afterwards one can determine

$$
\begin{aligned}
u_{t+\Delta t}^{(k)} &= u_{t+\Delta t}^{(k-1)} + \Delta u^{(k)} \\
\Delta F^{(k)} &= F^{(k)} - F^{(k-1)} + \left(\overline{K}_T - K_t\right)\Delta u^{(k)} \\
\Delta Q^{(k+1)} &= \Delta Q^{(k)} - \Delta F^{(k)}
\end{aligned}
\tag{2.43}
$$

4. Replacing k by $k + 1$ one can repeat the computations indicated by Eqs. (2.42) and (2.43).

It should be noted that in the above algorithm the tangent stiffness matrix $\overline{K}_T$ is computed only once during the iterative process.

In addition to the algorithms of Newmark, the algorithms of Houbolt and Wilson, mentioned in Sect. 2.2.2 for the linear case, can also be used for the present nonlinear case in conjunction with the algorithm of Newton-Raphson for iterations. At this point it should be stressed that, dynamic inelastic analyses, which are closer to reality than any other method of analysis described in this chapter, are used (i) to verify the degree of accuracy of the other methods, (ii) to determine very useful parameters of inelastic behavior used in seismic codes, such as ductility μ or the behavior factor q and (iii) to perform direct seismic design of structures by determining through them the mean value of the maximum values of response quantities like displacements or the base shear force obtained for a number of seismic motions. Finally, it should be noticed that there are some advanced methods of time integration of nonlinear equations especially designed for large structures. The most well-known of those methods is the reduction method (Chopra 2007), which works with a number of global approximation vectors (basis vectors) much smaller than the total number of degrees-of-freedom of the structure and thus succeeds in reducing the computational load. The approximation is based on the fact that the structural response is excited mainly by the lower modes and not by all of them.

With regard to computer programs for nonlinear dynamic analysis of steel framed buildings, one can mention the more general DRAIN-2DX & 3DX (Prakash et al. 1993, 1994), SAP 2000 (2020), ETABS (2020), Ruaumoko-2D & 3D (Carr 2005) and OpenSees (2020) and the more special IDARC (Valles et al. 1996) and ADAPTIC (Izzuddin and Elnashai 1989). It is worth noticing that SAP 2000 (2020) and ETABS (2020) can perform not only analysis of steel structures but design under static or seismic loads in accordance with the provisions of the most widely used codes. For example, they can perform static and seismic design of steel structures according to Eurocodes EC3 (2009) and EC8 (2004). Out of these programs, OpenSees (2020) is an open source program, while SAP 2000 (2020) and ETABS (2020) are commercial. Other commercial programs that can perform nonlinear seismic analysis of steel structures are, e.g., ABAQUS (2020), ADINA (2020) and ANSYS (2020). However, these programs, even though they are extremely powerful, they are not very practical for the analysis of framed building structures. A dynamic analysis of framed buildings that takes into account both material and geometric nonlinearities is usually named nonlinear time history (NLTH) analysis.

Finally, one should point out that damping matrix C, which is of the Rayleigh type, is constant in linear problems and usually assumed to be constant in nonlinear problems as well. However, since C depends on the stiffness K, as Eq. (2.24) indicates, and stiffness K depends on the deformation in nonlinear problems (decreases during the motion), C is not really constant in nonlinear problems. Ruaumoko-2D & 3D (Carr 2005) and OpenSees (2020) computer programs have two options for the user: either to use a constant matrix C, computed for the initial value of K from Eq. (2.24), or to use a variable with time matrix C computed for the tangent stiffness K at every time step from Eq. (2.24). It has been observed (Bernal 1994; Hall 2006; Charney 2008; Zareian and Medina 2010; Chopra and McKenna 2016; Puthanpurayil et al. 2016) that the use of the initial stiffness Rayleigh damping model in nonlinear dynamic problems results in responses with unrealistic damping forces, especially for high damping and inelasticity. Various modifications of the Rayleigh model, including the tangent stiffness one, have been proposed by the above authors. However, it seems there is no clear consensus on this issue as yet.

2.3.3 *Static Nonlinear Analysis*

This analysis is also known as pushover analysis and is used as an easy but approximate way of determining the seismic response of a structure. This method of analysis takes into account the inelastic behavior of the structure and phenomena P-Δ but applies the seismic loads statically. More specifically, the structure is loaded by vertical constant dead and live loads as in the dynamic case as well as by lateral loads, which increase progressively until the final structural collapse. The distribution of these lateral loads along the height of the structure approximates the distribution of the inertial seismic loads. Usually, this approximation is based on a displacement distribution following the first mode shape.

The above simple method of analysis can be used in a 2D or 3D framework for the discovery of probable strength weakness as well as for an easy verification of the various phases of the deformational behavior of the structure. It can also be used for design purposes as complementary method to construct the force-displacement relationship of a SDOF in the framework of the N2 method (EC8 2004; Fajfar 2000; Fajfar et al. 2005). However, it cannot replace the dynamic inelastic analysis as the best method of design verification, mainly because of the monotonic way of the lateral load imposition instead of the real cyclic one and the constant along the height load distribution instead of the real one that varies with time. In an effort to improve the accuracy of pushover analysis, the adaptive pushover (Antoniou and Pinho 2004; Kalkan and Kunnath 2006) or the modal pushover (Chopra and Goel 2002; Reyes and Chopra 2011) analyses have been developed at the expense of increasing complexity to the extent that its advantage of simplicity over the dynamic analysis to be lost. For more information on this method, its characteristics and its applications, one can consult the comprehensive articles of Krawinkler and Seneviratna (1998), Krawinkler (2006), Baros and Anagnostopoulos (2008), Fajfar (2018) and Peres et al. (2020).

2.3.4 *Dynamic Inelastic Spectrum Analysis*

As it has been pointed out in this and the previous chapter, modern seismic codes accept inelastic structural response to the design seismic motion in order to secure safety at a reasonable cost. An inelastic structure is dynamically analyzed for design purposes either directly, by employing a series of NLTH analyses, or indirectly by employing a static elastic analysis in conjunction with an inelastic spectrum. An inelastic response spectrum can be constructed on the basis of an inelastic SDOF system under specific ground motion analyzed by NLTH analysis. However, all inelastic design spectra used in seismic codes nowadays consist of elastic design spectra whose ordinates have been reduced by the behavior factor q, which takes into account approximately the effect of the inelastic behavior of the structure. Much more on this subject of response and design spectra can be found in the next chapter dealing with the force-based design method in the framework of EC8 (2004).

2.3.5 *Additional Remarks on Dynamic Nonlinear Analysis*

In this section, some additional aspects directly related to dynamic nonlinear structural analysis are briefly discussed for reasons of completeness and pertinent references are provided for a more in depth study. These aspects are the following:

1. *Hysteretic modeling for cyclic loading*

   The nature of earthquake loading is cyclic exhibiting many load reversals during the time interval of its action. As a result of that, the inelastic material behavior of a steel element is hysteretic exhibiting also many stress reversals and creating loops in the force-displacement diagram with an area that represents the energy of dissipation. These loops are stable if the element does not show any degradation or deterioration during the cyclic process (Fig. 2.5a). In reality, this happens only during the first few cycles and then the material exhibits stiffness and strength deterioration, which progressively increases from cycle to cycle until final failure, as shown in Fig. 2.5b. This figure, shows the original stable hysteretic loop (with the dashed line) and the subsequent deteriorating hysteretic loops (with solid lines), as well as the backbone curve (with the heavy solid line).

Figure 2.6 depicts a typical idealized backbone curve corresponding to the cyclic one of Fig. 2.5b, which is defined in terms of the three points A, B and C with coordinates (d_y, F_y), (d_u, F_u) and (d_m, 0), respectively. Empirical expressions for d_y, d_u and d_m can be obtained from experiments (Lignos and Krawinkler 2011; Lignos

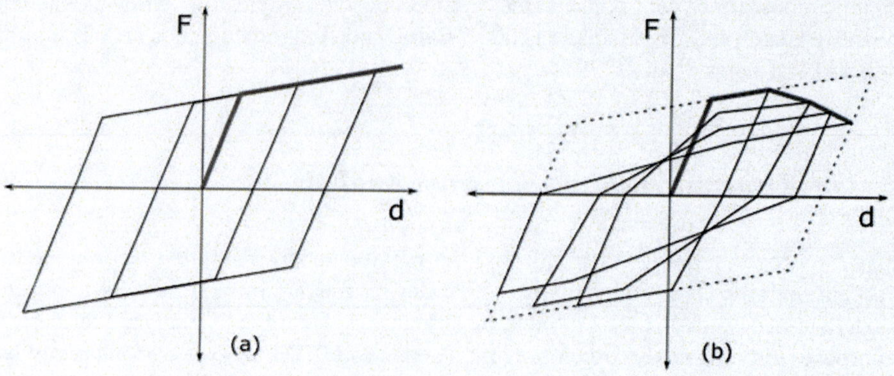

Fig. 2.5 Force-displacement hysteretic behavior without (**a**) and with (**b**) stiffness and strength deterioration

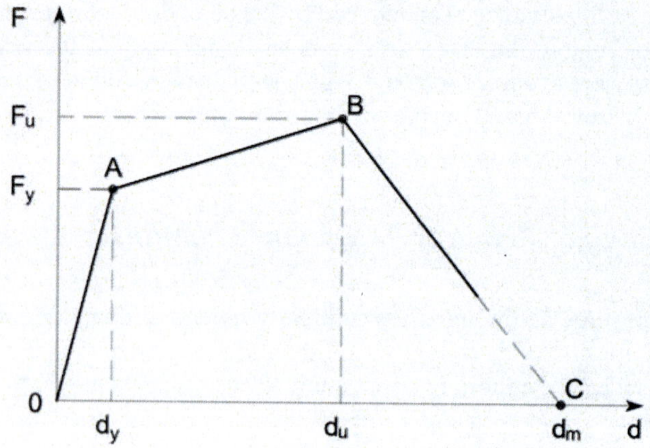

Fig. 2.6 Typical backbone curve for steel

et al. 2019). In accordance with Fig. 2.6, strength cyclic degradation can be simulated by establishing a functional relationship between yield strength reduction and ductility, which has as parameters the ductilities d_u/d_y and d_m/d_y corresponding to the beginning and end of the degradation process.

In the above described cyclic deterioration, strength loss occurs in subsequent cycles and not in the same loading cycle. In in-cycle strength deterioration, strength loss occurs within a cycle. These two kinds of strength deterioration are shown in Fig. 2.7.

In steel and steel members, the first kind occurs, e.g., due to the Bauschinger effect, while the second, e.g., due to local and/or lateral torsional buckling and fracture of steel. In-cycle deterioration is more damaging than cyclic one, even though both kinds of deterioration appear in practice.

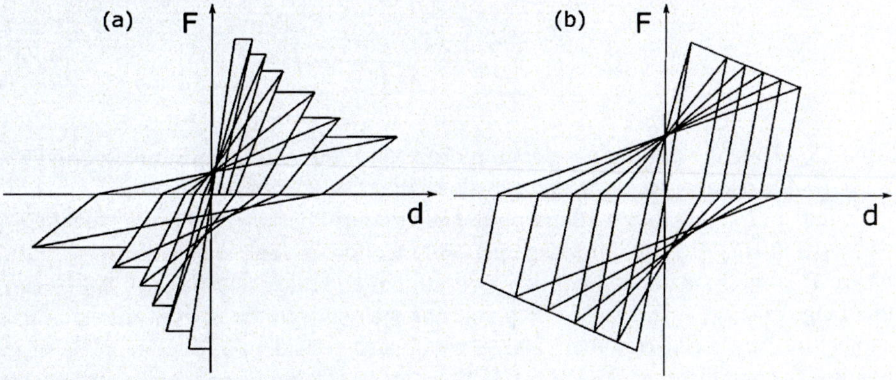

Fig. 2.7 The two kinds of strength deterioration: (**a**) cyclic deterioration; (**b**) in-cycle deterioration

More on this subject of hysteretic behavior of steel members can be found in Sivaselvan and Reinhorn (2000), Ibarra et al. (2005), FEMA 440 (2005), Lignos and Krawinkler (2011), Lignos et al. (2019) and the review of Deierlein et al. (2010).

2. *Selection and scaling of earthquakes*

   Nonlinear time history (NLTH) analyses are mainly used for design verification and performance assessment of structures. The earthquake ground motions used in such analysis should be selected from seismic databases like COSMOS (2020) or PEER (2020) and scaled in such a way so that they represent the seismic hazard of interest as accurately as possible.

   Selection of recorded ground motions can be done on the basis of the following criteria (Katsanos et al. 2010):

(a) Earthquake magnitude (M) and distance (R) in km of the fault from the site of interest
(b) Soil profile at the site of interest, strong motion duration, geophysical/seismological parameters (type of fault, rupture mechanism, directivity of seismic waves) and acceleration to velocity ratio of seismic waves
(c) Spectral matching, which is the main selection criterion by seismic codes and is based on establishing compatibility between the response spectra of the selected records and a target spectrum, which is the elastic (or design) spectrum, as defined by codes. It should be pointed out that, the generation of spectrum compatible artificial acceleration records of usually high energy content, is different from spectral matching. It is recommended to select seismic records first on the basis of magnitude and distance and then use those selected records for spectral matching. As an example of spectral matching, one can present the one by Iervolino et al. (2010), which defines the average spectrum deviation δ between a record set and the code spectrum as

$$\delta = \sqrt{\frac{1}{n}\sum_{i=1}^{n}\left[\frac{S_{aj}(T_i) - S_{ac}(T_i)}{S_{ac}(T_i)}\right]^2} \quad (2.44)$$

where $S_{\mathrm{aj}}(T_i)$ is the pseudo-acceleration ordinate of the real spectrum j at period T_i, $S_{\mathrm{ac}}(T_i)$ is the pseudo-acceleration ordinate of the code spectrum at period T_i and n is the number of values used within a predefined range of periods. According to ASCE/SEI 7-16 (2017), the recommended period range for far-fault sites is $0.2T_i - 1.5T_i$, where T_i is the fundamental period of the structure, while according to EC8 (2004) this range is $0.2T_i - 2.0T_i$. At this point, one may consult the review article on the subject by Beyer and Bommer (2007), which also considers bi-directional seismic loading, the Whittaker et al. (2011) NIST Report on selecting and scaling earthquake motions with a summary in Haselton et al. (2012), the article of Heo et al. (2011) comparing amplitude scaling with spectrum matching, the review article of Deierlein et al. (2010) for practicing engineers and the ASCE/SEI 7-16 (2017) document on minimum loads on structures.

In the context of EC8 (2004), concerning the minimum number of seismic records required for NLTH analyses this is three. When at least seven records are used, the mean structural response is acceptable. However, when three to six records are used, only the maximum structural response is acceptable (Katsanos et al. 2010).

3. *Geometric nonlinear effects*

Steel structures, because of their flexibility, can experience large displacements under small strains. Thus, gravity loads acting on the deformed structure can create second order moments at the member and structural levels, the so called P-δ and P-Δ effects. The former effects in buildings have to do with the lateral member displacements, while the latter with building lateral displacements, i.e., storey drifts. In buildings under seismic loading, P-δ effects do not have to be modeled in NLTH analyses (Deierlein et al. 2010). However, P-Δ effects have to be modeled since they reduce structural lateral stiffness and strength and eventually drive the structure to dynamic instability.

These effects are illustrated on the basis of the behavior of the one storey building under gravity load P and lateral load V shown in Fig. 2.8. One can easily prove (Bernal 1987) that the total lateral stiffness $K = K_f - K_g$, where $K_f = V_y/\Delta_y$ is the flexural and $K_g = P/h$ is the geometrical stiffness. This clearly indicates a stiffness reduction due to the P-Δ phenomenon. One can also easily prove that by introducing the stability coefficient $\theta = P\Delta_y/V_yh$, the relation $V_y^* = V_y(1-\theta)$ can be obtained (Bernal 1987), where V_y and V_y^* are the shear forces at yielding without and with P-Δ effects, respectively. This clearly indicates a strength reduction due to the P-Δ phenomenon. Figure 2.9 depicts the shear force V versus displacement Δ relation with and without the P-Δ effects. One can see from Fig. 2.9 the stiffness and strength reduction due to the P-Δ effect pictorially. Geometric nonlinearities involving large displacements and P-Δ effects can be included in the stepwise time integration algorithm of Newmark with the aid of the geometric stiffness matrix K_g at the

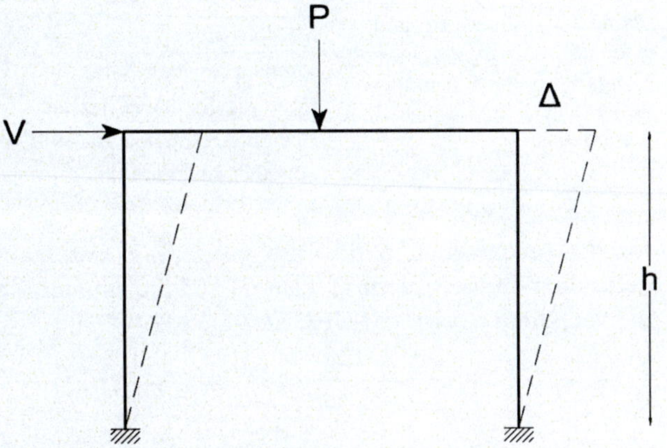

Fig. 2.8 One storey shear building under gravity and lateral forces

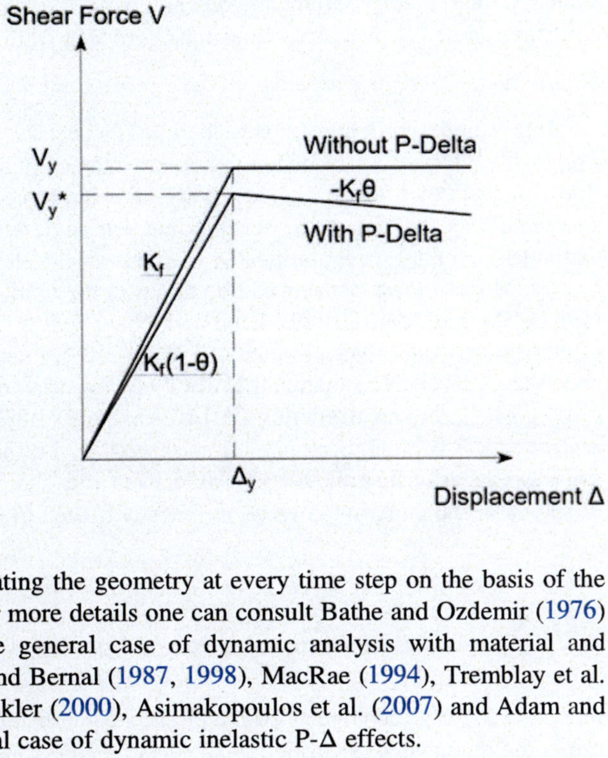

Fig. 2.9 Influence of P-Δ effects on shear force-displacement relation

element level and by updating the geometry at every time step on the basis of the current displacements. For more details one can consult Bathe and Ozdemir (1976) and Bathe (1996) for the general case of dynamic analysis with material and geometric nonlinearities and Bernal (1987, 1998), MacRae (1994), Tremblay et al. (1999), Gupta and Krawinkler (2000), Asimakopoulos et al. (2007) and Adam and Jager (2012) for the special case of dynamic inelastic P-Δ effects.

4. *Fundamental period of steel frames*

In order to determine seismic forces according to response spectrum analysis or to select the time step as a fraction of the fundamental period in time domain analysis by numerical integration, or to determine the seismic intensity measure defined as the spectral acceleration at the fundamental period, an approximate estimate of the fundamental period of the structure to be designed is required. Empirical expressions

Table 2.1 Values of C and x for steel buildings: values of C are in *ft* (outside parentheses) and in *m* (inside parentheses)

Structural type	C	x
Moment resisting frames	0.028 (0.0724)	0.8
Eccentrically braced frames	0.030 (0.0731)	0.75
Buckling-restrained braced frames	0.030 (0.0731)	0.75
All other structural types	0.020 (0.0488)	0.75

for the fundamental period are given in codes for various building structures depending on the structural material and the structural typology. All of them are of the very simple form (ASCE/SEI 7-16 2017)

$$T = CH^x \tag{2.45}$$

where T is the fundamental period in sec, H is the total height of the building in *ft* or *m* and C and x are coefficients depending on the type and the material of the structure. Table 2.1 provides values for C and x for various types of steel buildings.

5. *Additional modeling aspects*

More refined modeling of steel buildings can take into account various other aspects like the axial force-bending moment interaction in columns, the flexibility of floors in their own plane, the flexibility of beam-to-column joints, the torsion of space building structures, the seismic incident angle effect, the effect or repeated earthquakes, or the irregularities in plan view and elevation of plane and space frames. Most of these aspects can be found in the works, e.g., of Nakashima et al. (1984), Shi and Atluri (1989), Chui and Chan (1996), Tena-Colunga and Abrams (1996), Ju and Lin (1999), Foutch and Yun (2002), Fragiacomo et al. (2004), Zhao and Wong (2006), Krishnan and Hall (2006a, b), Liao et al. (2007), Loulelis et al. (2012), McCrum and Broderick (2014), Kostinakis et al. (2018) and the excellent review article of Anagnostopoulos et al. (2015). Furthermore, information about these aspects as well as about others associated with special cases, like base isolation or supplemental damping devices, are mostly treated in Chaps. 12 and 13.

2.3.6 *Elastic and Inelastic Design Spectra*

In Sect. 2.2.2 (2), the concept of the elastic response spectrum was defined and its use in the framework of elastic modal superposition was described for determining the maximum elastic structural response to a seismic excitation. However, a structure cannot be designed on the basis of just one elastic response spectrum corresponding to a single earthquake at its site. Furthermore, because a response spectrum is anomalous (Fig. 2.3), especially for small values of damping, and the position of its peaks and valleys change from one earthquake to another of similar characteristics, the structure has to be designed on the basis of smoothed or envelope response spectra, which take into account the possible variations between earthquakes that may occur at that site. Thus, for the seismic design of a structure at a particular site, an elastic design spectrum is constructed on the basis of estimated peak values of ground acceleration (*PGA*), velocity (*PGV*) and displacement (*PGD*) there. On the assumption that there is a number of seismic records available from the site of interest or other sites with the same seismological conditions, those records are normalized so as to have the same maximum ground acceleration. For these normalized ground motions, the corresponding response spectra are constructed and their mean spectral values are determined to produce the mean response spectra. The resulting curves are finally idealized by straight line segments thereby resulting in the elastic design spectra, like the pseudo-acceleration design spectra of Fig. 2.10.

A qualitative comparison between the acceleration response spectrum curves of Fig. 2.3a and the acceleration design spectrum curves of Fig. 2.10 clearly shows how the anomalous with peaks and valleys curves of Fig. 2.3a become the much smoother curves of Fig. 2.10.

An inelastic response spectrum can be constructed on the basis of the response of an inelastic SDOF system to a specific seismic motion. Assuming ideal elastoplastic material behavior with initial stiffness k, yield displacement u_y, maximum displacement u_m and maximum or yield force f_y (Fig. 2.11), one can write Eq. (2.34) for a SDOF system in the form

$$\ddot{u} + 2\xi\omega\dot{u} + f/m = -\ddot{u}_g \tag{2.46}$$

where $f = f(u, \dot{u})$ is the resistance force. For the particular case of zero damping and after defining the ductility μ as

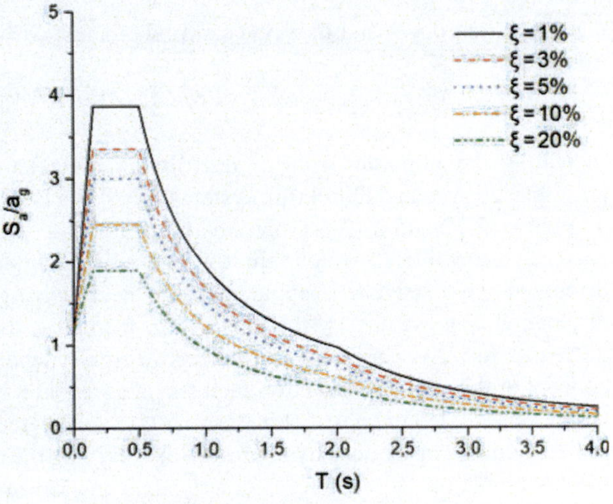

Fig. 2.10 Elastic pseudo-acceleration design spectrum of EC8 (2004) for various amounts of damping ratio ξ (Spectrum type 1, $a_g = 0.24$ g, soil type B and $\xi = 1\%, 3\%, 5\%, 10\%$ and 20%)

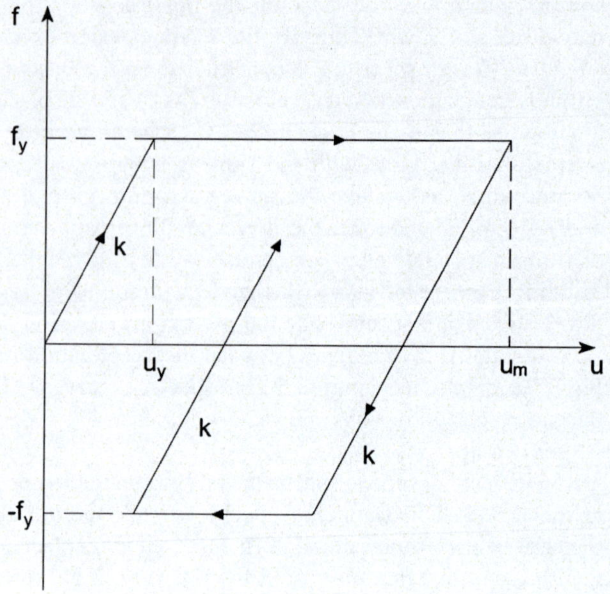

Fig. 2.11 Ideal elastoplastic material behavior

$$\mu = u_m / u_y \tag{2.47}$$

one can solve Eq. (2.46) by numerical stepwise time integration and obtain $u(t)$ and $\ddot{u}(t)$ and hence S_d and S_a. With reference Fig. 2.11, the relations

$$f_y = ku_y = m\omega^2 u_y, f_y = mS_a \tag{2.48}$$

can be written down and thus one can find

$$S_a = \omega^2 u_y = \omega^2 (u_m/\mu) = \omega^2 (S_d/\mu) \tag{2.49}$$

indicating the important role of ductility in the relation between S_a and S_d for elastoplastic systems. For elastic systems $\mu = 1$ and Eq. (2.49) reduces to Eq. (2.12).

Figures 2.12 and 2.13 present inelastic pseudo-acceleration response and design spectra, respectively, which are the inelastic counterparts of the elastic ones discussed previously. It is apparent from these two figures that as the energy of dissipation, represented by the ductility μ, increases, the strength f_y of the system decreases form its elastic value and this decrease depends on period T. Thus, the concept of the strength reduction factor R_y was introduced for elastoplastic systems, which can be used to construct inelastic pseudo-acceleration design spectra from the corresponding elastic ones by appropriately dividing the ordinates of the elastic ones by this reduction factor. This factor R_y is a function of μ and T. At this point the articles of Vidic et al. (1994) and Fajfar and Vidic (1994) on inelastic design spectra

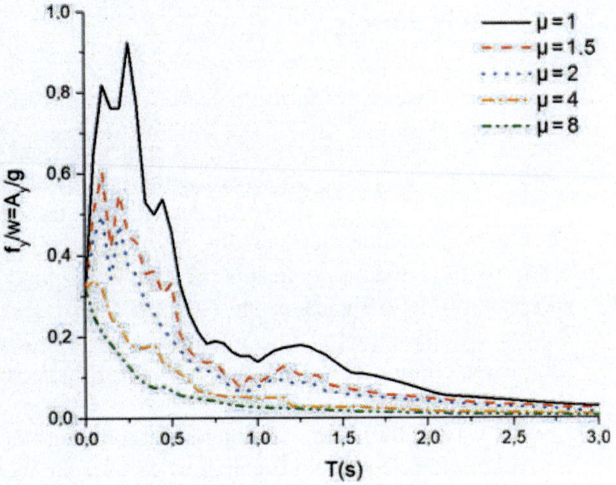

Fig. 2.12 Response spectrum for elastoplastic systems with 5% damping under the 1940 El Centro motion ($w = \text{mg}, A_y = \omega^2 u_y$)

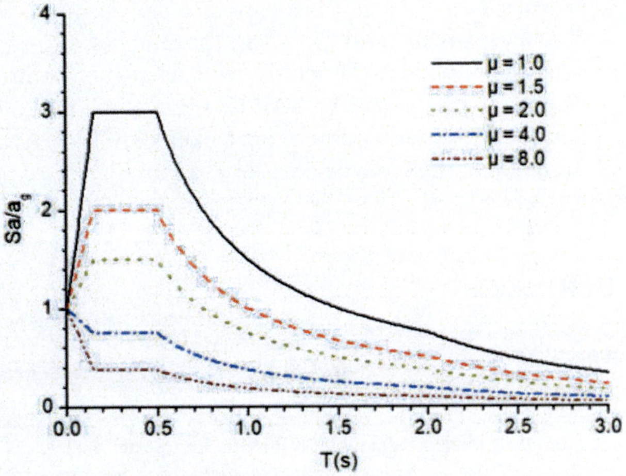

Fig. 2.13 Constant-ductility pseudo-acceleration spectrum of EC8 (2004), type 1, for various values of μ (ξ = 5%, *PGA* = 0.36 g, soil type B)

with explicit expressions of R in terms of μ and T for various types of material behavior should be mentioned. It should be stressed that the R (or q) factor for a structure as defined by Eq. (1.6) of Chap. 1 is higher than the corresponding one for a simple elastoplastic SDOF system as that of Fig. 2.11, exactly because it includes not only the ductility effect but also the one of overstrength. In seismic codes, the strength reduction factor R (or behavior factor q in EC8 2004), which includes the effect of overstrength, is assigned constant values for the various types of structures and their material. More on the subject of code-based inelastic design spectra can be found in the next chapter dealing with the description and application of EC8 (2004) to seismic design of steel building structures.

2.4 Conclusions

The presented discussion in the preceding sections of this chapter can lead to the following conclusions:

1. This chapter has briefly presented in the framework of the finite element method linear and nonlinear methods for determining the dynamic response of building structures to seismic motions.
2. Various fundamental concepts of dynamic analysis of building structures, like natural frequencies or periods, modal shapes, viscous damping, modal superposition, stepwise time integration or time history analysis, material and geometric nonlinearities, elastic and inelastic spectra, have been presented and discussed.
3. Special topics, like hysteretic material modeling under cyclic loading including deterioration, selection and scaling of earthquake records, dynamic inelastic P-Δ effects, pushover methods, structural modeling and empirical formulae for the fundamental period of various types of steel frames have been also presented and discussed.
4. The most widely used special and general purpose computer programs for seismic analysis of building structures have been very briefly presented. Special purpose computer programs like SAP/ETABS, Ruaumoko, or OpenSees are to be preferred over general purpose ones like ABAQUS, ANSYS, or ADINA because of simplicity, efficiency and easier availability.

References

ABAQUS (2020) Abaqus unified FEA, complete solutions for realistic simulation. Dassault Systems Simulia Corp, Johnston, RI

Adam C, Jager C (2012) Simplified collapse capacity assessment of earthquake excited regular framed structures vulnerable to P-delta. Eng Struct 44:159–171

ADINA (2020) Automatic dynamic incremental nonlinear analysis, premier simulation software for advanced analysis. ADINA R & D Inc, Watertown, MA

Anagnostopoulos SA (1997) Buildings. In: Beskos DE, Anagnostopoulos SA (eds) Computer analysis and design of earthquake resistant structures: a handbook. Computational Mechanics Publications, Southampton, pp 369–440

Anagnostopoulos SA, Kyrkos MT, Stathopoulos KG (2015) Earthquake induced torsion in buildings: critical review and state of the art. Earthq Struct 8:305–377

ANSYS (2020) ANalysis SYStems, engineering simulation & 3D design software. ANSYS Inc, Canonsburg, PA

Antoniou S, Pinho R (2004) Development and verification of a displacement based adaptive pushover procedure. J Earthq Eng 8:643–661

ASCE/SEI 7-16 (2017) Minimum design loads and associated criteria for buildings and other structures. American Society of Civil Engineers, Reston, VI

Asimakopoulos AV, Karabalis DL, Beskos DE (2007) Inclusion of P-Δ effect in displacement-based seismic design of steel moment resisting frames. Earthq Eng Struct Dyn 36:2171–2188

Baros D, Anagnostopoulos SA (2008) An assessment of static non-linear pushover analyses in 2-D and 3-D applications. In: Bento R, Pinho R (eds) 3D pushover 2008-nonlinear static methods for design/assessment of 3D structures. IST Press, Lisbon

Bathe KJ (1996) Finite element procedures. Prentice Hall, Englewood Cliffs, NJ

Bathe KJ, Ozdemir H (1976) Elastic-plastic large deformation static and dynamic analysis. Comput Struct 6:81–92

Bernal D (1987) Amplification factors for inelastic dynamic P–Δ effects in earthquake analysis. Earthq Eng Struct Dyn 15:635–651

Bernal D (1994) Viscous damping in inelastic structural response. J Struct Eng ASCE 120:1240–1254

Bernal D (1998) Instability of buildings during seismic response. Eng Struct 20:496–502

Beyer K, Bommer JJ (2007) Selection and scaling of real accelerograms for bi-directional loading: a review of current practice and code provisions. J Earthq Eng 11(S1):13–45

Bruneau M, Uang CM, Sabelli R (2011) Ductile design of steel structures, 2nd edn. McGraw Hill, New York

Carr AJ (2005) Ruaumoko 2D and 3D: programs for inelastic dynamic analysis. Theory and user guide to associated programs. Department of Civil Engineering, University of Canterbury, Christchurch

Challa VRM, Hall JF (1994) Earthquake collapse analysis of steel frames. Earthq Eng Struct Dyn 23:1199–1218

Charney FA (2008) Unintended consequences of modeling damping in structures. J Struct Eng ASCE 134:581–592

Charney FA, Marshall J (2006) A comparison of the Krawinkler and scissors models for including beam-column joint deformations in the analysis of moment-resisting steel frames. Eng J AISC 43:31–48

Chi WM, El-Tawil S, Deierlein GG, Abel JF (1998) Inelastic analysis of a 17-story steel framed building damaged during Northridge earthquake. Eng Struct 20:481–495

Chopra AK (2007) Dynamics of structures: theory and applications in earthquake engineering, 3rd edn. Prentice-Hall, Upper Saddle River, NJ

Chopra AK, Goel RK (2002) A modal pushover analysis procedure to estimate seismic demands for buildings. Earthq Eng Struct Dyn 31:561–582

Chopra AK, McKenna F (2016) Modeling viscous damping in nonlinear response history analysis of buildings for earthquake excitation. Earthq Eng Struct Dyn 45:193–211

Chui PPT, Chan SL (1996) Transient response of moment-resistant steel frames with flexible and hysteretic joints. J Constr Steel Res 39:221–243

COSMOS (2020) Consortium of organizations for strong motion observation systems, San Francisco. http://www.cosmos-eq.org/

Deierlein GG, Reinhorn AM, Wilford MR (2010) Nonlinear structural analysis for seismic design: a guide for practicing engineers, NEHRP seismic design technical brief No. 4. National Institute of Standards and Technology (NIST), NIST GCR 10-917-5, Gaithersburg, MD

EC3 (2009) Eurocode 3, Design of steel structures – Part 1–1: general rules and rules for buildings, EN 1993-1-1. European Committee for Standardization (CEN), Brussels

EC8 (2004) Eurocode 8, Design of structures for earthquake resistance, Part 1: general rules, seismic actions and rules for buildings, EN 1998-1-1. European Committee for Standardization (CEN), Brussels

Elnashai AS, Izzuddin BA (1993) Modeling of material non-linearities in steel structures subjected to transient dynamic loading. Earthq Eng Struct Dyn 22:509–532

ETABS (2020) Extended three dimensional analysis of building systems, Version 18, integrated analysis, design and drafting of building systems. Computers and Structures Inc., Walnut Creek, CA

Fajfar P (2000) A nonlinear analysis method for performance-based seismic design. Earthquake Spectra 16:573–592

Fajfar P (2018) Analysis in seismic provisions for buildings: past, present and future, the 5th Prof. Nicholas Ambraseys lecture. Bull Earthq Eng 16:2567–2608

Fajfar P, Vidic T (1994) Consistent inelastic design spectra: hysteretic and input energy. Earthq Eng Struct Dyn 23:523–537

Fajfar P, Marusic D, Perus I (2005) Torsional effects in pushover-based seismic analysis of buildings. J Earthq Eng 9:831–854

FEMA 356 (2000) Prestandard and commentary for the seismic rehabilitation of buildings. Federal Emergency Management Agency, Washington, DC

FEMA 440 (2005) Improvement of nonlinear static seismic analysis procedures. Federal Emergency Management Agency, Washington, DC

Foutch DA, Yun SY (2002) Modeling of steel moment frames for seismic loads. J Constr Steel Res 58:529–564

Fragiacomo M, Amadio C, Macorini L (2004) Seismic response of steel frames under repeated earthquake ground motions. Eng Struct 26:2021–2035

Gupta A, Krawinkler H (2000) Dynamic P-delta effects for flexible inelastic steel structures. J Struct Eng ASCE 126:145–154

Hall JF (2006) Problems encountered from the use (or misuse) of Rayleigh damping. Earthq Eng Struct Dyn 35:525–545

Haselton CB, Whittaker AS, Hortacsu A, Baker JW, Bray J, Grant DN (2012) Selecting and scaling earthquake ground motions for performing response-history analyses. In: Proceedings of 15th World Conference on Earthquake Engineering, Lisbon, Portugal

Heo YA, Kunnath SK, Abrahamson N (2011) Amplitude-scaled versus spectrum-matched ground motions for seismic performance assessment. J Struct Eng ASCE 137:278–288

Ibarra LF, Medina RA, Krawinkler H (2005) Hysteretic models that incorporate strength and stiffness deterioration. Earthq Eng Struct Dyn 34:1489–1511

Iervolino I, Galasso C, Cosenza E (2010) REXEL: computer aided record selection for code-based seismic structural analysis. Bull Earthq Eng 8:339–362

Izzuddin BA, Elnashai AS (1989) ADAPTIC: a program for the adaptive analysis of space frames, Report no ESEE-89/7. Imperial College, London

Ju SH, Lin MC (1999) Comparison of building analyses assuming rigid or flexible floors. J Struct Eng ASCE 125:25–31

Kalkan E, Kunnath SK (2006) Adaptive modal combination for nonlinear static analysis of building structures. J Struct Eng ASCE 132:1721–1731

Katsanos EI, Sextos AG, Manolis GD (2010) Selection of earthquake ground motion records: a state-of-the-art review from a structural engineering perspective. Soil Dyn Earthq Eng 30:157–169

Kostinakis KG, Manoukas GE, Athanatopoulou AM (2018) Influence of seismic incident angle on response of symmetric in plan buildings. KSCE J Civ Eng 22:725–735

Krawinkler H (1978) Shear in beam-column joints in seismic design of steel frames. AISC Eng J 15:82–91

Krawinkler H (2006) Importance of good nonlinear analysis. Struct Des Tall Spec Build 15:515–531

Krawinkler H, Seneviratna GDPK (1998) Pros and cons of a pushover analysis of seismic performance evaluation. Eng Struct 20:452–464

Krishnan S, Hall JF (2006a) Modeling steel frame buildings in three dimensions. I: panel zone and plastic hinge beam elements. J Eng Mech ASCE 132:345–358

Krishnan S, Hall JF (2006b) Modeling steel frame buildings in three dimensions. II: elastofiber beam element and examples. J Eng Mech ASCE 132:359–374

Liao KW, Wen YK, Foutch DA (2007) Evaluation of 3D steel moment frames under earthquake excitations. I: modeling, II: reliability and redundancy. J Struct Eng ASCE 133:462–470. 471–480

Lignos DG, Krawinkler H (2011) Deterioration modeling of steel components in support of collapse prediction of steel moment frames under earthquake loading. J Struct Eng ASCE 137:1291–1302

Lignos DG, Hartloper AR, Elkady A, Deierlein GG, Hamburger R (2019) Proposed updates to the ASCE 41 nonlinear modeling parameters for wide-flange steel columns in support of performance-based seismic engineering. J Struct Eng ASCE 145:04019083-1–04019083-13

Lopez OA, Cruz M (1996) Number of modes for the seismic design of buildings. Earthq Eng Struct Dyn 25:837–855

Loulelis DG, Hatzigeorgiou GD, Beskos DE (2012) Moment resisting steel frames under repeated earthquakes. Earthq Struct 3:231–248

MacRae GA (1994) P-Δ effects on single-degree-of-freedom structures in earthquakes. Earthquake Spectra 10:539–568

McCrum DP, Broderick BM (2014) Seismic assessment of a steel braced plan mass symmetric/asymmetric building structure. J Constr Steel Res 101:133–142

Nakashima M, Huang T, Lu LW (1984) Effect of diaphragm flexibility on seismic response of building structures. In: Proceedings of 8th World Conference on Earthquake Engineering, San Francisco, CA, vol 4, pp 735–742

OpenSees (2020) Open system for earthquake engineering simulation. Pacific Earthquake Engineering Research Center, University of California at Berkeley, CA. http://opensees.berkeley.edu/

PEER (2020) Pacific Earthquake Engineering Research Center, Strong Ground Motion Database, Berkeley, CA. http://peer.berkeley.edu/

Peres R, Bento R, Castro JM (2020) Nonlinear static performance assessment of plan-irregular steel structures. J Earthq Eng 24:226–253

Prakash V, Powell GH, Campbell S (1993) DRAIN-2DX, Base program description and user guide, Version 1.10, Report No UCB/SEMM-93/17. University of California, Berkeley, CA

Prakash V, Powell GH, Campbell S (1994) DRAIN-3DX, base program description and user guide, Version 1.10, Report No UCB/SEMM-94/08. University of California, Berkeley, CA

Puthanpurayil AM, Lavan O, Carr AJ, Dhakal RP (2016) Elemental damping formulation: an alternative modeling of inherent damping in nonlinear dynamic analysis. Bull Earthq Eng 14:2405–2434

Reyes JC, Chopra AK (2011) Three-dimensional modal pushover analysis of buildings subjected to two components of ground motion, including its evaluation for tall buildings. Earthq Eng Struct Dyn 40:789–806

SAP 2000 (2020) Structural analysis program 2000, integrated software for structural analysis and design, Version 22. Computers and Structures Inc., Walnut Creek, CA

Schneider SP, Amidi A (1998) Seismic behavior of steel frames with deformable panel zones. J Struct Eng ASCE 124:35–42

Shi G, Atluri SN (1989) Static and dynamic analysis of space frames with nonlinear flexible connections. Int J Numer Methods Eng 28:2635–2650

Sivaselvan MV, Reinhorn AM (2000) Hysteretic models for deteriorating inelastic structures. J Eng Mech ASCE 126:633–640

Tena-Colunga A, Abrams DP (1996) Seismic behavior of structures with flexible diaphragms. J Struct Eng ASCE 122:439–445

Tremblay R, Cote B, Leger P (1999) An evaluation of P-Δ amplification factors in multistory steel moment resisting frames. Can J Civ Eng 26:535–548

Valles RE, Reinhorn AM, Kunnath SK, Li C, Madan A (1996) IDARC 2D, Version 4.0, a computer program for the inelastic damage analysis of buildings, Technical Report NCEER-96-0010. National Center for Earthquake Engineering Research, State University of New York, Buffalo, NY

Vidic T, Fajfar P, Fischinger M (1994) Consistent inelastic design spectra: strength and displacement. Earthq Eng Struct Dyn 23:507–521

Villaverde R, Hanna MM (1992) Efficient mode superposition algorithm for seismic analysis of non-linear structures. Earthq Eng Struct Dyn 21:849–858

Whittaker A, Atkinson GM, Baker JW, Bray J, Grant DN, Hamburger R, Haselton C, Somerville P (2011) Selecting and scaling earthquake ground motions for performing response-history analyses. NEHRP Consultants Joint Venture, National Institute of Standards and Technology (NIST), NIST GCR 11-917-15, Gaithersburg, MD

Zareian F, Medina RA (2010) A practical method for proper modeling of structural damping in inelastic plane structural systems. Comput Struct 88:45–53

Zhao D, Wong KKF (2006) New approach for seismic nonlinear analysis of inelastic framed structures. J Eng Mech ASCE 132:959–966

Chapter 3
Force-Based Design of EC8

Abstract This chapter briefly describes the force-based design (FBD) method as applied and used for the seismic design of steel building structures in the framework of Eurocode 8 (EC8). The FBD method of EC8 uses forces as the main design parameters and performs the design in two steps involving the strength checking of the structure under the design basis earthquake and the displacement checking under the frequent occurred earthquake. Both the simple lateral force method and the modal response spectrum analysis method for determination of the seismic design forces are presented and discussed. The concepts of the behavior factor to account for inelastic effects in the framework of elastic analysis and that of capacity design in order to ensure high levels of ductility before collapse are also presented. The chapter consists of six sections describing, besides some introductory remarks, performance design requirements, seismic action and soil types, design rules for buildings in general and steel buildings in particular, illustration of the theory with characteristic numerical examples involving steel building frames of the moment resisting and the braced types and a number of conclusions at the end.

Keywords Force-based design · Eurocode 8 · Steel building structures · Seismic action · Design spectra · Behavior factor · Modal response spectrum analysis · Nonlinear time history analysis · Pushover analysis · Steel MRFs · Steel CBFs · Steel EBFs

3.1 Introduction

The force-based design (FBD) method that employs forces as the main design parameters is the oldest method for seismic design of structures. The method is also the most familiar and popular method to engineers because of its similarity with the conventional structural design under static loads. This is the reason that almost all current seismic design codes are based on the FBD method. For example, Eurocode 8 (EC8 2004) in Europe or the International Building Code (IBC 2018) in the USA use the FBD method for the seismic design of buildings and other structures made of

G. A. Papagiannopoulos et al., *Seismic Design Methods for Steel Building Structures*, Geotechnical, Geological and Earthquake Engineering 51,
https://doi.org/10.1007/978-3-030-80687-3_3

steel, reinforced concrete or other materials. This chapter briefly describes the FBD method for the seismic design of steel building structures in the framework of EC8 (2004).

Building seismic design according to EC8 (2004) is a two performance level design involving the no collapse and the damage limitation requirements. Thus, this design method consists of two steps involving the strength checking and the displacement checking. The structure is assumed to behave inelastically under the design earthquake and its seismic forces are usually determined by an elastic modal response spectrum analysis in conjunction with the behavior factor that takes care of inelasticity. With known seismic member forces one is then able to size those members on the basis of strength requirements using, for example, EC3 (2009) for steel structures. After this checking, the displacement checking follows in order to limit damage. In most of the cases, this second checking, which requires iterations, controls the design. Alternatively, one can determine seismic forces and displacements by a nonlinear dynamic analysis in the time domain of the structure subjected to design spectrum compatible seismic accelerograms. However, this is an approach not familiar to engineers due to its complexity and is usually reserved as a check on the accuracy of the results at the research level or as a tool for the design of very complex and important structures. On the other hand, for regular and rather simple buildings, whose response is mainly due to the contribution of their fundamental mode, one can employ the lateral force method, which determines the design seismic shear of that structure by considering it to be a single-degree-of-freedom system. After that, that shear force is distributed to the members of the building and the member sizing follows.

During the design process and especially during the member sizing, a care is exercised so that the strong column-weak beam rule of capacity design is followed. This rule is used so that the structure will collapse in a global mechanism and hence mobilize all its available ductility and avoid premature failure.

This chapter briefly describes the basic aspects of the seismic provisions of EC8 (2004) and is basically restricted to structural analysis and member sizing. Thus, discussion about connections, slabs, secondary systems and special structural characteristics is not provided here and the interested reader is advised to consult EC8 (2004) for details. Representative numerical examples illustrating the application of the FBD method of EC8 (2004) are presented in detail in Sect. 3.6. An assessment of the EC8 (2004) provisions for seismic design of steel structures can be found in the book edited by Landolfo (2013).

3.2 Performance Requirements

EC8 (2004) requires the satisfaction of two performance requirements, the no-collapse requirement and the damage limitation requirement. These two requirements are met by designing the structure for two limit states (or performance levels), the ultimate limit state and the damage limitation state. These two limit states correspond to the life safety (LS) and the immediate occupancy (IO), respectively, performance levels of performance-based design of Chap. 1 and are associated with their respective seismic hazards (intensities). Thus, LS level ensures structural integrity without any local or global collapse under the design basis earthquake (DBE) with a probability of exceedance 10% in 50 years or return period of 475 years. The IO level ensures no damage in the structure under the frequent occurred earthquake (FOE) with a probability of exceedance 50% in 30 years or a return period of 43 years.

Thus, the EC8 (2004) seismic design method is partially a performance-based design method with only two performance levels, the LS and the IO ones with the former level not associated with a deformation requirement. The collapse prevention (CP) performance level associated with a very strong and rare earthquake (maximum credible earthquake—MCE) is not explicitly considered in EC8 (2004) and its satisfaction is ensured only implicitly by imposing some specific design detailing rules during the design process.

3.3 Seismic Action and Soil Types

The seismic action at a given point of the ground is represented either in the form of an elastic ground acceleration response spectrum or in the form of acceleration time-histories. Three independent components of the seismic action are considered: two horizontal and one vertical along the orthogonal directions x, y and z, respectively.

The seismic action is influenced by the local ground conditions represented by the five ground (soil) types A, B, C, D and E shown in Table 3.1. The elastic acceleration response spectrum $S_e(T)$ for the two horizontal components is defined as

$$0 \leq T \leq T_B \quad S_e(T) = a_g S[1 + \frac{T}{T_B}(\eta 2.5 - 1)] \tag{3.1}$$

$$T_B \leq T \leq T_C \quad S_e(T) = a_g S \eta 2.5 \tag{3.2}$$

$$T_C \leq T \leq T_D \quad S_e(T) = a_g S \eta 2.5 \frac{T_C}{T} \tag{3.3}$$

Table 3.1 Ground (soil) types: $v_{s,30}$ = average shear wave velocity in the top 30 m

Soil type	Description of stratigraphic profile	$v_{s,30}$ (m/s)
A	Rock or rock-like formation	>800
B	Very dense sand or gravel, or very stiff clay	360–800
C	Medium-dense sand or gravel, or stiff clay	180–360
D	Loose-to-medium sand or gravel, or soft-to-firm clay	<180
E	C or D soil type of 5–20 m thick underlain by rock	–

Table 3.2 Importance factors for various building categories

Importance class	Buildings	γ_I
I	Buildings of minor importance (e.g., agricultural)	0.8
II	Ordinary buildings (not in other categories)	1.0
III	Buildings of importance (e.g., schools, assembly halls)	1.2
IV	Buildings of vital importance (e.g., hospitals, fire stations)	1.4

Table 3.3 Values of S and T_B, T_C and T_D for elastic spectrum Type 1

Ground type	S	T_B (s)	T_C (s)	T_D (s)
A	1.00	0.15	0.40	2.00
B	1.20	0.15	0.50	2.00
C	1.15	0.20	0.60	2.00
D	1.35	0.20	0.80	2.00
E	1.40	0.15	0.50	2.00

$$T_D \leq T \leq 4\text{ s} \quad S_e(T) = a_g S\eta 2.5 \frac{T_C T_D}{T^2} \tag{3.4}$$

where T is the vibration period of a linear single-degree-of-freedom (SDOF) system, α_g is the design ground acceleration on type A ground (provided in seismic zone maps for any specific country), S is the soil factor, η is the damping correction factor, T_B is the lower limit of the period of the constant spectral acceleration branch, T_C is the upper limit of the period of the constant spectral acceleration branch and T_D defines the beginning of the constant displacement response range of the spectrum.

In EC8 (2004) the seismic hazard is usually described in terms of the value of the reference peak ground acceleration (*PGA*) on soil type A symbolized as α_{gR}. This value corresponds to the 10% probability of exceedance in 50 years or the return period of 475 years associated with the DBE.

The design ground acceleration α_g on soil type A is given as $\alpha_g = \gamma_I\, \alpha_{gR}$, where γ_I is the importance factor with values 0.8–1.4 depending on the building category, as shown in Table 3.2.

The damping correction factor η has a value of $\eta = 1$ for 5% viscous damping, while it is given by

$$\eta = \sqrt{10/(5+\xi)} \geq 0.55 \tag{3.5}$$

for values of damping ratio ξ, other than 5%. The soil factor S and the periods T_B, T_C and T_D receive values depending on the ground type and the type of spectrum (type 1 or 2). Type 1 elastic spectrum is defined for earthquakes with surface-wave magnitude $M_s \geq 5.5$, while Type 2 for $M_s \leq 5.5$. Table 3.3 provides values of S and T_B, T_C and T_D in terms of the five soil types A, B, C, D and E for spectrum Type 1, while Fig. 3.1 the shapes of elastic spectra Type 1 for the five ground types. In Fig. 3.1 the $T_B = 0.15$ s, $T_C = 0.50$ s and $T_D = 2.00$ s for soil type B are indicated. Corresponding table and figure for elastic spectrum Type 2 can be found in EC8

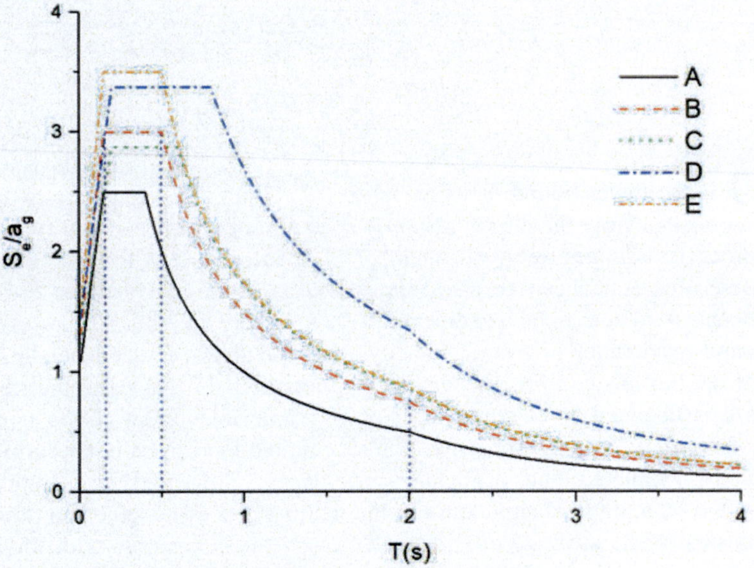

Fig. 3.1 Type 1 elastic spectra for ground types A to E and 5% damping

(2004). Explicit expressions for the vertical component of the elastic response spectrum can be also found in EC8 (2004).

According to EC8 (2004), the structure is seismically designed by accepting non-linear behavior under the design earthquake and hence energy dissipation. This energy of dissipation is taken into account by the introduction of the behavior factor q, which reduces the elastic response spectrum ordinates and creates the design spectrum. Use of the design spectrum in conjunction with a modal superposition elastic analysis leads to the design base shear force of the structure, which can be appropriately distributed to its members for their dimensioning.

The behavior factor q is defined as the ratio of its elastic behavior seismic forces (with 5% viscous damping) to its inelastic behavior design seismic forces (at first yield). The values of the behavior factor q, including the influence of viscous damping other than 5%, are given for various structural systems and steel material in terms of various ductility classes in Sect. 3.4.

The design spectrum $S_d(T)$ for the two horizontal seismic components is defined as

$$0 \leq T \leq T_B \quad S_d(T) = a_g S[\frac{2}{3} + \frac{T}{T_B}(\frac{2.5}{q} - \frac{2}{3})] \tag{3.6}$$

$$T_B \leq T \leq T_C \quad S_d(T) = a_g S \frac{2.5}{q} \tag{3.7}$$

$$T_C \le T \le T_D \quad S_d(T) = a_g S \frac{2.5}{q} \left(\frac{T_C}{T}\right) \ge \beta a_g \tag{3.8}$$

$$T_D \le T \quad S_d(T) = a_g S \frac{2.5}{q} \left(\frac{T_C T_D}{T^2}\right) \ge \beta a_g \tag{3.9}$$

where β is the lower bound factor equal to 0.2. Figure 3.2 shows the two design spectra obtained from the elastic response spectrum of type 1 and soil type B after division of its ordinates by $q_1 = 2$ and $q_2 = 4$ in accordance with Eqs. (3.6)–(3.9).

The seismic action can be also represented as a vector with three orthogonal components (two horizontal and one vertical), which are time-history expressions of the ground acceleration or ground velocity or ground displacement depending on the kind of application and on the available information. In that case, the structural response is obtained by a numerical stepwise time integration of the non-linear equations of motion of the structure. The seismic motion can be in the form of real (recorded), simulated and artificial accelerations. Simulated accelerations are constructed by a physical simulation of the faulting process, source and travel path mechanisms, while artificial ones so as to match (be compatible with) the elastic response spectra of EC8 (2004) with 5% viscous damping. A minimum of 3 real, simulated or artificial accelerograms should be used.

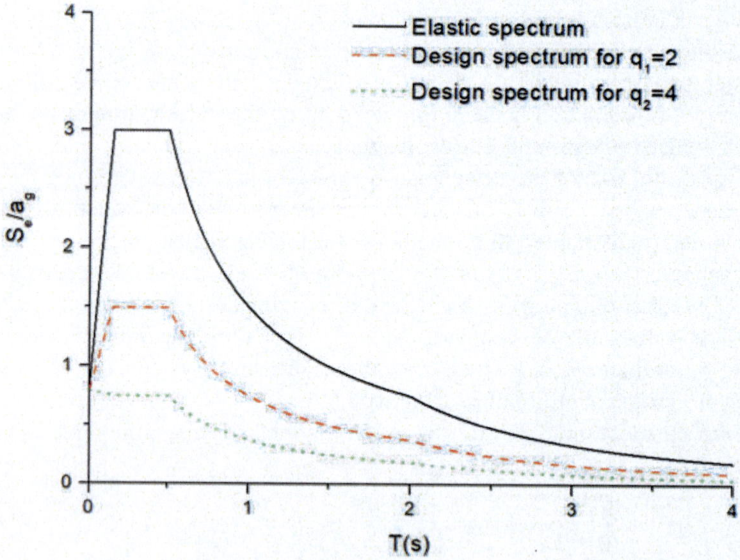

Fig. 3.2 Elastic response spectrum of type 1 and soil type B and corresponding design spectra resulting after dividing its ordinates by $q_1 = 2$ and $q_2 = 4$

3.4 Design of Buildings

Earthquake resistant buildings should satisfy the basic principles of: structural simplicity, uniformity, symmetry and redundancy; bi-directional resistance and stiffness; torsional resistance and stiffness; diaphragmatic behavior at storey level and adequate foundation. Satisfaction of the above basic principles results in regular buildings in plan view and elevation with a high seismic performance.

3.4.1 Primary and Secondary Seismic Members

The members (beams and/or columns) of a framed building structure are of two types depending on whether or not contribute to the seismic resisting system (Fig. 3.3). The primary seismic members are part of the seismic action resisting system of the building. The secondary ones are not part of that system and their strength and stiffness are not taken into account when considering seismic action. The secondary members and their connections are designed for gravity loads under seismic action. However, P-Δ effects are taken into account when designing those secondary members. The total contribution to lateral stiffness of all secondary seismic members should not exceed 15% of the stiffness due to the primary seismic members.

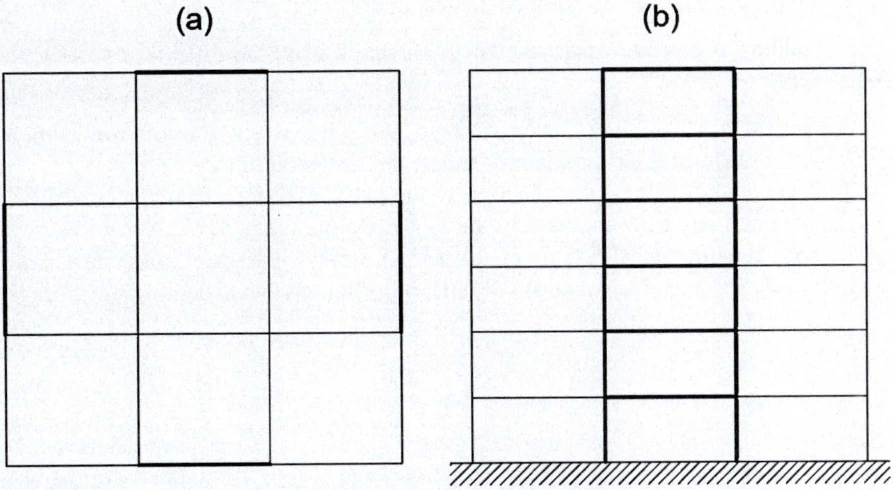

Fig. 3.3 Plan view (**a**) and elevation (**b**) of a steel moment resisting framed building with primary and secondary members shown by bold and light lines, respectively

3.4.2 Building Regularity

Building structures can be either regular (in plan and elevation) or non-regular. Regularity of structures has implications on their analysis and design, as shown in Table 3.4. Thus, for buildings which are regular in plan view and elevation, a planar model can be used and the analysis can be done by the lateral force method.

Reference values of the behavior factor q are given in the next section, while a brief description of the lateral force and modal analysis methods is provided in this section.

A building is considered as regular in plan if all of the following criteria are satisfied:

1. The structure is approximately symmetrical in plan with respect to the two orthogonal axes as far as its lateral stiffness and mass distributions are concerned.
2. Each floor is delimited by a polygonal convex line. If re-entrant corners or edges exist, these setbacks in plan should be very small.
3. The in-plan floor stiffness is sufficiently large in comparison with the lateral stiffness of the vertical structural elements.
4. The in-plan slenderness $\lambda = L_{max}/L_{min} \leq 4$, where L_{max} and L_{min} are the larger and smaller in plan dimension of the building along two orthogonal directions.
5. At each storey and for each main building plan direction x and y, the structural eccentricity e_O and the torsional radius r satisfy the two conditions $e_{ox,\, y} \leq 0.30 r_{x,\, y}$, $r_{x,\, y} \geq l_s$ for the analysis direction x and y, respectively, where l_s is the radius of gyration of the floor mass in plan. Figure 3.4 shows the eccentricity vector e_o between the center of mass (CM) and the center of stiffness (CS) of the plan view of a building and its two components e_{ox} and e_{oy}.

A building is considered as regular in elevation if all the following criteria are satisfied:

1. All lateral load resisting systems (cores, structural walls, frames) run without interruption from their foundations to the top of the building.
2. Both the lateral stiffness and the mass of the individual storeys remain constant or reduce gradually from the base to the top of the building.
3. In framed buildings the ratio of the actual storey resistance to the resistance required by the analysis do not vary disproportionally between adjacent storeys.

Table 3.4 Effects of structural regularity on seismic analysis and design

Regularity		Allowed simplification		Behavior factor
Plan	Elevation	Model	Linear elastic analysis	(for linear analysis)
Yes	Yes	Planar	Lateral force	Reference value
Yes	No	Planar	Modal analysis	Decreased value
No	Yes	Spatial	Lateral force	Reference value
No	No	Spatial	Modal analysis	Decreased value

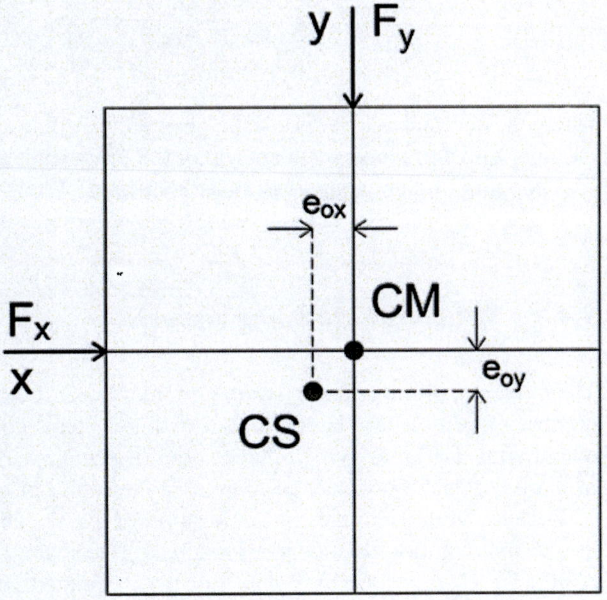

Fig. 3.4 Plan view of a building with eccentricities e_{ox} and e_{oy} between CM and CS under seismic forces F_x and F_y

4. When setbacks are present, specific rules are satisfied among the lateral widths of the structure in elevation.

For more details on all the above, one should consult Sect. 4.2.3.3 of EC8 (2004).

3.4.3 *Combination of Seismic Action with Other Actions*

Building structures are seismically designed on the basis of the seismic action combination

$$\sum_{j\geq1} G_{k,j} + A_{Ed} + \sum_{i\geq1} \varphi\psi_{2,i}Q_{k,i} \tag{3.10}$$

at the ultimate limit state (LS performance level), where $G_{k,\ j}$ is the permanent action j (self-weight and dead loads), $Q_{k,\ i}$ is the variable action i (live storey loads and snow), A_{Ed} is the design seismic action and $\psi_{2,\ i}$ is a coefficient equal to 0.3 for domestic, residential and office areas, 0.6 for congregation and shopping areas, 0.8 for storage areas and 0.2 for snow loads.

The inertial effects of the design seismic action come from the masses associated with the gravity loads in the combination

$$\sum_{j\geq 1} G_{k,j} + A_{Ed} + \sum_{i\geq 1} \varphi\psi_{2,i} Q_{k,i} \tag{3.11}$$

where the coefficient φ for domestic, residential, office and congregation areas is 0.5 (storeys with correlated occupancies) or 0.8 (independently occupied storeys), while for shopping or storage areas is 1.0.

3.4.4 Structural Modeling Aspects

The modeling of the building structure for its linear or non-linear analysis involves a number of vertical and lateral load resisting systems connected by horizontal floor diaphragms. For rigid in their planes floor diaphragms, the masses and the moments of inertia of each floor may be lumped at the center of gravity.

Because of uncertainties in the location of masses and in the spatial variation of the seismic motion, accidental torsional effects are considered even in regular buildings. This is done by displacing the calculated center of mass of each floor i by an accidental eccentricity.

$$e_{ai} = \pm 0.05 L_i \tag{3.12}$$

where L_i is the floor dimension perpendicular to the direction of the seismic motion.

3.4.5 Methods of Linear Seismic Analysis

Depending on the building structural characteristics (see Table 3.4), the following two types of linear elastic analyses may be used:

1. The lateral force method of analysis for buildings with predominantly first mode response along each principal direction. This method can be used only if both of the following criteria are met: (i) the regularity in elevation criterion and (ii) their fundamental period T_1 satisfies $T_1 \leq 4T_C$ and $T_1 \leq 2.0$ s. For buildings with heights up to 40 m, T_1 may be approximately calculated from

$$T_1 = C_t H^{3/4} \tag{3.13}$$

where $C_t = 0.085$ for moment resistant space steel frames, 0.075 for moment resistant space concrete frames and eccentrically braced steel frames and 0.050 for all other structures and H is the building height in m.

The seismic base shear force F_b for each horizontal direction is computed as

$$F_b = S_d(T_1)m\lambda \tag{3.14}$$

where m is the total mass of the building above the foundation or the top of a rigid basement and λ is a correction factor equal to $\lambda = 0.85$ if $T_1 \leq 2T_C$ and there are more than two stories or $\lambda = 1.0$ otherwise. This base shear is distributed at every floor i with mass m_i and height z_i from the foundation top or the rigid basement according to the relation

$$F_i = F_b \frac{z_i m_i}{\sum z_j m_j} \tag{3.15}$$

where F_i is the horizontal force at floor i, provided that the fundamental mode shape is approximated by horizontal displacements increasing linearly with the height.

Accidental eccentricity effects can be taken into account here by multiplying the action effects resulting from the application of F_i by a factor δ given as

$$\delta = 1 + 0.6(x/L_e) \tag{3.16}$$

where x is the distance of the element considered from the center of mass of the building in plan, along the direction perpendicular to that of the seismic action and L_e is the distance between the two outermost lateral load resisting elements, measured perpendicularly to the direction of the seismic action. If the analysis is done by using two planar models (for each main horizontal direction), Eq. (3.16) is used with 0.6 being replaced by 1.2.

2. The modal response spectrum analysis. This method of analysis should be applied to buildings irregular in elevation and regular or irregular in plan, as shown in Table 3.4. The sum of the effective modal masses for the considered modes significantly contributing to the response should be at least 90% of the total mass of the structure for each relevant direction. Furthermore, all modes with effective modal masses greater than 5% of the total mass should be taken into account. If these two conditions are not satisfied (e.g., in buildings with significant torsional modal contributions), the minimum number k of modes to be considered in a spatial analysis should satisfy both the following two conditions: $k \geq 3\sqrt{n}$ and $T_k \leq 0.20$ s, where n is the number of stories.

The independent modal responses E_{Ei} are combined to produce the maximum value E_E of the seismic action effect (force, displacement, etc.) according to the square root of the sum of the squares (SRRS) rule

$$E_E = \sqrt{\sum E_{Ei}^2} \tag{3.17}$$

The response in two vibration modes i and j (translational or torsional) are considered as independent to each other if their periods satisfy the relation $T_j \leq 0.9T_i$.

If this relation is not satisfied, more accurate combination rules than the SRSS of Eq. (3.17) should be used, e.g., the complete quadratic combination (CQC) rule.

When a spatial building model is used for the analysis, the accidental torsional effects may be determined as the envelope of the effects resulting from the application of static loading, consisting of sets of torsional moments M_{ai} about the vertical axis of each storey i of the form

$$M_{ai} = e_{ai} F_i \tag{3.18}$$

When two separate planar models are used for the analysis, the torsional effects may be taken into account by using (3.16) to the action effects computed in accordance with (3.18).

3.4.6 *Methods of Nonlinear Seismic Analysis*

Instead of using linear elastic methods of analysis in conjunction with the design spectrum/behavior approach, one can also use non-linear analysis methods. The model used for elastic analysis is extended here to include the strength of structural elements and their post-elastic behavior. At least a bilinear force-deformation relationship should be used to simulate material behavior. Two types of non-linear analysis can be used:

1. Non-linear static (pushover) analysis. The structural response is determined under constant gravity loads and monotonically increasing horizontal loads simulating the seismic motion. At least two vertical distributions of the lateral loads should be applied: a uniform and a modal distribution. With this analysis, one can do the following: (i) determine the overstrength ratio values a_u/a_1, (ii) determine the expected plastic mechanisms and distribution of damage, (iii) use it as an alternative to the design method based on linear elastic analysis with the behavior factor q. Using this analysis one can determine the relation between the base shear force and the control displacement (e.g., roof displacement) varying between 0–150% of the target displacement. This relation forms the capacity curve of the structure. More information on the subject can be found in Sect. 4.3.3.4.2 of EC8 (2004).
2. Non-linear time-history (NLTH) analysis. Using this type of analysis one can obtain the time domain response of the structure to the accelerograms representing the seismic motions by solving numerically the differential equations of its motion via a stepwise time integration. If the response is obtained from at least 7 non-linear time-history analyses involving recorded, simulated or artificial accelerograms, the average of the responses of these analyses should be used as the design value of the response E_d. Otherwise, the most unfavorable value of the response should be used as E_d.

3.4.7 Combination of the Effects of Seismic Actions

Concerning the horizontal components of the seismic action (i) in general, they should be taken as acting simultaneously; (ii) they may be combined as follows: the response to each component should be evaluated separately using the combination rule of modal responses; the maximum value of each action effect on the structure due to the two components may be estimated from the safe side by the SRSS rule involving the effects of those two components; (iii) alternatively, the responses E_{Edx} and E_{Edy} due to seismic actions along the x and y directions, respectively, may be combined as $E_{Edx} + 0.3E_{Edy}$ and $0.3E_{Edx} + E_{Edy}$ and select the maximum of the two. Concerning the vertical component of the seismic action one can consult Sect. 4.3.3.5.2 of EC8 (2004).

3.4.8 Computation of Displacements

1. When using linear elastic analysis, one determines first the displacement d_e by a linear analysis under the design seismic load and then the displacements d_s of the structure due to the design seismic action using the relation (equal displacement rule)

$$d_s = qd_e \tag{3.19}$$

   Torsional effects can be taken into account during the computation of d_e.

2. When using nonlinear static (pushover) or dynamic analyses, the displacements are those computed during these analyses without further modification.

At this point, one can comment that the approximate equal displacement rule overestimates or underestimates displacements depending on the fundamental period of the framed steel building (Gupta and Krawinkler 2000a, b).

3.4.9 Conditions Associated with the Two Limit States

As it has been stated earlier in Sect. 3.1, the two limit states of EC8 (2004) are the ultimate limit state and the damage limit state. In this section, the necessary requirements to be satisfied for these two states will be described.

(1) Ultimate limit state

The ultimate limit state under the seismic design action is satisfied if the following conditions associated with resistance, ductility and equilibrium are met:

(i) Resistance conditions

1. For all structural elements there should be

$$E_d \leq R_d \tag{3.20}$$

where E_d is the design value of the seismic action effect (including second order effects, if necessary) and R_d is the corresponding design resistance of the element depending on the material and computed according to the particular Eurocode (e.g., EC3 2009 for steel structures). For dissipative zones, designed and detailed for ductility, condition (3.20) should be satisfied in terms of member deformations (e.g., plastic hinge or chord rotations).

2. Second order (⍰-Δ) effects evaluated on the basis of interstorey drift sensitive coefficient θ defined for every storey of the framed building as

$$\theta = \frac{P_{tot} d_r}{V_{tot} h} \tag{3.21}$$

where P_{tot} is the total gravity load at and above the storey considered, d_r is the design interstorey drift, V_{tot} is the total seismic storey shear and h is the interstorey height. Thus, if $\theta \leq 0.10$, no second order effects are taken into account; if $0.10 \leq \theta \leq 0.20$, second order effects may be taken into account approximately by multiplying the relevant seismic action effects by a factor equal to $1/(1-\theta)$; the value of θ should not be greater than 0.30.

(ii) Global and local ductility conditions

Both the structural elements and the whole structure should have adequate ductility, which depends on the selected system and the behavior factor. Capacity design provisions should be also satisfied in order to achieve the hierarchy of resistance of the various structural components necessary for ensuring the desired pattern of plastic hinges and mechanism of collapse for avoiding brittle failure modes. Thus, at all joints of frame buildings the condition

$$\sum M_{Rc} \geq 1.3 \sum M_{Rb} \tag{3.22}$$

should be satisfied, where $\sum M_{Rc}$ is the sum of the design values of the moment of resistance of the columns framing the joint and $\sum M_{Rb}$ is the sum of the design values of the moments of resistance of the beams framing the joint. Expression (3.22) should be satisfied in two orthogonal vertical planes of bending for buildings with frames arranged in two orthogonal directions.

(iii) Equilibrium conditions

Building structures should be stable including overturning or sliding. Conditions regarding resistance of foundations are not considered here. They can be found in Sect. 4.4.2.6 of EC8 (2004).

(2) Damage limit state

The damage limit state is associated not with the design seismic action (DBE) but with the frequent occurred earthquake (FOE) and is equivalent with the IO performance level. This state is satisfied if the interstorey drift $d_r\nu$ satisfies the limitation requirement

$$d_r\nu \leq 0.005h \tag{3.23}$$

for buildings with non-structural elements of brittle materials

$$d_r\nu \leq 0.0075h \tag{3.24}$$

for buildings having ductile non-structural elements and

$$d_r\nu \leq 0.010h \tag{3.25}$$

for buildings with non-structural elements fixed so as not to interfere with structural deformations, or without non-structural elements, where h is the storey height and ν is a reduction factor equal to 0.4 for importance classes III and IV and to 0.5 for importance classes I and II.

Non-structural elements of buildings, such as mechanical appendages and equipment, curtain walls, partitions, railings etc. are not considered here. The interested reader can look at Sect. 4.3.5 of EC8 (2004).

3.5 Specific Rules for Steel Buildings

Steel buildings should be seismically designed as either structures with low dissipative behavior or structures with dissipative behavior corresponding to low or medium to high ductility class, respectively, as shown in Table 3.5 where the range of the reference values of the behavior factor q is also provided.

Design in accordance with low dissipative behavior is recommended only for low seismicity areas. Here, the seismic action effects may be calculated on the basis of an elastic design spectrum analysis with $1.5 \leq q < 2$, the structural members should be

Table 3.5 Structural ductility classes and values of q

Design concept	Structural ductility class	Values of q
Low dissipative structural behavior	DCL (low)	$\leq 1.5 - 2$
Dissipative structural behavior	DCM (medium)	≤ 4, also limited by values of Table 3.6
	DCH (high)	Only limited by values of Table 3.6

Table 3.6 Upper limit values of q for regular in elevation buildings

Structural type	Ductility class	
	DCM	DCH
Moment resisting frame	4	$5\alpha_u/\alpha_1$
Concentrically braced frame (i) Diagonal bracings (ii) V bracings	4 2	4 2.5
Eccentrically braced frame	4	$5\alpha_u/\alpha_1$
Moment resisting frame with concentric bracing	4	$4\alpha_u/\alpha_1$

of cross-sectional classes 1, 2 or 3 and member resistance should be evaluated in accordance with EC3 (2009) without additional requirements.

Design in accordance with dissipative behavior is recommended for medium to high seismicity areas. Here, parts of the structure (dissipative zones) resist seismic actions by inelastic behavior and this is taken into account by using higher values of q in the elastic design spectrum analysis (Table 3.6). These values of q depend on the ductility class (DCM or DCH) and the structural type. Depending on the ductility class, specific requirements on the class of steel sections should be satisfied. Adoption of this type of design requires the satisfaction of a number of requirements concerning the materials, the structural types and behavior factors, the structural analysis, the dissipative structural behavior and the design of various types of frames, such as moment resisting frames or braced frames.

3.5.1 *Structural Types and Behavior Factors*

Steel buildings are of the following three main structural types in accordance with their primary resisting system under seismic actions:

(a) Moment resisting frames (MRF), which resist horizontal forces mainly by members acting essentially in a flexural manner
(b) Concentrically braced frames (CBF), which resist horizontal forces mainly by members subjected to axial forces
(c) Eccentrically braced frames (EBF), which resist horizontal forces mainly by axially loaded members but with an eccentricity of the layout permitting dissipation of energy in seismic links by either cyclic bending or cyclic shear.
(d) Moment resisting frames combined with concentric bracings (MRF-CBF), which resist horizontal forces by members acting in flexure and by members subjected to axial forces.

Additional not very common, structural types can be found in EC8 (2004).

In MRF, the dissipative zones should be mainly located in plastic hinges at the ends of beams or beam-column joints where energy is dissipated by cyclic bending.

The dissipative zones may also be located at the bottom of columns. Figure 3.5 shows the location of such zones in MRF.

In CBF the dissipative zones should be mainly located in the tensile diagonals (Fig. 3.6) or in the tensile and compression diagonals (Fig. 3.7). In EBF the dissipative zones should be mainly located in bending or shear links (Fig. 3.8). In combined MRF-CBF the dissipative zones are the ends of beams or beam-column joints and the tensile diagonals (Fig. 3.9).

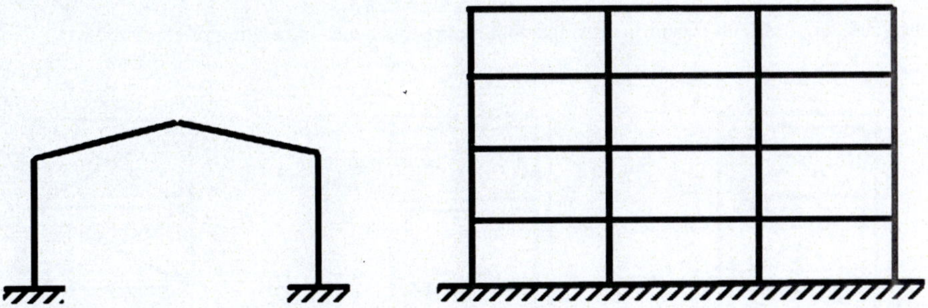

Fig. 3.5 Moment resisting frames with dissipative zones at the ends of beams and at the bottom of columns

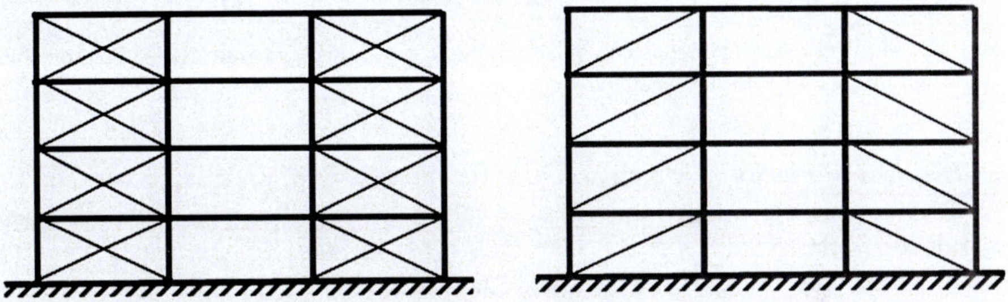

Fig. 3.6 Frames with concentric diagonal bracings with dissipative zones in tension diagonals only

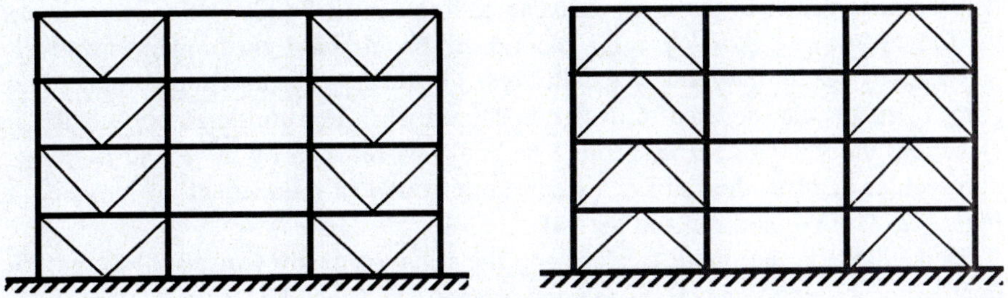

Fig. 3.7 Frames with concentric V-bracings with dissipative zones in tension and compression diagonals

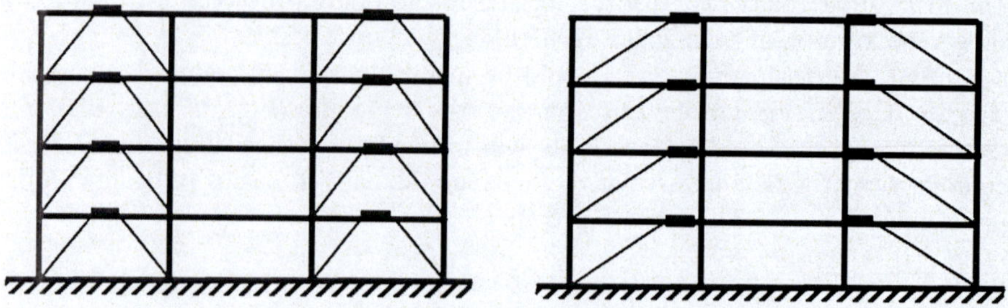

Fig. 3.8 Frames with eccentric bracings with dissipative zones in bending or shear links

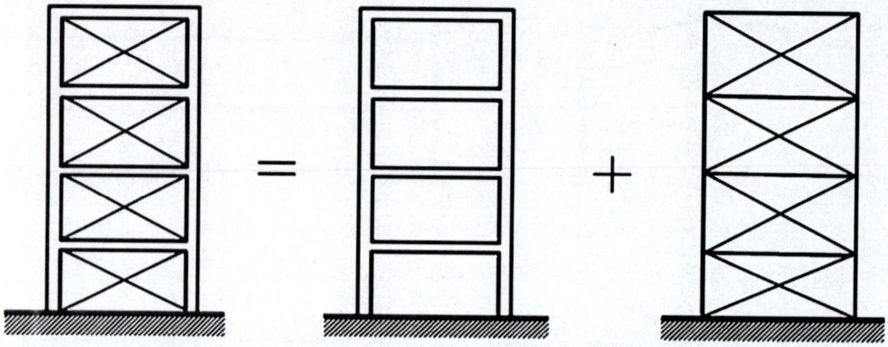

Fig. 3.9 Moment resisting frames combined with concentric bracings with dissipative zones in moment frame and in tension diagonals

The behavior factor q, which accounts for the energy of dissipation capacity of the structure, has upper limits the reference values for various regular structural types and ductility classes of Table 3.6. The ratio α_u/α_1 in Figs. 3.5, 3.8 and 3.9 and in Table 3.6 accounts for the overstrength of the structure due to its redundancy, where α_1 and α_u are multipliers of the horizontal design seismic action at formation of first plastic hinge in the system and of global plastic mechanism, respectively. Factors α_u and α_1 can be determined by static inelastic (pushover) analysis of the structure. Approximate values of the α_u/α_1 ratio can be found in EC8 (2004). For example, for the frames of Fig. 3.5, $\alpha_u/\alpha_1 = 1.1$ and 1.3 for the left and right frame, respectively and $\alpha_u/\alpha_1 = 1.2$ for the frames of both Figs. 3.8 and 3.9. The maximum value of the ratio α_u/α_1 that can be used in design is equal to 1.6. For buildings non-regular in elevation, the values of q in Table 3.6 should be reduced by 20%. For buildings non-regular in plan, the ratio α_u/α_1 can be obtained as the average of its value in Figs. 3.2, 3.3, 3.4, and 3.5 and the value of 1.0.

Depending on the ductility class and the behavior factor q used in design, the required cross-sectional class is the one provided in Table 3.7.

Table 3.7 Required cross-sectional class of dissipative elements depending on ductility class and behavior factor q

Ductility class	Reference value of behavior factor q	Required cross-sectional class
DCM	$1.5 < q \leq 2$	Class 1, 2 and 3
	$2 < q \leq 4$	Class 1 and 2
DCH	$q > 4$	Class 1

3.5.2 Specific Design Rules

Specific design rules of EC8 (2004) for MRF, CBF, EBF and combined MRF-CBF regarding to their members and connections are summarized in the following:

Starting with the MRF, plastic hinges can form in beams or in beam-to-column connections but not in columns except those at the base of the frame, at single storey structures and at the top level of multi-storey structures. The desired plastic hinge formation pattern is achieved on the basis of the following rules for beams and columns. More specifically, beams that possess sufficient resistance against lateral-torsional buckling may exhibit plastic hinge at their ends. For beams of cross-sectional class 1 or 2 the conditions that should be satisfied where plastic hinge formation is expected are

$$M_{Ed}/M_{pl.Rd} \leq 1 \tag{3.26}$$

$$N_{Ed}/N_{pl.Rd} \leq 0.15 \tag{3.27}$$

$$V_{Ed}/V_{pl.Rd} \leq 0.5 \tag{3.28}$$

where N_{Ed}, V_{Ed} and M_{Ed} are the design axial force, shear force and bending moment, respectively and $N_{pl.\,Rd}$, $V_{pl.\,Rd}$ and $M_{pl.\,Rd}$ are the associated with axial force, shear force and bending moment plastic design resistances. If Eq. (3.27) is not satisfied, a reduction of plastic moment resistance and rotation capacity at the location where plastic hinge formation is expected should be performed. In Eq. (3.28), $V_{Ed} = V_{Ed,\,G} + V_{Ed,\,M}$, where $V_{Ed,\,G}$ and $V_{Ed,\,M}$ are the shear forces due to non-seismic action and to plastic moments with opposite signs at the ends of the beam, respectively. For beam of cross-section class 3, the aforementioned design resistances are the elastic ones.

Columns in MRF are verified for the most unfavorable combination of bending moments M_{Ed}, shear forces V_{Ed} and axial force N_{Ed} of the form

$$M_{Ed} = M_{Ed,G} + 1.1\gamma_{ov}\Omega M_{Ed,E} \tag{3.29}$$

$$V_{Ed} = V_{Ed,G} + 1.1\gamma_{ov}\Omega V_{Ed,E} \tag{3.30}$$

$$N_{Ed} = N_{Ed,G} + 1.1\gamma_{ov}\Omega N_{Ed,E} \tag{3.31}$$

which are characterized by amplification of the column moments $M_{Ed,\ E}$ and forces $V_{Ed,\ E}$ and $N_{Ed,\ E}$ due to the design seismic action by the magnification coefficient Ω in order to achieve the capacity design rule 'strong columns-weak beams', which leads to a global collapse mechanism. This magnification coefficient is defined as

$$\Omega = min\left(M_{pl.Rd.i}/M_{Ed.i}\right) \leq 1 \tag{3.32}$$

where $M_{Ed.\ i}$ is the design value of the bending moment in beam i and $M_{pl.\ Rd.\ i}$ is the corresponding plastic moment. In Eqs. (3.29)–(3.31), γ_{ov} is the material overstrength factor with a recommended value equal to 1.25, while $M_{Ed,\ G}$, $V_{Ed,\ G}$ and $N_{Ed,\ G}$ are the column forces due to the non-seismic actions. The most unfavorable shear force V_{Ed} should be less than 50% of the ultimate shear resistance of the column cross-section.

Web panels at beam-column joints (Fig. 3.10) are required to be designed against shear and buckling on the basis of the expressions

$$V_{wp.Ed}/V_{wp.Rd} \leq 1.0 \tag{3.33}$$

$$V_{wp.Ed}/V_{wb.Rd} \leq 1.0 \tag{3.34}$$

where $V_{wp.\ Ed}$ is the design shear force in the web panel taking into account the plastic resistance of adjacent beams and connections, $V_{wp.\ Rd}$ is the plastic shear resistance of the web panel (not necessarily considering the effects of axial and bending stresses) and $V_{wb.\ Rd}$ is the shear buckling resistance of the web panel. If the resistance of the web panel has to be improved, this is accomplished by either adding stiffeners and/or welding additional plates to the column panel zone (shaded area in Fig. 3.10).

The rotation capacity θ_p of the plastic hinge region in connections should be $\geq$35 mrad for DCH and $\geq$25 mrad for DCM ductility class cases. The plastic hinge rotation θ_p is determined as the ratio of the beam deflection at mid-span over 0.5

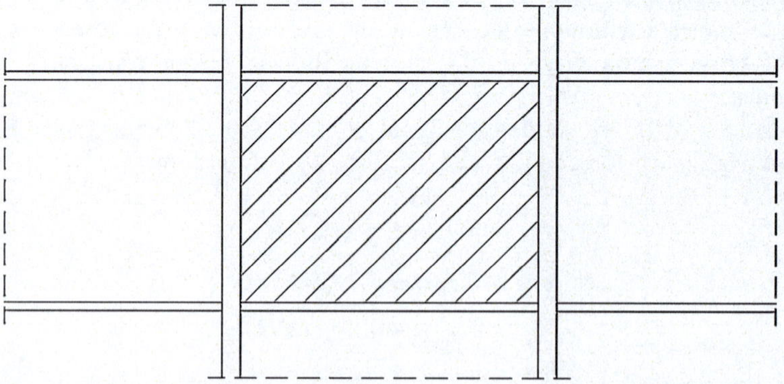

Fig. 3.10 Web panel with stiffeners and additional plates

times the span of the beam. Effects of strength and stiffness degradation as well as web panel shear should be checked so as not to affect significantly θ_p. The elastic deformation of the column is not included in θ_p.

Continuing now with CBFs, design should allow yielding of the bracing members in tension to occur before yielding or buckling of beams or columns and before failure of the connections. The bracing members should be configured in such a way that undesirable response effects are avoided, particularly under seismic load reversals. This is accomplished by enforcing at every storey the rule

$$\frac{|A^+ - A^-|}{(A^+ + A^-)} \leq 0.05 \tag{3.35}$$

where A^+ and A^- are the areas of the horizontal projections of the cross-sections of the tension diagonals for positive and negative seismic action direction, respectively.

For elastic analysis of CBF with diagonal bracings, only the tension ones are taken into account, whereas for CBF with V bracings, both tension and compression diagonal are considered. For non-linear static or dynamic analysis, both tension and compression diagonals of the CBF are taken into account as well as pre- and post-buckling situations are considered.

The design of the diagonal members is performed on the basis of non-dimensional slenderness $\bar{\lambda}$. More specifically, in CBF with diagonal bracings in X configuration $\bar{\lambda}$ is limited by $1.3 < \bar{\lambda} \leq 2.0$. For CBF with diagonal bracings not in X configuration as well as for CBF with V bracings, $\bar{\lambda}$ is less than or equal to 2.0. No limit on $\bar{\lambda}$ is applied to structures of up to two storeys, essentially permitting the use of tension elements (i.e., cables) as bracings. To satisfy a uniform dissipative behavior of the diagonal bracings, the maximum overstrength Ω_i of each diagonal defined as

$$\Omega_i = \left(N_{pl.Rd.i}/N_{Ed.i}\right) \tag{3.36}$$

should not differ from the minimum value of Ω_i by more than 25%. In Eq. (3.36), $N_{pl.Rd.\,i}$ and $N_{Ed.\,i}$ are the design resistance and the design axial force for seismic action, respectively, of the diagonal i. Bracing connections can be of any type under the condition that they can transfer the tension or compression effectively. The in- or out-of-plane behavior of the connection has to be modelled in non-linear time history analysis because it has impact on the global drift of the structure.

Beams and columns of the CBF should satisfy

$$N_{pl.Rd}(M_{Ed}) \geq N_{Ed,G} + 1.1\gamma_{ov}\Omega N_{Ed,E} \tag{3.37}$$

where $N_{pl.\,Rd}(M_{Ed})$ is the design buckling resistance of the beam or column taking account the interaction with bending moment M_{Ed} coming from the seismic action, $N_{Ed,\,G}$ is the axial force of the beam or column due to non-seismic action, $N_{Ed,\,E}$ is the axial force of the beam or column due to seismic action, γ_{ov} is the material overstrength factor (recommended value is 1.25) and Ω is the minimum value of Ω_i. Beams in CBF with V bracings should be designed (i) to resist non-seismic action without considering the support provided by the diagonals and (ii) against the unbalanced vertical seismic action effect occurring after buckling of the compression diagonal. This vertical action effect is considered using $N_{pl.\ Rd}$ for the brace in tension and $0.3N_{pl.\ Rd}$ for the brace in compression. Finally, for the case of diagonal bracings where there is no intersection between tension and compression diagonals, the design should take into account the tensile and compression forces developed in columns adjacent to the diagonals in compression assuming a compression force equal to the design buckling resistance of the diagonals.

For the case of an EBF, as shown in Fig. 3.5, the bracing members intersect the beam and, thus, forces are transmitted by shear and bending. The seismic link element formed horizontally or vertically by this intersection, dissipates energy by a plastic shear and/or bending mechanism. A uniform dissipative behavior of the seismic links has to be ensured and yielding of seismic links should take place before any other yielding of steel members.

Seismic links are classified into three categories according to the plastic mechanism developed: (i) short links that yield in shear, (ii) long links that yield in bending and (iii) intermediate links that yield in combined shear and bending. If $N_{Ed}/N_{pl.\ Rd} \leq 0.15$ the link should satisfy

$$V_{Ed} \leq V_{p.link} \tag{3.38}$$

$$M_{Ed} \leq M_{p.link} \tag{3.39}$$

where $M_{p.\ link}$ and $V_{p.\ link}$ are the design resistances of the link (EC8 2004) and N_{Ed}, V_{Ed} and M_{Ed} are the design axial force, shear force and bending moment, respectively, at both ends of the link. If $N_{Ed}/N_{pl.\ Rd} \geq 0.15$, Eqs. (3.38) and (3.39) should be satisfied for reduced values of $M_{p.\ link}$ and $V_{p.\ link}$ that take into account the axial force (EC8 2004). For I-sections, links of equal length e and equal moments occurring at both ends are classified as

$$short: e < e_s = \frac{1.6M_{p.link}}{V_{p.link}} \tag{3.40}$$

$$long: e > e_L = \frac{3.0M_{p.link}}{V_{p.link}} \tag{3.41}$$

$$intermediate: e_s < e < e_L \tag{3.42}$$

If one plastic hinge forms at one end in I-sections, links of length e are classified as

$$short: e < e_s = 0.8(1+\alpha)\frac{M_{p.link}}{V_{p.link}} \tag{3.43}$$

$$long : e > e_L = 1.5(1 + \alpha)\frac{M_{p.link}}{V_{p.link}} \tag{3.44}$$

$$intermediate : e_s < e < e_L \tag{3.45}$$

where α is the ratio of the absolute value of the smaller-to-larger bending moments at the two ends of the link. The link rotation angle θ_p should not exceed 0.08 rads for short links and 0.02 rads for long links, whilst linear interpolation can be used to determine θ_p for the case of intermediate links. EC8 (2004) stipulates specific rules for the design of stiffeners and their associated welds in short, long and intermediate links. Lateral supports at the ends of the link should be also provided.

The members not containing links, i.e., columns and diagonals for the case of horizontal links and beams for the case of vertical links, should be designed in accordance with

$$N_{Rd}(M_{Ed}, V_{Ed}) \geq N_{Ed,G} + 1.1\gamma_{ov}\Omega N_{Ed,E} \tag{3.46}$$

where $N_{Rd}(M_{Ed}, V_{Ed})$ is the design axial resistance of column or the diagonal member taking account the interaction with bending moment M_{Ed} and shear force V_{Ed} coming from the seismic action, $N_{Ed,\ G}$ is the compression force in the column or the diagonal member due to non-seismic action, $N_{Ed,\ E}$ is the compression force in the column or the diagonal member due to seismic action, γ_{ov} is the material overstrength factor (recommended value is 1.25) and Ω is a factor which is the minimum of the following: (i) the minimum value of $\Omega_i = 1.5V_{p.\ link.\ i}/V_{Ed.\ i}$ among all short links and (ii) the minimum value of $\Omega_i = 1.5M_{p.\ link.\ i}/M_{Ed.\ i}$ among all intermediate and long links, where $V_{Ed.\ i}$ and $M_{Ed.\ i}$ are the design values of shear force and bending moment in link i for seismic action, whereas $V_{p.\ link.\ i}$ and $M_{p.\ link.\ i}$ are the shear and bending plastic design resistances of link i.

If the structure is designed to dissipate energy in the seismic links, the connections of the links or the elements containing the links should be capacity designed (EC8 2004).

The case of partial strength connections or semi-rigid connections is permitted in MRF, CBF and EBF on the basis of the design requirements and the satisfaction of certain conditions mentioned in EC8 (2004).

Dual (combined) MRF-CBF acting in the same direction should be designed for a single value of the q factor. The horizontal forces are distributed to the MRF and the CBF according to their elastic stiffness. The design rules for MRF and CBF are also employed for the design of dual MRF-CBF.

It should be finally stressed that for all types of steel structures presented above, the use of an actual maximum value of steel yield strength in dissipative zones is considered (Sect. 6.2 of EC8 2004). This value of $f_{y,\ max} \leq 1.1\gamma_{ov}f_y$ can be substantially higher than the nominal one f_y, since the overstrength factor $\gamma_{ov} = 1.25$.

3.6 Numerical Examples

In this section three numerical examples are presented in order to show how the EC8 (2004) provisions are applied to the seismic design of steel buildings consisting of moment resisting and braced frames. The interested reader can look at Plumier (2012) and Landolfo (2014) for additional examples dealing with MRFs and CBFs.

3.6.1 Seismic Design of Steel Building with MRFs by Lateral Force Method

Consider the six-storey steel office building of Fig. 3.11 consisting of five and four moment resisting frames (MRFs) along the x and y horizontal directions, respectively. The orientation of the columns, shown in Fig. 3.12, is selected in order to have (i) the same number of strong and weak axis column bending in both x and y directions and (ii) the strong axis column bending in interior beam-column connection points along the x direction where longer spans result in deeper beams for satisfaction of the strong column-weak beam rule at those points. The grade of steel is assumed to be S355 for both beams and columns. The connections are assumed to be rigid and the floors are assumed to exhibit diaphragmatic action. The gravity load consists of dead load G and live load $Q = 2.5$ kN/m^2. Load G is 5 kN/m^2 on floors and 3 kN/m on walls.

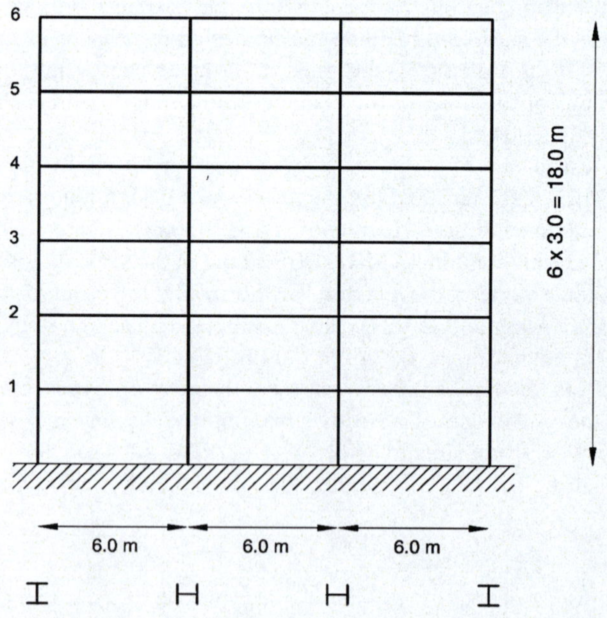

Fig. 3.11 Elevation of six-storey steel building

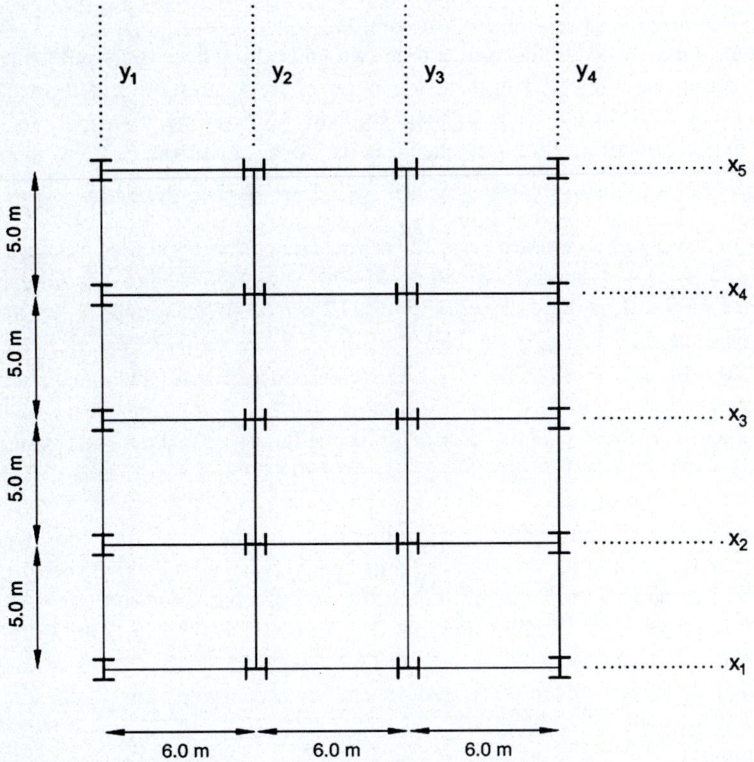

Fig. 3.12 Plan view and column orientation along horizontal directions x and y of six-storey steel building

The office building is assumed to be founded in a site with $a_{gR} = 0.24$ g, soil type C and design spectrum of type 1. A preliminary design of this building in accordance with EC8 (2004) will be presented here by following the procedure of the worked example in Plumier (2012). On the basis of the above data, one has form Table 3.2 that $\gamma_I = 1.0$ implying $a_g = 1.0 \cdot 0.24 \cdot 9.81 = 2.35$ m/s^2 and from Table 3.3 that $S = 1.15$, $T_B = 0.20$ s, $T_C = 0.60$ s and $T_D = 2.0$ s. The maximum behavior factor q for this type of building is from Fig. 3.5 and Table 3.6, $q = 5 \cdot (a_u/a_1) = 5 \cdot 1.3 = 6.5$. However, in order to avoid many iterations during the deformation checking phase, which usually controls the seismic design, a value of $q = 4$ is chosen here. Use is made here of the lateral force method and its results are compared against those of the modal response spectrum analysis method but without giving details of the latter method as this is used with details in the next two examples. The designs by these two methods here are finally analyzed by NLTH analyses for further comparison and validation. More specifically, the design procedure here consists of the following steps:

(1) Beam section selection for vertical loads

Beams are assumed to be fixed at their two ends along both the x and y directions. Their sections are selected first on the basis of their strength and those sections are then checked on the basis of their stiffness (deflection). Consider first beams along the x direction. Strength checking is done under the vertical static load combination $1.35G + 1.5Q$, where the dead load G consists of the floor weight $5\,\mathrm{m} \cdot 5\,\mathrm{kN/m}^2 = 25\,\mathrm{kN/m}$ and the wall weight given as 3 kN/m, while the live load Q is equal to $5\,\mathrm{m} \cdot 2.5\,\mathrm{kN/m}^2 = 12.5\,\mathrm{kN/m}$. Hence the combined vertical load is $1.35 \cdot (25 + 3) + 1.5 \cdot 12.5 = 56.55\,\mathrm{kN/m}$. Thus, the design bending moment $M_{sd} = (1.35G + 1.5Q)L^2/12 = 56.55 \cdot 6^2/12 = 169.65\,\mathrm{kNm}$ and hence the minimum required $W_{pl} = M_{sd}/f_y = 169.65 \cdot 10^5/355 \cdot 10^2 = 477.88\,\mathrm{cm}^3$. Selection of a IPE270 with $W_{pl} = 484\,\mathrm{cm}^3 > 477.88\,\mathrm{cm}^3$ is acceptable. This beam section is now checked for stiffness by examining if the maximum beam deflection $f_1 = (G + Q)L^4/384EI \leq L/300$. For IPE270 one has $I = 5790\,\mathrm{cm}^4$ and hence $f_1 = [(25 + 3 + 12.5) \cdot 6^4 \cdot 10^{12}]/(384 \cdot 2 \cdot 10^5 \cdot 5790 \cdot 10^4) = 11.80\,\mathrm{mm} < L/300 = 6000/300 = 20\,\mathrm{mm}$.

Consider now beams along the y direction. In this case the floor weight is equal to $6\,\mathrm{m} \cdot 5\,\mathrm{kN/m}^2 = 30\,\mathrm{kN/m}$ and hence $G = 30 + 3 = 33\,\mathrm{kN/m}$, while $Q = 6\,\mathrm{m} \cdot 2.5 = 15\,\mathrm{kN/m}$. Thus, $1.35G + 1.5Q = 67.05\,\mathrm{kN/m}$, while $G + Q = 33 + 15 = 48\,\mathrm{kN/m}$. The design bending moment $M_{sd} = 67.05 \cdot 5^2/12 = 139.69\,\mathrm{kNm}$ and hence the minimum required $W_{pl} = 139.69 \cdot 10^5/355 \cdot 10^2 = 393.48\,\mathrm{cm}^3$. The IPE270 with $W_{pl} = 484\,\mathrm{cm}^3 > 393.48\,\mathrm{cm}^3$ is acceptable. The maximum beam deflection $f_1 = (48 \cdot 5^4 \cdot 10^{12})/(384 \cdot 2 \cdot 10^5 \cdot 5790 \cdot 10^4) = 6.75\,\mathrm{mm} < 5000/300 = 16.67\,\mathrm{mm}$.

In conclusion, the IPE270 can be used in both directions under the gravity load combination. With this section selection for beams, one can proceed for sizing beams and columns under lateral (seismic) loading. In order to avoid many iterations, the subsequent calculations are done for a greater size of beam. A IPE400 beam section is finally selected with $W_{pl} = 1307\,\mathrm{cm}^3$ and $I = 23{,}130\,\mathrm{cm}^4$ with respect to strong axis. The beams are considered to support secondary beams per 1.5 m in order to mitigate the effect of lateral-torsional buckling.

(2) Column section selection on the basis of strong column-weak beam rule

According to EC8 (2004), the strong column-weak beam rule described by Eq. (3.22), should be satisfied at every joint of the frame. Since it has been assumed here that the steel grade in both beams and columns is S355 and taking into account that $M_{Rc} = f_y W_{plc}$ and $M_{Rb} = f_y W_{plb}$, one can obtain from Eq. (3.22), the relation

$$\sum W_{plc} \geq 1.3 \sum W_{plb} \tag{3.47}$$

At interior joints of the frame, there are 2 beams and 2 columns framing into those joints and Eq. (3.47) takes the form

$$W_{plc} \geq 1.3 W_{plb} \tag{3.48}$$

At exterior joints of the frame, there is 1 beam and 2 columns framing into those joints and Eq. (3.47) takes the form

$$2W_{plc} \geq 1.3W_{plb} \tag{3.49}$$

Before proceeding with the checks indicated by Eqs. (3.48) and (3.49), which involve W_{plc} values with respect to both the strong and the weak section axis, a column selection is made. A section HEM340 is selected with $W_{plcs} = 4718$ cm^3, $W_{plcw} = 1953$ cm^3, $I_{cs} = 76370$ cm^4, $I_{cw} = 19710$ cm^4 and $A = 316$ cm^2, where the subscripts *s* and *w* stand for strong and weak axis, respectively. Taking into account the different column orientations in Fig. 3.11, one has to do on the basis of Eqs. (3.48) and (3.49) the following checks:

1. line y_2, interior joint: Eq. (3.48) becomes $W_{plcw} \geq 1.3W_{plb}$, or 1953 cm^3 > 1.3 · 1307 = 1700 cm^3, which is satisfied
2. line y_2 exterior joint: Eq. (3.49) becomes $2W_{plcw} \geq 1.3W_{plb}$, or 2 · 1953 = 3906 cm^3 > 1.3 · 1307 = 1700 cm^3, which is satisfied
3. line y_1, interior or exterior joint: Here one has cases like the previous ones with W_{plcs} instead of W_{plcw} and thus Eqs. (3.48) and (3.49) are satisfied
4. line x_2, interior joint: Eq. (3.48) becomes $W_{plcs} \geq 1.3W_{plb}$, or 4718 cm^3 > 1.3 · 1307 = 1700 cm^3, which is satisfied
5. line x_2, exterior joint: Eq. (3.49) becomes $2W_{plcw} \geq 1.3W_{plb}$, or 2 · 1953 = 3906 cm^3 > 1.3 · 1307 = 1700 cm^3, which is satisfied

In conclusion, the strong column-weak beam rule described by Eq. (3.33) is satisfied with the chosen beam and column sections.

(3) Interior column under axial compression

For a loaded area 5 · 6 = 30 m^2, a floor weight 5 kN/m^2, a wall weight 3 kN/m and live load 2.5 kN/m^2, one has G = 30 · 5 + (6 + 5) · 3 = 183 kN/*storey* and Q = 30 · 2.5 = 75 kN/*storey*. Thus, the static vertical load combination is $1.35G + 1.5Q$ = 1.35 · 183 + 1.5 · 75 = 359.55 kN/*storey* and the axial compressive force at the ground level is 6 · 359.55 = 2157.3 kN. Assuming a buckling length equal to the storey height of 3.0 m, the slenderness of the HEM340 column with respect to its weak axis is $\lambda = kL/r_z$ = 1.0 · 300/7.9 = 37.97, while the Euler slenderness for S355 steel is $\lambda_E = 93.9\varepsilon = 93.9\sqrt{235/355} = 76.40$. Thus, the dimensionless slenderness $\bar{\lambda} = \lambda/\lambda_E = 37.97/76.4 = 0.497$ and hence $\chi = 0.86$. Finally, for A_c = 316 cm^2 one has $N_{b.\,Rd} = \chi A_c f_y$ = 0.86 · 316 · 355 · 10^2 = 9647480 N = 9647.48 kN > 2157.3 kN. In conclusion, the section selection of the frame consists of IPE400 sections for all beams and HEM340 sections for all columns.

(4) Computation of the seismic mass

A floor area of the building is (4 · 5)(3 · 6) = 20 · 18 = 360 m^2. Total length for partitions and facade at one storey is 4 · 20 + 5 · 18 = 170 m. Since the mass has units of kg, for all the given weights, the corresponding masses have to be determined. Thus, assuming $g \approx 10$ m/s^2, a weight W in kN corresponds to a mass $m = W/g = 10^2 W$ in kg. Hence the floor dead load mass is equal to $5 \cdot 10^2$ (kg/m^2) $\cdot$ 360 m^2 = 180,000 kg/*storey*, the floor live load mass is equal to $3 \cdot 10^2$ (kg/m^2) $\cdot$ 360 m^2 = 108,000 kg/*storey* and the mass of partitions and facade is $3 \cdot 10^2$ (kg/m) $\cdot$ 170 m = 51,000 kg/*storey*. At the roof there are various pieces of equipment (air conditioning, water tanks, elevator rooms etc.) of an assumed mass of 51,000 kg. The total mass of the steel skeleton per storey is calculated as follows: Columns HEM340: $3 \cdot (5 \cdot 4) \cdot 248$ kg/m = 14,480 kg, Beams IPE400: $((5 \cdot 18) + (4 \cdot 20)) \cdot 66.3$ kg/m = 11,271 kg. Total mass of steel skeleton per storey is 14,480 + 11,271 = 25,751 kg/*storey*. Finally, one can determine the total seismic mass per storey as 180,000 + $0.5 \cdot 0.3 \cdot$ 108,000 + 51,000 + 25,751 = 272,951 kg/*storey* and the total seismic mass for the whole building as m = $6 \cdot 272951$ = 1,637,706 kg. In the above, the 0.5 stands for the coefficient φ of Eq. (3.11). One can observe that the steel skeleton represents only 9.43% of the total seismic mass, while the floor represents 71.88% of the total seismic mass implying that if a weight reduction is desirable, this has to be done with respect to the floors.

(5) Seismic design shear forces by the lateral force method

The approximate method of lateral force can be used here because the building is regular in elevation and its fundamental period T_1 satisfies the relations $T_1 \le 4T_c$ and $T_1 \le 2.0$ s. Indeed, by using Eq. (3.13) one has $T_1 = 0.085 \cdot 18^{3/4} = 0.74$ s and $T_1 = 0.74$ s < 2.0 s $< 4T_c = 4 \cdot 0.6 = 2.4$ s. Since $T_C = 0.6$ s $< T_1 = 0.74$ s $< T_D = 2.0$ s, one can determine the design pseudo-acceleration $S_d(T)$ from Eq. (3.8) as $S_d(T) = 2.35 \cdot 1.15 \cdot (2.5/4.0) \cdot (0.6/0.74) = 1.37$ m/s^2. Thus, the seismic design shear F_b can be determined from Eq. (3.14) with $\lambda = 0.85$ (since $T_1 = 0.74$ s $< 2T_c = 2 \cdot 0.6 = 1.2$ s) as $F_b = mS_d(T)\lambda = 1{,}637{,}706 \cdot 1.37 \cdot 0.85 \cdot 10^{-3} = 1907$ kN. This force can be applied to any direction of the building because T depends only on the height of the building. Here, results are presented for frames in the x direction, which includes five identical frames with floor diaphragms. Thus, the seismic design shear F_{bx} in one frame is $F_{bx} = 1907/5 = 381.4$ kN. In this regular and symmetrical with respect to x and y directions building where the center of mass and the center of rigidity coincide on every floor, there are no natural eccentricities. Thus, torsional effects are only due to accidental eccentricities. These can be taken into account in the present case by amplification of F_{bx} by $\delta = 1 + 0.6(x/L_e)$ of Eq. (3.16), where $L_e = 20$ m is the horizontal dimension of the building perpendicular to the earthquake direction x and $x = 0.5L_e = 10$ m is the greatest distance from the center of rigidity to the frame considered. This leads to $\delta = 1 + 0.6(10/20) = 1.3$ and to an amplified $F_{bx} = 1.3 \cdot 381.4 = 495.82$ kN. This design seismic shear force is distributed to the rigid in-plane floor levels in an inverted triangular manner as described by Eq. (3.15) and since floor masses are equal, storey shear forces $F_i = F_{bx}(z_i/\sum z_i) = 495.82(z_i/63)$. More specifically, $F_1 = 23.61$ kN, $F_2 = 47.22$ kN, $F_3 = 70.83$ kN, $F_4 = 94.44$ kN, $F_5 = 118.05$ kN and $F_6 = 141.66$ kN. These lateral seismic forces together with the vertical loads of $G + 0.3Q$ on beams constitute the seismic load combination for the single frame considered along the x_2 line. On the basis of the seismic mass calculations, one has that the mass of the $G + 0.3Q$ load is equal to 180,000 + $0.3 \cdot$ 108,000 +

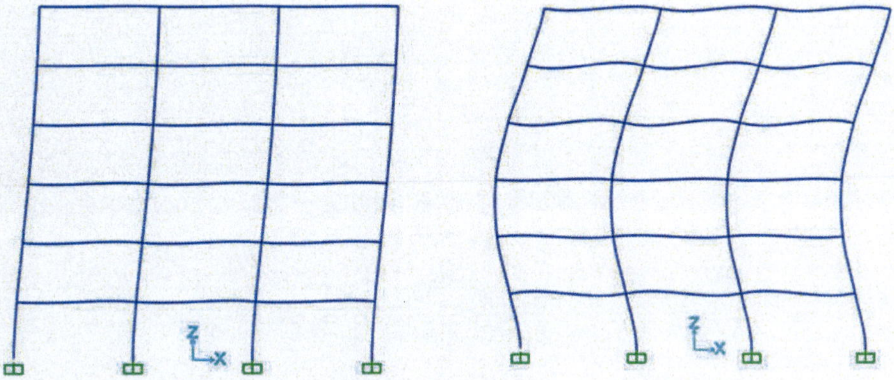

Fig. 3.13 Modal shapes for modes 1 (left) and 2 (right)

51,000 + 25,751 = 289,151 kg/*storey*. Frames along the lines x_1 and x_5 carry 1/8 of this mass, while frames along the lines x_2, x_3 and x_4 carry 1/4 of this mass. Thus, the frame along line x_2 (considered here) has a seismic design mass in one beam equal to (289,151/4) kg/18 m = 4015.99 kg/m, which corresponds to $G + 0.3Q$ load equal to 40.16 kN/m.

(6) Seismic design shear forces by modal response spectrum analysis

The plane single frame along line x_2 is analyzed by the modal response spectrum analysis. The design peak ground acceleration is $a_g = 0.24 \cdot 9.81 = 2.35$ m/s^2. Torsional effects can be taken into account by amplifying seismic lateral forces by $\delta = 1 + 0.6 \cdot (10/20) = 1.30$, as it is mentioned in the relevant paragraph of EC8 (2004). Hence seismic forces to be obtained by spectrum analysis are computed on the basis of an amplified $a_g = 1.3 \cdot 2.35 = 3.055$ m/s^2. The spectrum analysis of the plane frame under study is performed in SAP 2000 (2010) and one finds the first two modal periods to be $T_1 = 1.247$ s and $T_2 = 0.384$ s. The modal shapes associated with the first two modes are shown in Fig. 3.13.

(7) Analysis and design results and comparisons

The response results from the lateral force and the modal response spectrum analysis methods are shown in Tables 3.8 and 3.9, respectively. As shown in Tables 3.8 and 3.9, the inter-storey drift sensitivity coefficient θ exceeds 0.1 and its maximum value is 0.159 and 0.157 for the lateral force and response spectrum methods, respectively. Therefore, for design purposes, the analysis results have to be multiplied by $1/(1 - \theta)$, which turns out to be almost the same for both methods and equal to 1.19.

Tables 3.10 and 3.11, corresponding to the lateral force method and the modal response spectrum analysis, respectively, present the non-multiplied by 1.19 maximum values of axial force (N), shear force (V) and bending moment (M) of the critical beam and column elements. These critical elements are the internal column of the 1st storey and the middle beam of the 2nd storey and are highlighted with an ellipse in Fig. 3.14. The values for N, V, M are given separately for the $G + 0.3Q$ and E loads and for both the left and right ends of the critical elements. It is observed that

Table 3.8 Response results from lateral force method

Seismic load combination $G + 0.3Q + E$ with $G + 0.3Q = 40.16$ kN/m							
Storey	Storey displacement d_i (m)	Design interstorey displacement $q\,(d_i - d_{i-1})$ (m)	Storey lateral forces F_i (kN)	Storey shear force V_i (kN)	Storey cumulative gravity load P_i (kN)	Storey height h_i (m)	Interstorey sensitivity coefficient θ
1	0.0092	0.0368	23.61	495.82	4337.28	3.0	0.107
2	0.0248	0.0624	47.22	472.20	3614.40	3.0	0.159
3	0.0406	0.0632	70.83	424.98	2891.52	3.0	0.143
4	0.0541	0.0540	94.44	354.15	2168.64	3.0	0.110
5	0.0643	0.0408	118.05	259.71	1445.76	3.0	0.076
6	0.0707	0.0256	141.66	141.66	722.88	3.0	0.044

Table 3.9 Response results from modal response spectrum analysis method

Seismic load combination $G + 0.3Q + E$ with $G + 0.3Q = 40.16$ kN/m

Storey	Storey displacement d_i (m)	Design interstorey displacement $q\ (d_i - d_{i-1})$ (m)	Storey lateral forces F_i (kN)	Storey shear force V_i (kN)	Storey cumulative gravity load P_i (kN)	Storey height h_i (m)	Interstorey sensitivity coefficient θ
1	0.0073	0.0292	23.74	398.63	4337.28	3.0	0.106
2	0.0195	0.0488	41.75	374.89	3614.40	3.0	0.157
3	0.0316	0.0484	52.13	333.14	2891.52	3.0	0.140
4	0.0418	0.0408	64.31	281.01	2168.64	3.0	0.105
5	0.0494	0.0304	85.83	216.70	1445.76	3.0	0.068
6	0.0542	0.0192	130.87	130.87	722.88	3.0	0.036

Table 3.10 M, V, N values of the critical beam and column elements obtained by the lateral force method

Beam	Left end		Right end	
	G + 0.3*Q*	*E*	*G* + 0.3*Q*	*E*
M (kNm)	−120.59	207.89	120.59	−207.89
V (kN)	−120.48	69.30	120.48	69.30
N (kN)	0	0	0	0
Column	**Bottom end**		**Top end**	
	G + 0.3*Q*	*E*	*G* + 0.3*Q*	*E*
M (kNm)	1.41	476.55	−3.66	−65.03
V (kN)	1.69	180.53	1.69	180.53
N (kN)	−1467.50	16.07	−1467.50	16.07

Table 3.11 M, V, N values of the critical beam and column elements obtained by response spectrum analysis method

Beam	Left end		Right end	
	G + 0.3*Q*	*E*	*G* + 0.3*Q*	*E*
M (kNm)	−120.59	161.77	120.59	161.77
V (kN)	−120.48	53.93	120.48	53.93
N (kN)	0	0	0	0
Column	**Bottom end**		**Top end**	
	G + 0.3*Q*	*E*	*G* + 0.3*Q*	*E*
M (kNm)	1.41	379.26	−3.66	61.14
V (kN)	1.69	145.29	1.69	145.29
N (kN)	−1467.50	12.43	−1467.50	12.43

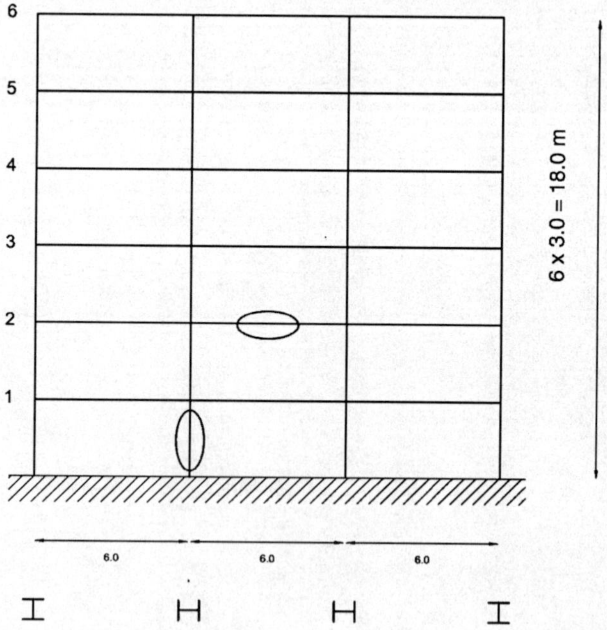

Fig. 3.14 Critical beam and column elements (highlighted by an ellipse)

the values in Table 3.11 are different than the corresponding ones in Table 3.10. This is because the correct value of the fundamental period $T_1 = 1.247$ s from the dynamic analysis is greater than the approximate value of $T_1 = 0.74$ s from the lateral force method and leads to a lower pseudo-acceleration $S_d(T)$ and hence to a lower seismic action values and responses. The dynamic analysis has shown that the first modal mass is 80.1% of the total seismic mass m, while the second modal mass is 10.8% of the total seismic mass m. Thus, according to EC8 (2004), the contribution to the response of the first two modes is enough for design purposes since $80.1 + 10.8 = 90.9\% > 90.0\%$.

The section and stability design checks for the critical beam and column of Fig. 3.14, after EC3 (2009), are presented in the following on the basis of the response results of Table 3.11. It is recalled that the values of both Tables 3.10 and 3.11 have to be multiplied by $1/(1 - \theta)$. For illustration purposes of the design procedure, the θ values of Table 3.9 are employed for the aforementioned design checks, i.e., $1/(1 - 0.157) = 1.19$ for the beam and $1/(1 - 0.106) = 1.12$ for the column.

In particular, the seismic demands for the critical beam following Sect. 3.5.1 are: $M_{Ed} = (-120.59 - 1.19 \cdot 161.77) = -313$ kNm, $N_{Ed} = 0$ and $V_{Ed} = 120.48 + 1.19$ $(463.99 + 463.99)/6 = 304.52$ kN, where 463.99 kNm $= M_{pl.\ Rd}$ for IPE400. Then, section checks are performed on the basis of Eqs. (3.26)–(3.28) and one has $M_{Ed}/M_{pl.\ Rd} = 313/463.99 = 0.674 < 1$, $V_{Ed}/V_{pl.\ Rd} = 304.52/875.81 = 0.35 < 0.5$ (interaction of M-V and M-N does not take place, hence the plastic moment capacity of the beam remains $M_{pl.\ Rd} = 463.99$ kNm). Regarding stability checks, only the one corresponding to lateral torsional buckling is needed and for curve b and $a_{LT} = 0.34$, one calculates $\bar{\lambda}_{LT} = 0.391$, $\Phi_{LT} = 0.609$, $\chi_{LT} = 0.93$ and finally $M_{b.\ Rd} = 431.29$ kNm > 313 kNm. Thus, the IPE400 beam satisfies all checks. It is stressed that the presence of a composite slab essentially renders redundant the lateral torsional buckling check of the beam because due to the effective width considered, $M_{b.\ Rd}$ would be higher than the one calculated above.

The design checks are now performed for the bottom end of the critical column on the basis of Eqs. (3.29)–(3.31), where $\gamma_{ov} = 1.25$ and $\Omega = 463.99/313 = 1.48$ from Eq. (3.32). Therefore, one obtains $M_{Ed} = (M_{Ed,G} + 1.12 \cdot 1.1\gamma_{ov}\Omega M_{Ed,E}) = (1.41 + 2.28$ $(379.26)) = 865.82$ kNm, $V_{Ed} = (1.69 + 2.28(145.29)) = 332.95$ kN, $N_{Ed} = (-1467.5 - 2.28(12.43)) = -1496$ kN. Only interaction of M-N takes place leading to a plastic moment capacity of $M_{N.\ pl.\ Rd} = 1606.75$ kNm and, thus, one has $M_{Ed}/M_{N.\ pl.\ Rd} = 865.82/1606.75 = 0.539 < 1$, $V_{Ed}/V_{pl.\ Rd} = 332.95/2025 = 0.164 < 0.5$ and $N_{Ed}/N_{pl.\ Rd} = 1496/11218 = 0.133 < 1$. Column stability checks for flexural buckling are critical for minor axis and considering curve b and $a = 0.34$, one calculates $\bar{\lambda}_z = 0.994$, $\Phi_z = 1.129$, $\chi_z = 0.601$ and $N_{b.\ z.\ Rd} = 6739.19$ kN > 1496 kN. Regarding lateral-torsional buckling and considering curve a and $a = 0.21$, one has $\bar{\lambda}_{LT} = 0.488$, $\Phi_{LT} = 0.649$, $\chi_{LT} = 0.928$ and $M_{b.\ Rd} = 1554.24$ kNm > 865.82 kNm. Finally the member interaction equations of EC3 (2009) are checked and using $C_{my} = 0.66$, $C_{mz} = 1$, $C_{mLT} = 0.66$, $k_{yy} = 0.658$, $k_{yz} = 0.612$, $k_{zy} = 0.948$ and $k_{zz} = 1.016$, the demand to capacity ratio for the column is found to 0.66. Thus, the HEM340 column satisfies all checks. It should be also noted that the maximum moment due to seismic action for a column in the 2nd storey is $M_{Ed,\,E} = 226.63kNm$. Therefore, even with the 1/(1 − 0.157) = 1.19 increase, M_{Ed} calculated by Eq. (3.29) is lower than the previous value of 865.82 kNm found for the critical column of the 1st storey.

Every joint satisfies Eq. (3.22) and the panel zone conditions of Eqs. (3.33) and (3.34) are also satisfied and no doubler plates are required. Finally, for the damage limitation (serviceability) limit state, one makes use of the response spectrum method and from Table 3.9 obtains that the maximum design inter-storey displacement is 0.0488 m. Hence from Eq. (3.25) and assuming $v = 0.5$, one has $0.0488 \cdot 0.5 = 0.0244 < 0.01 \cdot 3 = 0.03$. Thus, the damage limitation criterion is satisfied under the assumption of the interstorey drift limit of 0.01, i.e., the non-structural components are fixed to the structure and do not interfere with its deformation. For the cases of non-ductile or brittle non-structural elements, this limit is reduced to 0.0075 and 0.005, respectively, and for the steel frame presented herein, the damage limitation criterion obviously cannot be satisfied. Similarly, the damage limitation criterion is not satisfied for the limit 0.01 (not to mention the other two aforementioned limits) if the maximum inter-storey drift value that comes from the lateral force method is utilized. In cases of violation of the damage limitation criterion, the whole design procedure described up to here has to be repeated by assuming heavier sections.

After having satisfied the interstorey drift limit, attention is focused on the validity of the design methods presented previously. Therefore, non-linear dynamic analysis with the aid of SAP 2000 (2010) of the steel frame of Figs. 3.11 and 3.12 is performed employing 10 seismic motions that are compatible to the design spectrum mentioned above. Mean values for the yield base shear (corresponding to the first plastic hinge), maximum base shear and inter-storey drift ratio (*IDR*) were found to be 1163 kN, 1955 kN and 2.52%, respectively. Under the assumption that 2.5% is an acceptable value for the *IDR* of a steel MRF with respect to the life-safety seismic performance level of EC8 (2004) (see Sect. 3.2), it is concluded that the mean *IDR* value 2.52% obtained from non-linear dynamic analyses is marginally higher than 2.50% and hence acceptable. It should be also noted that the plastic hinge distribution was found to follow the expected from design pattern without formation of storey mechanism. Of course, if a lower than 2.5% *IDR* value has to be achieved, one should repeat the whole process until convergence between the targeted and obtained *IDR* values.

3.6.2 Seismic Design of a 3D Steel Building with MRFs by Response Spectrum Analysis

The six-storey steel office building of Figs. 3.11 and 3.12 is now designed employing a three-dimensional seismic analysis. The grade of steel is S355 for both beams and columns. The connections are assumed to be rigid and for each floor, a composite steel/concrete slab having a thickness of 0.14 m is considered. The composite slab is modeled using standard shell elements and according to EC4 (2004) provides a diaphragmatic action in each floor plane. The gravity load consists of dead load G and live load $Q = 2.5$ kN/m^2. Load G is 5 kN/m^2 on floors and 3 kN/m on walls. The beam and column sections are the same with those of the previous example, i.e., IPE400 and HEM340, respectively. In addition, secondary IPE330 beams pinned to the main IPE400 beams are considered per 2.0 m along the x direction of the structure of Fig. 3.12.

Frame global imperfections are calculated as $\varphi = a_h \cdot a_m \cdot \varphi_0$, where $\varphi_0 = 1/200$, $a_h = 2/\sqrt{h}$, but $2/3 \leq a_h \leq 1.0$ and $a_m = \sqrt{0.5(1 + 1/m)}$. In the a_h and a_m expressions, h and m are the height of the structure and the number of columns, respectively. These frame imperfections have to be considered either in the x or in the y direction of the structure of Fig. 3.12. Therefore, for $h = 18m$ and $m = 20$ one has $a_h = 0.667$, $a_m = 0.724$ and hence $\varphi = 0.0024$. For the conversion of frame imperfections into equivalent horizontal loads, a vertical loading coming from the combination $G + 0.3Q = 5 \cdot 20 \cdot 18 + 15 \cdot 3 \cdot 6 + 16 \cdot 3 \cdot 5 + 0.3 \cdot 2.5 \cdot 20 \cdot 18 = 2580$ kN is considered. The equivalent horizontal loads per storey due to frame imperfections are along both the x and y directions equal to $F = 0.0024 \cdot 2580 = 6.23$ kN. Frame imperfections can be disregarded if $H_{ed} \geq 0.15V_{ed}$ where H_{ed} is the design value of the storey shear and V_{ed} is the total design vertical load on the structure at the bottom of the storey.

As demonstrated in Sect. 3.6.1, the strong column-weak beam rule is satisfied at every joint of the frame. The office building is assumed to be founded in a site with $a_{gR} = 0.24$ g, soil type C and design spectrum of type 1. An accidental eccentricity of 5% in both horizontal directions is considered according to EC8 (2004). A value of $q = 4$ for the behavior factor is also assumed. A higher value of q could have been selected, such as $q = 6.5$, but since deformation usually controls the design, high values of q usually result in more iterations when trying to satisfy deformation requirements. Loading combinations have as follows: $1.35G + 1.5Q$, $G + 0.3Q \pm I_x \pm E_x \pm 0.3E_y$ and $G + 0.3Q \pm I_y \pm 0.3E_x \pm E_y$, where I_x, I_y are the equivalent loads that represent the imperfections and E_x, E_y are the seismic loads along directions x and y, respectively.

The modal response spectrum analysis of the structure under study is performed in SAP 2000 (2010), which also provides the option of taking into account accidental eccentricity automatically. The first eight modal periods and their associated participating mass ratios are shown in Table 3.12. It is observed that for translational along the x and y directions and rotational about the z direction motions, the sums of participating masses are 96.4%, 92.6% and 90.1%, respectively, all higher than the EC8 (2004) limit of 90%. These first eight modes are utilized in modal superposition. The deformed building shapes associated with the first three modes are shown in Fig. 3.15.

Results from modal response spectrum method in order to determine the inter-storey drift sensitivity coefficient θ in both directions x and y of the structure are shown in Tables 3.13 and 3.14. The results for θ are nearly the same irrespectively of the sign used in the seismic combinations ($\pm I_x \pm E_x \pm 0.3E_y$, $\pm I_y \pm 0.3E_x \pm E_y$).

Table 3.12 First eight modal periods and participating mass ratios

Mode number	Period (s)	Participating mass %		
		Translational x	Translational y	Rotational z
1	1.139	81.8	0	0
2	1.061	0	82.3	0
3	0.579	0	0	85.6
4	0.360	10.4	0	0
5	0.341	0	10.3	0
6	0.199	4.2	0	0
7	0.195	0	0	0.4
8	0.193	0	0	4.1
Sum of participating mass %		96.4	92.6	90.1

More specifically, θ_x exceeds 0.1 at the 2nd and 3rd stories and its maximum value is 0.132, whereas θ_y exceeds 0.1 at the 2nd storey and its maximum value is 0.114. Therefore, for design purposes where either θ_x and/or θ_y exceeds 0.1, the analysis results have to be multiplied by $1/(1 - \theta_{max})$, where θ_{max} is the maximum value of θ_x and θ_y with respect to the same storey.

Table 3.15 presents the maximum values for axial force (N), shear force (V) and bending moment (M) of the critical beam and column elements. These critical elements are the internal column of the 1st storey and the outer beam of the 2nd storey, both highlighted with heavy lines of red color in Fig. 3.16. The values for N, V, M are given for both the left and right ends of the critical elements and they are provided separately for the $G + 0.3Q$, $E_x + 0.3E_y$, $0.3E_x + E_y$ loading combinations. The corresponding values for axial force (N), shear force (V) and bending moment (M) due to imperfections I_x and I_y have been included in the seismic combinations $E_x + 0.3E_y$ and $0.3E_x + E_y$, respectively.

The section and stability design checks for the critical beam and column of Fig. 3.16, after EC3 (2009), are presented in the following. The θ value to be included in the aforementioned design checks is $1/(1 - 0.132) = 1.15$ for the beam whereas for the column the θ value of the first storey is lower than 0.10 and thus it does not enter in the design calculations.

The seismic demands on the critical beam following Sect. 3.5.1 are: $M_{Ed} = (-10.98 - 1.15 \cdot 105.43) = -132.22$ kNm, $N_{Ed} = 24.97$ kN and $V_{Ed} = 10.52 + 1.15$ $(463.99 + 463.99)/5 = 223.96$ kN. Then, section checks are performed on the basis of Eqs. (3.26)–(3.28) and one has $M_{Ed}/M_{pl.\ Rd} = 132.22/463.99 = 0.284 < 1$, $V_{Ed}/V_{pl.\ Rd} = 223.96/875.81 = 0.255 < 0.5$ (interaction of M-V and M-N does not take place, hence the plastic moment capacity of the beam remains $M_{pl.\ Rd} = 463.99$ kNm). Regarding stability checks: (i) for flexural buckling in minor axis, which is the critical axis, one has for curve b and $a_z = 0.34$ that $\bar{\lambda}_z = 1.66$, $\Phi_z = 2.12$, $\chi_z = 0.29$ and finally $N_{b.\ z.\ Rd} = 871.17$ kN > 24.97 kN; (ii) for lateral torsional buckling one has for curve b and $a_{LT} = 0.34$ that $\bar{\lambda}_{LT} = 0.787$, $\Phi_{LT} = 0.91$, $\chi_{LT} = 0.732$ and finally $M_{b.\ Rd} = 339.8$ kNm > 132.22 kNm. Thus, the IPE400 beam satisfies all checks. It is recalled that the presence of a composite slab

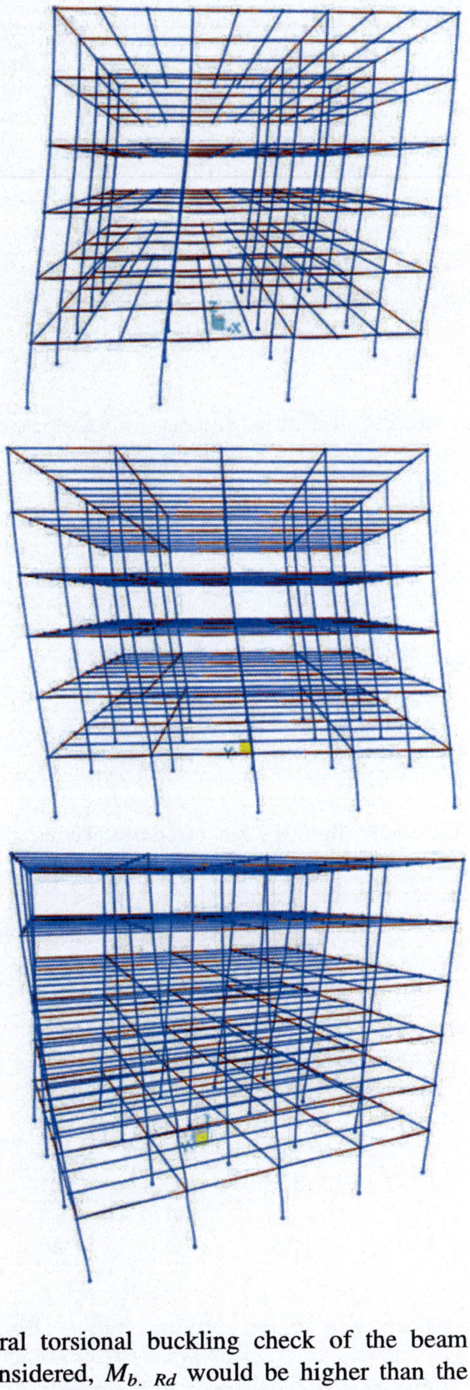

Fig. 3.15 Deformed building shapes for modes 1 (up), 2 (middle) and 3 (bottom)

essentially renders redundant the lateral torsional buckling check of the beam because due to the effective width considered, $M_{b.\ Rd}$ would be higher than the one calculated above. On the other hand, consideration of the effective width for the calculation of $M_{pl.\ Rd}$ of the beam may lead to violation of the strong-column weak-

Table 3.13 Inter-storey drift sensitivity coefficient θ_x

Storey	Storey displacement d_{xi} (m)	Design interstorey displacement $q\ (d_{xi} - d_{xi-1})$ (m)	Storey shear force V_{xi} (kN)	Storey cumulative gravity load P_i (kN)	Storey height h_i (m)	Interstorey sensitivity coefficient θ_x
1	0.0050	0.0200	1700.16	24,938.80	3.0	0.097
2	0.0126	0.0304	1585.66	20,784.71	3.0	0.132
3	0.0198	0.0288	1400.78	16,639.91	3.0	0.114
4	0.0258	0.0240	1187.26	12,476.89	3.0	0.084
5	0.0302	0.0176	925.48	8312.57	3.0	0.052
6	0.0328	0.0104	558.64	4168.95	3.0	0.026

Table 3.14 Inter-storey drift sensitivity coefficient θ_y

Storey	Storey displacement d_{yi} (m)	Design interstorey displacement $q\ (d_{yi} - d_{yi-1})$ (m)	Storey shear force V_{yi} (kN)	Storey cumulative gravity load P_i (kN)	Storey height h_i (m)	Interstorey sensitivity coefficient θ_y
1	0.0015	0.0060	543.90	24,938.80	3.0	0.092
2	0.0036	0.0084	508.86	20,784.71	3.0	0.114
3	0.0056	0.0080	449.74	16,639.91	3.0	0.099
4	0.0072	0.0064	378.95	12,476.89	3.0	0.070
5	0.0084	0.0048	291.40	8312.57	3.0	0.046
6	0.0091	0.0028	171.19	4168.95	3.0	0.023

Table 3.15 M, V, N values of the critical beam and column elements obtained by modal response spectrum analysis method

Beam	**Left end**			**Right end**		
	$G + 0.3Q$	$E_x + 0.3E_y$	$0.3E_x + E_y$	$G + 0.3Q$	$E_x + 0.3E_y$	$0.3E_x + E_y$
M (kNm)	−4.03	29.09	97.08	−10.98	31.60	105.43
V (kN)	−7.63	12.14	40.50	10.52	12.14	40.53
N (kN)	0	7.51	24.97	0	7.51	24.97
Column	**Bottom end**			**Top end**		
	$G + 0.3Q$	$E_x + 0.3E_y$	$0.3E_x + E_y$	$G + 0.3Q$	$E_x + 0.3E_y$	$0.3E_x + E_y$
M_2 (kNm)	1.41	25.02	105.85	−3.66	19.20	64.07
M_3 (kNm)	0.58	308.85	95.56	−1.67	104.11	31.20
V_2 (kN)	0.75	140.72	42.17	0.75	140.72	42.17
V_3 (kN)	0	16.97	56.63	0	16.97	56.63
N (kN)	−1983	18.15	5.44	−1976	18.15	5.44

beam capacity design criterion. If this is the case, one has to size the columns on the basis of $M_{pl.\ Rd}$ of the beam obtained by inclusion of an effective width.

The design checks are now performed for the left (bottom) end of the column on the basis of Eqs. (3.29)–(3.31), where $\gamma_{ov} = 1.25$ and $\Omega = 463.99/132.22 = 3.51$

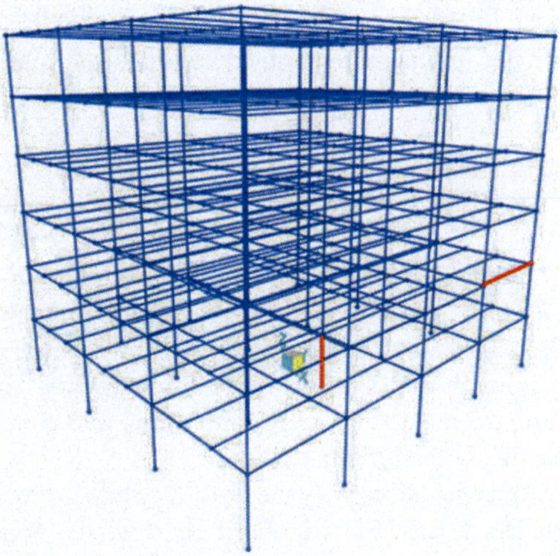

Fig. 3.16 Critical beam and column elements (highlighted with heavy lines of red color)

from Eq. (3.32). Taking into account that an IPE400 section has been employed for all beams in all stories, the value of 132.22 is the maximum one leading essentially to the minimum value of Ω. Therefore, upon consideration of the $E_x + 0.3E_y$ combination, one has that (indices 2 and 3 correspond to minor and major axis, respectively): $M_{3.\ Ed} = M_{Ed,\ G} + 1.1\gamma_{ov}\Omega M_{Ed,\ E} = 0.58 + 4.83 \cdot 308.85 = 1492.32$ kNm, $M_{2.\ Ed} = M_{Ed,\ G} + 1.1\gamma_{ov}\Omega M_{Ed,\ E} = 1.41 + 4.83 \cdot 25.02 = 122.26$ kNm, $V_{3.\ Ed} = 0 + 4.83 \cdot 16.97 = 81.97$ kN, $V_{2.\ Ed} = 0.75 + 4.83 \cdot 140.72 = 680.42$ kN and $N_{Ed} = -1983 - 4.83 \cdot 18.15 = -2070.66$ kN. For the aforementioned N, V, M design values and because interaction of M-V and M-N does not take place, the chosen section HEM340 marginally satisfies the biaxial moment section criterion, i.e., $(M_{3.\ Ed}/M_{3.\ Rd})^2 + (M_{2.\ Ed}/M_{2.\ Rd}) = (1492.32/1674.90)^2 + 122.26/693.32 = 0.793 + 0.176 = 0.969$, whereas shear and axial stress ratio are satisfied, i.e., $V_{2.\ Ed}/V_{2.\ pl.\ Rd} = 122.26/2025 = 0.06 < 0.5$, $V_{3.\ Ed}/V_{3.\ pl.\ Rd} = 81.97/5198 = 0.016 < 0.5$, $N_{Ed}/N_{pl.\ Rd} = 2070.66/11218 = 0.184 < 1$.

Similar considerations can be made by using now the $0.3E_x + E_y$ combination where $M_{3.\ Ed} = M_{Ed,\ G} + 1.1\gamma_{ov}\Omega M_{Ed,\ E} = 0.58 + 4.83 \cdot 95.56 = 462.13$ kNm, $M_{2.\ Ed} = M_{Ed,\ G} + 1.1\gamma_{ov}\Omega M_{Ed,\ E} = 1.41 + 4.83 \cdot 105.85 = 512.67$ kNm and, thus, the biaxial moment section criterion, i.e., $(M_{3.\ Ed}/M_{3.\ Rd})^2 + (M_{2.\ Ed}/M_{2.\ Rd}) = (462.13/1674.90)^2 + 512.67/693.32 = 0.076 + 0.739 = 0.815$ is satisfied. Shear and axial stress ratio are satisfied too.

Column stability checks for flexural buckling are of importance only for minor axis (since $\chi_y = 1.0$) and by considering curve b and $a = 0.34$, one has $\overline{\lambda}_z = 0.603$, $\Phi_z = 0.75$, $\chi_z = 0.836$ and $N_{b.\ z.\ Rd} = 9374.52$ kN > 2070.66 kN. Regarding lateral-torsional buckling, for curve a and $a = 0.21$, one obtains $\overline{\lambda}_{LT} = 0.375$, $\Phi_{LT} = 0.589$, $\chi_{LT} = 0.959$ and $M_{b.\ Rd} = 1606.55$ kNm > 1492.32 kNm. Finally, the member interaction equations of EC3 (2009) are checked for the $E_x + 0.3E_y$ combination. Thus, considering $C_{my} = 0.723$, $C_{mz} = 0.784$, $C_{mLT} = 0.723$, $k_{yy} = 0.711$, $k_{yz} = 0.482$, $k_{zy} = 0.919$ and $k_{zz} = 0.803$, the demand to capacity ratio for the column is found to be 1.14. This is something expected in view of the fact that, as shown above, the biaxial moment section criterion was marginally satisfied. Similarly, for the $0.3E_x + E_y$ combination and using $C_{my} = 0.705$, $C_{mz} = 0.823$, $C_{mLT} = 0.705$, $k_{yy} = 0.694$, $k_{yz} = 0.506$, $k_{zy} = 0.919$ and $k_{zz} = 0.843$, the demand to capacity ratio for the column is found to be 1.10.

It should be noted that the maximum moment (major axis) due to the $E_x + 0.3E_y$ combination at the most stressed column in the 2nd storey is $M_{Ed,\,E} = 202.18$ kNm. Therefore, even with the $\theta = 1.15$ increase, M_{Ed} calculated by Eq. (3.29) is found to be 1122.14 kNm, which is lower than the value of 1492.32 kNm corresponding to the critical column of the 1st storey. Similar conclusion can be reached for the maximum moments (major and minor axes) of the $0.3E_x + E_y$ combination. Therefore, the most stressed column of the 2nd storey satisfies the corresponding stability checks with sufficient margin.

In conclusion, only the critical column of the 1st storey fails to satisfy the stability checks at least for two out of eight seismic loading combinations, essentially due to the large value of Ω that enters in the calculations. Had the Ω value been lower, the critical column would pass all checks. A reduction of the value of Ω can be obtained by changing the section of the beams of some stories. From another point of view, the aforementioned deficiency in the design of the critical column demonstrates the potential problems behind the use of simplified 2D models (Sect. 3.6.1) instead of the exact 3D ones. In the 2D model, the critical column passed all checks, something that did not happen when the 3D model was employed.

Just for illustration purposes of the whole design process, it is assumed that the critical column of the structure under study marginally satisfies all checks and, hence, a redesign is not needed. To check the damage limitation (serviceability) limit state, one makes use of Tables 3.13 and 3.14 and obtains that the maximum design inter-storey displacement is 0.0304 m. Hence from Eq. (3.25) and assuming $\nu = 0.5$, one has $0.0304 \cdot 0.5 = 0.0152 < 0.01 \cdot 3 = 0.03$. Thus, the damage limitation criterion is satisfied under the assumption of the interstorey drift limit of 0.01, valid for non-structural components fixed to the structure and not interfering with its deformation. The damage limitation criterion is also satisfied when the interstorey drift limit is 0.0075 but it is violated when this limit is 0.005. It is recalled that in cases of violation of the damage limitation criterion, the whole design procedure described up to here has to be repeated by assuming heavier sections.

After having satisfied the interstorey drift limit, attention is focused on the validity of the design methods presented previously. Therefore, non-linear dynamic analysis with the aid of SAP 2000 (2010) of the steel frame of Figs. 3.8 and 3.9 is performed employing 10 seismic motions that are compatible to the design spectrum mentioned above. Mean values for the yield base shear V_y (corresponding to the first plastic hinge), maximum base shear V_{max} and inter-storey drift IDR were found to be $V_{y,\,x} = 6864.61$ kN, $V_{y,\,y} = 9982.75$ kN, $V_{max,\,x} = 11196.30$ kN, $V_{max,\,y} = 10881.40$ kN, $IDR_x = 1.69\%$ and $IDR_y = 1.46\%$. Under the assumption that 2.5% is an acceptable value for the IDR of a steel MRF structure with respect to the life-safety seismic performance level of EC8 (2004) (see also Sect. 3.2 and SEAOC 1999), it is concluded that the aforementioned mean IDR value of 1.69%

obtained from non-linear dynamic analyses is lower and thus form the conservative side. Moreover, the plastic hinge distribution was found to follow the pattern expected from design without formation of a storey mechanism.

3.6.3 *Seismic Design of a 3D Steel Building with MRFs and CBFs by Response Spectrum Analysis*

The six-storey steel office building of Figs. 3.17, 3.18, and 3.19 is designed employing a three-dimensional seismic analysis. According to the column orientation shown in Fig. 3.17, this building consists of moment resisting frames in the *x* direction and concentrically braced frames in the *y* direction, shown by dashed lines. The elevation per horizontal directions *x* and *y*, respectively, is shown in Figs. 3.18 and 3.19, respectively. As depicted in Fig. 3.19, the braces are diagonally configured at the outer bays of all stories. The braces are assumed to intersect at their middle (cross-braces) forming a diagonal X type of braces in which one of the braces is continuous along the diagonal, whereas the other one is spliced. End conditions for both the continuous and the spliced braces are assumed to be fixed in-plane and

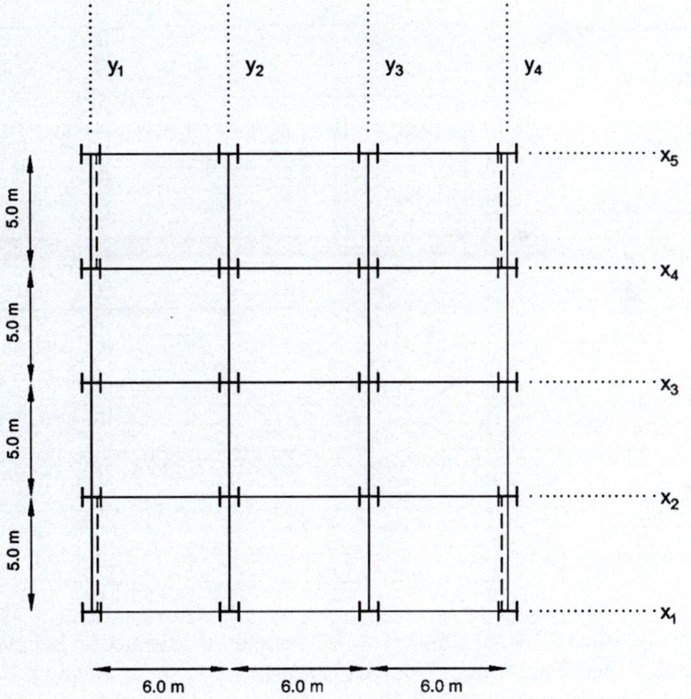

Fig. 3.17 Column orientation along horizontal directions *x* and *y* of six-storey steel building

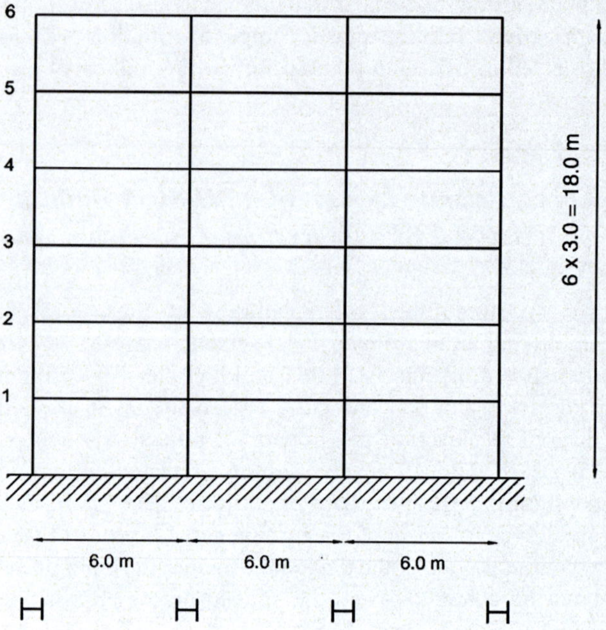

Fig. 3.18 Elevation along the *x* direction of the six-storey steel building

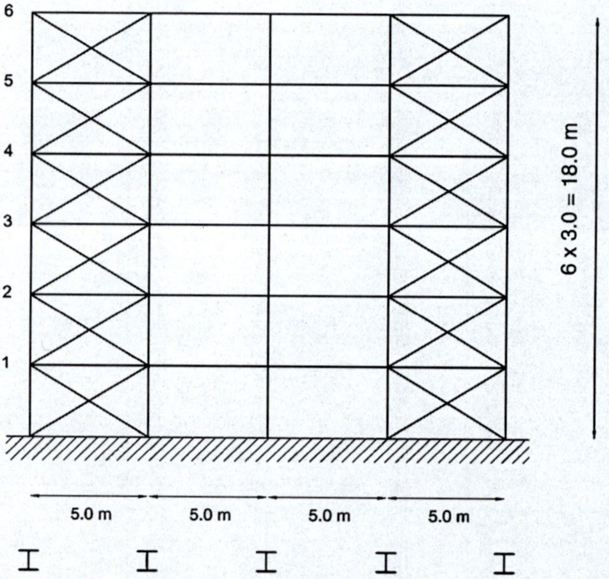

Fig. 3.19 Elevation along the *y* direction of the six-storey steel building

pinned out-of-plane (SEAOC 2009). The splice detail is assumed to behave as fixed in-plane and pinned out-of-plane (SEAOC 2009).

The grade of steel is S355 for both beams and columns and S275 for the braces. Column-base and beam-to-column (moment resisting frames) connections are assumed to be rigid, whereas those of the braces, of the beams in braced frames and of the secondary beams to main beams are assumed to be pinned. For each floor, a composite slab having a thickness of 0.14 m is considered, which is modeled using standard shell elements and provides a diaphragmatic action in each floor plane, as in the previous example. The gravity load consists of dead load G and live load $Q = 2.5$ kN/m^2. Load G is 5 kN/m^2 on floors and 3 kN/m on walls. The beam and column sections selected are IPE330 and HEM320, respectively. Secondary IPE270 beams pinned to the main IPE330 beams are considered per 2.0 m along the x direction of the structure of Fig. 3.17. The sections of the braces vary at each floor due to the Ω restriction mentioned in the following. In particular, the selected per storey sections for braces are: SHS 120 x 120 x 12.5 (1st–3rd stories), SHS 120 x 120 x 8 (4th storey), SHS 100 x 100 x 10 (5th storey) and SHS 90 x 90 x 5 (6th storey).

Frame imperfections either in the x or in the y direction of the structure are the same as those calculated in Sect. 3.6.2. Thus, the equivalent horizontal loads per storey due to frame imperfections are $0.0024 \cdot 2580 = 6.23$ kN, where 2580 kN is the total vertical load due to the $G + 0.3Q$ loading combination.

The office building is assumed to be founded in a site with $a_{gR} = 0.24$ g, soil type C and design spectrum of type 1. An accidental eccentricity of 5% in both horizontal directions and a common value of $q = 4$ is considered in both horizontal directions. Loading combinations have as follows: $1.35G + 1.5Q$, $G + 0.3Q \pm I_x \pm E_x \pm 0.3E_y$ and $G + 0.3Q \pm I_y \pm 0.3E_x \pm E_y$, where I_x, I_y are the equivalent loads that represent the imperfections and E_x, E_y are the seismic loads along x and y directions, respectively. The modal response spectrum analysis of the structure under study is performed in SAP 2000 (2010), which also provides the option of taking into account accidental eccentricity automatically. The first ten modal periods and their associated participating mass ratios are shown in Table 3.16. These first ten modes are utilized in modal superposition. The deformed building shapes associated with the first three modes are shown in Fig. 3.20.

Table 3.16 First ten modal periods and participating mass ratio

		Participating mass %		
Mode number	Period (s)	Translational x	Translational y	Rotational z
1	1.214	79.8	0	0
2	0.609	0	77.3	0
3	0.443	0	0	81.2
4	0.370	11.1	0	0
5	0.215	0	14.5	0
6	0.193	4.7	0	0
7	0.165	0	2.6	4.5
8	0.159	0	0.8	3.1
9	0.154	0	0.5	0.9
10	0.151	0	0	0.4
Sum of participating mass %		95.6	95.7	90.1

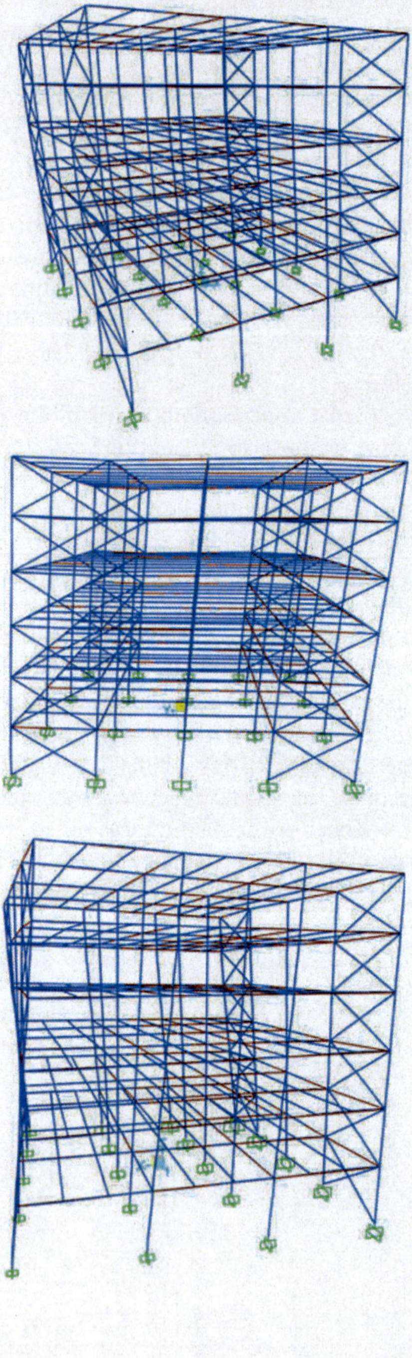

Fig. 3.20 Deformed building shapes for modes 1 (up), 2 (middle) and 3 (bottom)

Results from modal response spectrum analysis for the determination of the inter-storey drift sensitivity coefficient θ in both directions x and y of the structure are shown in Tables 3.17 and 3.18 for the seismic combinations $\pm E_x \pm 0.3E_y$ and $\pm 0.3E_x \pm E_y$, respectively. In these tables the values in parentheses correspond to the negative sign in both terms of the above seismic combinations. The corresponding values for axial force (N), shear force (V) and bending moment (M) due to imperfections I_x and I_y have been included in the seismic combinations $E_x + 0.3E_y$ and $0.3E_x + E_y$, respectively.

From the θ results presented in Tables 3.17 and 3.18, it is concluded that the most unfavorable seismic combination is $+E_x + 0.3E_y$ because θ_x exceeds 0.1 at the 2nd, 3rd and 4th stories and θ_y exceeds 0.1 at the 2nd and 4th stories. Therefore, for design purposes, the seismic actions have to be multiplied by the factor $1/(1 - \theta_{max})$, where θ_{max} is the maximum value of θ_x and θ_y with respect to the same storey. These multiplication factors read $1/(1 - 0.152) = 1.18$ for the 2nd storey, $1/(1 - 0.136) = 1.16$ for the 3rd storey and $1/(1 - 0.105) = 1.12$ for the 4th storey. No multiplication factors have to be applied to the rest of the stories.

Table 3.19 presents the maximum values for axial force (N), shear force (V) and bending moment (M) of the critical beams (pinned and rigidly connected), brace and column elements. These critical elements are highlighted with heavy lines of red color in Fig. 3.21. The values for N, V, M are provided separately for the $G + 0.3Q$, $\pm E_x \pm 0.3E_y$ and $\pm 0.3E_x \pm E_y$ loading combinations.

The section and stability design checks for the critical elements of Fig. 3.21, after EC3 (2009), are presented in the following making use of Table 3.19. Since all critical elements are in the 2nd storey, the seismic action has to be multiplied by 1.18. It should be noted that the stress ratio to the column just below the critical one, due to the total action (dead, live and seismic loads), is very close to the corresponding one of the critical column.

Equation (3.35) is satisfied because all braces have been symmetrically configured and the horizontal projection of the cross-section of the tension diagonal for positive and negative direction (the terms in the nominator of Eq. (3.35)) is the same. The design action for the critical brace is $N_{Ed} = 45.34 + 1.18 \cdot 408.23 = 527.05\,\text{kN} < N_{pl.\,Rd} = 1478.31$ kN. To check against flexural buckling one has for curve a and $a_z = 0.21$ that $\bar{\lambda}_z = 1.52\ \Phi_z = 1.79$, $\chi_z = 0.36$ and finally $N_{b.\,z.\,Rd} = 537.98\,\text{kN} > 527.05$ kN. Thus, the stability criterion is satisfied. In addition to the checks presented above, EC8 (2004) stipulates for each brace $1.3 \leq \bar{\lambda} \leq 2.0$ and the maximum brace overstrength $\Omega = N_{pl.\,Rd}/N_{Ed}$ not to differ from the minimum value by more than 25%, i.e., all Ω values not to surpass $1.25\Omega_{min}$. Table 3.20 provides the design checks in terms of $\bar{\lambda}$, $N_{pl.\,Rd}$, N_{Ed}, $N_{b.\,Rd}$ and Ω values for the most stressed braces at each storey. It is noted that N_{Ed} is the axial force due to the seismic combination taking into account, where needed, a multiplication of the axial force of the seismic action due to second-order effects. On the basis of the Ω values presented in Table 3.20, $\Omega_{min} = 2.80$ and $1.25\Omega_{min} = 3.50$, thus, all Ω values do not surpass 3.50 (the 6th storey satisfies the Ω limit marginally). To perform capacity design to beams and columns, one makes use of $\Omega_{min} = 2.80$ which is derived from the critical brace of the 2nd storey.

Table 3.17 Inter-storey drift sensitivity coefficients θ_x and θ_y for $\pm E_x \pm 0.3E_y$

Storey	Design interstorey displacement $q\,(d_{xi} - d_{xi-1})$ (m)	Design interstorey displacement $q\,(d_{yi} - d_{yi-1})$ (m)	Storey shear force V_{xi} (kN)	Storey shear force V_{yi} (kN)	Storey cumulative gravity load P_i (kN)	Storey height h_i (m)	Interstorey sensitivity coefficient θ_x	Interstorey sensitivity coefficient θ_y
1	0.020	0.0012	1716.54 (1800.39)	142.24 (169.59)	24,604.74	3.0	0.096 (0.091)	0.069 (0.058)
2	0.036	0.0017	1612.77 (1706.74)	79.46 (103.48)	20,503.95	3.0	0.152 (0.144)	0.146 (0.112)
3	0.036	0.0017	1445.69 (1501.61)	95.53 (107.31)	16,403.16	3.0	0.136 (0.131)	0.097 (0.087)
4	0.0308	0.0020	1227.44 (1222.21)	77.82 (140.36)	12,302.37	3.0	0.102 (0.103)	0.105 (0.058)
5	0.0232	0.0017	956.66 (984.64)	103.53 (113.38)	8201.58	3.0	0.066 (0.064)	0.045 (0.041)
6	0.0152	0.0015	700.52 (436.11)	78.62 (124.70)	4100.79	3.0	0.030 (0.048)	0.026 (0.016)

Table 3.18 Inter-storey drift sensitivity coefficients θ_x and θ_y for $\pm 0.3E_x \pm E_y$

Storey	Design interstorey displacement $q\,(d_{xi} - d_{xi-1})$ (m)	Design interstorey displacement $q\,(d_{yi} - d_{yi-1})$ (m)	Storey shear force V_{xi} (kN)	Storey shear force V_{yi} (kN)	Storey cumulative gravity load P_i (kN)	Storey height h_i (m)	Interstorey sensitivity coefficient θ_x	Interstorey sensitivity coefficient θ_y
1	0.006	0.0042	491.74 (457.90)	388.85 (533.89)	24,604.74	3.0	0.075 (0.081)	0.089 (0.065)
2	0.0108	0.0053	516.86 (586.76)	309.42 (322.09)	20,503.95	3.0	0.143 (0.126)	0.116 (0.111)
3	0.0108	0.0056	527.74 (583.67)	332.33 (344.11)	16,403.16	3.0	0.112 (0.101)	0.091 (0.088)
4	0.0092	0.0063	519.61 (493.28)	365.88 (449.54)	12,302.37	3.0	0.073 (0.076)	0.071 (0.057)
5	0.0072	0.0056	477.38 (505.36)	356.77 (366.61)	8201.58	3.0	0.041 (0.039)	0.043 (0.041)
6	0.0044	0.0053	452.29 (438.30)	433.74 (422.63)	4100.79	3.0	0.013 (0.014)	0.017 (0.017)

Table 3.19 M, V, N values of the critical beam, brace and column elements obtained by response spectrum analysis method

Beam (rigidly connected)	**Left end**			**Right end**		
	$G + 0.3Q$	$\pm E_x \pm 0.3E_y$	$\pm 0.3E_x \pm E_y$	$G + 0.3Q$	$\pm E_x \pm 0.3E_y$	$\pm 0.3E_x \pm E_y$
M (kNm)	−10.73	−60.68	−18.19	−14.63	−63.02	−18.89
V (kN)	−11.56	−20.61	−6.18	12.86	−20.61	−6.18
N (kN)	−1.17	−12.21	−3.72	−1.17	−11.48	−3.50
Column	**Bottom end**			**Top end**		
	$G + 0.3Q$	$\pm E_x \pm 0.3E_y$	$\pm 0.3E_x \pm E_y$	$G + 0.3Q$	$\pm E_x \pm 0.3E_y$	$\pm 0.3E_x \pm E_y$
M_2 (kNm)	−2.36	−6.99	−23.32	2.15	−6.70	−22.31
M_3 (kNm)	10.95	−116.58	−39.74	−11.73	−78.85	−33.53
V_2 (kN)	7.56	−64.09	−23.91	7.56	−64.09	−23.91
V_3 (kN)	−1.51	−4.56	−15.19	−1.51	−4.56	−15.19
N (kN)	−818.02	−318.28	−920.05	−810.82	−318.28	−920.05
Brace	**Left end**			**Right end**		
	$G + 0.3Q$	$\pm E_x \pm 0.3E_y$	$\pm 0.3E_x \pm E_y$	$G + 0.3Q$	$\pm E_x \pm 0.3E_y$	$\pm 0.3E_x \pm E_y$
N (kN)	−45.41	−127.74	−413.46	−46.04	−127.74	−413.46
Beam (pinned)	**Mid-span**					
	$G + 0.3Q$	$\pm E_x \pm 0.3E_y$	$\pm 0.3E_x \pm E_y$			
M (kNm)	10.88	0	0			
V (kN)	0	0	0			
N (kN)	−0.11	−4.28	−13.74			

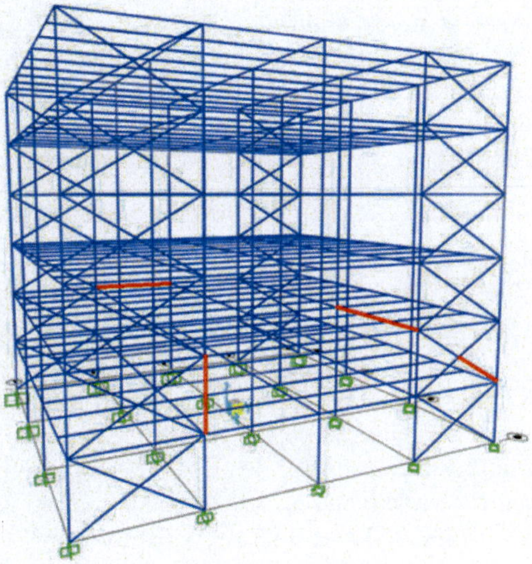

Fig. 3.21 Critical beam, brace and column elements (highlighted with heavy lines of red color) for the steel building

Table 3.20 Design checks for braces

Storey	Section	$\bar{\lambda}$	$N_{pl.\ Rd}$ (kN)	N_{Ed} (kN)	$N_{b.\ Rd}$ (kN)	Ω
1	SHS 120x120x12.5	1.52	1478.13	434.77	537.98	3.40
2	SHS 120x120x12.5	1.52	1478.13	527.05	537.98	2.80
3	SHS 120x120x12.5	1.52	1478.13	471.17	537.98	3.14
4	SHS 120x120x8	1.47	985.60	351.28	381.92	2.81
5	SHS 100x100x7.1	1.46	881.65	263.98	345.98	3.34
6	SHS 100x100x4	1.73	417.72	119.56	121.14	3.49

One important thing to mention is the effective length considered for the out-of-plane buckling of the continuous brace employed herein. According to Kitipornchai and Finch (1986) and Stoman (1989), due to the stabilizing force offered by the tension brace, this effective length can conservatively be assumed equal to 1.0 (i.e., equal to the length of the continuous brace) even though lower than 1.0 values are also reported by them (by altering the end-fixity conditions and/or the splice detail). If the value of 1.0 is adopted, the $N_{b.\ Rd}$ values of Table 3.20 are doubled and the only problem that obviously remains is that for all braces the condition $\bar{\lambda} < 1.3$ is not satisfied. In general, the violation of this 1.3 (lower) limit of $\bar{\lambda}$ has been reported by various researchers, e.g., Brandonisio et al. (2012) and Bosco et al. (2017), as a major drawback of the seismic design process of EC8 (2004) regarding CBFs. In fact, satisfaction of this 1.3 limit is in most cases difficult in praxis, reducing the realistic design possibilities. To take into account this commonly met 1.3 limit violation, design of beams and columns is performed in accordance with Brandonisio et al. (2012) and Bosco et al. (2017), i.e., by using Eq. (3.37) with Ω being replaced by $\Omega^* = \max\ (2\chi_i\Omega_i)$ where χ_i is the buckling reduction factor of the brace at storey i and Ω_i is the brace overstrength at each storey (see Table 3.20). Thus, for the steel structure under study the maximum value of Ω^* comes from the brace of the 5st storey and reads $\Omega^* = 2\chi\Omega = 2 \cdot 0.39 \cdot 3.34 = 2.61$. Just for reasons of illustration of this Ω^* approach, the design checks for the critical beams and columns are firstly performed by using the minimum Ω as stipulated by current version of EC8 (2004) and then by Ω^*.

The seismic demands on the critical rigidly connected beam (Sect. 3.5.1) are: $M_{Ed} = -14.63 + 1.18 \cdot (-63.02) = -88.99$ kNm, $N_{Ed} = -1.17 + 1.18 \cdot (-11.48) = -14.71$ kNm and $V_{Ed} = 12.86 + 1.18 \cdot (285.42 + 285.42)/6 = 125.12$ kN. Then, section checks are performed on the basis of Eqs. (3.26)–(3.28) and one has $M_{Ed}/M_{pl.\ Rd} = 88.99/285.42 = 0.311 < 1$, $V_{Ed}/V_{pl.\ Rd} = 125.12/875.81 = 0.143 < 0.5$ (interaction of M-V and M-N does not take place, hence the plastic moment capacity of the beam remains $M_{pl.\ Rd} = 285.42$ kNm). Regarding stability checks: (i) for flexural buckling in minor axis, which is the critical axis, one has for curve b and $a_z = 0.34$ that $\bar{\lambda}_z = 0.65$, $\Phi_z = 0.79$, $\chi_z = 0.81$ and finally $N_{b.\ z.\ Rd} = 1397.06$ kN > 14.71 kN; (ii) for lateral torsional buckling one has for curve b and $a_{LT} = 0.34$ that $\bar{\lambda}_{LT} = 0.35$, $\Phi_{LT} = 0.59$, $\chi_{LT} = 0.94$ and finally $M_{b.\ Rd} = 208.72$ kNm $> M_{Ed} = 88.99$ kNm. Considering $C_{my} = 0.4$, $C_{mz} = 0.541$, $C_{mLT} = 0.4$, $k_{yy} = 0.404$, $k_{yz} = 0.333$, $k_{zy} = 0.983$ and $k_{zz} = 0.555$, the demand to capacity ratio reads $14.71/1397.06 + 0.983 \cdot 88.99/208.72 = 0.01 + 0.42 = 0.43 < 1.0$ and, thus, it is satisfied. It is recalled that the presence of a composite slab essentially increases $M_{b.\ Rd}$ and thus the above criterion results in a smaller capacity ratio.

Regarding the critical section of the pinned beam (braced frame), using Eq. (3.37) one has $N_{Ed} = (-0.11 - 1.18 \cdot 1.1 \cdot 1.25 \cdot 2.80 \cdot 13.74) = -62.53$ kN. Regarding stability checks: (i) for flexural buckling in minor axis, which is the critical axis, one has for curve b and $a_z = 0.34$ that $\bar{\lambda}_z = 1.62$, $\Phi_z = 2.06$, $\chi_z = 0.30$ and finally $N_{b.\ z.\ Rd} = 517.43$ kN > 62.53 kN; (ii) for lateral torsional buckling one has for curve b and $a_{LT} = 0.34$ that $\bar{\lambda}_{LT} = 1.10$, $\Phi_{LT} = 1.26$, $\chi_{LT} = 0.54$ and finally $M_{b.\ Rd} = 118.45$ kNm $> M_{Ed} = 10.88 + 1.18 \cdot 1.1 \cdot 1.25 \cdot 2.80 \cdot 0 = 10.88$ kNm. The demand to capacity ratio reads $62.53/517.43 + 0.582 \cdot 10.88/118.45 = 0.12 + 0.05 = 0.17 < 1.0$ and, thus, it is satisfied. Using now Ω^* instead of Ω in Eq. (3.37), one has $N_{Ed} = (-0.11 - 1.18 \cdot 1.1 \cdot 1.25 \cdot 2{,}61 \cdot 13.74) = -58.30$ kN and the corresponding demand to capacity ratio reads $58.30/517.43 + 0.582 \cdot 10.88/118.45 = 0.11 + 0.05 = 0.16 < 1.0$ and, thus, it is satisfied.

The design checks are now performed for the bottom end of the critical column, as part of the braced frame, on the basis of Eq. (3.37) and the $G + 0.3Q \pm 0.3E_x \pm E_y$ combinations. More specifically, one has $N_{Ed} = (-818.02 - 1.18 \cdot 1.1 \cdot 1.25 \cdot 2.80 \cdot 920.05) = -4997.81$ kN, $M_{3.\ Ed} = 10.95 + 1.18 \cdot 39.74 = 57.84$ kNm, $M_{2.\ Ed} = -2.36 + 1.18 \cdot 23.32 = 25.17$ kNm, $V_{3.\ Ed} = -1.51 + 1.18 \cdot (-15.19) = 19.43$ kN and $V_{2.\ Ed} = 7.56 + 1.18 \cdot (-23.91)$ kN $= -20.65$ kN. For the aforementioned N, V, M design values only M-N interaction takes place but the chosen section HEM320 sufficiently satisfies the biaxial moment section criterion, i.e., $(M_{3.\ Ed}/M_{3.\ Rd})^2 + (M_{2.\ Ed}/M_{2.\ Rd}) = (57.84/981.55)^2 + 25.16/632.39 = 0.043 < 1.0$, as well as the shear and axial stress ratio, i.e., $V_{2.\ Ed}/V_{2.\ pl.\ Rd} = 20.65/1943 = 0.01 < 0.5$, $V_{3.\ Ed}/V_{3.\ pl.\ Rd} = 19.43/5194 = 0.003 < 0.5$, $N_{Ed}/N_{pl.\ Rd} = 4997.81/11076 = 0.45 < 1$.

Column stability checks for flexural buckling are of importance only for major axis and by considering curve b and $a = 0.34$, one has $\overline{\lambda}_y = 0.531$, $\Phi_z = 0.698$, $\chi_z = 0.87$ and $N_{b.\ y.\ Rd} = 9636.58$ kN > 4997.81 kN. Regarding lateral-torsional buckling, for curve a and $a = 0.21$, one obtains $\overline{\lambda}_{LT} = 0.377$, $\Phi_{LT} = 0.59$, $\chi_{LT} = 0.959$ and $M_{b.\ Rd} = 1509.55$ kNm > 57.85 kNm. Finally, the member interaction equations of EC3 (2009) are checked where for $C_{my} = 0.756$, $C_{mz} = 0.928$, $C_{mLT} = 0.756$, $k_{yy} = 0.776$, $k_{yz} = 0.669$, $k_{zy} = 0.949$ and $k_{zz} = 1.116$ corresponding demand to capacity ratio criterion reads $4997.81/9636.58 + 0.776 \cdot 57.85/1509.55 + 0.669 \cdot 25.16/632.39 = 0.518 + 0.03 + 0.027 = 0.576 < 1.0$ and, thus, it is satisfied. Finally, using Ω^* instead of Ω in Eq. (3.37), one has $N_{Ed} = (-818.02 - 1.18 \cdot 1.1 \cdot 1.25 \cdot 2.61 \cdot 920.05) = -4714.18$ kN and the corresponding demand to capacity ratio reads $4714.18/9636.58 + 0.776 \cdot 57.85/1509.55 + 0.669 \cdot 25.16/632.39 = 0.489 + 0.03 + 0.027 = 0.546 < 1.0$ and, thus, it is satisfied.

The same critical column has to be checked, as part of a MRF, employing Eqs. (3.29)–(3.32). For reasons of brevity, these checks are omitted and the reader may consult the previous example 3.6.2.

To check the damage limitation (serviceability) limit state, one makes use of Tables 3.17 and 3.18 and finds that the maximum design inter-storey displacement for the directions x and y (see Fig. 3.17) is 0.036 and 0.0063, respectively. Hence from Eq. (3.25) and assuming $v = 0.5$, one has $0.036 \cdot 0.5 = 0.018 < 0.01 \cdot 3 = 0.03$ for the x direction and $0.0063 \cdot 0.5 = 0.003 < 0.01 \cdot 3 = 0.03$ for the y direction. Thus, the damage limitation criterion is satisfied under the assumption of the interstorey drift limit being equal to 0.01, valid for non-structural components being fixed to the structure and not interfering with its deformation. The damage limitation criterion is also satisfied if one considers the interstorey drift limit to be 0.0075 but is violated in the x direction when this limit is 0.005.

In the following, a validation of the previously presented design method is presented by comparing its results against those of a non-linear dynamic analysis done with the aid of SAP 2000 (2010). Thus, the steel frame of Figs. 3.17, 3.18 and 3.19 is analyzed by employing 10 seismic motions that are compatible to the design spectrum mentioned above. Mean values for the yield base shear V_y (corresponding to the first plastic hinge), maximum base shear V_{max} and inter-storey drift IDR were found to be $V_{y,\ x} = 1908.79$ kN, $V_{y,\ y} = 2832.28$ kN, $V_{max,\ x} = 10{,}278$ kN, $V_{max,\ y} = 10{,}560.40$ kN, $IDR_x = 1.76\%$ and $IDR_y = 0.81\%$.

It is assumed that 2.5% and 1.0% are the limit values for *IDR*s of a steel MRF (direction x of the structure) and a steel CBF (direction y of the structure), respectively, regarding the life-safety (LS) seismic performance level (see Sect. 3.2 and SEAOC 1999). The maximum *IDR* values found from NLTH analysis, were 2.14% for the x direction (MRF) and 0.94% for the y direction (CBF), i.e., lower than the aforementioned limit *IDR* values. Nevertheless, it should be stressed that in 3 out of 10 seismic motions, undesirable plastic hinges were triggered at the bottom end of two columns (adjacent to braced bays) of the 2nd storey. In the rest of the cases, the plastic hinge distributions followed the pattern expected from design and no formation of storey mechanism took place.

3.7 Conclusions

Closing this chapter and having in mind all the material presented in the previous sections, one can state the following conclusions:

1. The force-based seismic design method of Eurocode 8 as applied to steel building structures has been briefly presented. The method uses forces as the main design parameters and assumes that the structure will experience inelastic deformations during the design earthquake. For an initial section selection, the method performs design in two steps: the strength checking and the deformation checking with the second one usually controlling the design.
2. The first step of the method starts with the determination of the seismic base shear force with the aid of an elastic modal response spectrum analysis in conjunction with the inelastic design acceleration spectrum. This inelastic spectrum comes from the elastic one through division of its ordinates by the behavior factor q, which takes care of inelasticity in an approximate manner. This seismic base shear force is distributed along the building height and the structure is elastically analyzed for gravity and seismic forces for member design force determination and subsequent dimensioning through strength checking.
3. The second step of the method starts with the determination of elastic displacements of the structure designed for strength and their conversion to inelastic ones on the basis of the equal displacement rule. These displacements are then checked against values for the damage limit case and member sizes are adjusted appropriately until an iterative satisfaction of these limit values. During the design process care is exercised in order to satisfy capacity design rules in order to ensure full exploitation of ductility and a global collapse mechanism.
4. In this chapter the method has been applied to three numerical examples for illustration purposes. The first example deals with a plane steel MRF treated by the simple lateral force method as well as the modal response spectrum analysis. The other two examples dealing with a space steel MRF and a space steel MRF/CBF are treated by the modal response spectrum analysis. The seismic response of all design cases is determined by nonlinear time history analyses and its maximum values are checked against performance limits. All three examples are solved in great detail for the benefit of the reader of this book.
5. The force-based design method enjoys universal acceptance and all current seismic design codes are based on this method. It follows the procedure of classical structural design under static loads and this makes the method familiar to engineers. It is supported by a large documentation and is continuously

updated and improved because the great majority of the research effort is devoted to its improvements.

6. The present version of the method has some noticeable disadvantages. The first has to do with the use of crude and constant values for the behavior factor q, which really depends on deformation and the structural period. The second has to do with the employment of the equal displacement rule for the determination of inelastic displacements from elastic ones, even though this rule underestimates or overestimates displacements depending on the fundamental structural period. The third has to do with the use of some equations for the satisfaction of the capacity design requirements in MRFs, which even though ensure the strong column-weak beam rule satisfaction, fail in some cases to drive the structure to a global collapse mechanism. Other shortcomings of the method have to do with the non-uniform distribution of plastic hinges above some stories of CBFs and the initiation of a soft storey mechanism and the absence of design recommendations for buckling-restrained braces or fracture and low-cycle fatigue criteria for ordinary braces.

References

Bosco M, Brandonisio G, Marino EM, Mele E, De Luca A (2017) Ω* method: an alternative to Eurocode 8 procedure for seismic design of X-CBFs. J Constr Steel Res 134:135–147

Brandonisio G, Toreno M, Grande E, Mele E, De Luca A (2012) Seismic design of concentric braced frames. J Constr Steel Res 78:22–37

EC3 (2009) Eurocode 3, Design of steel structures – Part 1–1: general rules and rules for buildings, EN 1993-1-1. European Committee for Standardization (CEN), Brussels

EC4 (2004) Eurocode 4, Design of composite steel and concrete structures – Part 1–1: general rules and rules for buildings, EN 1994-1-1. European Committee for Standardization (CEN), Brussels

EC8 (2004) Eurocode 8, Design of structures for earthquake resistance, Part 1: general rules, seismic actions and rules for buildings, EN 1998-1-1. European Committee for Standardization (CEN), Brussels

Gupta A, Krawinkler H (2000a) Dynamic P-delta effects for flexible inelastic steel structures. J Struct Eng ASCE 126:145–154

Gupta A, Krawinkler H (2000b) Estimation of seismic drift demands for frame structures. Earthq Eng Struct Dyn 29:1287–1305

IBC (2018) International building code. International Code Council, Washington, DC

Kitipornchai S, Finch DL (1986) Stiffness requirements for cross bracing. J Struct Eng ASCE 112:2702–2707

Landolfo R (2013) Editor, assessment of EC8 provisions for seismic design of steel structures. ECCS/Wiley/Ernst & Sohn, Berlin

Landolfo R (2014) Seismic design of steel structures-chapter 6, in Eurocodes-design of steel buildings with worked examples, CEN/TC 250/SC3. European Commission, Brussels

Plumier A (2012) Specific rules for the design and detailing of steel buildings: (i) steel moment resisting frames. In: Acun B, Athanasopoulou A, Pinto A, Carvalho E, Fardis M (eds) Eurocode 8: seismic design of buildings, worked examples. JRC Scientific and Technical Reports, EUR 25204EN-2012, European Union, Luxembourg, pp 105–128

SAP 2000 (2010) Structural analysis program 2000, static and dynamic finite element analysis of structures, Version 14. Computers and Structures Inc., Berkeley, CA
SEAOC (1999) Recommended lateral force requirements and commentary, 7th edn. Structural Engineers Association of California, Sacramento
SEAOC (2009) Seismic design recommendations. Structural Engineers Association of California, Sacramento, CA
Stoman SH (1989) Effective length spectra for cross bracings. J Struct Eng ASCE 115:3112–3122

Chapter 4
Direct Displacement-Based Design

Abstract The direct displacement-based seismic design method as applied to steel framed buildings in accordance with the Model Code DBD12 is presented. Moment resisting, eccentrically braced and concentrically braced frames are mainly considered. The method employs displacements as the main design parameters of the problem and succeeds in effectively controlling damage. It is based on the construction of an equivalent linear single-degree-of-freedom system to the original nonlinear frame and a displacement design spectrum with high amounts of viscous damping. Thus, by assuming a target interstorey drift ratio and determining the design displacement and the equivalent damping of the single-degree-of freedom system, one can obtain from the displacement spectrum the required period and hence stiffness and design base shear necessary for the structure to achieve the assumed deformation. Using the computed design base shear, one can distribute it along the height of the frame and dimension beams and columns in conformity with the capacity design rule. Numerical examples involving steel moment resisting and braced (eccentrically and concentrically) frames are presented for illustration purposes and demonstration of the advantages of the method. New developments of the method pertaining to various improvements and further applications are also briefly discussed.

Keywords Displacement-based design · Steel framed buildings · Equivalent damping · Displacement design spectrum · Target interstorey drift ratio · Moment resisting frames · Braced frames

4.1 Introduction

As it was mentioned in the previous chapter, current seismic codes, such as EC8 (2004), employ the force-based design (FBD) method, which uses forces as the main design parameters. This method performs design in two steps: the first step involves a strength checking, while the second one a displacement checking, usually accomplished iteratively. During the last 25 years or so, the displacement-based design

G. A. Papagiannopoulos et al., *Seismic Design Methods for Steel Building Structures*, Geotechnical, Geological and Earthquake Engineering 51,
https://doi.org/10.1007/978-3-030-80687-3_4

(DBD) method has emerged as a viable alternative of the FBD method. The DBD method uses displacements as the main design parameters and since displacements are more intimately related to damage than forces, can more effectively control damage. DBD requires only one step during the design process, i.e., a strength checking, since the displacement checking is automatically satisfied.

Among the most important proposed DBD methods, one can mention those of Moehle (1992), Panagiotakos and Fardis (2001), Chopra and Goel (2001) and Priestley and Kowalsky (2000) for reinforced concrete (R/C) structures and Medhekar and Kennedy (2000a, b) for steel structures. The most widely known DBD method is the direct DBD (DDBD) method, which, after an extensive research and development effort, resulting in the publication of many articles, a book by Priestley et al. (2007) and the model code DBD12 by Sullivan et al. (2012), has gained a wide acceptance. A comparison of the DDBD method against other DBD methods has been reported by Sullivan et al. (2003). The DDBD method can successfully handle seismic design of R/C, steel, masonry and timber buildings as well as buildings with isolation and supplemental damping devices. Among the existing works on the DDBD method as applied to steel structures one can mention those of Calvi et al. (2015) and Roldan et al. (2016) on moment resisting frames, Wijesundara and Rajeev (2012) and Sahoo and Prakash (2019) on concentrically braced frames and Sullivan (2012, 2013) on eccentrically braced frames.

The most important problem with the DDBD method is the replacement of the original nonlinear multi-degree-of-freedom (MDOF) building by an equivalent single-degree-of-freedom (SDOF) system in accordance with the substitute structure concept of Shibata and Sozen (1976). This replacement simplifies considerably the method at the expense of losing modeling accuracy as one goes from the MDOF system to the equivalent SDOF one. Thus, higher mode and P-Δ effects are lost because of this simplification, which is based on an assumed first mode displacement profile of the building. These problems of the method have been detected by its developers and corrected later by adding correction terms in the expressions for the lateral displacement profile and the design base shear and its distribution to take into account higher mode and P-Δ effects (Sullivan et al. 2012). However, the replacement of the MDOF system by an equivalent SDOF system not only prevents one to take into account directly higher mode and P-Δ effects but also to take into account the real distribution of deformation and damage at the member level and hence local failures. In connection with steel structures, various improvements and extensions of the method have been made after the publication of the DBD12 Model Code (Sullivan et al. 2012) and most of them are discussed in Sect. 4.4.

The DDBD method follows very briefly the following steps: It starts by assuming a target interstorey drift ratio and expresses the storey lateral displacements in terms of this deformation target; then it determines the design displacement, the effective mass and the equivalent viscous damping ratio of the equivalent linear SDOF system in terms of the storey lateral displacements and ductility; with known design displacement and equivalent damping ratio, the required natural period of the SDOF system is obtained from a displacement design spectrum with high amounts of damping; finally, with known natural period, the required stiffness is obtained and hence the required seismic base shear, which is distributed to the members of the original building for their dimensioning.

This chapter presents the basic steps of the method as applied to steel building structures mainly on the basis of the provisions of the model code DBD12 authored by Sullivan et al. (2012) and illustrates it with the aid of three examples dealing with plane steel moment resisting frames (MRFs), concentrically braced frames (CBFs) and eccentrically braced frames (EBFs).

4.2 Performance and Capacity Design Requirements

This section describes performance and capacity design requirements for steel framed buildings in accordance with the model code DBD12 of Sullivan et al. (2012).

For buildings in moderate to high seismicity areas (Zone A), performance criteria for levels of seismic design intensity 1, 2 and 3, corresponding to no damage, repairable damage and no collapse, respectively, should be satisfied. For buildings in very high seismicity areas (Zone B), performance criteria only for level of seismic design intensity 3 should be satisfied. Table 4.1 provides the seismic design intensity levels 1, 2 and 3 as functions of the probability of exceedance and the class of building importance (I, II, III, IV) according to EC8 (2004).

In DDBD, the main deformation/damage measure is the interstorey drift ratio (*IDR*). Table 4.2 provides limit values for *IDR* for the aforementioned intensity levels 1 and 2. No limits are given for level 3 intensity because the performance criterion for that level is no collapse. The special non-structural building elements of Table 4.2 refer to elements detailed to sustain building displacements. Residual *IDR* limits are also provided in the DBD12 model code (Sullivan et al. 2012) since residual deformations may render a building un-operational, or un-safe or

Table 4.1 Seismic intensity levels

Class	Level 1	Level 2	Level 3
I	Not required	50% in 50 years	10% in 50 years
II	50% in 50 years	10% in 50 years	2% in 50 years
III	20% in 50 years	4% in 50 years	1% in 50 years
IV	10% in 50 years	2% in 50 years	1% in 50 years

Table 4.2 Limit values of *IDR* for buildings with various non-structural elements and two intensity levels

IDR limit	Level 1	Level 2
Brittle non-structural elements	0.004	0.025
Ductile non-structural elements	0.007	0.025
Special non-structural elements	0.010	0.025

Table 4.3 Strain limits for plastic hinges of steel structural elements

Strain limit	Level 1	Level 2	Level 3
Class 1 section, Flexural plastic hinges	0.010	No limit	No limit
Class 2, 3 sections, Flexural plastic hinges	ε_y	ε_y	ε_y
Steel braces	$\chi_{br}\,\varepsilon_y$	$0.25\mu_f\,\varepsilon_y$	$0.5\mu_f\,\varepsilon_y$

Table 4.4 Chord rotation limits for steel structural elements

Chord rotation limit	Level 1	Level 2	Level 3
EBF short links	γ_y	$\gamma_y + 0.08$	$\gamma_y + 0.10$
EBF long links	γ_y	$\gamma_y + 0.02$	$\gamma_y + 0.025$

unacceptable for occupants. The limit values of residual *IDR* for buildings are given as 0.002 and 0.005 for design intensity levels 1 and 2, respectively.

While *IDR* limits refer to non-structural deformation/damage, material strain limits at plastic hinges or chord rotations of structural members refer to structural deformation/damage. In principle, a good design should take into account both non-structural and structural deformation/damage measures. However, it has been observed, that at least for regular framed buildings, non-structural deformation/damage measures govern the design. The DBD12 model code (Sullivan et al. 2012) presents limit values for both types of structural measures, even though it recognizes that use of chord rotation limits are more convenient than the material strain ones. However, no limit values in terms of chord rotations are given in the model code for steel MRFs and only limit values concerning short and long links in EBFs are provided. Of course, one can use chord limit values for steel buildings from the ASCE 41-17 (2017) provisions.

Table 4.3, taken form DBD12 (Sullivan et al. 2012), provides the strain or target limit values ε_T for plastic hinges of steel structural elements. In that table ε_y is the steel yield strain, $\chi_{br} = N_{b,R}/f_yA$ and $\mu_f = 2.4 + 8.3\bar{\lambda}$, where $N_{b,R}$ is the buckling strength for axial compression, f_y is the steel yielding strength, A is the cross-sectional area of the brace and $\bar{\lambda}$ is the slenderness ratio

$$\bar{\lambda} = \left(L_e/r_g\right)\sqrt{f_y/\pi^2 E} \qquad (4.1)$$

with L_e and r_g being the effective length and the cross-sectional radius of gyration of the brace element, respectively and E is the modulus of elasticity. Table 4.4, taken from DBD12 (Sullivan et al. 2012), provides the chord rotation limits for steel structural elements. In that table γ_y is the yield chord rotation.

Finally, the model code of DBD12 (Sullivan et al. 2012) provides the following steel material strengths for plastic hinge regions

$$
\begin{aligned}
f_{ye} &= 1.10 f_y \; for\ S355\&S450 \\
f_{ye} &= 1.15 f_y \; for\ S275 \\
f_{ye} &= 1.20 f_y \; for\ S235
\end{aligned} \tag{4.2}
$$

and the following steel material feasible strengths (overstrengths) of plastic hinge regions for capacity design

$$
\begin{aligned}
f_{yo} &= 1.30 f_y \; for\ S355\&S450 \\
f_{yo} &= 1.35 f_y \; for\ S275 \\
f_{yo} &= 1.40 f_y \; for\ S235
\end{aligned} \tag{4.3}
$$

where f_y is the nominal yield strength of structural steel.

Capacity design, as it was also mentioned in previous Chaps. 1 and 3, has the goal of not permitting the structure to develop undesirable inelastic mechanisms, which could lead to a premature collapse, i.e., collapse (partial or full) before the structure has mobilized all its strength capacity. The desirable distribution of plastic hinges only at the ends of beams and bases of columns in order to create a global collapse mechanism requires strong columns and weak beams. Thus, design moments and shears have to be amplified in columns in order for them to remain elastic. Amplified moments and shears may be determined through NLTH analyses, or by the Effective modal superposition method or an approximate method, which are briefly described as follows (DBD12 of Sullivan et al. 2012):

1. NLTH analyses are used to determine capacity design force levels in a structural model with resistance of plastic hinge actions increased for overstrength by using Eqs. (4.2) and (4.3), while all other elements and actions will be elastic. The capacity design forces will be the mean response values to a set of spectrum compatible accelerograms.
2. Effective modal superposition method is used to determine capacity design moments and shears by combining modal moments and shears obtained from a modal analysis based on effective member stiffness at maximum displacement response by a modal combination rule, e.g., the SRSS rule. The first mode response may be multiplied for overstrength by 1.25.
3. The approximate method is applicable to certain structural types and actions. For more details one can consult the model code DBD12 (Sullivan et al. 2012).

4.3 Basic Steps of DDBD Method for Steel Building Frames

This section provides a brief description to the DDBD method, restricted to plane structures for presentation convenience, by following the model code DB12 (Sullivan et al. 2012).

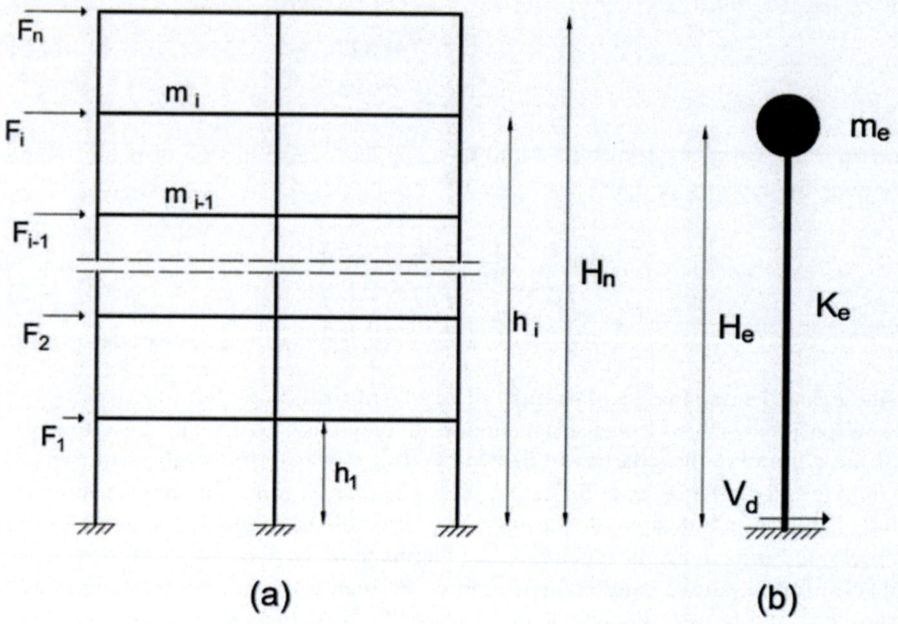

Fig. 4.1 Characteristics of the MDOF structure (**a**) and its SDOF representation (**b**)

Consider a plane moment resisting frame of n stories under a horizontal seismic motion that has to be designed by the DDBD method, as shown in Fig. 4.1a. The basic idea of the method is to determine the required seismic design base shear V_d for this frame that will ensure that its displacements will not exceed the target displacements. This is accomplished by constructing an equivalent linear SDOF system to the MDOF frame under consideration as it is shown in Fig. 4.1b. To this end, one has to determine the design displacement u_d, the effective mass m_e and the equivalent viscous damping ξ_{eq} of this equivalent SDOF system.

Assuming a target or limit design inter-storey drift ratio (IDR) and that the inelastic displacement profile corresponds to the fundamental mode, the lateral displacement u_i at the storey i ($i = 1,2, \ldots, n$) can be obtained from the expression

$$u_i = \omega_\theta IDR_T h_i \frac{(4H_n - h_i)}{(4H_n - h_1)} \tag{4.4}$$

where ω_θ is a reduction factor (taking values from 1 to 0.85 decreasing linearly from a frame of 6 stories to one of 16 stories) introduced to take into account higher mode effects and H_n and h_i are the total height of the frame and the height at storey i from the base, respectively.

The effective mass m_e of the equivalent SDOF system is evaluated from

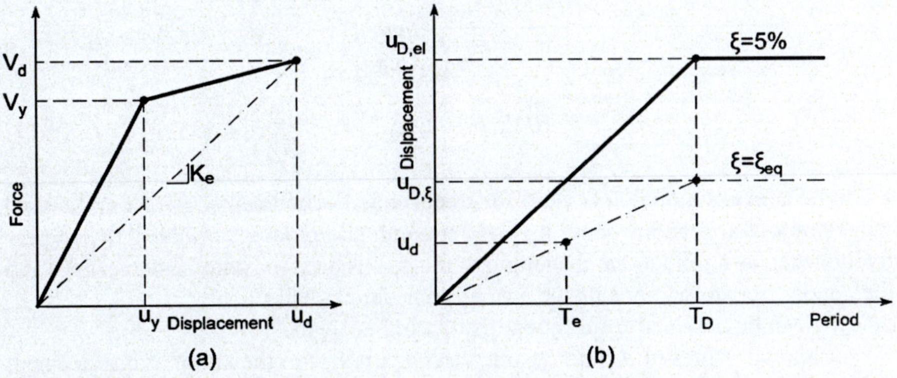

Fig. 4.2 The effective stiffness at the design target displacement u_d (**a**) and the design displacement spectra for equivalent damping ξ_{eq} (**b**)

$$m_e = \sum_{i=1}^{n} (m_i u_i)/u_d \tag{4.5}$$

where m_i is the total mass at storey i and u_d is the characteristic or design displacement shown in Fig. 4.2a and given by

$$u_d = \frac{\sum_{i=1}^{n} \left(m_i u_i^2\right)}{\sum_{i=1}^{n} (m_i u_i)} \tag{4.6}$$

The estimation of the equivalent damping ξ_{eq} for the case of steel MRFs can be obtained from the relation

$$\xi_{eq} = 0.05 + 0.577\left(\frac{\mu - 1}{\mu\pi}\right) \tag{4.7}$$

where μ is the displacement ductility, defined as

$$\mu = u_d/u_y \tag{4.8}$$

with u_y being the yield displacement shown in Fig. 4.2a. The yield displacement can be approximated as

$$u_y = H_e IDR_y \tag{4.9}$$

where H_e and IDR_y are the effective height (see Fig. 4.1b) and yield drift of the SDOF structure, respectively, expressed as

$$H_e = \frac{\sum_{i=1}^{n}(m_i u_i h_i)}{\sum_{i=1}^{n}(m_i u_i)} \tag{4.10}$$

$$IDR_y = 0.65\varepsilon_y \frac{L_b}{h_b} \tag{4.11}$$

In the above, Eq. (4.11) is valid for steel framed structures, $\varepsilon_y = f_y/E$ is the steel yield strain with f_y and E being its yield strength and modulus of elasticity, respectively, whereas L_b and h_b are the length of the beams between column centerlines and the depth of beam sections of the frames considered, respectively. Alternatively, u_y or IDR_y can be obtained from rational structural analysis.

For known values of u_d and ξ_{eq} one can calculate with the aid of a displacement design spectrum like the one shown in Fig. 4.2b, the corresponding damped design displacement $u_{D,\ \xi}$ and then the effective period T_e as

$$T_e = \frac{u_d}{u_{D,\xi}} T_D \tag{4.12}$$

where T_D and $u_{D,\ \xi}$ is the corner period of the displacement design spectrum and the design displacement with damping $\xi = \xi_{eq}$ at T_D, respectively.

Then, the effective stiffness is determined by

$$K_e = 4\pi^2 m_e / T_e^2 \tag{4.13}$$

and finally, the design base shear by

$$V_d = K_e u_d + c\sum_{i=1}^{n} P_i u_i / H_e \tag{4.14}$$

where the second term takes care of P-Δ effects with P_i denoting the total gravity load on storey level i and $c = 1.0$ for steel structures. One can easily prove that the second term of the right hand side of Eq. (4.14) can be replaced by the term $cm_e g u_d/H_e$. Related to P-Δ effects, the stability coefficient $\theta_{P-\Delta,\ i}$ for all stories i of the form

$$\theta_{P-\Delta,i} = \frac{P_i(u_i - u_{i-1})}{V_{di}(h_i - h_{i-1})} \tag{4.15}$$

where V_{di} is the shear force at level i, should not exceed 0.30.

The above design base shear can be distributed to the floor masses of the frame by using the relations

$$\begin{aligned} &\mathit{Floors}\ 1\ \mathit{to}\ n-1: F_i = kV_d(m_i u_i)/\sum_{i=1}^{n}(m_i u_i) \\ &\mathit{Roof}\ (\mathit{floor}\ n): F_n = (1-k)V_d + kV_d(m_n u_n)/\sum_{i=1}^{n}(m_i u_i) \end{aligned} \tag{4.16}$$

where for framed structures $k = 0.9$. The above force distribution assumes that 10% of the base shear is additionally applied at roof level in order to take care of higher-modes effects.

Using the above lateral forces, one can dimension the frame members. It is apparent, that the DDBD method is a one step method as requiring only a strength check and not two checks for strength and deformation as it is the case with the FBD method and this is because in the former method deformation requirements are automatically satisfied.

For the case of steel braced frames, one should use expressions for ξ_{eq} different than that of Eq. (4.7) and determine the ductility $\mu = u_d/u_y$ with u_y obtained from rational structural analysis or approximate expressions. Thus, according to the DDBD model code (Sullivan et al. 2012), one has the following expressions for ξ_{eq}, u_y and u_i for various types of braced frames:

1. For steel concentrically braced frames (CBFs)

$$\begin{aligned} \xi_{eq} &= 0.03 + \left(0.23 - \overline{\lambda}/15\right)(\mu - 1)\, for\ \mu \leq 2 \\ \xi_{eq} &= 0.03 + \left(0.23 - \overline{\lambda}/15\right) for\ \mu > 2 \end{aligned} \tag{4.17}$$

where $\overline{\lambda}$ is the slenderness ratio defined by Eq. (4.1). For CBFs with braces yielding in tension before buckling in compression, the approximate expression for u_y of the form

$$u_y = 2H_e\left[\frac{\varepsilon_y}{\sin 2a_{br}} + \frac{H_e k_{col} \varepsilon_y}{L_b}\right] \tag{4.18}$$

can be used, where a_{br} is the inclination angle of the brace with respect to the horizontal direction, L_b is the bay length and k_{col} is the average ratio of the design stress to the yield stress in the column section, which can be computed as

$$k_{col} = \frac{1}{n}\sum_{i=1}^{n}\left(\frac{N_{E,col}}{N_{R,col}}\right)_i \tag{4.19}$$

where $N_{E,\,col}$ is the seismic axial force in the column at the design deformation limit state, $N_{R,\,col}$ is the column section resistance and n is the number of storeys. A value of $k_{col} = 0.25$ is recommended as an initial estimate.

Concerning the design displacement profile of tension yielding CBFs, IDR_T in Eq. (4.4) is replaced by the minimum of IDR_T for non-structural elements (Table 4.2) and the structural storey drift limit IDR_c given as

$$IDR_c = 2\varepsilon_c / \sin 2a_{br} \tag{4.20}$$

where ε_c is the brace strain at the design limit state as described in Table 4.3. For the case of CBFs of the chevron type for which brace buckling is the inelastic deformation mechanism, design displacement profile, ξ_{eq} and u_y should be determined through rational structural analysis.

2. For steel eccentrically braced frames (EBFs)

$$\xi_{eq} = 0.05 + 0.519\left(\frac{\mu - 1}{\mu\pi}\right) \tag{4.21}$$

where the yield displacement u_y can be obtained from rational structural analysis. In addition, Eq. (4.4) is replaced by

$$u_i = \omega_\theta\left[IDR_y h_i + \left(IDR_c - IDR_y\right)h_i\frac{(2H_n - h_i)}{(2H_n - h_1)}\right] \tag{4.22}$$

where $\omega_\theta = 1.0$ for $n \leq 6$ and $\omega_\theta = 0.85$ for $n \geq 16$ with a linear interpolation in between and IDR_c is the critical storey drift limit. The critical storey drift limit IDR_c is defined as the minimum value between the non-structural drift limit $IDR_{c,\,ns}$ and the structural storey drift limit $IDR_{c,\,str}$. Sullivan (2012) suggests $IDR_{c,\,ns} = 2.0\%$ to 2.5% for the ultimate limit state and 0.5% to 1.0% for the serviceability limit state. The $IDR_{c,\,str}$ at level i can be determined from rational structural analysis or from the expression

$$IDR_{c,str,i} = IDR_{y,i} + IDR_{p.i} \tag{4.23}$$

where $IDR_{y,\,i}$ and $IDR_{p.\,i}$ are the yield and plastic storey drifts, respectively at level i. For these two IDRs explicit expressions can be found in Sullivan (2012, 2013) in terms of various geometrical and material parameters as well as the link plastic rotation γ_p, with limit values of 0.08 and 0.02 for short and long links, respectively, at the ultimate limit state (EC8 2004).

3. For steel frames with buckling-restrained braces (BRBs)

The ξ_{eq} is given again by Eq. (4.20), while the yield displacement u_y can be obtained from rational structural analysis or with the aid of the yield drift profile

$$IDR_{iy} = \frac{2\varepsilon_y/\gamma_i}{sin\,2a_i} + \rho\varepsilon_y\,tan\,a_i + 2\rho\varepsilon_y\sum_{i=1}^{n-1} h_i/L \tag{4.24}$$

where ρ is a safety factor against column buckling ($\rho < 1$), γ is the brace effective yield coefficient, a_i is the angle of the brace with the horizontal direction at level i and L is the bay length. Suggested values for ρ and γ are 0.4 and 1.14 (Maley et al. 2010).

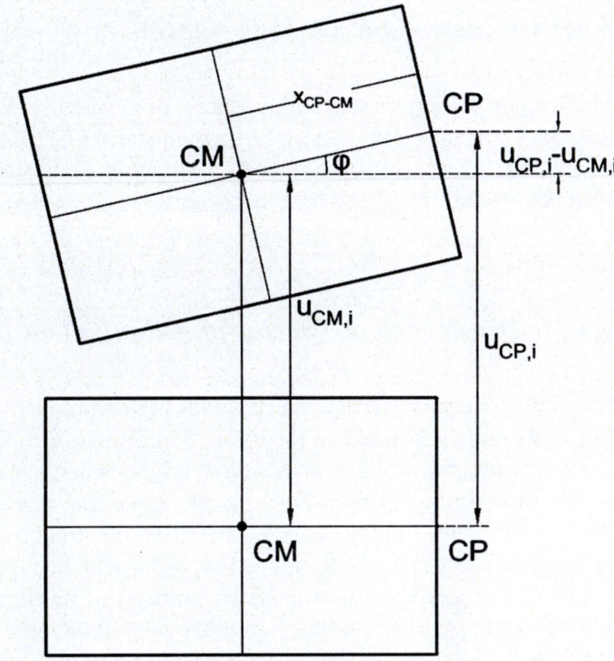

Fig. 4.3 Plan view of framed building under torsion and translational displacements $u_{CM,\,i}$ and $u_{CP,\,i}$ at storey level i

Finally, it should be noted that, in case the framed building is three-dimensional (3D) and there is a possibility of torsion due to eccentricity (no coincidence of centers of mass and stiffness at storey levels), Eq. (4.4) has to be replaced by

$$u_i = \omega_\theta IDR_T h_i \frac{(2H_n - h_i)}{(2H_n - h_1)} - (u_{CP,i} - u_{CM,i}) \tag{4.25}$$

where $u_{CP,\,i}$ is the peak displacement of the critical point (CP) in the structure and $u_{CM,\,i}$ the peak displacement at the center of mass (CM) both at storey i, as shown in Fig. 4.3. The additional displacement $u_{CP,\,i} - u_{CM,\,i}$ may be positive or negative depending on the angle of twist or torsion φ. The DDBD method controls the displacement at the center of mass and this displacement does not depend on the angle of twist φ. The critical point usually corresponds to a point at the perimeter of the building where the effect of torsion becomes maximum. One can observe from Fig. 4.3 that $u_{CP,\,i} - u_{CM,\,i} = \varphi x_{CP\,-\,CM}$. For more information on the subject of torsion in 3D buildings, one can consult Priestley et al. (2007).

4.4 Discussion on Various Aspects of the Method

This section elaborates on various aspects of the DBD method described in Sect. 4.3, discusses new developments of the method after 2012 (publication year of the Model Code DBD12) and just informs the reader about some additional applications of the method.

4.4.1 *Displacement Spectra and Equivalent Damping Ratios*

Elastic displacement design spectra with high amounts of damping like the one in Fig. 4.2b, are constructed on the basis of information concerning the seismicity and soil characteristics of the site as well as of the specific shape of EC8 (2004) elastic acceleration design spectra. For example, assuming that the peak ground acceleration of a site with soil type B is $a_{gR} = 0.30$ g, the importance factor of the structure (of height of 18.0 m) to be designed is $\gamma_I = 1.0$ and the type of the elastic spectrum is 1, one can proceed as follows: (i) an estimate of the fundamental period T of the structure using the approximate formula $T_1 = 0.085H^{3/4}$ of Eq. (3.13) is obtained equal to $T_1 = 0.74$ s; (ii) for soil type B, one can obtain from Table 3.3 $S = 1.20$, $T_C = 0.5$ s and $T_D = 2.0$ s and observe that $T_C = 0.5 < T_1 = 0.74 < T_D = 2.0$ s; (iii) for $T_C < T_1 < T_D$ and $a_g = \gamma_I a_{gR} = 1.0 \cdot 0.30$ g $= 2.943$ m/s^2, one can obtain from Eq. (3.3) of the elastic acceleration spectrum with 5% damping ($\eta = 1$), $S_a(T_D) = a_g S\eta 2.5(T_C/T_D) = 2.943 \cdot 1.20 \cdot 1.0 \cdot 2.5(0.5/2.0) = 2.207$ m/s^2; (iv) for small amounts of damping (like 5%), $S_d = S_a(T/2\pi)^2$ and thus $S_d(T_D) = S_a(T_D)(T_D/2\pi)^2 = 2.207 \cdot (2/2\pi)^2 = 0.224$ m. Thus, with reference Fig. 4.2b, one has that for $T_D = 2.0s$, $u_{D,\,el} = 0.224$ m for 5%.

With the pair ($u_{D,\,el}$, T_D) at $\xi = 5\%$ known, one can obtain the ($u_{D,\,\xi}$, T_D) pair for any value of $\xi = \xi_{eq} > 5\%$ with the aid of the damping reduction factor η of the form

$$\eta = \left[0.10/\left(0.05 + \xi_{eq}\right)\right]^{0.5} \tag{4.26}$$

For example, for ξ_{eq} obtained from Eq. (4.7) for steel MRFs and found to be equal to 8%, one can determine from Eq. (4.26) $\eta = 0.877$ and hence $u_{D,\,\xi} = u_{D,\,el} \cdot \eta = 0.224 \cdot 0.877 = 0.196$ m. Thus, on the assumption that the design displacement has been found to be $u_d = 0.180$ m, one can easily compute T_e from Eq. (4.12) and find $T_e = (u_d/u_{D,\,\xi})T_D = (0.18/0.196)2.0 = 1.837$ s.

In connection with the above, some comments should be made, which read as follows:

1. The value of 2.0 s for the corner periods (the same for all soil types) has been taken from EC8 (2004) in connection with acceleration spectra. However, it has been observed by Boore and Bommer (2005) and Priestley et al. (2007) that $T_D = 2.0$ s underestimates displacement spectral values and that $T_D = 8.0$ s would

be a better choice. Many authors have followed this suggestion in their applications, e.g., Maley et al. (2010), Garcia et al. (2010), Morelli (2014), Nievas and Sullivan (2015) and Roldan et al. (2016) and obtained reasonable results.

2. The expression of Eq. (4.26) for the damping reduction factor has been adopted in the DBD12 Model Code (Sullivan et al. 2012) from EC8 (2004) and is the most widely used in applications. Dwairi et al. (2007) through extensive parametric studies involving a large number of real seismic records verified the validity of Eq. (4.26). The Model Code DBD12 (Sullivan et al. 2012) also offers the option of using an expression for η of the form

$$\eta = \left[0.07/\left(0.02 + \xi_{eq}\right)\right]^{0.5} \tag{4.27}$$

which had been initially proposed in the old version of EC8 (1994). This expression was verified by Grant et al. (2005) through extensive parametric studies involving artificial seismic records. Studies by Pennucci et al. (2011a) also shown in Sullivan et al. (2012) have verified the results of both Dwairi et al. (2007) and Grant et al. (2005) and shown that the maximum difference between the two η expressions (new and old) is about 0.06 in the range 15%–30% of damping, with the new one to give the higher values.

3. Instead of using the equivalent damping ξ_{eq} in conjunction with the damping reduction factor η for constructing elastic displacement spectra for high amounts of damping, Pennucci et al. (2011b) proposed the employment of the displacement reduction factor η_{Δ}, which is given as a function of the displacement ductility μ. Thus, inelastic displacement spectra are constructed in one step instead of the two required for the case of elastic displacement spectra with high amounts of damping and this likely leads to higher accuracy. This displacement reduction factor can be thought of as the counterpart of the behavior or strength reduction factor used in acceleration spectra in the framework of the FBD.
4. In Sect. 4.3, various expressions for the equivalent damping ratio ξ_{eq} of the equivalent SDOF system to the original MDOF nonlinear structure have been presented for various steel frames (MRFs, CBFs, EBFs). This representation of the original MDOF structure by an equivalent SDOF system has the advantage of simplicity but it cannot take into account higher mode and P-Δ effects. The developers of the DDBD method have detected that and corrected it rather artificially by adding appropriate terms in the expression for the inelastic displacement profile and the base shear. Very recently, Muho et al. (2020) and Kalapodis et al. (2021) working on plane reinforced concrete and steel frames, respectively, have proposed an improved DDBD method, which replaces the original nonlinear MDOF structure by an equivalent linear MDOF structure with the same mass and elastic stiffness as in the original one. This is accomplished with the aid of equivalent modal damping ratios for the first few modes

significantly contributing to the response. This approach succeeds in taking into account higher mode and P-Δ effects in a rational and more accurate manner.

4.4.2 *Expressions for the Yield Displacement*

The yield displacement u_y in the Model Code DBD12 (Sullivan et al. 2012) is determined by using Eqs. (4.9) and (4.11) or by rational structural analysis. The only problem with the approximate Eq. (4.11) is that the designer does not know at the beginning the depth of the beam cross-section h_b. Thus, one has to assume a value of h_b and proceed with iterations during the design process.

For plane steel MRFs Garcia et al. (2010) and Maley et al. (2010) have proposed for IDR_y the expression

$$IDR_y = (1/6)\left(\varphi_{y,b}L_b + 0.9\varphi_{y,c}h_c\right) \tag{4.28}$$

where the subscripts b and c refer to beams and columns, respectively, L_b and h_c are the bay or beam length and the interstorey height, respectively, and the yield curvature φ_y for beams or columns is given by

$$\varphi_y = \left(W_{pl}/I\right)\varepsilon_y \tag{4.29}$$

In the above equation, W_{pl} and I are the section plastic modulus (cm^3) and moment of inertia (cm^4), respectively and $\varepsilon_y = f_y/E$ is the yield strain of steel with f_y and E being the yield strength and modulus of elasticity, respectively, of the steel material. The advantage of Eq. (4.28) over Eq. (4.11) is that the former takes into account the effects of both beams and columns instead of just the effect of beams taken by the latter equation. Table 4.5 provides the plastic section moduli W_{pls} and W_{plw} with respect to the strong and weak sectional axes, respectively, of some commonly used in praxis HEM and IPE sections for columns and beams, respectively. This table can be used in conjunction with the capacity design requirement for section selection in order to use Eq. (4.28) for the estimation of IDR_y.

Table 4.5 Strong and weak axis plastic section moduli W_{pl} of some HEM and IPE sections

Section	W_{pls} (cm^3)	W_{plw} (cm^3)	Section	W_{pls} (cm^3)	W_{plw} (cm^3)
HEM280	2966	1397	IPE270	484	96.95
HEM300	4078	1913	IPE300	628.4	125.2
HEM320	4435	1951	IPE330	804.3	153.7
HEM340	4718	1953	IPE360	1019	191.1
HEM360	4989	1942	IPE400	1307	229.0
HEM400	5571	1934	IPE450	1702	276.4
HEM450	6331	1939	IPE500	2194	335.9
HEM500	7094	1932	IPE550	2787	400.5

Table 4.6 Coefficients $b_1 - b_4$ for Eq. (4.30)

n_s	b_1	b_2	b_3	b_4
$n_s \leq 10$	−3.396	4.401	4.209	−1.197
$10 < n_s \leq 20$	−3.559	4.536	4.415	−1.207

Table 4.7 Coefficients $b_1 - b_4$ for Eq. (4.31)

n_s	b_1	b_2	b_3	b_4
$n_s \leq 10$	−4.520	5.680	5.459	−0.890
$10 < n_s \leq 20$	−2.682	3.899	2.923	−0.946

For plane steel MRFs and CBFs of the X-brace type, Dimopoulos et al. (2012) developed empirical expressions for the yield displacement u_{yi} at storey level i through extensive parametric studies involving nonlinear dynamic analyses of 36 MRFs and 36 CBFs of X-brace type under 84 far-fault seismic motions. The NLTH analyses were conducted by the Ruaumoko computer program (Carr 2005) in conjunction with the Remennikov and Walpole (1997) model for braces and having in mind the work of Tremblay (2002) on the seismic behavior of braces. These expressions have the form

$$u_{y,i} = h_i^{b_i}(h_i/H_n)^{b_2} n_s^{b_3} e^{b_4} \tag{4.30}$$

for MRFs

$$u_{y,i} = h_i^{b_i}(h_i/H_n)^{b_2} n_s^{b_3} e^{b_4} \tag{4.31}$$

for CBFs of the X-brace type. In the above equations, h_i is the interstorey height, H_n is the total height, n_s is the number of storeys, $e = 235/f_y$ with f_y being the steel yield strength in N/mm^2 and $b_1 - b_4$ are coefficients given in terms of n_s in Tables 4.6 and 4.7 for MRFs and CBFs of the X-brace type, respectively. The above expressions, in spite of their simplicity (not requiring design estimates) have good accuracy. More complicated expressions (requiring design estimates like stiffness and strength ratios or slenderness ratio) with higher accuracy can be found in Dimopoulos et al. (2012).

4.4.3 *Additional Information on the DDBD Method*

The Model Code DBD12 (Sullivan et al. 2012), in addition to the plane steel CBFs and EBFs discussed in Sect. 4.3, also discusses the case of plane steel buckling-restrained braced frames (BRBFs). Some other general subjects indirectly related to steel frames, such as flexible foundations, base isolation devices and supplemental dampers are also discussed in the above Model Code.

During the last 10 years or so, several publications on the DDBD method as applied to plane steel frames have appeared in the literature. Among those one can mention the works of Macedo and Castro (2012) on MRFs, Nievas and Sullivan

(2015) and Roldan et al. (2016) on MRFs with setbacks and partially-restrained joints, respectively, Medhekar and Kennedy (2000a, b), Wijesundara et al. (2011), Wijesundara and Rajeev (2012), Salawdeh and Goggins (2016) and Sahoo and Prakash (2019) on CBFs, Sullivan (2012, 2013) on EBFs and Maley et al. (2010) and Garcia et al. (2010) on dual systems consisting of MRFs and BRBFs and MRFs and RC walls, respectively. A displacement-based seismic design method for plane steel CBFs of chevron type has also been developed by Della Corte and Mazzolani (2008) and Della Corte et al. (2010). Finally, the comprehensive report DiSTEEL by Calvi et al. (2015) on the application of the DDBD method to steel MRFs should be mentioned.

4.5 Numerical Examples

In this section, three numerical examples are presented in detail to illustrate the method and demonstrate its merits. These examples deal with plane steel frames of the MRF, EBF and CBF type.

4.5.1 *Seismic Design of Steel Building with MRFs*

Consider the six-storey steel office building of Figs. 3.11 and 3.12 of the previous chapter reproduced here as Figs. 4.4 and 4.5 for reasons of convenience and easy reference. This building consists of five and four moment resisting frames (MRFs) along the x and y horizontal directions, respectively. The three-bay six-storey MRF along the x_2 direction that was designed in Sect. 3.5.1 by the FBD method of EC8 (2004) is designed here again by the DDBD method assuming $IDR_T = 2.5\%$ and steel grade S355 for both beams and columns. The gravity and seismic load details of Sect. 3.5.1 are summarized here and read as follows: seismic peak acceleration $a_g = 2.35$ m/s^2 associated with soil type C and gravity load for the seismic load combination acting on beams $G + 0.3Q = 40.16$ kN/m.

According to Eq. (4.4), the lateral displacement u_i for the assumed $IDR_T = 2.5\%$, takes the form $u_i = 1.0 \cdot 0.025 \cdot h_i \cdot (4 \cdot 18 - h_i)/(4 \cdot 18 - 3) = 0.3623 \cdot 10^{-3} \cdot h_i \cdot (72 - h_i)$. Thus, the design displacement u_d, which is given by Eq. (4.6), taking into account that storey masses are the same in all the stories, finally becomes $u_d = \sum_{i=1}^{6} u_i^2 / \sum_{i=1}^{6} u_i = 0.3623 \cdot 10^{-3} \cdot (2715363/3717) = 0.265$ m. The effective mass m_e of the equivalent linear SDOF system can now be computed from Eq. (4.5) with a storey mass $m_i = 272951$ kg/5 $= 54590.2 \rightarrow 54590$ kg, where the value of 272951 kg taken from Sect. 3.5.1 corresponds to the whole building consisting of

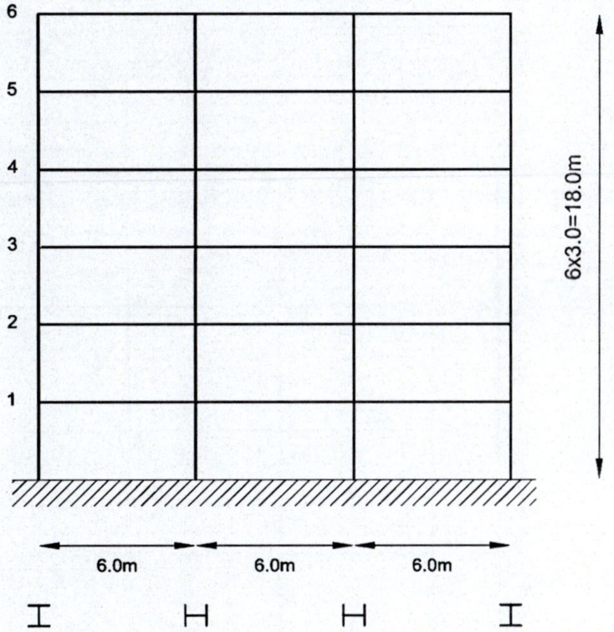

Fig. 4.4 Elevation of six-storey steel building

5 frames like the one under design here. Thus, $m_e = m_i\left(\sum_{i=1}^{6} u_i\right)/u_d =$ $54590 \cdot 0.3623 \cdot 10^{-3} \cdot 3717/0.265 = 277414$ kg.

Next the yield displacement u_y is determined by using Eqs. (4.9) and (4.11). From Eq. (4.11), assuming $h_b = 0.40$ m and using $\varepsilon_y = f_y/E = 355/210{,}000 = 0.0017$ one has $IDR_y = 0.65 \cdot 0.0017 \cdot 6/0.40 = 0.0165$. This $h_b = 0.40m$ corresponds to a IPE400 section for beams for which the plastic moment of resistance $M_{pl.\ Rd} = W_{pl}f_y/$ $1.0 = 1307 \cdot 10^{-3} \cdot 355/1 = 463.99$ kNm is larger than the design moment $M_{sd} =$ $(G + 0.3Q)L_b^2/10 = 40.16 \cdot 6^2/10 = 144.58$ kNm for the frame beams assumed to be designed only for gravity loads. The section selection for beams and columns can be done with the aid of Table 4.5 in such a way so as to satisfy capacity design requirements. Taking into account that for the frame studied herein the steel grade for beams and columns is the same, one selects a HEM340 section for columns and an IPE400 section for beams. On the basis of the values provided in Table 4.5 and column orientation depicted in Fig. 4.5, this section selection obviously satisfies the criteria $W_{plc} \geq 1.3W_{plb}$ at an interior joint and $2W_{plc} \geq 1.3W_{plb}$ at an exterior joint, where the subscripts b and c denote beam and column, respectively.

By selecting the aforementioned sections and with the aid of Eq. (4.29), one computes IDR_y from Eq. (4.28) where $W_{pl}/I = 6.18$ m^{-1} for columns (HEM340) and $W_{pl}/I = 5.65$ m^{-1} for beams (IPE400). In particular, from Eq. (4.29) one has $\varphi_{y,\ c} = 6.18 \cdot 0.0017 = 10.51 \cdot 10^{-3}$ m^{-1} and $\varphi_{y,\ b} = 5.65 \cdot 0.0017 = 9.61 \cdot 10^{-3}$ m^{-1} and with $h_c = 3.0$ m and $L_b = 6.0$ m, Eq. (4.28) yields $IDR_y = (1/6)$ $(9.61 \cdot 10^{-3} \cdot 6.0 + 0.9 \cdot 10.51 \cdot 10^{-3} \cdot 3) = 0.0143$. The above found two values of IDR_y can be used in Eq. (4.9) to determine the corresponding values of u_y,

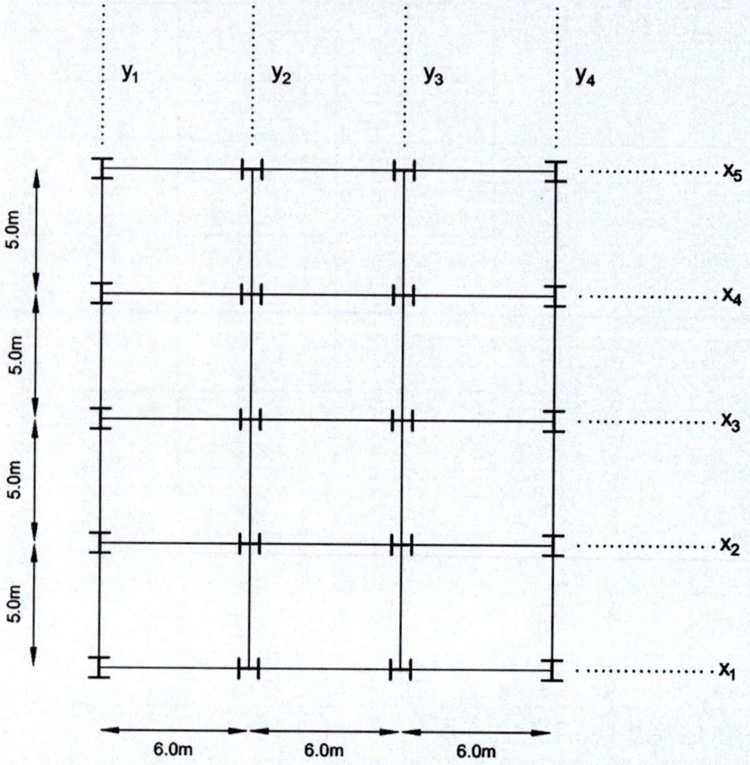

Fig. 4.5 Plan view and column orientation along horizontal directions x and y of six-storey steel building

provided the effective H_e is known. Use of Eq. (4.10) with m_i the same in all stories provides $H_e = \sum_{i=1}^{n}(m_i u_i h_i)/\sum_{i=1}^{n}(m_i u_i) = 47061/3717 = 12.66$ m. Thus, Eq. (4.9) gives $u_y = 0.0165 \cdot 12.66 = 0.209$ m and $u_y = 0.0143 \cdot 12.66 = 0.181$ m for the two values of IDR_y. Finally, u_y is also estimated on the basis of Eqs. (4.30) and (4.6). Using Table 4.6, one has from Eq. (4.30) $u_{y,i} = \left(h_i^{-3.396+4.401}/18^{4.401}\right)6^{4.209}(235/355)^{-1.197} = 0.00923 h_i^{1.005}$ and from Eq. (4.6) $u_y = 0.00923 \cdot \sum_{i=1}^{n} h_i^{2.01}/\sum_{i=1}^{n} h_i^{1.005} = 0.122$ m. Selecting as u_y the intermediate value of $u_y = 0.181$ m, the displacement ductility μ can be found from Eq. (4.8) to be $\mu = u_d/u_y = 0.265/0.181 = 1.46$. With this value of μ, Eq. (4.7) gives $\xi_{eq} = 0.05 + 0.577 \cdot [(1.46 - 1)/1.46\pi] = 0.109$.

For the steel frame of Fig. 4.4 with total height $H_n = 3 \cdot 6 = 18.0$ m, an estimate of its fundamental period can be found from Eq. (3.13) to be $T_1 = 0.085 \cdot 18^{3/4} = 0.74$ s. For soil type C one can obtain from Table 3.3 that $S = 1.15$, $T_B = 0.20$ s, $T_C = 0.60$ s and $T_D = 2.0$ s. However, on the basis of the discussion in Sect. 4.4.1 the more reasonable value of $T_D = 8.0$ s is used here instead of the value of $T_D = 2.0$ s suggested by EC8 (2004) in connection with acceleration spectra. Since $T_C = 0.6 \text{ s} \leq T_1 = 0.74 \text{ s} < T_D = 8.0$ s, one can obtain from Eq. (3.3)

$S_a(T_D) = a_g S\eta 2.5(T_C/T_D) = 2.35 \cdot 1.15 \cdot 1.0 \cdot 2.5 \cdot (0.6/8) = 0.507$ m/s^2, where use was made of $\eta = 1.0$ because damping is 5%. For small amounts of damping (5%), the spectral displacement at $T_D = 8.0$ s is $S_d(T_D) = S_a(T_D) \cdot (T_D/2\pi)^2$ or $S_d(8.0) = S_a(8.0) \cdot (8.0/2\pi)^2 = 0.507 \cdot 1.623 = 0.823$ m.

The damping reduction factor η is then computed from Eq. (4.27) and found to be $\eta = [0.10/(0.05 + 0.109)]^{0.5} = 0.793$. Thus, with reference Fig. 4.2b, one finds $u_{D,\,\xi}$ to be $u_{D,\,\xi} = S_d(8s)_{\xi\,=\,5\%} \cdot \eta = 0.823 \cdot 0.793 = 0.653$ m and from Eq. (4.12) the effective period $T_e = (u_d/u_{D,\,\xi}) \cdot T_D = (0.265/0.653) \cdot 8 = 3.25$ s. With this effective period T_e and known effective mass m_e one can determine from Eq. (4.13) the effective stiffness K_e of the equivalent SDOF system as $K_e = 4\pi^2 m_e/T_e^2 = 4\pi^2 \cdot 277{,}414/3.25^2 = 1{,}035{,}812\ \text{N/m} = 1035.81\ \text{kN/m}$.

The design base shear of the frame can be determined on the basis of Eq. (4.14) with $P_i = (G + 0.3Q) \cdot 3L_b = 40.16 \cdot 3 \cdot 6 = 722.88$ kN and $\sum_{i=1}^{6} P_i u_i = P_i \sum_{i=1}^{6} u_i = 722.88 \cdot 0.3623 \cdot 10^{-3} \cdot 3717 = 973.48$ kN. Thus, $V_d = 1035.81 \cdot 0.265 + 1.0 \cdot 973.48/12.66 = 351.38$ kN. This design base shear is distributed as lateral forces F_i to the storey levels in accordance with Eq. (4.16). These forces for m_i being the same in all storeys take the form $F_i = 0.9V_d\left(u_i/\sum_{j=1}^{6} u_j\right)$ for $i = 1 - 5$ and $F_6 = (1 - 0.9)V_d + 0.9V_d\left(u_n/\sum_{j=1}^{6} u_j\right)$. Thus, since $u_i/\sum_{j=1}^{6} u_j = h_i(72 - h_i)/3717$ these seismic lateral forces are: $F_1 = 0.9 \cdot 351.38 \cdot 3 \cdot (72 - 3)/3717 = 17.61$ kN, $F_2 = 0.9 \cdot 351.38 \cdot 6 \cdot (72 - 6)/3717 = 33.69$ kN, $F_3 = 0.9 \cdot 351.38 \cdot 9 \cdot (72 - 9)/3717 = 48.24$ kN, $F_4 = 0.9 \cdot 351.38 \cdot 12 \cdot (72 - 12)/3717 = 61.26$ kN, $F_5 = 0.9 \cdot 351.38 \cdot 15 \cdot (72 - 15)/3717 = 72.74$ kN and $F_6 = 0.1 \cdot 351.38 + 0.9 \cdot 351.38 \cdot 18 \cdot (72 - 18)/3717 = 117.84$ kN. One can observe that $\sum_{i=1}^{i=6} F_i = 351.38\ \text{kN} = V_d$.

Using the above seismic forces and $G + 0.3Q$ gravity load on beams equal to 40.16 kN/m (Sect. 3.6.1, (5)), one can determine member forces and moments through an elastic analysis with the aid of SAP 2000 (2010) and finally reach a section selection consisting of HEM340 for all columns and IPE400 for all beams, i.e., the same as the one in the frame designed by the FBD in Sect. 3.6.1 of Chap. 3. The deformation response results (lateral drifts d_i and interstorey sensitivity coefficients θ_i of Eq. (4.15)) of the above designed frame are shown in Table 4.8 and found to be almost the same with those presented in Table 3.9 for the same frame analyzed by the modal response spectrum analysis method of EC8 (2004). It is observed that according to the Model Code DBD12 (Sullivan et al. 2012), all θ_i values satisfy the requirement to be less than 0.30. Therefore, following DBD12 (Sullivan et al. 2012), the analysis results have to be multiplied by $1/(1 - \theta) = 1/(1 - 0.16) = 1.19$ for the design of beams and columns according to EC3 (2009). Taking into account that the N_{Ed}, V_{Ed}, M_{Ed} values found here for the most critical beam and column (see Fig. 3.14) were lower than those presented in Tables 3.10 and 3.11, section and stability checks of EC3 (2009) are essentially satisfied and there is no need to repeat them herein. For the stability checks, the out-of-plane buckling length is assumed to be 6/4 = 1.5 m for each beam due to the presence of secondary beams.

Table 4.8 Response results to the lateral forces obtained by the DDBD method (HEM340/IPE400)

Seismic load combination $G + 0.3Q + E$ with $G + 0.3Q = 40.16$ kN/m							
Storey	Storey displacement d_i (m)	Design interstorey displacement $q(d_i - d_{i-1})$ (m)	Storey lateral forces F_i(kN)	Storey shear force V_i(kN)	Storey cumulative gravity load P_i(kN)	Storey height h_i (m)	Interstorey sensitivity coefficient θ
1	0.0065	0.0260	17.61	351.38	4337.28	3.0	0.107
2	0.0176	0.0444	33.69	333.76	3614.40	3.0	0.160
3	0.0287	0.0444	48.24	300.07	2891.52	3.0	0.143
4	0.0385	0.0392	61.26	251.65	2168.64	3.0	0.113
5	0.0459	0.0296	72.74	190.57	1445.76	3.0	0.075
6	0.0509	0.0200	117.84	117.84	722.88	3.0	0.041

It should be finally noted that unlike the case of EC8 (2004), no guidance is provided in Model Code DBD12 (Sullivan et al. 2012) regarding any amplification of the obtained lateral force distribution in a regular and symmetrical frame, as the one studied herein, due to accidental eccentricities. Thus, the amplification factor δ of EC8 (2004) that essentially leads to increased lateral forces is not employed in the context of the design of the present steel frame by the DDBD method.

Even though the Model Code DBD12 (Sullivan et al. 2012) does not explicitly require any drift checking for damage limitation, as it is the case with EC8 (2004), it does provide drift limit values for damage limitation (Level 1 in Table 4.2), which are almost the same as those of EC8 (2004). Thus, from Table 4.8 one reads the maximum design inter-storey displacement to be 0.0444 m. Hence from Eq. (3.25) and assuming $v = 0.5$, one has $0.0444 \cdot 0.5 < 0.01 \cdot 3 = 0.03$. Thus, the damage limitation criterion of EC8 (2004) is satisfied under the assumption of the limit value 0.01 (case of non-structural components being fixed to the structure and not interfering with its deformation).

The steel MRF is then subjected to non-linear time-history (NLTH) analysis with the aid of SAP 2000 (2010) employing 10 seismic motions that are compatible to the design spectrum mentioned above. Mean values for the yield base shear (corresponding to the first plastic hinge) and inter-storey drift ratio (*IDR*) were found to be 1203 kN and 2.04%, respectively. This *IDR* value is smaller than the 2.5% value initially targeted and taking into account that the plastic hinge distribution follows the expected from design pattern without formation of a soft storey mechanism, one can accept this design as the final one. However, in an effort to achieve an *IDR* value closer to the 2.5% value, the whole process is repeated here for a lighter section selection.

One chooses now sections HEM300 for columns and IPE360 for beams. On the basis of Table 4.5, this section selection satisfies the capacity design checks of Eqs. (3.48) and (3.49). The design displacement u_d remains equal to 0.265 m. The total mass of the steel skeleton per storey is calculated as follows: one has for columns HEM300 that $3 \cdot (5 \cdot 4) \cdot 238$ kg/m = 14280 kg and for beams IPE360 that $((5 \cdot 18) + (4 \cdot 20)) \cdot 57.1$ kg/m = 9707 kg; thus, the total mass of steel skeleton per storey is 14280 + 9707 = 23987 kg/*storey* and the total seismic mass per storey is 180,000 + 0.5 · 0.3 · 108,000 + 51,000 + 23,987 = 271,187 kg/*storey*. The effective mass m_e of the equivalent linear SDOF system can now be computed from Eq. (4.5) with a storey mass m_i = 271,187 kg/5 = 54,237.4→54,238 kg, where the number 5 denotes the 5 frames like the one under design comprising the whole building. Thus, $m_e = m_i\left(\sum_{i=1}^{6} u_i\right)/u_d = 54{,}238 \cdot 0.3623 \cdot 10^{-3} \cdot 3717/0.265 = 275{,}615$ kg. Thus, the frame along line x_2 (considered here) of Fig. 4.5 has a seismic design mass in one beam equal to (275,615/4) kg/18 m = 3828.12 kg/m, which corresponds to $G + 0.3Q$ load equal to 38.28 kN/m.

To determine the new yield displacement u_y, one employs Eq. (4.11). Assuming $h_b = 0.36$ m and using $\varepsilon_y = f_y/E = 355/210{,}000 = 0.0017$ one has $IDR_y = 0.65 \cdot 0.0017 \cdot 6/0.36 = 0.0184$. This $h_b = 0.36$ m corresponds to the IPE360 section for beams for which the plastic moment of resistance $M_{pl.\,Rd} = W_{pl}f_y/$

$1.0 = 1019 \cdot 10^{-3} \cdot 355/1 = 361.75$ kNm is larger than the design moment $M_{sd} = (G + 0.3Q)L_b^2/10 = 38.28 \cdot 6^2/10 = 137.81$ kNm for the frame beams assumed to be designed only for gravity loads. Then, one computes IDR_y from Eq. (4.28) where $W_{pl}/I = 6.89$ m^{-1} for columns (HEM300) and $W_{pl}/I = 6.26$ for beams (IPE360). In particular, from Eq. (4.29) one has $\varphi_{y,\,c} = 6.89 \cdot 0.0017 = 11.71 \cdot 10^{-3}$ m^{-1} and $\varphi_{y,\,b} = 6.26 \cdot 0.0017 = 10.64 \cdot 10^{-3}$ m^{-1} and with $h_c = 3.0$ m and $L_b = 6.0$ m, Eq. (4.28) yields $IDR_y = (1/6)(10.64 \cdot 10^{-3} \cdot 6.0 + 0.9 \cdot 11.71 \cdot 10^{-3} \cdot 3) = 0.0159$. The value of H_e remains equal to 12.66 m and since m_i is the same in all stories, Eq. (4.9) gives $u_y = 0.0184 \cdot 12.66 = 0.233$ m and $u_y = 0.0159 \cdot 12.66 = 0.201$ m for the two values of IDR_y. It is recalled that u_y can also be estimated on the basis of Eqs. (4.30) and (4.6) leading to the aforementioned value of 0.122 m. Selecting as u_y the intermediate value of $u_y = 0.201$ m, the displacement ductility μ from Eq. (4.8) is $\mu = u_d/u_y = 0.265/0.201 = 1.32$. With this value of μ, Eq. (4.7) gives $\xi_{eq} = 0.05 + 0.577 \cdot [(1.32 - 1)/1.32\pi] = 0.095$.

The spectral displacement remains the same as before and reads $S_d(8.0) = 0.823$ m. The damping reduction factor η is then computed from Eq. (4.27) and found to be $\eta = [0.10/(0.05 + 0.095)]^{0.5} = 0.831$. Thus, with reference Fig. 4.2b, one finds $u_{D,\,\xi}$ to be $u_{D,\,\xi} = S_d(8s)_{\xi\,=\,5\%} \cdot \eta = 0.823 \cdot 0.831 = 0.684$ m and from Eq. (4.12) the effective period $T_e = (u_d/u_{D,\,\xi}) \cdot T_D = (0.265/0.684) \cdot 8 = 3.10$ s. With this effective period T_e and known effective mass m_e one can determine from Eq. (4.13) the effective stiffness K_e of the equivalent SDOF system as $K_e = 4\pi^2 m_e/T_e^2 = 4\pi^2 \cdot 275{,}610/3.10^2 = 1131074$ N/m $= 1131.08$ kN/m.

The design base shear of the frame can be determined on the basis of Eq. (4.14) with $P_i = (G + 0.3Q) \cdot 3L_b = 38.28 \cdot 3 \cdot 6 = 689.04$ kN and $\sum_{i=1}^{6} P_i u_i = P_i \sum_{i=1}^{6} u_i = 689.04 \cdot 0.3623 \cdot 10^{-3} \cdot 3717 = 927.91$ kN. Thus, $V_d = 1131.08 \cdot 0.265 + 1.0 \cdot 927.91/12.66 = 373.03$ kN. This design base shear is distributed as lateral forces F_i to the storey levels in accordance with Eq. (4.16). These forces for m_i being the same in all storeys take the form $F_i = 0.9V_d\left(u_i/\sum_{j=1}^{6} u_j\right)$ for $i = 1 - 5$ and $F_6 = (1 - 0.9)V_d + 0.9V_d\left(u_n/\sum_{j=1}^{6} u_j\right)$. Thus, since $u_i/\sum_{j=1}^{6} u_j = h_i(72 - h_i)/3717$ these seismic lateral forces are: $F_1 = 0.9 \cdot 373.03 \cdot 3 \cdot (72 - 3)/3717 = 18.70$ kN, $F_2 = 0.9 \cdot 373.03 \cdot 6 \cdot (72 - 6)/3717 = 35.77$ kN, $F_3 = 0.9 \cdot 373.03 \cdot 9 \cdot (72 - 9)/3717 = 51.21$ kN, $F_4 = 0.9 \cdot 373.03 \cdot 12 \cdot (72 - 12)/3717 = 65.03$ kN, $F_5 = 0.9 \cdot 373.03 \cdot 15 \cdot (72 - 15)/3717 = 77.23$ kN and $F_6 = 0.1 \cdot 373.03 + 0.9 \cdot 373.03 \cdot 18 \cdot (72 - 18)/3717 = 125.11$ kN. One can observe that $\sum_{i=1}^{i=6} F_i = 373.03$ kN $= V_d$.

Using the above seismic forces and $G + 0.3Q$ gravity load on beams equal to 38.28 kN/m, one can determine member forces and moments through an elastic analysis with the aid of SAP 2000 (2010). The deformation response results (lateral drifts d_i and interstorey sensitivity coefficients θ_i of Eq. (4.15) of the above designed frame are shown in Table 4.9. It is observed that according to the Model Code DBD12 (Sullivan et al. 2012), all θ_i values satisfy the requirement to be less than 0.30. Since $\theta_{max} = 0.202$, the analysis results have to be multiplied by $1/(1 - \theta) = 1/(1 - 0.202) = 1.26$ for the design of beams and columns according to EC3 (2009).

Table 4.9 Response results to the lateral forces obtained by the DDBD method (HEM300/IPE360)

Seismic load combination $G + 0.3Q + E$ with $G + 0.3Q = 38.28$ kN/m

Storey	Storey displacement d_i (m)	Design interstorey displacement $q(d_i - d_{i-1})$ (m)	Storey lateral forces F_i(kN)	Storey shear force V_i(kN)	Storey cumulative gravity load P_i(kN)	Storey height h_i (m)	Interstorey sensitivity coefficient θ
1	0.0089	0.0356	18.70	373.05	4134.24	3.0	0.132
2	0.0245	0.0624	35.77	354.35	3445.20	3.0	0.202
3	0.0404	0.0636	51.21	318.58	2756.16	3.0	0.183
4	0.0543	0.0556	65.03	267.37	2067.12	3.0	0.143
5	0.0651	0.0432	77.23	202.34	1378.08	3.0	0.098
6	0.0723	0.0288	125.11	125.11	689.04	3.0	0.053

The corresponding section and stability checks of EC3 (2009) for beams are presented in the following having in mind that the out-of-plane buckling length is assumed to be 6/4 = 1.5 m for each beam due to the presence of secondary beams. For the critical beam (it is the same as in Fig. 3.14) one has that $M_{Ed} = 1.26 \cdot 265.87 = 335$ kNm, $N_{Ed} = 0$ and $V_{Ed} = 1.26 \cdot 165.16 = 208.10$ kN. Section checks are performed on the basis of Eqs. (3.26)–(3.28) and one has $M_{Ed}/M_{pl.\ Rd} = 335/361.75 = 0.926 < 1$, $V_{Ed}/V_{pl.\ Rd} = 208.10/719.57 = 0.29 < 0.5$ (interaction of M-V and M-N does not take place, hence the plastic moment capacity of the beam remains $M_{pl.\ Rd} = 361.75$ kNm). Regarding beam stability checks, only the one corresponding to lateral torsional buckling is needed and for curve b and $a_{LT} = 0.34$, one calculates $\bar{\lambda}_{LT} = 0.388$, $\Phi_{LT} = 0.607$, $\chi_{LT} = 0.931$ and finally $M_{b.\ Rd} = 336.66$ kNm > 335 kNm. Thus, the IPE360 beam marginally satisfies the stability check. However, it is recalled that due to the presence of the composite slab, $M_{b.\ Rd}$ would be higher than the one calculated above.

For the bottom end of the critical column (it is the same as in Fig. 3.14), one has $M_{Ed} = 1.26 \cdot 358.17 = 451.29$ kNm, $N_{Ed} = 1.26 \cdot 1388.05 = 1748.94$ kN and $V_{Ed} = 1.26 \cdot 132.42 = 166.85$ kN. Only interaction of M-N takes place leading to a plastic moment capacity of $M_{N.\ pl.\ Rd} = 1402.52$ kNm and, thus, one has $M_{Ed}/M_{N.\ pl.\ Rd} = 451.29/1402.52 = 0.322 < 1$, $V_{Ed}/V_{pl.\ Rd} = 166.85/1853.86 = 0.09 < 0.5$ and $N_{Ed}/N_{pl.\ Rd} = 1748.94/10756.50 = 0.163 < 1$. Column stability checks for flexural buckling are critical for minor axis and considering curve c and $a = 0.49$, one calculates $\bar{\lambda}_z = 0.981$, $\Phi_z = 1.173$, $\chi_z = 0.551$ and $N_{b.\ z.\ Rd} = 5925.86$ kN > 1748.94 kN. Regarding lateral-torsional buckling and considering curve a and $a = 0.21$, one has $\bar{\lambda}_{LT} = 0.424$, $\Phi_{LT} = 0.614$, $\chi_{LT} = 0.946$ and $M_{b.\ Rd} = 1369.93$ kNm > 451.29 kNm. Finally the member interaction equations of EC3 (2009) are checked and using $C_{my} = 0.56$, $C_{mz} = 1$, $C_{mLT} = 0.56$, $k_{yy} = 0.566$, $k_{yz} = 0.607$, $k_{zy} = 0.943$ and $k_{zz} = 1.012$ the demand to capacity ratio for the column is found to 0.605. Thus, the HEM300 column satisfies all checks.

From Table 4.9 one reads the maximum design inter-storey displacement to be 0.0636 m. Hence from Eq. (3.25) and assuming $v = 0.5$, one has $0.0636 \cdot 0.5 = 0.0318 > 0.01 \cdot 3 = 0.03$. Thus, the damage limitation criterion of EC8 (2004) is not satisfied under the assumption of the limit value 0.01 (case of non-structural components being fixed to the structure and not interfering with its deformation).

The steel MRF is then subjected to non-linear time-history (NLTH) analysis with the aid of SAP 2000 (2010) employing 10 seismic motions that are compatible to the design spectrum mentioned above. Mean values for the yield base shear (corresponding to the first plastic hinge) and inter-storey drift ratio (*IDR*) were found to be 727.14 kN and 2.63%, respectively. This *IDR* value is larger than the 2.5% value initially targeted and certainly from the non-conservative side. One particular thing to note is that in 8 out of the 10 NLTH analyses, the maximum value of *IDR* surpassed the target 2.5% value, reaching 2.92%. On the other hand, the plastic hinge distribution followed the expected from design pattern without formation of a soft storey mechanism.

In conclusion, the frame with HEM340 sections for columns and IPE400 sections for beams is the one which satisfies all checks, the same with that coming from the EC8 (2004) design. A final observation is that the DDBD method, at least on the basis of the present example, even though is theoretically supposed to automatically satisfy displacement requirements, it appears to be not so successful about that in applications, most probably because of inadequate proposed expressions for the yield displacement, the equivalent viscous damping and the damping reduction factor.

4.5.2 *Seismic Design of Steel Building with CBFs*

This example has been taken from Sahoo and Prakash (2019) and deals with the seismic design of a 3-storey steel building structure with four identical CBFs of the chevron type for resisting seismic loads in each of the two horizontal directions, as shown in Fig. 4.6. Figure 4.6a shows the plan view of the building, while Fig. 4.6b its elevation with the results of the two designs considered here: the original force-based design by Sabelli (2000) using the NEHRP provisions (FEMA 273 1997) and the present DDBD.

The building is in a site class D (FEMA 273 1997) and is designed for both the design basis earthquake (DBE) and the maximum considered earthquake (MCE) with target interstorey drift ratios (IDR_T) at the first storey equal to 1.5% and 3.0%, respectively. The seismic forces are assumed to be undertaken by the perimeter CBFs. All the other frames are designed for gravity load only. Steel yield stress f_y in braces and columns/beams is assumed to be 317 MPa and 345 MPa, respectively, with corresponding yield strain ε_y equal to 0.158% and 0.172%. The seismic action for the DBE is based on design spectral accelerations $S_{DS} = 1.395$ g and $S_{D1} = 0.77$ g for short-period and one-second, respectively (ASCE 7-10 2010). Effective seismic weights are 9390, 9390 and 10159kN for first to third floors for all four CBFs.

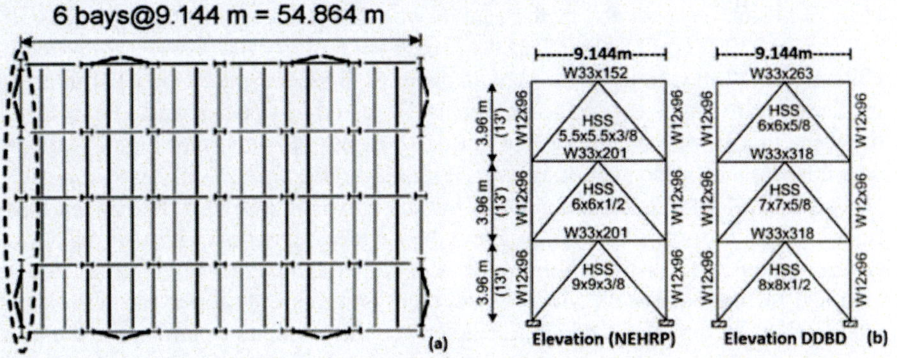

Fig. 4.6 Geometry in plan view (**a**) and elevation (**b**) of the 3-storey CBF building (after Prakash 2014, reprinted with permission from IITDCE)

Table 4.10 Determination of $u_{y,\,i}$ and u_i

i	h_i(m)	$u_{sy,\,i}$(m)	$u_{ry,\,i}$(m)	$u_{y,\,i}$(m)	δ_i	u_i (m)
3	3.96	0.0127	0.0044	0.0171	1.00	0.1783
2	3.96	0.0127	0.0044	0.0171	0.67	0.1189
1	3.96	0.0127	0.0044	0.0171	0.33	0.0594

Design base shear for the NEHRP design is 3355kN for the CBFs (Bayat et al. 2010).

The present design starts with the determination of the storey displacements from the equations (Priestley et al. 2007)

$$u_i = \delta_i(u_1/\delta_1) \tag{4.32}$$

$$\begin{aligned} \delta_i &= (h_i/H_n)\ for\ n \leq 4 \\ \delta_i &= 4/3(h_i/H_n)(1 - h_i/4H_n)\ for\ n > 4 \end{aligned} \tag{4.33}$$

where $u_1 = h_1 IDR_{T1}$ with $IDR_{T1} = 1.5\%$.

The storey yield displacement $u_{y,\,i}$ is the sum of the sway mechanism displacement u_{sy} and the rigid-body rotational displacement u_{ry} and reads (Wijesundara and Rajeev 2012)

$$u_{y,i} = u_{sy,i} + u_{ry,i} = \left(\frac{2\varepsilon_y}{sin\ 2a_{br}}\right)h_i + \left(\beta\varepsilon_{yc}h_i\right)\ tan\ a_{br} \tag{4.34}$$

where h_i is the interstorey height at level i, a_{br} is the angle between the brace and the horizontal direction, β is the ratio of the design axial force N_E to the yielding force N_R of the column and $\varepsilon_y = 0.158\%$ and $\varepsilon_{yc} = 0.172\%$ are the yield strains of the brace and column material, respectively. A value of $\beta = 0.75$ was assumed and used here. Table 4.10 shows the computations pertaining to u_i and $u_{y,\,i}$.

Using Eqs. (4.5),(4.6) and (4.10), one is able to determine the design displacement $u_d = 0.140$ m, the effective mass $m_e = 2{,}537{,}924.74$ kg and the effective height $H_e = 9.344$ m, respectively, of the equivalent SDOF system with the aid of masses $m_1 = m_2 = 9390 \cdot 10^3/9.81 = 957{,}186$ kg and $m_3 = 10{,}159 \cdot 10^3/9.81 = 1{,}035{,}576$ kg. Utilizing Eq. (4.6) with $u_{y,\,i}$ instead of u_i, the yield displacement $u_y = 0.0171$ m of the whole structure can be determined. Of course, u_y could be also determined directly by using Eq. (4.18), which is equivalent to Eqs. (4.6) and (4.34). Use of Eq. (4.8) finally determines the displacement ductility $\mu = u_d/u_y = 8.19$.

The equivalent viscous damping ξ_{eq} of the equivalent SDOF is determined from Eq. (4.17) in terms of the ductility μ and the slenderness ratio $\bar{\lambda}$ given by Eq. (4.1). Determination of $\bar{\lambda}$ for the whole frame is done iteratively. A value of $\bar{\lambda}$ is initially assumed for the equivalent SDOF system and braces are designed for the seismic lateral forces. A new value of $\bar{\lambda}$ is calculated and the calculations are repeated until a value of $\bar{\lambda}$ for the structure is found to be almost equal to that of the SDOF equivalent system. The design displacement spectrum for 5% damping is then constructed by using the information that $S_a(1) = a_g \cdot 1.15 \cdot 2.5 \cdot (0.6/1) = 0.77$ g for soil C of EC8

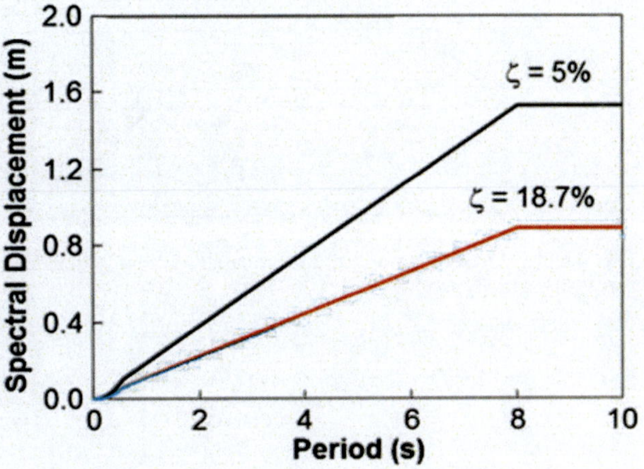

Fig. 4.7 Design displacement spectrum

(2004) that corresponds to site class D of NEHRP (FEMA 273 1997) and the relations $S_d(8) = S_a(8) \cdot (8/2\pi)^2$ and $S_a(8) = a_g \cdot 1.15 \cdot 2.5 \cdot (0.6/8)$. Thus, one finds $a_g = 0.446$ g $= 4.38$ m/s^2 and $S_d(8) = 1.53$ m, as shown in Fig. 4.7.

Use of Eqs. (4.1) and (4.17) yields $\xi_{eq} = 18.7\%$ for which $u_{D,\ \xi}$ can be found with the aid of the damping reduction factor η. Employment of Eq. (4.28) provides $\eta = [0.07/(0.02 + 0.187)]^{0.5} = 0.581$ and thus $u_{D,\ \xi} = 1.53 \cdot 0.581 = 0.889$ m. With $u_d = 0.140$ m one can determine from Eq. (4.12) the effective period $T_e = (u_d/u_{D,\ \xi})T_D = (0.140/0.889)8.0 = 1.26$ s.

The effective stiffness K_e of the SDOF system is then computed from Eq. (4.13) as $K_e = 4\pi^2 m_e / T_e^2 = 4\pi^2 2{,}537{,}925/1.26^2 = 63{,}046$ kN/m and finally the base shear force from Eq. (4.14) as $V_d = K_e u_d = 63{,}046 \cdot 0.140 = 8826.44$ kN for all four CBFs of the building. The P-Δ term in Eq. (4.14) has been omitted because of the presence of braces and the low height of the frame. The above design base shear is distributed along the height of the frame to lateral seismic forces F_i in accordance with the equation (ASCE 7-10 2010)

$$F_i = w_i h_i^k V_d / \sum\nolimits_{j=1}^{3} w_j h_j^k \tag{4.35}$$

where $w_i = m_i g$ and factor k varies linearly from 1 to 2 for corresponding values of structural period between 0.5 to 2.5 s. The structural period for CBFs is approximately given (ASCE 7-10 2010) by $T = 0.0488 H_n^{0.75}$. Thus for $H_n = 3.96 \cdot 3 = 11.88$ m, one has $T = 0.31$ s and hence $k = 1$. Finally, one obtains for one CBF, $V_d = 8826.44/4 = 2207$ kN, which after distribution along the height results in $F_1 = 263$ kN, $F_2 = 675$ kN and $F_3 = 1271$ kN.

The design proceeds with the dimensioning of the members of the frame. The design of braces at every storey is controlled by their maximum compressive forces due to seismic and gravity forces. On the assumption that braces share equally the storey shear, the axial compressive force in each brace C_i of storey i is given by

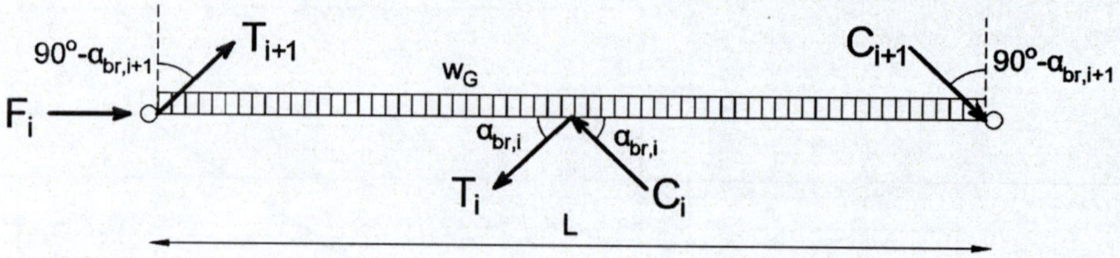

Fig. 4.8 Free body diagram of a beam at storey i under tensile T and compressive C brace axial forces and gravity load w_G

$$C_i = \frac{V_i}{2\cos a_{br}} + \frac{1}{\sin a_{br}}\left(\frac{5w_G L}{16}\right) \tag{4.36}$$

where V_i is the shear force at storey i, w_G is the uniform gravity load and L is the length of the beam. The brace size of the brace is determined iteratively so as to have its slenderness ratio $\bar{\lambda}$ to be the same with the assumed value for the SDOF system. The effective length factor for the out-of-plane buckling of braces is assumed to be $k = 0.85$. The satisfaction of no brace fracture and delayed brace fracture for the DBE and MCE cases, respectively, is secured by the imposed IDR_T and the width-to-thickness ratio limits (Goel and Chao 2008).

Design of beams and columns is accomplished in accordance with capacity design rules. Beams are designed on the basis of the ultimate demand from braces in tension and compression (pre- and post-buckling cases). Figure 4.8 depicts the free body diagram of a beam of the frame, pinned at both ends, under the action of the brace axial forces in tension T and compression C as well as the uniformly distributed gravity loading w_G.

In Fig. 4.8, $T_i = R_y f_y A_{g,\ i}$ is the brace strength in tension and $C_i = 1.14 P_{cr} A_{g,\ i}$ or $0.3 P_{cr} A_{g,\ i}$ are the brace strengths in buckling (pre-buckling or post-buckling, respectively), where $R_y = 1.1$ is the steel material overstrength factor, f_y is the yield strength, P_{cr} is the critical buckling load and A_g is the gross cross-section of the brace members. Thus, on the basis of Fig. 4.8, one can determine the design bending moment $M_{b,\ i}$ and the axial force $C_{b,\ i}$ at the storey i as

$$M_{b,i} = (T_i - C_i)\sin a_{br,i+1}\frac{L}{4} + \frac{w_{G,i}L^2}{8} \tag{4.37}$$

$$C_{b,i} = T_{i+1}\cos a_{br,i+1} + 0.5F_i \tag{4.38}$$

for the design of the beams of the CBF. Columns in the frame are designed for axial compressive forces coming from gravity loads G, cumulative vertical components of compressive brace capacity and half the net vertical resultant of the brace loads acting at mid-span of the beams. Thus, the column design axial compressive force at level i has the form

$$C_{col,i} = C_{col,G,i} + \sum_{j=i+1}^{n} C_j \sin a_{br,j} + \sum_{j=i}^{n} (T_j - C_j) \sin a_{br,j} \quad (4.39)$$

Figure 4.6b shows the final member sections for the present DDBD method and the NEHRP method. For more details on the design of CBF members one can consult the book of Goel and Chao (2008) and the report of Bayat et al. (2010) as well as the example on CBFs of Chap. 7.

The seismic performance of the two designs of Fig. 4.6b is assessed with the aid of NLTH analyses conducted by the computer program OpenSees (McKenna et al. 2007). Inelastic material behavior and large deformations are considered in the simulation. The frames are subjected to 40 earthquakes LA1 to LA40 (earthquakes in greater Los Angeles area) with the first 20 of them spectrum compatible with the DBE and the rest 20 of them spectrum compatible with the MCE with their spectra associated with 5% damping. In the nonlinear analyses, Rayleigh damping of the tangent stiffness type was used with its coefficients computed on the basis of first and third natural frequencies and 3% damping. Newmark's integration algorithm was used for the solution of the nonlinear equations in the time domain. As it was mentioned at the beginning of this example, IDR_T values of 1.5% and 2.0% for the DBE and MCE levels were assumed. Corresponding residual IDR_T values were 0.5% and 2.0% (Sahoo and Prakash 2019).

Figure 4.9 presents maximum response results along the height of the two CBFs designed by NEHRP and DDBD methods for the DBE hazard level. Response results include peak storey displacements, peak *IDR*s, peak storey shears and peak residual *IDR*s (*RIDR*s) as obtained from the mean and mean + SD of the LA01-LA20 motions associated with the DBE hazard level. One can observe that the mean values of storey displacements are lower than the assumed design displacements profiles and hence conservative. However, their mean + SD values seem to be close to the design displacement profiles. It is also observed that both *IDR*s and *RIDR*s do not exceed their limit values with the mean + SD values to be closer to those limits. Peak storey shears appear to be much higher than the design base shear of 8826.44/ 2 = 4413 kN for the two CBFs. In general, displacements are lower in the DDBD design than those in the NEHRP and storey shears are higher in the DDBD design than those in the NEHRP design (4413 kN versus 3353 kN for the two CBFs), as expected. This has as a result the DDBD design to be heavier than the NEHRP one (Fig. 4.6b). However, if one looks at Figs. 4.10 and 4.11 will realize that the NEHRP design has serious problems of damage and undesirable mechanisms of collapse, while the present DDBD design is free of those problems.

The damage index in Fig. 4.10 refers to braces and is calculated using Miner's rule in damage accumulation under low-cycle fatigue during the response simulation. It is observed that for all 20 seismic motions brace damage is lower for the DDBD case than for the NEHRP case. Furthermore, it is also observed that for LA01, LAO2 and LA20 motions, the damage index for the NEHRP frame becomes equal to one implying brace failure. In conclusion, the design according to the

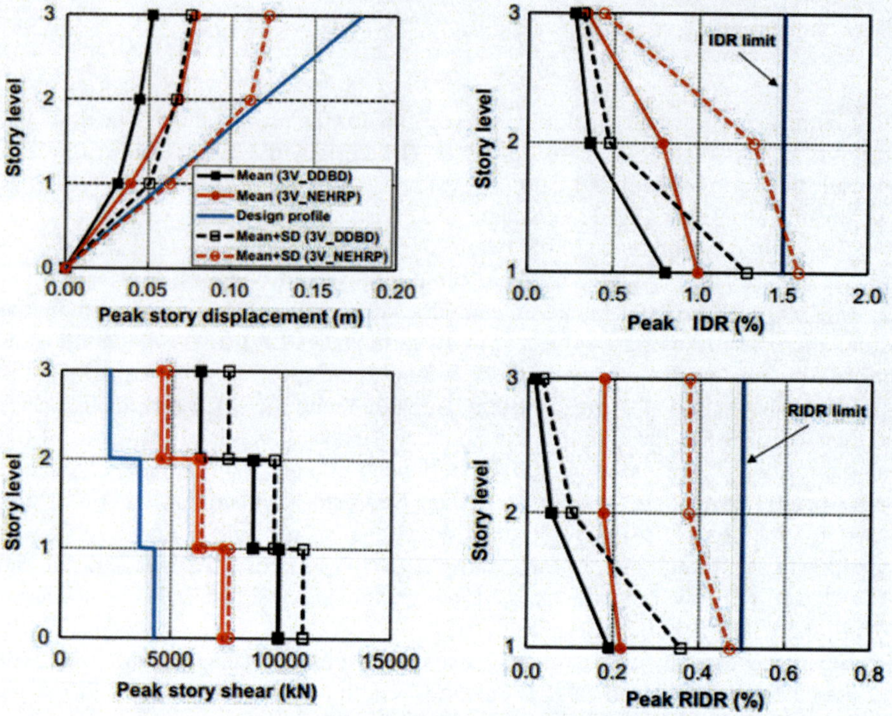

Fig. 4.9 Maximum response results for the NEHRP and DDBD designed CBFs along their height for the DBE hazard level (after Prakash 2014, reprinted with permission from IITDCE)

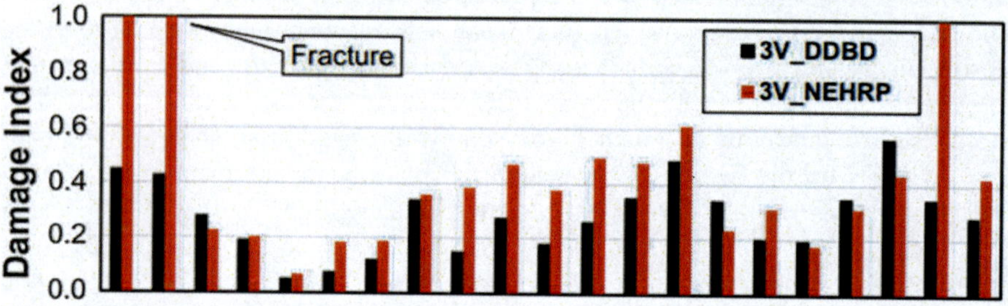

Fig. 4.10 Brace damage index for the two designed CBFs for 20 seismic motions spectrum compatible with the DBE (after Prakash 2014, reprinted with permission from IITDCE)

DDBD method results in a heavier but safer frame structure. Analogous results to be found in Sahoo and Prakash (2019) have been obtained for the MCE hazard level.

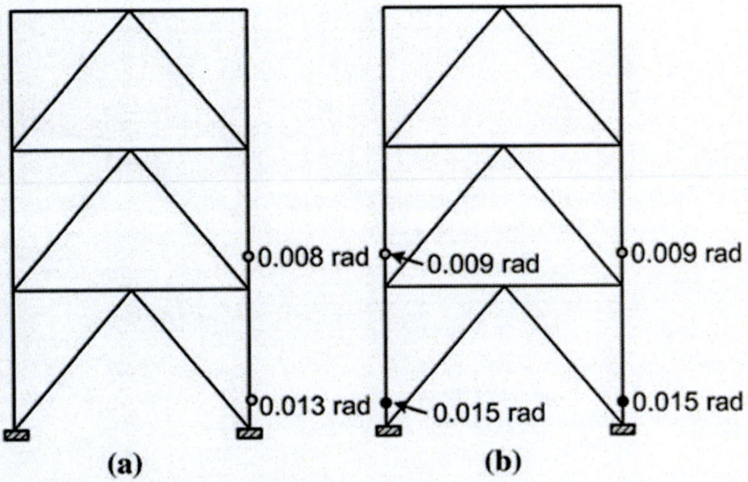

Fig. 4.11 Plastic hinge formation for the two designed (**a**: DDBD, **b**: NEHRP) CBFs for the DBE hazard level (black and white circles correspond to plastic hinges and yielding hinges, respectively) (after Prakash 2014, reprinted with permission from IITDCE)

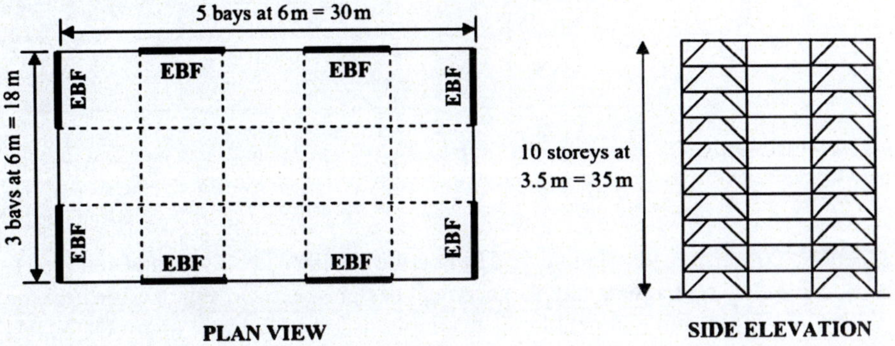

Fig. 4.12 Geometry in plan view and elevation of the 10-storey EBF building (after Sullivan 2012, reprinted with permission from PAEE)

4.5.3 *Seismic Design of Steel Building with EBFs*

This example has been taken from Sullivan (2012) and deals with the seismic design of a 10-storey steel building structure with four identical EBFs for resisting seismic loads in each of the two horizontal directions, as shown in Fig. 4.12. The bay lengths and storey heights are everywhere the same and equal to 6.0 m and 3.5 m, respectively. The steel grade and the modulus of elasticity are assumed to be S450 and $E = 205{,}000$ N/mm^2, respectively. The yield strength is then 440 N/mm$^2 \cdot 1.2 =$ 528 N/mm^2, where 1.2 is an overstrength factor. The storey weight for the building is 4600 kN implying that for any of the four EBFs the storey weight is $4600/4 =$

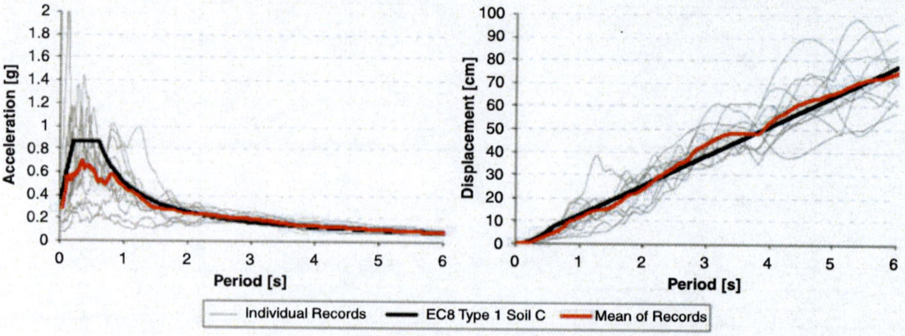

Fig. 4.13 Design acceleration and displacement spectra for the building of Fig. 4.12 including the average and the individual spectra of 10 spectrum-compatible seismic motions used for NLTH analyses (after Sullivan 2012, reprinted with permission from PAEE)

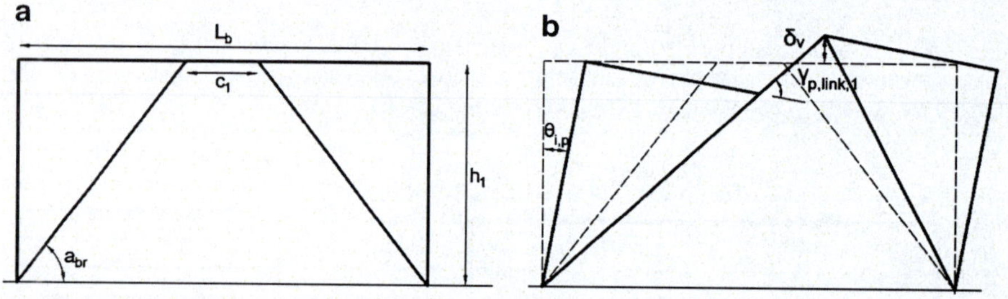

Fig. 4.14 Geometry of a one-storey EBF with central link (**a**) before deformation and (**b**) after plastic deformation

1150 kN. Concerning the seismic load, it is assumed that $a_g = 0.30$ g and soil type C resulting in the acceleration and displacement design spectra of Fig. 4.13, where the spectra of 10 spectrum-compatible seismic motions to be used later on for the NLTH analyses are also shown. The EBFs are assumed to have central links (Fig. 4.12b) of variable length along the height for design optimization and thus the design plastic rotation of the links is equal to 0.08 rad. A non-structural interstorey drift ratio limit (IDR_T) of 2.5% for the ultimate limit state is also assumed.

The design starts with an initially assumed section size and length of the links in order to determine IDR_c from Eq. (4.23) to be used in Eq. (4.22) for the determination of the displacement u_i at every level i. Thus, one determines for every storey i

$$IDR_{p,i} = e_i \gamma_{p,i} / L_b \tag{4.40}$$

$$IDR_{y,i} = \frac{2\delta_{v.i}}{L_b - e_i} + \frac{2k_{br,i}\varepsilon_y}{\sin 2a_{br,i}} + \frac{2k_{cols,i}\varepsilon_y(h_i - h_s)}{L_b} \tag{4.41}$$

and then $IDR_{c,\,i} = IDR_{p,\,i} + IDR_{y,\,i}$ (Eq. (4.23)). In the above, h_s is the storey height, L_b is the bay length, e is the link length, γ_p is the link plastic rotation (Fig. 4.14), k_{br} is the ratio of the design stress to the yield stress of the brace section, k_{cols} is the average ratio of the design stress to the yield stress in the column section and δ_v is the elastic vertical displacement of the end of the link. Explicit expressions for k_{br}, k_{cols} and δ_v at storey level i read as

$$k_{br,i} = \frac{N_{E,br,i}}{N_{Rs,br,i}} \quad , \quad k_{cols,i-1} = \frac{1}{i-1}\sum_{j=1}^{i-1}\frac{N_{E,col,j}}{N_{Rs,col,j}} \tag{4.42}$$

$$\delta^s_{v,i} = 0.577 f_y A_{v,i}\left[\frac{e_i^2(L_b - e_i)}{24EI_i} + \frac{e_i}{2GA_{v,i}}\right] \tag{4.43}$$

$$\delta^l_{v,i} = M_{p,i}\left[\frac{e_i(L_b - e_i)}{12EI_i} + \frac{1}{GA_{v,i}}\right] \tag{4.44}$$

where $N_{E,\ br}$ is the design axial force in the brace, $N_{Rs,\ br}$ is the brace section resistance, $N_{E,\ col}$ is the design axial force in the column, $N_{Rs,\ col}$ is the column section resistance, I and $A_v = \left(1/\sqrt{3}\right)t_w\left(d - t_f\right)$ are the moment of inertia and shear area of the link, respectively, E and G are the elastic and shear modulus of steel, respectively and M_p is the plastic moment of resistance of the link equal to $M_p = f_y b t_f (d - t_f)$ with b being the flange width, d the section depth and t_f and t_w, the flange and web thickness, respectively of the link cross section.

Sullivan et al. (2012) suggests initial estimates of $k_{br,\ i}$ and $k_{cols,\ i-1}$ equal to 0.25. In conclusion, use is made of Eqs. (4.40) and (4.41) to compute IDR_c from Eq. (4.23) and then Eq. (4.22) is employed with $\omega_\theta = 0.84$ for the present case of 10 stories and $IDR_T = 2.5\%$ to determine the displacements u_i. Once u_i are known, one can easily determine IDR_i and with the aid of Eq. (4.41) compute the storey ductilities μ_i as

$$\mu_i = IDR_i / IDR_{y,i} \tag{4.45}$$

For determining the ductility of the whole structure μ, Sullivan et al. (2012) suggests the weighted average type of expression

$$\mu = \sum_{i=1}^{10} \mu_i V_i IDR_i / \sum_{i=1}^{10} V_i IDR_i \tag{4.46}$$

Design shear values V_i are not known at this stage. However, only relative shear proportions are really needed in Eq. (4.46). Thus, by assuming a design base shear $V_b = 1$, one can obtain the lateral forces F_i from Eq. (4.16) with $k = 1$ and from there the V_i needed in Eq. (4.46).

With known μ, one can easily compute the equivalent damping ξ_{eq} and the damping reduction factor η. In this example, Sullivan et al. (2012) uses for ξ_{eq} the expression

$$\xi_{eq} = 0.07\left[1 + \frac{1.17(\mu - 1)}{1 + e^{\left(\sqrt{\mu} - 1\right)}}\right]^2 \tag{4.47}$$

instead of the one by (4.21) and for η the expression (4.27) instead of the one by Eq. (4.26). Use of Eqs. (4.12) and (4.13) finally leads to the value of the design base shear V_b, which is distributed to the storey levels of the frame with the aid of Eq. (4.16) providing the lateral forces F_i. With known forces F_i, the design storey shears V_i and hence the design shear forces in the links $V_{link,\ i}$ can be easily determined as

$$V_{link,i} = V_i h_1 / L_b \tag{4.48}$$

Thus, the initially selected link sizes are checked by comparing $V_{link,\ i}$ against the plastic resistance

$$V_{p,link} = \left(f_y/\sqrt{3}\right) t_w (d - t_f) \tag{4.49}$$

for the case of short links. Column and brace sizes are then selected using amplified design forces in conformity with capacity design and EC3 (2009) provisions. With selected column and brace sections and the new link sizes, the initially assumed values of $k_{br,\ i}$, $k_{cols,\ i-1}$ and $\delta_{v,\ i}$ are updated by using Eqs. (4.42)–(4.44) and iterations continue till convergence. Tables 4.11 and 4.12 present intermediate and final, respectively, sections for links, braces and columns. The final design is associated with a ductility $\mu = 3.40$, $\xi_{eq} = 21.6\%$ and a base shear force of $V_b = 964$ kN for the EBF. Furthermore, the ratios of the storey shear resistance over the storey shear demand are not different by more than a factor of 1.25 overall or by more than a factor of 1.15 in successive stories.

The designed here EBF was subjected to the 10 EC8 (2004) spectrum compatible seismic motions associated with the spectra of Fig. 4.10 and its dynamic response to those motions was recorded through NLTH analyses conducted with the aid of the computer program Ruaumoko (Carr 2009). Columns, braces and beam parts outside the links are considered to be elastic, while links are considered to be inelastic. Floors are assumed to act as rigid diaphragms but links are assumed to be unconstrained. Large displacement analyses including P-Δ effects with Rayleigh damping of the tangent-stiffness type are conducted (taking 3% damping for the second mode and 1.37% damping for the first mode). Figure 4.15 depicts maximum storey displacements and *IDR* along the height of the EBF as obtained from the NLTH analyses of the designed frame. It is observed that the average design *IDR* profile is close to the design *IDR* capacity and both less than the $IDR_T = 2.5\%$. It is also observed that the first mode displacement and *IDR* profiles are much lower than the average displacement and *IDR* profiles, indicating significant higher mode effects.

Table 4.11 Intermediate design values for the 10-storey EBF structure (after Sullivan 2012, reprinted with permission from PAEE)

Level	Height, h_i (m)	Mass, m_i	Yield Drift	Drift capacity	Δ_i	θ_i	$m_i.\Delta_i$	$m_i.\Delta_i^2$	$m_i.\Delta_i.h_i$	Δ_d (m)	m_e (T)	H_e (m)
10	35.0	117.2	0.87%	1.54%	0.306	0.46%	35.9	11.0	1256.7	0.232	989	23.5
9	31.5	117.2	0.86%	1.66%	0.290	0.55%	34.0	9.9	1072.1			
8	28.0	117.2	0.79%	1.59%	0.271	0.64%	31.8	8.6	890.0			
7	24.5	117.2	0.77%	1.70%	0.249	0.74%	29.2	7.2	714.2			
6	21.0	117.2	0.78%	1.71%	0.223	0.83%	26.1	5.8	548.8			
5	17.5	117.2	0.67%	1.60%	0.194	0.92%	22.7	4.4	397.9			
4	14.0	117.2	0.65%	1.72%	0.162	1.02%	19.0	3.1	265.4			
3	10.5	117.2	0.58%	1.64%	0.126	1.11%	14.8	1.9	155.3			
2	7.0	117.2	0.57%	1.77%	0.087	1.20%	10.2	0.9	71.7			
1	3.5	117.2	0.49%	1.69%	0.045	1.29%	5.3	0.2	18.6			
						Total:	229.0	53.0	5390.7			

Table 4.12 Final section sizes, ductility demands, shear demands and shear ratios for the 10-storey EBF structure (after Sullivan 2012, reprinted with permission from PAEE)

Level	Brace section	Column section	Link section	Link length (m)	μ_i	$V_{d,i}$(kN)	$V_{R,i}/V_{d,i}$
10	HE 180 A	HE 180 A	HE 120 A	0.50	1.49	222	1.36
9	HE 180 A	HE 180 A	HE 160 A	0.60	2.03	346	1.48
8	HE 180 A	HE 240 A	HE 180 A	0.60	3.00	461	1.33
7	HE 200 A	HE 240 A	HE 200 A	0.70	3.04	567	1.31
6	HE 200 A	HE 300 B	HE 180 B	0.70	3.03	661	1.38
5	HE 200 B	HE 300 B	HE 180 B	0.70	3.37	744	1.27
4	HE 200 B	HE 450 B	HE 200 B	0.80	3.51	812	1.36
3	HE 200 B	HE 450 B	HE 200 B	0.80	3.83	866	1.31
2	HE 200 B	HE 600 B	HE 200 B	0.90	3.34	903	1.23
1	HE 200 B	HE 600 B	HE 200 B	0.90	3.60	922	1.24

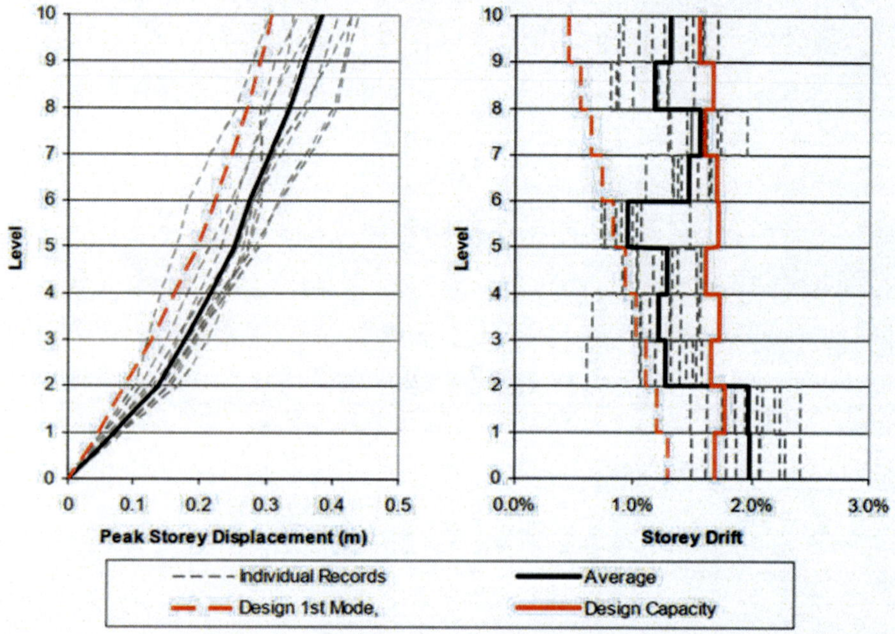

Fig. 4.15 Maximum storey displacements and interstorey drift ratios from DDBD methods and NLTH analyses of the building of Fig. 4.12 (after Sullivan 2012, reprinted with permission from PAEE)

Closing the discussion on this EBF example, one should mention that more details on the theory and especially more details on examples can be found in Sullivan (2013) where three building examples dealing with 5, 10 and 15 storey EBFs are presented.

4.6 Conclusions

Based on the material presented in the preceding sections of this chapter, the following conclusions can be stated:

1. The direct displacement-based design (DDBD) method for the seismic design of steel building frames has been presented. The method employs displacements as the main design parameters for better damage control. Use is made of the equivalent linear SDOF system to the original nonlinear MDOF structure and the displacement design spectrum with high amounts of damping. The SDOF is characterized by its equivalent mass and equivalent damping for taking care of the inelastic energy of dissipation. Thus, one determines the equivalent period of this SDOF system with the aid of the displacement design spectrum and from there the effective stiffness and design base shear force required for resisting the assumed design displacement. Using this base shear the structure can be dimensioned for strength.
2. The method has been successfully applied to a variety of steel building structures involving MRFs, CBFs, EBFs, BRBFs, MRFs with setbacks or flexible joints and MRFs in combination with RC walls. In this chapter, numerical examples dealing with MRFs, CBFs and EBFs have been presented in detail in order to illustrate the method and demonstrate its advantages.

   The DDBD method is a one-step design method involving only a strength checking and not a two-steps design method involving a strength and a deformation checking as it is the case with the force-based design (FBD) method. In the DDBD method no deformation checking is needed because this is automatically satisfied from the start when the designer decides about the design displacement of the structure.
3. The two main problems with DDBD method are (i) the use of the equivalent SDOF system, which results in a loss of modeling accuracy reflecting in its inability to take into account higher order mode and P-Δ effects as well as local failures and (ii) the employment of a displacement design spectrum, which is not so familiar to engineers as it is the acceleration design spectrum associated with the FBD method. However, the first problem has been practically almost solved by adding corrective terms in the expressions for the design displacements and the design base shear, while the second one will be solved over time as engineers will become accustomed with the displacement design spectrum.
4. The DDBD method, as it is currently used for the seismic design of steel structures, has not reached as yet a level of maturity as it is the case with the FBD method used in current codes. The main problem is that no definitive and reliable expressions for design displacements, yield displacements, equivalent damping, damping reduction factors, spectral period T_D and distribution of lateral forces over the frame height have been accepted as yet. Much more research work is required in order to (i) to cover all kinds of steel frames including three-dimensional ones and (ii) to improve and unify the aforementioned design

expressions and include appropriately the findings in a future edition of the Model Code DBD12 (Sullivan et al. 2012).

References

ASCE/SEI 41-17 (2017) Seismic evaluation and retrofit of existing buildings. American Society of Civil Engineers, Reston, VI

ASCE/SEI 7-10 (2010) Minimum design loads and associated criteria for buildings and other structures. American Society of Civil Engineers, Reston, VI

Bayat MR, Goel SC, Chao SH (2010) Performance-based plastic design of earthquake resistant concentrically braced steel frames, Research Report UMCEE 10-02. Department of Civil and Environmental Engineering, The University of Michigan, Ann Arbor, MI

Boore DM, Bommer JJ (2005) Processing of strong-motion accelerograms: needs, options and consequences. Soil Dyn Earthq Eng 25:93–115

Calvi GM, Sullivan T, Roldan R, O'Reilly G, da Silva LS, Rebelo C, Castro M, Agusto H, Landolfo R, Della Corte G, Terraciano G, Salvatore W, Morelli F (2015) Displacement based seismic design of steel moment resisting frame structures (DiSTEEL). Directorate-General for Research and Innovation, EUR 27157 EN, Brussels

Carr AJ (2005) Ruaumoko 2D and 3D: programs for inelastic dynamic analysis. Theory and user guide to associated programs. Department of Civil Engineering, University of Canterbury, Christchurch

Carr AJ (2009) Ruaumoko 2D and 3D: programs for inelastic dynamic analysis. Theory and user guide to associated programs. Department of Civil Engineering, University of Canterbury, Christchurch

Chopra AK, Goel RK (2001) Direct displacement-based design: use of inelastic vs. elastic design spectra. Earthquake Spectra 17:47–64

Della Corte G, Mazzolani FM (2008) Theoretical developments and numerical verification of a displacement-based design procedure for steel braced structures. In: Proceedings of 14th World Conference on Earthquake Engineering, Beijing, China

Della Corte G, Landolfo R, Mazzolani FM (2010) Displacement-based seismic design of braced steel structures. Steel Construct 3:134–139

Dimopoulos AI, Bazeos N, Beskos DE (2012) Seismic yield displacements of plane moment resisting and X-braced steel frames. Soil Dyn Earthq Eng 41:128–140

Dwairi HM, Kowalsky MJ, Nau JM (2007) Equivalent damping in support of direct displacement-based design. J Earthq Eng 11:512–530

EC3 (2009a) Eurocode 3, Design of steel structures – Part 1–1: general rules and rules for buildings, EN 1993-1-1. European Committee for Standardization (CEN), Brussels

EC8 (1994) Eurocode 8, Design of structures for earthquake resistance, Part 1: general rules, seismic actions and rules for buildings. European Committee for Standardization (CEN), Brussels

EC8 (2004) Eurocode 8, Design of structures for earthquake resistance, Part 1: general rules, seismic actions and rules for buildings, EN 1998-1-1. European Committee for Standardization (CEN), Brussels

FEMA 273 (1997) NEHRP guidelines for the seismic rehabilitation of buildings. Federal Emergency Management Agency, Washington, DC

Garcia R, Sullivan TJ, Della Corte G (2010) Development of a displacement-based design method for steel frame-RC wall buildings. J Earthq Eng 14:252–277

Goel SC, Chao SH (2008) Performance-based plastic design: earthquake resistant steel structures. International Code Council, Washington DC

Grant DN, Blandon CA, Priestley MJN (2005) Modeling inelastic response in direct displacement-based design, Research Report ROSE-2005/03. IUSS Press, Pavia

Kalapodis NA, Muho EV, Beskos DE (2021) Seismic design of plane steel MRFs, EBFs and CBFs by improved direct displacement-based design method. Soil Dyn Earthq Eng (submitted)

Macedo L, Castro JM (2012) Direct displacement-based seismic design of steel moment frames. In: Proceedings of 15th World Conference on Earthquake Engineering, Lisbon, Portugal

Maley TJ, Sullivan TJ, Della Corte G (2010) Development of a displacement-based design method for steel dual systems with buckling-restrained braces and moment resisting frames. J Earthq Eng 14:106–140

McKenna F, Fenves GL, Jeremic B, Scott MH (2007) Open system for earthquake simulation (OpenSees). University of California, Berkeley, CA

Medhekar MS, Kennedy DJL (2000a) Displacement-based seismic design of buildings-theory. Eng Struct 22:201–209

Medhekar MS, Kennedy DJL (2000b) Displacement-based seismic design of buildings-application. Eng Struct 22:210–221

Moehle J (1992) Displacement-based design of RC structures subjected to earthquakes. Earthquake Spectra 8:403–428

Morelli F (2014) Displacement-based seismic design of steel moment resisting frames. Ph.D. Thesis, Department of Civil Engineering, University of Pisa, Pisa, Italy

Muho EV, Qian J, Beskos DE (2020) A direct displacement-based seismic design method using a MDOF equivalent system: application to R/C framed structures. Bull Earthq Eng 18:4157–4188

Nievas CI, Sullivan TJ (2015) Applicability of the direct displacement-based design method to steel moment resisting frames with setbacks. Bull Earthq Eng 13:3841–3870

Panagiotakos TB, Fardis MN (2001) A displacement-based seismic design procedure for RC buildings and comparison with EC8. Earthq Eng Struct Dyn 30:1439–1462

Pennucci D, Sullivan TJ, Calvi GM (2011a) Performance-based seismic design of tall RC wall buildings, Research Report ROSE 2011/02. IUSS Press, Pavia

Pennucci D, Sullivan TJ, Calvi GM (2011b) Displacement reduction factors for the design of medium and long period structures. J Earthq Eng 15(S1):1–29

Prakash A (2014) Performance of steel concentric braced frames designed using direct-displacement based design approach. M.S. Thesis. Department of Civil Engineering, Indian Institute of Technology Delhi, Delhi, India

Priestley MJN, Kowalsky MJ (2000) Direct displacement-based seismic design of concrete buildings. Bull N Z Soc Earthq Eng 33:421–443

Priestley MJN, Calvi GM, Kowalsky MJ (2007) Displacement-based seismic design of structures. IUSS Press, Pavia

Remennikov AM, Walpole WR (1997) Analytical prediction of seismic behavior for concentrically-braced steel systems. Earthq Eng Struct Dyn 26:859–874

Roldan R, Sullivan TJ, Della Corte G (2016) Displacement-based design of steel moment resisting frames with partially-restrained beam-to-column joints. Bull Earthq Eng 14:1017–1046

Sabelli R (2000) Research on improving the design and analysis of earthquake resistant steel braced frames. FEMA/EERI Report. Federal Emergency Management Agency, Washington DC

Sahoo DR, Prakash A (2019) Seismic behavior of concentrically braced frames designed using direct displacement-based method. Int J Steel Struct 19:96–109

Salawdeh S, Goggins J (2016) Direct displacement-based seismic design for single storey steel concentrically braced frames. Earthq Struct 10:1125–1141

SAP 2000 (2010a) Structural analysis program 2000, static and dynamic finite element analysis of structures, Version 14. Computers and Structures Inc, Berkeley, CA

Shibata A, Sozen MA (1976) Substitute structure method for seismic design in reinforced concrete. J Struct Div ASCE 102:1–18

Sullivan TJ (2012) Formulation of a direct displacement-based design procedure for steel eccentrically braced frame structures. In: Proceedings of 15th World Conference on Earthquake Engineering, Lisbon, Portugal, Paper No 2121

Sullivan TJ (2013) Direct displacement-based seismic design of steel eccentrically braced frame structures. Bull Earthq Eng 11:2197–2231

Sullivan TJ, Calvi GM, Priestley MJN, Kowalsky MJ (2003) The limitations and performances of different displacement based design methods. J Earthq Eng 7:201–241
Sullivan TJ, Priestley MJN, Calvi GM (2012) A model code for the displacement-based seismic design of structures, DBD12. IUSS Press, Pavia
Tremblay R (2002) Inelastic seismic response of steel bracing members. J Constr Steel Res 58:665–701
Wijesundara KK, Rajeev P (2012) Direct displacement-based design of steel concentric braced frame structures. Aust J Struct Eng 13:243–257
Wijesundara K, Nascimbene R, Sullivan TJ (2011) Equivalent viscous damping of steel concentrically braced frame structures. Bull Earthq Eng 9:1535–1558

Chapter 5
Hybrid Force-Displacement Design

Abstract The hybrid force/displacement (HFD) seismic design method for plane and space, regular and irregular, unbraced and braced steel building frames is presented. The method combines the advantages of the force-based design (FBD) and displacement-based design (DBD) methods. The HFD design method starts by considering both non-structural and structural target deformations in the form of maximum interstorey drift ratios (*IDR*) and member rotational ductilities μ_θ, respectively. These target deformations are transformed to a target roof displacement. Thus, the behavior (strength reduction) factor q is determined as a function of the target roof displacement ductility. After that, the HFD method proceeds as a FBD approach for strength checking using the pseudo-acceleration design spectrum analysis for seismic force determination. The necessary explicit empirical expressions needed for the determination of q are obtained by regression analysis on response databanks created for the various types of frames considered here. These databanks are obtained with the aid of extensive parametric studies involving many frames under many seismic motions analyzed by nonlinear time history analyses. The HFD method is constructed to be a performance-based seismic design method. Numerical examples are presented to illustrate the application of the method and demonstrate its advantages over the FBD and DBD seismic design methods.

Keywords Hybrid design method · Steel structures · Target deformations · Behavior factor · Design spectrum analysis · Performance-based seismic design

5.1 Introduction

Current seismic design codes for building structures, such as EC8 (2004), employ the force-based design (FBD) method, which uses seismic forces as the basic design parameters, as explained in Chap. 3. EC8 (2004) works with two limit states, the ultimate limit state (ULS) and the damage limit state (DLS). The structure is first designed for strength on the basis of the ULS associated with the design earthquake (475 years return period). Seismic design forces for member dimensioning are

G. A. Papagiannopoulos et al., *Seismic Design Methods for Steel Building Structures*, Geotechnical, Geological and Earthquake Engineering 51,
https://doi.org/10.1007/978-3-030-80687-3_5

determined by using the pseudo-acceleration elastic design spectrum with ordinates divided by the behavior (strength reduction) factor q to take care of inelasticity. Afterwards, the so-designed structure is checked for deformation on the basis of the DLS associated with the frequent earthquake (72 years return period). Satisfaction of deformation requirements is accomplished iteratively usually leading to a stiffer structure than the originally selected.

In an effort to more effectively control structural seismic damage, displacement-based design (DBD) methods for seismic design have been also developed, as explained in Chap. 4. Here, displacements rather than forces are the main design parameters. The direct displacement-based design (DDBD) method (Priestley et al. 2007) is the most widely used one among the existing DBD methods. According to this method, the multi degree of freedom (MDOF) original inelastic structure is substituted by a single degree of freedom (SDOF) equivalent linear structure with high amount of damping. Using this equivalent SDOF structure in conjunction with a displacement design spectrum with high amounts of damping, one can determine the seismic base shear force required for the structure to experience the desired deformation in the form of the target interstorey drift ratio (*IDR*). Both structural and non-structural damage are controlled by imposing limits only on *IDR*.

The hybrid force/displacement (HFD) design method for plane and space steel building structures (Tzimas et al. 2013, 2017) has recently emerged as a good alternative of both the FBD and the DDBD methods as combining the advantages of both of them in a hybrid force/displacement design scheme. This method is the subject of the present chapter. The method has been originally developed for plane unbraced and braced regular and irregular steel frames (Karavasilis et al. 2006, 2007a, b, 2008a, b, 2009; Zotos and Bazeos 2009; Stamatopoulos and Bazeos 2011). A comparison of the FBD, DDBD and HFD methods for plane steel moment resisting frames (MRF) done by Bazeos (2009) has revealed the advantages of the HFD. All the aforementioned works on the HFD are associated with far-fault ordinary ground motions. Karavasilis et al. (2010) extended the HFD method to the case of near-fault pulse-like ground motions. The HFD starts by adopting both *IDR* and member ductility target values μ_θ in order to control both non-structural and structural damage, respectively. Then, transforms them to a target roof displacement in order to obtain a behavior factor q that takes both of them into account. Seismic design forces are determined with the aid of response spectrum analysis and use of an acceleration design spectrum as in the FBD method. Therefore, the engineer works with familiar to him concepts and tools and avoids the employment of a highly damped displacement design spectrum, which is used by the DDBD. On the other hand, the employed q leads to a design satisfying the two predefined deformation limits and no deformation checking is required as it is the case with the FBD method.

In this chapter, the basic steps of the HFD method are presented as applied to regular in plan view and elevation space steel MRFs under far-fault ground motions for illustration purposes. Torsional effects are taken into account by considering accidental eccentricity. The required for the method explicit empirical expressions involving deformational transformations and the behavior factor q are obtained by regressing the results of a response databank obtained through extensive parametric studies. These studies involve nonlinear time history (NLTH) analyses of 38 space frames with three values of eccentricity (0%, 5% and 10%) subjected to 42 pairs of horizontal ground motions and four different performance levels. Explicit empirical expressions are also provided in this chapter for irregular space steel MRF obtained in a similar way. These frames are irregular in plan view and elevation (with setbacks and mass irregularity). Furthermore, explicit empirical design equations for the HFD design method as applied to plane regular steel moment resisting and concentrically braced frames as well as to plane irregular steel MRFs (with setbacks and mass irregularity) are also given in here. Detailed seismic design examples covering various frame typologies are finally presented to illustrate the HFD design method and demonstrate its merits over other methods of seismic design.

5.2 Basic Steps of the HFD Design Method

This section briefly describes the basic steps of the HFD seismic design method as applied to steel space MRFs with rectangular plan view and regular in elevation. The same steps apply to all other types of steel frames considered here provided that the design equations in steps (5) and (6) are replaced by the corresponding ones for the other types of frames presented in Sects. 5.4–5.7. Thus, the HFD seismic design method for steel space MRFs with rectangular plan view and regular in elevation (Fig. 5.1) consists of the following steps (Tzimas et al. 2017):

1. Definition of the main geometrical frame features, such as the number of stories, n_s, the number of bays, n_b, the storey heights, h, and bay widths, b.
2. Definition of the considered performance levels: IO under the FOE; LS under the DBE and CP under the MCE with seismic intensity levels provided by the corresponding elastic response spectra.
3. Definition of input limit (target) values for the performance metrics of maximum interstorey drift ratio (*IDR*) and maximum member rotational ductility μ_θ as in the DBD method. Table 5.1 provides those limits for the aforementioned three performance levels considered here (FEMA 273 1997), where local ductility values represent limits on flange and web slenderness and limits on axial force of columns, values in parentheses are in accordance with ASCE 41-13 (2014) and $\overline{\lambda}$ denotes brace slenderness ratio.
4. Estimation of the yield roof displacement $u_{r,\,y}$ and the fundamental period, T, of the frame. Initial estimates of these variables may be obtained by designing the frame only for strength requirements under the FOE by assuming elastic behavior. Good estimates for $u_{r,\,y}$ can also be obtained by the empirical expressions of Dimopoulos et al. (2012) for steel plane MRFs and CBFs of the X-braced type. A good first estimate of T can also be obtained from the empirical formula

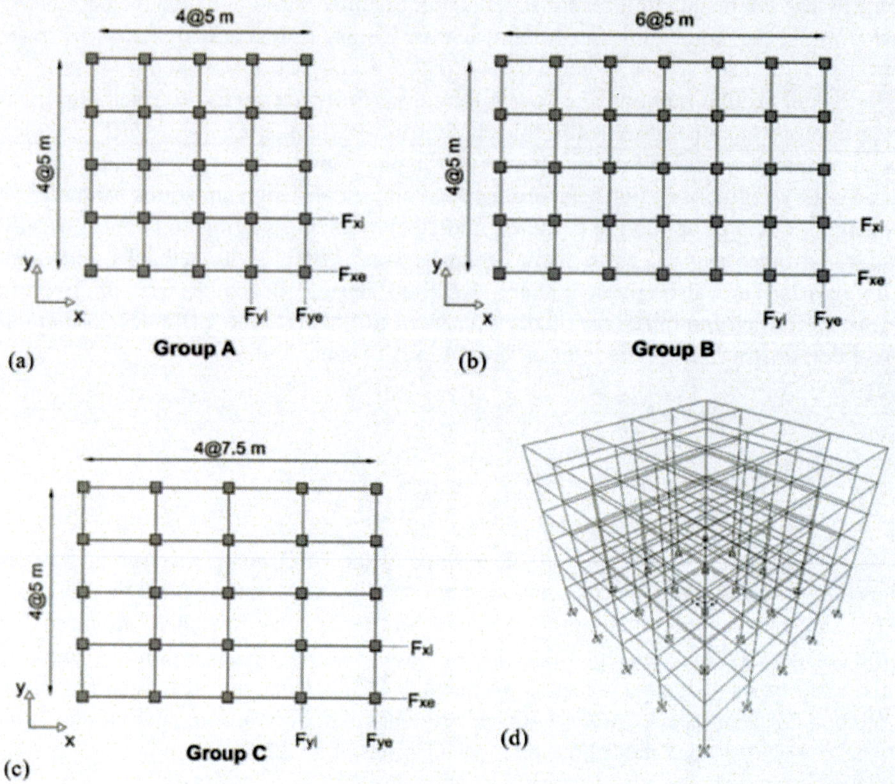

Fig. 5.1 Groups of space steel frames considered here: **(a–c)** plan view of frames of groups A, B, C; **(d)** perspective view of a six storey frame of group A (after Tzimas 2013, reprinted with permission from UPCE)

Table 5.1 Target values of performance metrics for three performance levels

	Immediate occupancy		Life safety		Collapse prevention	
	Inter-storey drift	Local ductility	Inter-storey drift	Local ductility	Inter-storey drift	Local ductility
MRFs	0.7% transient; negligible permanent	1.00	2.5% transient; 1.0% permanent	7.00 (9.00)	5% transient or permanent	9.00 (11.00)
CBFs	0.5% transient; negligible permanent	$0.2(2.4 + 8.3\bar{\lambda})$	1.5% transient; 0.4% permanent	$0.75(2.4 + 8.3\bar{\lambda})$	2% transient or permanent	$2.4 + 8.3\bar{\lambda}$

$T = 0.116 \times H^{0.8}$, where H is the total height of the frame in m (ASCE/SEI 7-16 2017).

Table 5.2 Values of parameters b_1 and b_2 of Eq. (5.1) for regular steel space frames

	IDR_y–$IDR_{1.8\%}$		$IDR_{1.8\%}$–$IDR_{3.2\%}$		$IDR > IDR_{3.2\%}$	
Number of storeys	b_1	b_2	b_1	b_2	b_1	b_2
3	0.84	1.01	1.03	1.06	0.99	1.05
6	0.37	0.88	0.93	1.11	1.51	1.25
9	0.29	0.88	2.07	1.37	2.38	1.41
12	0.28	0.91	1.46	1.32	5.58	1.71
15	0.22	0.89	5.04	1.67	6.88	1.76

5. Transformation of performance metrics to the target roof displacement. Transformation of the maximum (target) *IDR* to maximum (target) roof displacement, $u_{r,\max(IDR)}$, is done by the relation

$$u_{r,\max(IDR)} = b_1 H (IDR)^{b_2} \tag{5.1}$$

where H is the total height of the frame (in m) and b_1 and b_2 parameters given by Table 5.2 in terms of the number of stories, n_s, and the level of maximum *IDR*.

Transformation of the maximum (target) μ_θ, to maximum (target) roof displacement, $u_{r,\max(\mu)}$, is done by the relation

$$u_{r,\max(\mu)} = \mu_{r,\theta} u_{r,y} \tag{5.2}$$

where $\mu_{r,\theta}$ is the maximum rotational roof ductility given in terms of the maximum local ductility μ_θ as.

$$\begin{aligned} \mu_{r,\theta} &= 1 + 0.81(\mu_\theta - 1)\, for\ \mu_\theta \leq 4.68 \\ \mu_{r,\theta} &= 2.58 + 0.38(\mu_\theta - 1)\, for\ \mu_\theta > 4.68 \end{aligned} \tag{5.3}$$

Thus, the maximum design roof displacement $u_{r,\max(d)}$ is obtained as

$$u_{r,\max(d)} = min\left(u_{r,\max(IDR)}, u_{r,\max(\mu)}\right) \tag{5.4}$$

6. Computation of the behavior factor q of the frame from

$$\begin{aligned} q &= 1 + 1.35\left(\mu_{r,d} - 1\right) for\ e \neq 0 \\ q &= 1 + 1.30\left(\mu_{r,d} - 1\right) for\ e = 0 \end{aligned} \tag{5.5}$$

where e is the accidental eccentricity and $\mu_{r,d}$ is the design roof displacement ductility expressed as

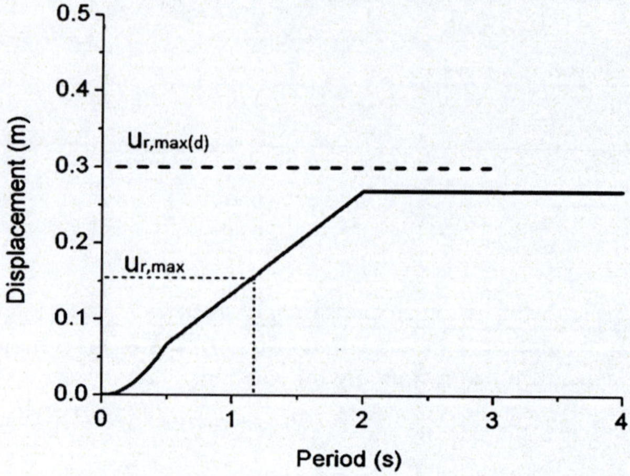

Fig. 5.2 Displacement design spectrum of EC8 (2004) for $PGA = 0.36$ g and soil type B

$$\mu_{r,d} = \frac{u_{r,\max(d)}}{u_{r,y}} \tag{5.6}$$

7. Design of the frame for strength by using elastic response spectrum analysis based on an elastic spectrum with ordinates divided by q as in the FBD method. Capacity and ductile design rules of seismic codes (EC8 2004) are also observed. Deformation checking is not required since this is automatically satisfied because q depends on deformation.
8. Iterations with respect to the input variable $u_{r,\ y}$ are required with their number depending on the initial estimate.

In some cases, the above seismic design procedure may experience problems of convergence when the demand value of the $u_{r,\ \max}$ is lower than the target value $u_{r,\ \max\ (d)}$. This $u_{r,\ \max}$ is determined from the displacement design spectrum for a SDOF system with a period equal to the fundamental period of the MDOF structure under design. The displacement design spectrum is constructed form the corresponding acceleration one of EC8 (2004). If $u_{r,\ \max}$ is lower than $u_{r,\ \max\ (d)}$, as shown in Fig. 5.2, a new target value $u_{r,\ \max\ (d)}$ equal to $u_{r,\ \max}$ is adopted and the method proceeds without any problem. On the basis of the new $u_{r,\ \max\ (d)}$ value, the corresponding new target values IDR_{max} and $\mu_{\theta,\ max}$ can be computed with the aid of the design Eqs. (5.1–5.4).

In order to determine the roof displacement of a MDOF structure from the spectral displacement of its equivalent SDOF system more realistically the latter displacement has to be multiplied by a modification factor equal to 1.0, 1.2, 1.3, 1.4 and 1.5 for buildings with 1, 2, 3, 5 and $\geq$ 10 number of floors, respectively (ASCE 41-13 2014).

5.3 Design Equations for Space Regular Steel MRFs

In this section, the derivation of the basic Eqs. (5.1)–(5.6) for space regular steel MRFs is provided in some detail for illustration purposes. This includes the geometry, sections and seismic design of the considered frames, the nonlinear modeling of those frames and the seismic motions considered for their nonlinear dynamic analysis, the creation of a response databank through parametric analyses of the frames and finally, the derivation of the basic design equations through regression analysis of the results in the response databank.

5.3.1 *Geometry and Seismic Design of Frames Considered*

In here, 38 steel space MRFs with rectangular base and regular in plan view and elevation are considered. The storey height is 3.0 m, while the bay span is either 5.0 m or 7.5 m. The frames are classified in three groups A, B and C, as shown in Fig. 5.1a–c. Groups A and B consist of 5 frames with 3, 6, 9, 12 and 15 stories, while group C of 4 frames with 3, 6, 9 and 12 stories. The bay dimensions along the two x and y horizontal directions of every group are shown in Fig. 5.1a–c. Figure 5.1d shows in three-dimensional (3D) perspective view a typical 6 stories and 4 bays (in both directions) space frame with square plan view. In all frame groups accidental eccentricities of 0% and 5% (along both directions x and y separately) are considered. In addition, for group A, an accidental eccentricity of 10% as well as different steel grades in beams are considered.

The frames were designed according to the provisions of EC3 (2009) and EC8 (2004) and the use of the computer program SAP 2000 (2010). The steel grade was assumed to be S235 for beams and S355 for columns.

The dead and live design loads were assumed to be $G = 6.5$ kN/m^2 and $Q = 2.0$ kN/m^2 resulting in a G + 0.3Q combination equal to 7.1 kN/m^2. Structural self-weight is not included in G as it is added during the analysis. Type 1 elastic design spectrum of EC8 (2004) for soil type B and peak ground acceleration (*PGA*) equal to 0.24 g, where g = 9.81 m/s^2 is the acceleration of gravity constitutes the seismic action. The behavior factor q was assumed equal to 6.5 for both x and y building directions. The design action combinations were assumed to be $G + 0.3Q \pm E_x \pm 0.3E_y$, $G + 0.3Q \pm E_y \pm 0.3E_x$ and $1.35G + 1.5Q$, where E_x and E_y denote the seismic actions along the x and y directions, respectively.

The steel sections were assumed to be IPE for the beams and square hollow sections (SHS) for the columns, both of class 1. SHS columns were assumed to be under bidirectional bending and axial load, while beam-to-column joints were assumed to be rigid and of full strength. The final sections of the frames are shown in Table 5.3. The designed frames have low values of the stiffness parameter ρ and high values of the strength parameter α, defined as (Chopra 2007)

Table 5.3 Sections of space steel frames considered in the parametric studies (the indices *i* and *e* mean internal and external beam/column sections, respectively) (after Tzimas 2013, reprinted with permission from UPCE)

Three storey space MRF						
Floor	Group A–Group B			Group C		
	IPE		SHS	IPE		SHS
	F_{xe}–F_{ye}	F_{xi}–F_{yi}	C_i–C_e	F_{xe}–F_{ye}	F_{xi}–F_{yi}	C_i–C_e
1	270	300	250 × 16	330–300	400–300	300 × 16
2	270	300	250 × 16	330–300	400–300	300 × 16
3	270	300	250 × 16	330–300	400–300	300 × 16
Six storey space MRF						
Floor	Group A–Group B			Group C		
	IPE		SHS	IPE		SHS
	F_{xe}–F_{ye}	F_{xi}–F_{yi}	C_i–C_e	F_{xe}–F_{ye}	F_{xi}–F_{yi}	C_i–C_e
1	330	360	350 × 16	330–300	400–360	400 × 16
2	330	360	300 × 16	360–330	450–400	400 × 16
3	330	360	300 × 16	360–330	450–400	350 × 16
4	300	330	300 × 12	330–300	400–360	350 × 16
5	300	330	300 × 12	330–300	360–330	300 × 16
6	300	330	300 × 12	330–300	360–330	300 × 16
Nine storey space MRF						
Floor	Group A–Group B			Group C		
	IPE		SHS	IPE		SHS
	F_{xe}–F_{ye}	F_{xi}–F_{yi}	C_i–C_e	F_{xe}–F_{ye}	F_{xi}–F_{yi}	C_i–C_e
1	300	360	400 × 16	400–400	500–500	400 × 20
2	330	400	400 × 16	450–400	500–500	400 × 20
3	330	400	400 × 16	400–360	500–450	400 × 20
4	330	400	400 × 16	400–360	500–450	400 × 16
5	330	400	400 × 16	400–360	450–400	400 × 16
6	300	360	350 × 16	360–330	400–360	350 × 16
7	270	330	350 × 16	360–330	400–360	350 × 16
8	270	300	300 × 12	330–300	400–330	300 × 16
9	270	300	350 × 12	330–300	400–330	300 × 16
Twelve storey space MRF						
Floor	Group A–Group B			Group C		
	IPE		SHS	IPE		SHS
	F_{xe}–F_{ye}	F_{xi}–F_{yi}	C_i–C_e	F_{xe}–F_{ye}	F_{xi}–F_{yi}	C_i–C_e
1	360	450	400 × 20	500–450	550–500	400 × 20
2	400	450	400 × 20	500–450	600–550	400 × 20
3	400	450	400 × 20	500–450	600–550	400 × 20
4	360	450	400 × 20	500–400	550–500	400 × 20
5	360	400	400 × 20	500–400	550–500	400 × 20
6	360	400	400 × 16	450–400	500–500	400 × 20
7	330	400	400 × 16	400–360	500–450	400 × 16
8	330	360	350 × 16	400–360	500–450	400 × 16

(continued)

Table 5.3 (continued)

9	300	360	300 × 16	360–330	450–400	350 × 16
10	300	360	300 × 16	360–300	450–360	350 × 16
11	270	330	300 × 12	360–300	450–360	300 × 16
12	270	330	300 × 12	360–300	450–360	300 × 16
Fifteen storey space MRF						
Floor	Group A–Group B			Group C		
	IPE		SHS	IPE		SHS
	F_{xe}–F_{ye}	F_{xi}–F_{yi}	C_i–C_e	F_{xe}–F_{ye}	F_{xi}–F_{yi}	C_i–C_e
1	450	500	400 × 20	–	–	–
2	450	500	400 × 20	–	–	–
3	450	500	400 × 20	–	–	–
4	400	500	400 × 20	–	–	–
5	400	450	400 × 20	–	–	–
6	400	450	400 × 16	–	–	–
7	360	450	400 × 16	–	–	–
8	360	450	400 × 16	–	–	–
9	330	400	400 × 16	–	–	–
10	330	400	350 × 16	–	–	–
11	300	400	350 × 16	–	–	–
12	300	360	350 × 16	–	–	–
13	270	360	300 × 16	–	–	–
14	270	330	300 × 16	–	–	–
15	270	330	300 × 16	–	–	–

$$\rho = \frac{\sum (I/L)_b}{\sum (I/L)_c}, \quad a = \frac{M_{RC,1,av}}{M_{RB,av}} \tag{5.7}$$

In the above, I and L are the second moment of inertia and length of steel member (column c or beam b), $M_{RC,\,1,\,av}$ is the average of the plastic moments of resistance of the columns of the first storey and $M_{RB,\,av}$ is the average of the plastic moments of resistance of the beams in all stories of the frame. The stiffness parameter ρ is evaluated at the storey more closely to the midheight of the frame. It is observed that the sections of the frames of groups A and B were found to be the same. It should be noticed that in Table 5.3, F stands for frame and C for column, while subscripts x and y denote the respective directions and e and i stand for the words exterior and interior, respectively. Furthermore, Table 5.4 provides the three first natural periods of the frames with zero accidental eccentricity.

Table 5.4 Vibration periods of the space steel frames considered in the parametric studies

Number of stories	Group	Period (s)		
		T_1	T_2	T_3
3	A	0.81	0.81	0.72
3	B	0.81	0.81	0.72
3	C	0.85	0.75	0.72
6	A	1.23	1.23	1.10
6	B	1.23	1.23	1.10
6	C	1.32	1.31	1.24
9	A	1.62	1.62	1.52
9	B	1.62	1.62	1.52
9	C	1.71	1.67	1.57
12	A	1.98	1.98	1.83
12	B	1.98	1.98	1.83
12	C	1.97	1.94	1.84
15	A	2.23	2.23	2.05
15	B	2.23	2.23	2.06

5.3.2 *Nonlinear Modeling and Seismic Motions Considered*

Nonlinear time history (NLTH) analyses of the frames considered here were performed with the aid of the Ruaumoko 3D computer program (Carr 2005). Both material and geometric nonlinearities were considered. The hysteretic behavior of beams and columns is simulated by bilinear elastoplastic hinges at their ends. Strain hardening in the moment-rotation relation was assumed equal to 3% (Gupta and Krawinkler 1999). Considering plastic hinge formation in columns, the effect of the axial force on the biaxial plastic moment strength is taken into account through the corresponding interaction formulae (Carr 2005).

The frames are modelled by considering a center-line representation of their inelastic members and rigid beam-column and column-base connections. Material deterioration and panel zone effects are not taken into account. Diaphragmatic action at the level of every floor of the frames is assumed. The mass and the mass moment of inertia of each floor is considered to be concentrated at the center of mass of that floor. Accidental eccentricity can be introduced by moving the center of mass of each floor in each horizontal direction, as defined in EC8 (2004). Damping is assumed to be of the Rayleigh type in conjunction with the tangent stiffness matrix (Carr 2005). The mass and stiffness coefficients of the resulting secant damping matrix are obtained by assuming 3% of critical damping in the first and n_{th} mode, where n is the number of the stories of the frame.

The space frames are subjected to 42 pairs of far-fault earthquake ground motions taken from PEER (2009) and listed in Tzimas (2013). Figure 5.3 depicts the elastic spectra for the two horizontal components of the ground motions. The selection of ground motions was based on the comparison between the spectral ordinates of each ground motion against the spectral ordinates of the design basis earthquake at the fundamental period of each frame. In that way the scaling factor of each ground

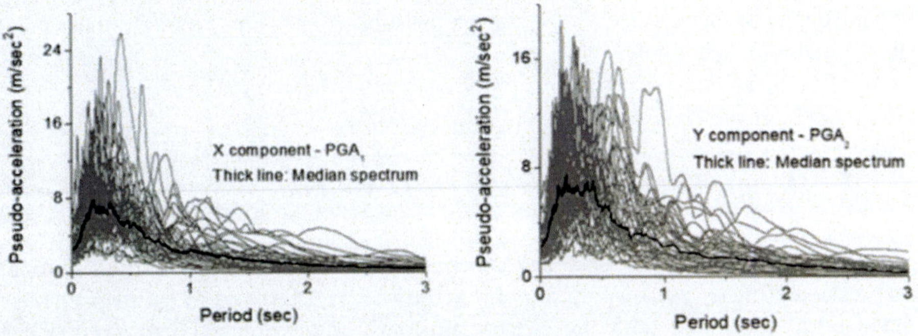

Fig. 5.3 Pseudo-acceleration response spectra for the two components of the seismic excitations (after Tzimas 2013, reprinted with permission from UPCE)

motion can be controlled in order not to take excessive values at higher performance levels as discussed in the following section.

5.3.3 *Parametric Analyses and Creation of Response Databank*

For the creation of the seismic response databank, 12,432 NLTH analyses were conducted in the framework of incremental dynamic analysis (IDA) (Vamvatsikos and Cornell 2002). Thus, a structure is repeatedly subjected to a single ground motion by scaling its amplitude for the construction of the peak structural response versus seismic intensity curve. The 42 pairs of the seismic motions in Tzimas (2013) are used twice for every frame by alternating the x and y directions. The determination of the appropriate scaling factor (SF) for driving the structure to every performance level (defined by its IDR_{max}) is done by the bisection method (Tzimas et al. 2017).

Although scaling of earthquake motions has been criticized, especially when the SF exceeds a certain large value, e.g., SF = 10 or 12 (De Luca et al. 2009), its use becomes practically necessary as it is difficult to find natural seismic records that can drive the structure to high performance levels. In this work, the highest value of SF is taken to be 8. Use of NLTH analysis for a specific performance level leads to the determination of the seismic response of the structure in the form of $u_{r, max}$, IDR_{max}, $\mu_{\theta, max}$, $\mu_{r, max}$ and q. It should be noted that (i) $u_{r, max}$ is the maximum roof displacement; (ii) IDR_{max} is the maximum interstorey drift ratio; (iii) the maximum local rotational ductility $\mu_{\theta, max} = 1 + \theta_p/\theta_y$, where θ_p and θ_y are the plastic rotation and the yield chord rotation at the ends of the member, respectively; (iv) the maximum roof displacement ductility, $\mu_{r, max} = u_{r, max}/u_{r, max, y1}$, where $u_{r, max, y1}$ is the maximum roof displacement at the appearance of the first plastic hinge; and

(v) the behavior factor $q = SF_t/SF_{y1}$ where the subscripts t and y_1 stand for target and first yielding, respectively.

5.3.4 Design Equations for the HFD Method

Using the created response databank described in the previous section, the empirical expressions for $u_{r,\max(IDR)}$, $\mu_{r,\theta}$ and q in Eqs. (5.1), (5.3) and (5.5), respectively, have been developed with the aid of non-linear regression analysis (MATLAB 2009).

Figure 5.4 provides graphically the distribution of the ratio $R_1 = u^{app}_{r,max,(IDR)}/u^{ex}_{r,max,(IDR)}$ as obtained in Tzimas et al. (2017) (for space frames) and in Karavasilis et al. (2008a) (for plane frames) for known *IDR* values. The superscript "app" stands for approximate and is associated with the value obtained by using the proposed empirical equations, (5.1) in Tzimas et al. (2017) and (5.4) in Karavasilis et al. (2008a), while "ex" is associated with the value of the databank obtained by NLTH analysis. The results of Fig. 5.4a include space frames with 0%, 5% and 10% accidental eccentricities. Figure 5.4 also provides the mean, median and standard deviation (Stdev) values for the two (a) and (b) cases. It is observed that Eq. (5.1) gives very good results, while Eq. (5.4) of Karavasilis et al. (2008a) gives results with higher standard deviation. Furthermore, Eq. (5.1) is much simpler than Eq. (5.4) of Karavasilis et al. (2008a) because it does not include the stiffness and strength parameters ρ and α of Eq. (5.7). This is because the ρ values of the space frames in Tzimas et al. (2017) correspond to the lower values used in Karavasilis et al. (2008a), whereas the α values correspond to the higher values used in Karavasilis et al. (2008a). In general, the lower the ρ value is, the higher the α value becomes (Karavasilis et al. 2008a). However, as was shown in Karavasilis et al. (2008a), both the ρ and α parameters affect the structural behavior and thus

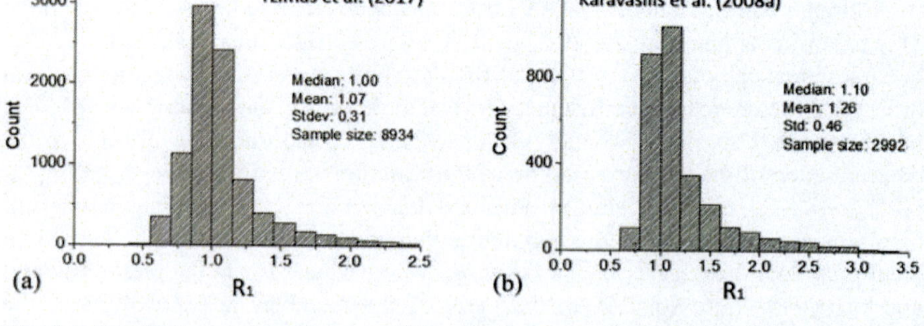

Fig. 5.4 Distribution of the ratio $R_1 = u^{app}_{r,max,(IDR)}/u^{ex}_{r,max,(IDR)}$ on the basis of **(a)** Eq. (5.1) of Tzimas et al. (2017) with known *IDR*; and **(b)** Eq. (5.4) of Karavasilis et al. (2008a) relation with known *IDR* (after Tzimas 2013, reprinted with permission from UPCE)

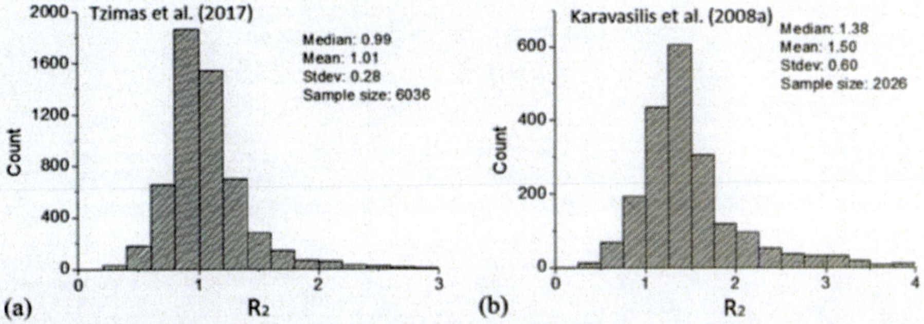

Fig. 5.5 Distribution of the ratio $R_2 = \mu_{r,\theta}^{app}/\mu_{r,\theta}^{ex}$ on the basis of (**a**) Eq. (5.3) with known μ_θ; and (**b**) Eq. (5.11) of Karavasilis et al. (2008a) relation with known μ_θ (after Tzimas 2013, reprinted with permission from UPCE)

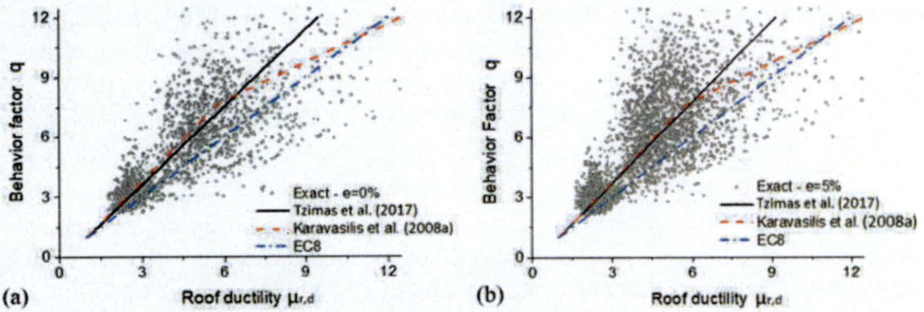

Fig. 5.6 Behavior factor q versus roof ductility $\mu_{r,\,d}$ as given by the space frame response databank, Eq. (5.5), the plane frame Karavasilis et al. (2008a) relation and the equal-displacement rule (EC8 2004) for accidental eccentricity of (a) 0% and (b) 5% (after Tzimas 2013, reprinted with permission from UPCE)

further investigation is needed using frames with a wider range of ρ and α values, to possibly include these parameters in the expressions for the space frames.

Figure 5.5 provides graphically the distribution of the ratio $R_2 = \mu_{r,\theta}^{app}/\mu_{r,\theta}^{ex}$ for known μ_θ values and using Eqs. (5.3) and (5.11) of Karavasilis et al. (2008a) for the computation of $\mu_{r,\theta}^{app}$ for space (with zero accidental eccentricity) and plane frames, respectively. It is observed that Eq. (5.3) gives very good results and certainly better than those of Eq. (5.11) in Karavasilis et al. (2008a).

Figure 5.6 shows graphically Eq. (5.5) for space frames of groups A, B and C and $e = 0\%$ and 5% together with the databank results, the relations of Karavasilis et al. (2008a) for plane frames and those based on the equal displacement rule (EC8 2004). Figure 5.6 exhibits considerable variability in the q value at higher roof ductility values, which can be related to record-to-record variability. It is observed from Fig. 5.6 that space frame results from Eq. (5.5) and $e = 0\%$ are very close to the plane frame results of Karavasilis et al. (2008a) for up to about $q = 6$. On the other hand, since values of $q > 8$ are not practically realistic, any comparison of the results

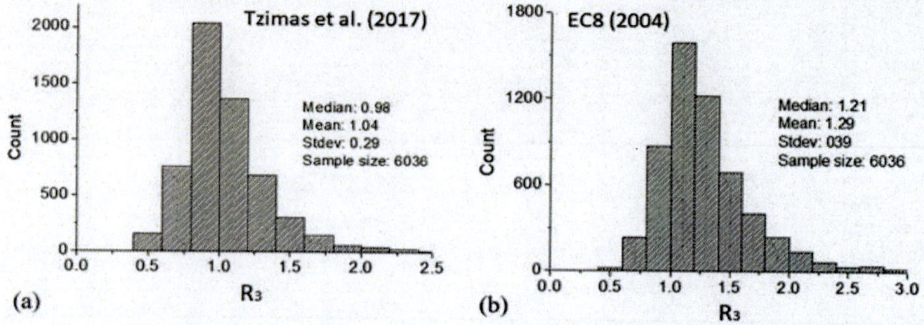

Fig. 5.7 Distribution of the ratio $R_3 = \mu_{r,d}^{app}/\mu_{r,d}^{ex}$ on the basis of (**a**) Eq. (5.6) with known q; and (**b**) EC8 (2004) relation with known q (after Tzimas 2013, reprinted with permission from UPCE)

with the Karavasilis et al. (2008a) plane frame results is practically meaningless. Figure 5.6 in conjunction with databank results also show that ductility values of the top floor and intermediate floors are smaller than the behavior factor q provided in EC8 (2004). Figure 5.7 provides the distribution of the values of the ratio $R_3 = \mu_{r,d}^{app}/\mu_{r,d}^{ex}$ with known q for the cases of using Eq. (5.6) and EC8 (2004) for determining $\mu_{r,d}^{app}$ and databank results for determining $\mu_{r,d}^{ex}$ for space frames with rectangular plan view. The same figure also provides median, mean and standard deviation values for the two aforementioned cases. One can observe that the Tzimas et al. (2017) results are more accurate than the EC8 (2004) ones, which over-predict ductility demands.

5.4 Design Equations for Space Irregular MRFs

In this section, space irregular steel MRFs are considered. Their irregularity may be with respect to the plan view (Fig. 5.8) or to elevation (Figs. 5.9 and 5.10). The material of this section comes from the work of Tzimas et al. (2020), where one can look at for more details.

Figure 5.8 depicts 6 types of steel framed buildings regular in height and irregular in plan view (L-shaped plan). Each type has 3, 6, 9, 12 and 15 storeys, hence there are $6 \times 5 = 30$ such MRF buildings. The beams consist of IPE steel sections of grade S235, while the columns of steel square hollow section (SHS) of grade S355. According to EC8 (2004), the buildings considered here have a low to moderate irregularity in plan view, since their slenderness ratio L_{max}/L_{min} with L_{max} and L_{min} being the largest and smallest in plan dimensions, respectively, varies between 1.33 and 4.0. Only natural eccentricities of these frames are considered.

Following the same procedure as in the previous Sect. 5.3 for regular frames, one is able to derive the design equations for the present irregular in plan steel space MRFs of Fig. 5.8. According to Tzimas et al. 2020, these design equations are the

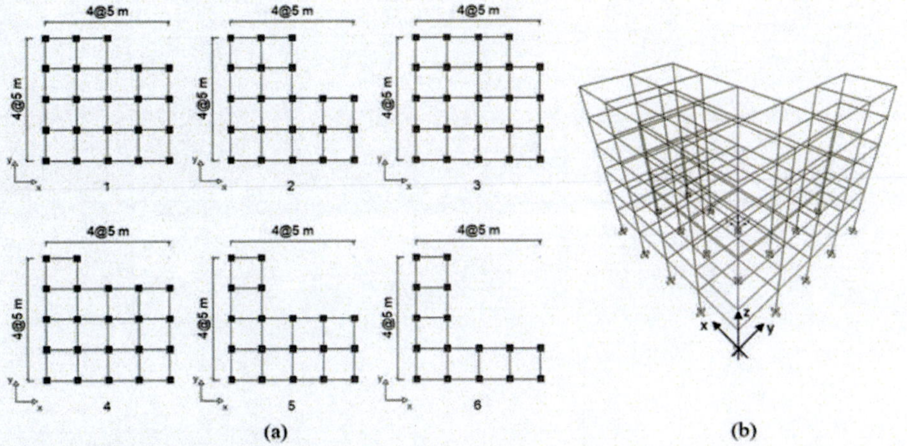

Fig. 5.8 Plan irregular space steel frames: (**a**) frames plan view (Type 1–6); (**b**) perspective view of the six-storeys four-bays space MRF (Type 5 in plan view) (after Tzimas 2013, reprinted with permission from UPCE)

same with those for the space regular MRFs of Sect. 5.2, at least for frame types 1–5 (Fig. 5.8a).

Figure 5.9 shows all 40 types of space steel building frames with setbacks along their elevation. Steel beam and column sections are as in the previous case of irregular in plan frames. The geometrical irregularity introduced by setbacks is quantified through the indices Φ_s and Φ_b, which are expressed in the form

$$\Phi_s = \frac{1}{n_s - 1} \sum_{i=1}^{i=n_s-1} \frac{L_i}{L_{i+1}}, \quad \Phi_b = \frac{1}{n_b - 1} \sum_{i=1}^{i=n_b-1} \frac{H_i}{H_{i+1}} \tag{5.8}$$

where n_s is the number of storeys, n_b is the number of bays of the first storey and H_i and L_i are defined as in Fig. 5.9a. Following the same procedure as in the previous Sect. 5.3 for regular frames, the following design equations for the present frames with setbacks along their height as in Fig. 5.9 are derived (Tzimas et al. 2020):

$$u_{r,\max\,(IDR)} = \beta H(IDR) \tag{5.9}$$

where the parameter β is computed from

$$\beta = 1 - b_1 (n_s - 1)^{b_2} \Phi_s^{b_3} \Phi_b^{b_4} \tag{5.10}$$

with the parameters b_1, b_2, b_3 and b_4 given in Table 5.5 for motions along the x and y directions,

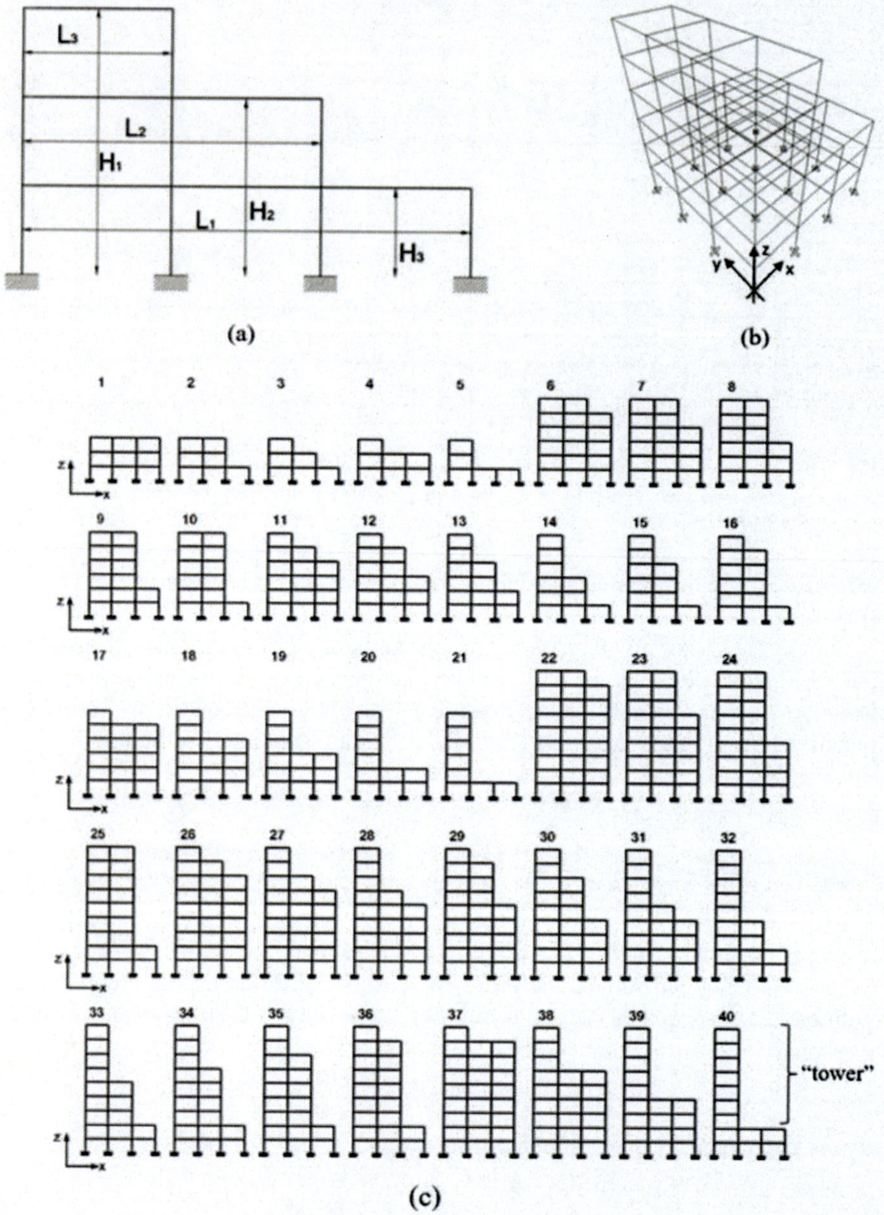

Fig. 5.9 (**a**) Geometry of frames with setbacks; (**b**) perspective view of a six-storey with setbacks along the height; (**c**) all the frames with setbacks considered (Frame type 1–Frame type 40) (after Tzimas 2013, reprinted with permission from UPCE)

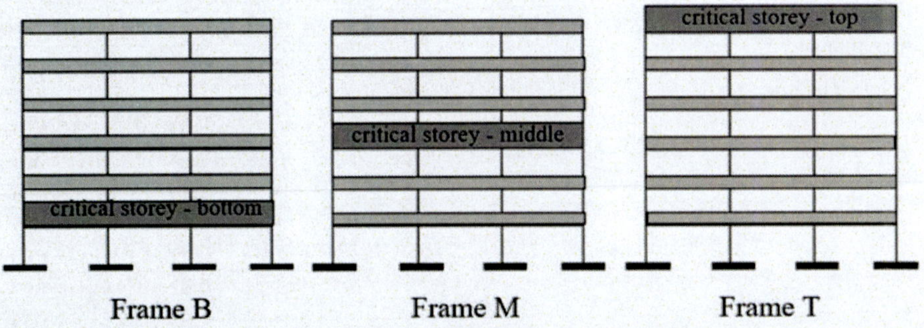

Fig. 5.10 The location of mass discontinuity considered in this study (after Tzimas 2013, reprinted with permission from UPCE)

Table 5.5 Values of parameters b_1–b_4 of Eq. (5.10) for steel space frames with setbacks

Direction	b_1	b_2	b_3	b_4
x	0.18	0.36	0.85	0.24
y	0.18	0.36	0.17	0.10

Table 5.6 Values of parameters b_1–b_3 of Eq. (5.13) for steel space frames with setbacks

Direction	b_1	b_2	b_3
x	1.39	1.05	−0.06
y	1.39	1.05	−0.34

$$u_{r,\max(\mu)} = \mu_{r,\theta} u_{r,y} \tag{5.11}$$

where $\mu_{r,\theta}$ is computed from

$$\mu_{r,\theta} = 1 + 0.79(\mu_\theta - 1) \tag{5.12}$$

$$q = 1 + b_1 \left(\mu_{r,d} - 1\right)^{b_2} \Phi_s^{b_3} \tag{5.13}$$

where the parameters b_1, b_2 and b_3 are obtained from Table 5.6 and $\mu_{r,d}$ is defined by Eqs. (5.4) and (5.6).

The above design equations are appropriately used in the 8 steps procedure of the HFD method of Sect. 5.2.

Figure 5.10 shows the three types of space framed buildings with mass discontinuity at the first (B), the middle (M) and the top (T) storey. The mass discontinuity is quantified by the mass ratio m_r, i.e., the ratio of the mass of the storey that sustains the large weight to the smaller mass of the masses of the adjacent storeys.

According to ASCE 7-10 (2010) a building structure is irregular with respect to the vertical mass discontinuity when $m_r > 1.5$. In the cases considered here, m_r equals 2 and 3. The frames considered here have 3, 6 and 9 stories with 4 bays and are regular in plan view. Thus, there are $3 \times 3 \times 2 = 18$ frame buildings in total with

Table 5.7 Values of parameters b_1 and b_2 of Eq. (5.14) for steel space frames with mass discontinuities

Number of storeys	Location of mass discontinuity	b_1	b_2
3	B	0.83	1.01
	M	0.83	1.01
	T	1.26	1.11
6	B	1.16	1.17
	M	0.72	1.03
	T	0.82	1.07
9	B	1.95	1.34
	M	1.47	1.24
	T	1.38	1.27

Table 5.8 Values of parameters b_1 and b_2 of Eq. (5.16) for steel space frames with mass discontinuities

Location of mass discontinuity	b_1	b_2
B	1.22	0.60
M	1.17	0.59
T	1.26	0.61

beams and columns of the same type as in the case of regular ones discussed in Sect. 5.3. The frames were designed according to EC8 (2004) with 5% accidental eccentricity. Following the same procedure as in Sect. 5.3 for regular frames, the following design equations for the present frames with mass discontinuity along their height as in Fig. 5.10 are derived (Tzimas et al. 2020):

$$u_{r,\max(IDR)} = b_1 H (IDR)^{b_2} \tag{5.14}$$

where values of the parameters b_1 and b_2 can be obtained from Table 5.7 in terms of the number of stories and the location of mass discontinuity,

$$u_{r,\max(\mu)} = \mu_{r,\theta} u_{r,y} \tag{5.15}$$

where

$$\mu_{r,\theta} = 1 + b_1(\mu_\theta - 1)^{b_2} \tag{5.16}$$

with the parameters b_1 and b_2 given in Table 5.8,

$$q = 1 + b_1(\mu_{r,d} - 1) \tag{5.17}$$

where the parameter b_1 has the values 1.38, 1.50 and 1.43 for the location of mass discontinuity at B, M and T, respectively and $\mu_{r,d}$ is obtained from Eqs. (5.4), (5.6) and (5.14)–(5.16).

The above design equations are appropriately used in the 8 steps procedure of the HFD design method of Sect. 5.2.

5.5 Design Equations for Plane Regular MRFs

In this section, plane regular MRFs are considered, as shown in Fig. 5.11. The frames have a number of stories n_s, a number of bays n_b, storey heights equal to 3.0 m and bay widths equal to 6.0 m. The beams and columns are of steel sections IPE and HEB, respectively. The modeling of these plane frames is the same as in the case of space frames (Sect. 5.3.2). The material of this section comes from the work of Tzimas et al. (2013) and Karavasilis et al. (2006, 2007a, 2008a), where more details can be found.

Following the same procedure as in Sect. 5.3 for space regular MRFs, the following design equations for plane regular MRFs (Fig. 5.11) can be derived (Karavasilis et al. 2008a):

$$\text{(i) } u_{r,\max\,(IDR)} = \beta H(IDR) \tag{5.18}$$

where the parameter β depends on the frame properties and is calculated by

$$\beta = 1 - 0.19(n_s - 1)^{0.54}\rho^{0.14}\alpha^{-0.19} \tag{5.19}$$

with ρ and a being the stiffness and strength ratios defined by Eq. (5.7)

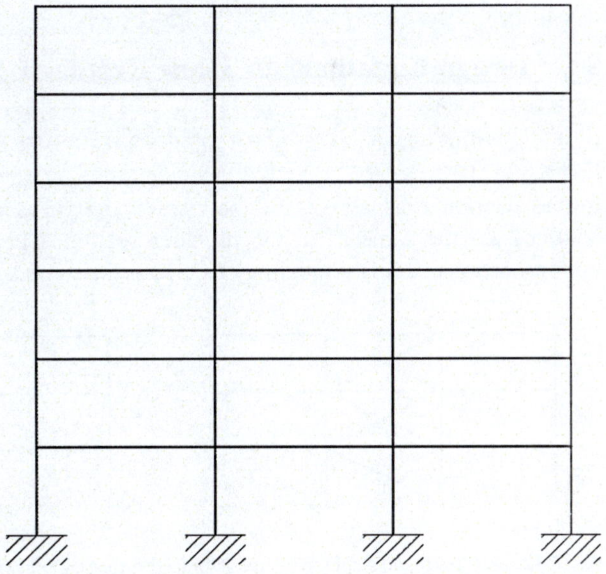

Fig. 5.11 Typical plane regular multi-storey multi-bay MRF

$$\text{(ii)}\ u_{r,\max(\mu)} = \mu_{r,\theta} u_{r,y} \tag{5.20}$$

where $\mu_{r,\theta}$ is computed by

$$\mu_{r,\theta} = 1 + 1.35(\mu_\theta - 1)^{0.86} \alpha^{0.43} n_s^{-0.31} \tag{5.21}$$

$$\text{(iii)}\ \begin{array}{l} q = 1 + 1.39(\mu_{r,d} - 1)\ for\ \mu_{r,d} \le 5.8 \\ q = 1 + 8.84(\mu_{r,d}^{0.32} - 1)\ for\ \mu_{r,d} > 5.8 \end{array} \tag{5.22}$$

with $\mu_{r,d}$ obtained by using Eqs. (5.4) and (5.6). The above design equations are appropriately used in the 8 steps procedure of the HFD design method of Sect. 5.2.

A comparison of design Eqs. (5.18), (5.19) and (5.21) for plane regular MRFs against the corresponding Eqs. (5.1) and (5.3) for space regular MRFs clearly shows the complexity of the former equations, especially due to the presence there of the stiffness and strength ratios ρ and α, respectively. These two ratios, defined by Eq. (5.7), require a knowledge of the sections of the frame to be designed for their computation. The absence of these ratios in the design equations for space regular MRFs is because their ranges defined on the basis of the frames considered for the derivation of those design equations are narrower in space regular MRFs than in plane regular MRFs. Indeed, one has $0.11 \le \rho \le 0.54$ and $1.30 \le \alpha \le 4.76$ for plane regular MRFs (Karavasilis et al. 2008a) and $0.15 \le \rho \le 0.34$ and $2.65 \le \alpha \le 6.31$ for space regular MRFs (Tzimas et al. 2017). Thus, no significant effect of ρ and α was identified in the response of the space regular MRFs as it can be also observed in Figs. 5.4, 5.5 and 5.6. Use of wider ranges of ratios ρ and α for space regular MRFs, would have most probably resulted in design equations dependent on these ratios.

5.6 Design Equations for Plane Regular CBFs

In this section, two types of plane regular steel CBFs are considered as shown in Fig. 5.12. The first type is the X-braced and the second type is the chevron type of frames. Geometrical aspects of the frames and types of sections for beams and columns are the same as in the previous section. The braces consist of circular hollow sections (CHS). Modeling of CBFs is based on the assumption that columns

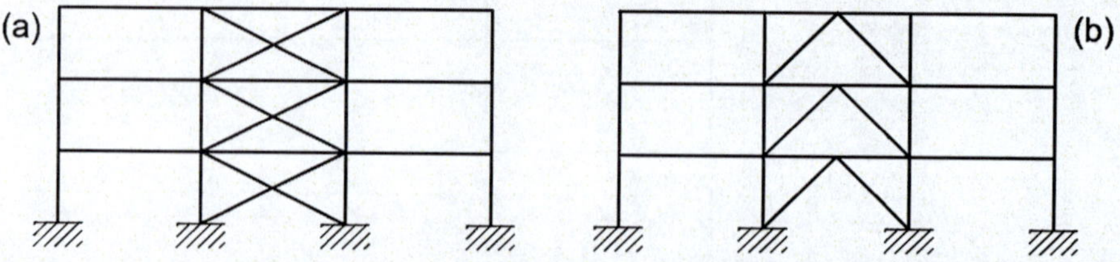

Fig. 5.12 Typical plane steel CBFs: **(a)** X-braced frame; **(b)** chevron braced frame

carry moments along the whole height of the frame. In addition, beams and diagonals are pin-connected to columns in X-braced frames, while beams are pin-connected to columns and diagonals are pin-connected to beams and columns in chevron braced frames. Finally, columns are assumed to be pinned and fixed at their base for X-braced and chevron frames, respectively. More details can be found in Tzimas et al. (2013), Karavasilis et al. (2007b), Zotos and Bazeos (2009) and Stamatopoulos and Bazeos (2011).

Following the same procedure as in Sect. 5.3 for space regular MRFs, the following design equations for plane regular CBFs (Fig. 5.12) can be derived:

(i) $u_{r,\,\max\,(IDR)}$ is given again by Eq. (5.18) with β as

$$\beta = 1 - 0.12(n_s - 1)^{0.31}\bar{\lambda}^{-0.11}\mathrm{a}^{-0.19}(T/T_c)^{0.14} \tag{5.23}$$

for X-braced CBFs and

$$\beta = 1 - 0.542(n_s - 1)^{-0.013}\bar{\lambda}^{-0.061}\mathrm{a}^{-0.03}T^{0.09} \tag{5.24}$$

for chevron CBFs, where T is the fundamental period of the frame, T_c is the period corresponding to the transition from the constant acceleration to the constant velocity regime of the design spectrum, $\bar{\lambda}$ is the brace slenderness ratio and a is the ratio of the contribution of the columns over that of the diagonal braces to the storey stiffness. Ratios $\bar{\lambda}$ and a, defined at the storey closest to the mid-height of the frame, are given as

$$\bar{\lambda} = \frac{L_b}{\pi r}\sqrt{\frac{f_y}{E}}\,, \quad \mathrm{a} = \frac{n_c I_c L_d}{n_d A_d h^3 \cos^2\varphi} \tag{5.25}$$

where L_b, L_d, A_d and r are the buckling length, the length, the cross-sectional area and the radius of gyration of the cross-section of the diagonal brace, respectively, I_c is the second moment of inertia of the column, h is the storey height, φ is the angle between diagonals and beams, n_c and n_d are the numbers of columns and diagonals in the storey, respectively and f_y and E are the yield strength and the modulus of elasticity of the steel material, respectively.

(ii) $u_{r,\,\max\,(\mu)}$ is given again by Eq. (5.20) with $\mu_{r,\,\theta}$ as

$$\mu_{r,\theta} = 1 + 1.51(\mu_{cb} - 1)^{0.73} n_s^{-0.18}\bar{\lambda}^{1.90}(T/T_c)^{0.28} \tag{5.26}$$

for X-braced CBFs and

$$\mu_{r,\theta} = 1 + 4.15(\mu_{cb} - 1)^{0.65} n_s^{-1.19} \overline{\lambda}^{-0.39} \mathrm{a}^{0.09} T^{0.48} \quad (5.27)$$

for chevron CBFs, where μ_{cb} is the local cyclic brace ductility given as a function of the slenderness ratio $\overline{\lambda}$ in Table 5.1.

$$\text{(iii)}\ q = 1 + 0.86(\mu_{r,d} - 1)^{0.62} n_s^{0.34} \overline{\lambda}^{0.70} \mathrm{a}^{-0.10} (T/T_c)^{0.24} \quad (5.28)$$

and for X-braced CBFs and

$$q = 1 + 4.96(\mu_{r,d} - 1)^{0.84} n_s^{-0.40} \overline{\lambda}^{0.03} \mathrm{a}^{0.11} T^{0.58} \quad (5.29)$$

for chevron CBFs, where $\mu_{r,\,d}$ is determined by using Eqs. (5.4) and (5.6).

The above design equations are appropriately used in the 8 steps procedure of the HFD design method of Sect. 5.2.

5.7 Design Equations for Plane Irregular MRFs

In this section, two types of plane irregular steel MRFs are considered, as shown in Fig. 5.13. The first type is a frame with setbacks and the second is a frame with mass discontinuity at its bottom (B), mid-height (M), or top (T). Geometrical aspects of the frames and types of sections for beams and columns are the same as in Sect. 5.5. More details can be found in Tzimas et al. (2013) and Karavasilis et al. (2008b, 2009)

Following the same procedure as in Sect. 5.3 for space regular MRFs, the following design equations for plane irregular steel MRFs (Fig. 5.13) can be derived:

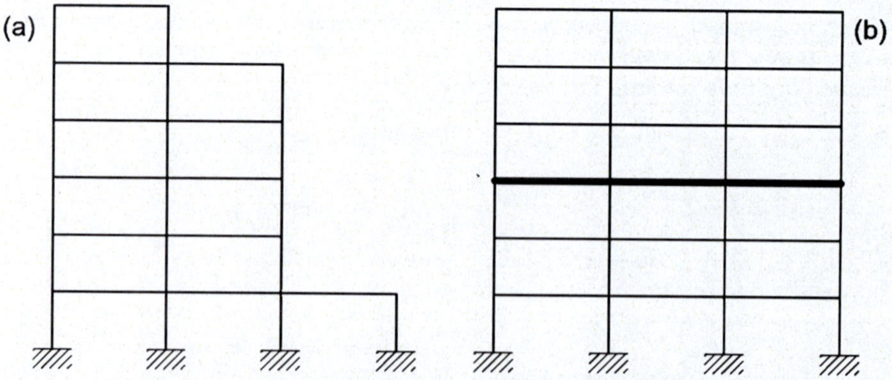

Fig. 5.13 Typical plane irregular steel MRFs: (**a**) with setbacks; (**b**) with mass discontinuity

(i) $u_{r,\max(IDR)}$ is given again by Eq. (5.18) with β as

$$\beta = 1 - 0.13(n_s - 1)^{0.52}\Phi_s^{0.38}\Phi_b^{0.14} \tag{5.30}$$

for plane frames with setbacks, where Φ_s and Φ_b are defined in Eq. (5.8) and

$$\begin{aligned} \beta &= 1 - 0.18(n_s - 1)^{0.4}a^{-0.13} \quad \textit{for B mass location} \\ \beta &= 1 - 0.17(n_s - 1)^{0.64}a^{-0.47} \quad \textit{for M mass location} \\ \beta &= 1 - 0.12(n_s - 1)^{0.66}a^{-0.28} \quad \textit{for T mass location} \end{aligned} \tag{5.31}$$

for plane frames with mass discontinuity.

(ii) $u_{r,\max(\mu)}$ is given again by Eq. (5.20) with $\mu_{r,\theta}$ as

$$\mu_{r,\theta} = 1 + 0.44(\mu_\theta - 1)^{1.26}a^{0.26} \tag{5.32}$$

for plane MRFs with setbacks and

$$\begin{aligned} \mu_{r,\theta} &= 1 + 0.50(\mu_\theta - 1)^{1.12}a^{0.36} \quad \textit{for B mass location} \\ \mu_{r,\theta} &= 1 + 0.78(\mu_\theta - 1) \quad \textit{for M mass location} \\ \mu_{r,\theta} &= 1 + 0.50(\mu_\theta - 1)^{1.18}a^{0.39} \quad \textit{for T mass location} \end{aligned} \tag{5.33}$$

for plane frames with mass discontinuity.

$$\text{(iii)}\ q = 1 + 1.92\left(\mu_{r,d} - 1\right)^{0.85}\Phi_s^{-0.17} \tag{5.34}$$

for plane MRFs with setbacks and

$$\begin{aligned} q &= 1 + 2.26(\mu_{r,d} - 1)^{0.69} \quad \textit{for B mass location} \\ q &= 1 + 2.42(\mu_{r,d} - 1)^{0.68} \quad \textit{for M mass location} \\ q &= 1 + 2.45(\mu_{r,d} - 1)^{0.60} \quad \textit{for T mass location} \end{aligned} \tag{5.35}$$

for plane frames with mass discontinuity, where $\mu_{r,d}$ is determined by using Eqs. (5.4) and (5.6).

The above design equations are appropriately used in the 8 steps procedure of the HFD design method of Sect. 5.2.

5.8 Seismic Design Examples for Steel Space MRFs

This section serves to illustrate the HFD seismic design method and demonstrate its merits by means of two representative examples dealing with one regular and one irregular (with setbacks) steel space MRFs. Additional examples dealing with space regular MRFs and MRFs exhibiting irregularities in plan view and in mass along the height can be found in Tzimas et al. (2017, 2020).

5.8.1 *Six-Storey Four-Bay Regular Steel Space MRF*

Consider a regular steel space MRF with six stories of height 3.0 m each and a square plan view with four bays of 6.0 m each in both directions, as shown in Fig. 5.1a. It is assumed that this space frame has an accidental eccentricity of 5%. The grade of steel is assumed S235 and S355 for beams and columns, respectively. The column sections are square hollow ones (SHS), while those of beams IPE. The dead and live design loads are assumed to be $G = 6.5$ kN/m^2 and $Q = 2.0$ kN/m^2, respectively, resulting in $G + 0.3Q$ load combination of 7.1 kN/m^2. The above loads do not include structural self-weight, which is taken into account separately. Furthermore, it is assumed that the IO under the FOE, the LS under the DBE and the CP under the MCE are the appropriate performance levels for seismic design. The FOE, DBE and MCE are expressed through the Type 1 elastic design spectra of EC8 (2004) for soil type B and peak ground acceleration under DBE (PGA_{DBE}) equal to 0.36 g, as shown in Fig. 5.14b. The peak ground acceleration under the FOE and the MCE are equal to $0.3 \times PGA_{DBE} = 0.3 \cdot 0.36$ g $= 0.108$ g and $1.5 \times PGA_{DBE} = 1.5 \cdot 0.36$ g $= 0.54$ g, respectively. Table 5.1 provides limit values for performance metrics according to FEMA 273 (1997) and ASCE 41-13 (2014).

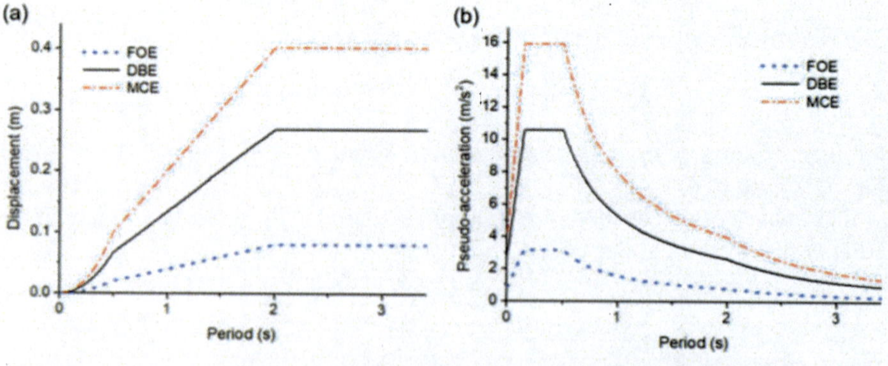

Fig. 5.14 Design spectra for three performance levels based on the DBE spectra of EC8 (2004) for soil type B and $PGA = 0.36$ g for: **(a)** Displacement S_d and **(b)** Pseudo-acceleration S_a

Table 5.9 Column and beam sections and first three natural periods of the six-storey space steel frames designed by the HFD and FBD methods (indices "i" and "e" stand for internal and external, respectively)

	HFD method			FBD (EC8) method		
	IPE		SHS	IPE		SHS
Floor	F_{xe}–F_{ye}	F_{xi}–F_{yi}	C_i–C_e	F_{xe}–F_{ye}	F_{xi}–F_{yi}	C_i–C_e
1	330	450	340 × 20	400	450	350 × 16
2	330	500	340 × 20	400	500	350 × 16
3	330	450	320 × 20	400	500	350 × 16
4	300	400	320 × 20	360	450	300 × 16
5	300	360	300 × 20	330	360	300 × 16
6	300	360	300 × 20	330	360	300 × 16
	$T_1 = 1.23$ s—$T_2 = 1.23$ s—$T_3 = 1.23$ s			$T_1 = 1.15$ s—$T_2 = 1.15$ s—$T_3 = 1.07$ s		

The frame is initially designed elastically for the FOE and its resulting dimensions are given in Table 5.9. This table also provides the first three natural periods of the frame with its fundamental period being translational and equal to 1.23 s. The roof displacement $u_{r,\,y}$ and the IDR_y under the FOE are found to be 0.091 m and 0.69%, respectively. It is observed that 0.69% < 0.70% indicating that this design satisfies the IO demands of Table 5.1.

From Table 5.1, the target values of maximum IDR and μ_θ for the LS performance level under the DBE are 2.5% and 9, respectively. Use of Eq. (5.1) with $b_1 = 0.93$ and $b_2 = 1.11$ from Table 5.2 results in $u_{r,\,\max\,(IDR)} = 0.93 \cdot (3 \cdot 6) \cdot 0.025^{1.11} = 0.279$ m. Use of Eq. (5.3) provides $\mu_{r,\,\theta} = 2.58 + 0.38(9 - 1) = 5.62$ and hence one has from Eq. (5.2) $u_{r,\,\max\,(\mu)} = 5.62 \cdot 0.091 = 0.511$ m. Thus, Eq. (5.4) yields $u_{r,\,\max\,(d)} = \min\,(0.279\text{ m}, 0.511\text{ m}) = 0.279$ m and indicates that IDR controls the LS design. In accordance with the displacement design spectrum of Fig. 5.14a, for a SDOF system with the same period as the fundamental period $T_1 = 1.23$ s of the MRF, its maximum displacement for the DBE is $0.165 \cdot 1.42 = 0.234$ m, where the factor 1.42 comes from ASCE 41-13 (2014) for the present case of the six-storey building. It is observed that the demand displacement of 0.234 m is lower than the $u_{r,\,\max\,(d)} = u_{r,\,\max\,(IDR)} = 0.279$ m and thus, one has to adopt a new $u_{r,\,\max\,(d)} = 0.234$ m and revise the IDR and μ_θ maximum target values. Solving Eq. (5.1) for IDR with $u_{r,\,\max\,(IDR)} = 0.234$ m, one obtains $IDR_{max} = 2.13\%$. On the other hand, one can obtain from Eq. (5.2) with $u_{r,\,\max\,(\mu)} = 0.234$ m, $\mu_{r,\,\theta} = 0.234/0.091 = 2.57$ and hence solve Eq. (5.3) for μ_θ and obtain $\mu_{\theta,\,max} = 2.94$. Eq. (5.6) yields $\mu_{r,\,d} = 0.234/0.091 = 2.57$ and thus Eq. (5.5) finally provides $q = 3.12$.

From Table 5.1, the target values of maximum IDR and μ_θ for the CP performance level under the MCE are 5% and 11, respectively. Use of Eq. (5.1) with $b_1 = 1.51$ and $b_2 = 1.25$ from Table 5.2 results in $u_{r,\,\max\,(IDR)} = 1.51 \cdot (3 \cdot 6) \cdot 0.05^{1.25} = 0.643$ m. Use of Eq. (5.3) provides $\mu_{r,\,\theta} = 2.58 + 0.38 \cdot (11 - 1) = 6.38$ and hence one has from Eq. (5.2) $u_{r,\,\max\,(\mu)} = 6.38 \cdot 0.091 = 0.581$ m. Thus, Eq. (5.4) yields $u_{r,\,\max\,(d)} = \min\,(0.643\text{ m}, 0.581\text{ m}) = 0.581$ m and indicates that local ductility controls the CP design. In accordance with the displacement design spectrum of

Fig. 5.14a for a SDOF system with the same period as the fundamental period $T_1 = 1.23$ s of the MRF, its maximum displacement for the MCE is $0.248 \cdot 1.42 = 0.352$ m, where the factor 1.42 comes from ASCE 41-17 (2017) for the present case of the six-storey building. It is observed that the demand displacement of 0.352 m is lower than the $u_{r,\,\max\,(d)} = u_{r,\,\max\,(\mu)} = 0.581$ m and thus, one has to adopt a new $u_{r,\,\max\,(d)} = 0.352$ m and revise the IDR and μ_θ maximum target values. Solving Eq. (5.1) for IDR with $u_{r,\,\max\,(IDR)} = 0.352$ m, one obtains $IDR_{max} = 3.08\%$. On the other hand, one can obtain from Eq. (5.2) with $u_{r,\,\max\,(\mu)} = 0.352$ m, $\mu_{r,\,\theta} = 0.352/0.091 = 3.87$ and hence solve Eq. (5.3) for μ_θ and obtain $\mu_{\theta,\,max} = 4.54$. Eq. (5.6) yields $\mu_{r,\,d} = 0.352/0.091 = 3.87$ and thus Eq. (5.5) finally provides $q = 4.87$.

From the pseudo-acceleration design spectra of Fig. 5.14b and $T_1 = 1.23$ s one can determine the design base shears V for the three performance levels IO, LS and CP as follows:

$$V_{IO}/M = S_\alpha^{IO}/q_{IO} = 1.30/1.00 = 1.30\,\mathrm{m/s^2}$$

$$V_{LS}/M = S_\alpha^{LS}/q_{LS} = 4.28/3.12 = 1.29\,\mathrm{m/s^2}$$

$$V_{CP}/M = S_\alpha^{CP}/q_{CP} = 6.30/4.87 = 1.29\,\mathrm{m/s^2}$$

where M is the mass of the framed building.

The above results clearly indicate that the IO performance level controls the design and thus the member dimensions provided in Table 5.9 are the final dimensions of the present MRF according to the HFD method.

For comparison purposes, the present MRF was also seismically designed by the FBD method of EC8 (2004) with the aid of the computer program SAP 2000 (2010) using a behavior factor $q = 6.5$. The resulting member dimensions are also provided in Table 5.9. The designed by the EC8 (2004) space MRF remains elastic under the FOE and experiences $u_{r,\,y} = 0.088$ m, $IDR_{max} = 0.66\%$ and $\mu_\theta = 1$. Under the DBE, the MRF experiences $u_{r,\,max} = 0.088 \cdot (PGA_{DBE}/PGA_{FOE}) = 0.088/0.3 = 0.293$ m and $IDR_{max} = 0.66\ \%\ /0.3 = 2.20\%$ while under the MCE the corresponding response values are $u_{r,\,max} = 0.293 \cdot (PGA_{MCE}/PGA_{DBE}) = 0.293 \cdot 1.5 = 0.440$ m and $IDR_{max} = 2.20\ \%\ \cdot 1.5 = 3.30\%$.

The results of the two designed MRFs by the HFD and FBD (EC8 2004) methods are compared with those obtained by NLTH analyses of the two frames by the Ruaumoko computer program (Carr 2005). Five pairs of artificial accelerograms compatible with the elastic design spectrum of EC8 (2004) with soil type B and PGA equal to 0.36 g (Fig. 5.15) were used for the analyses. These accelerograms were constructed by using a specialized software (Karabalis et al. 1993).

Table 5.10 provides the response results of the NLTH analyses together with those obtained by the two design methods. It is observed that the HFD method controls better the deformation/damage compared to the FBD (EC8 2004) method. This is because the HFD method uses a deformation dependent q factor, while the

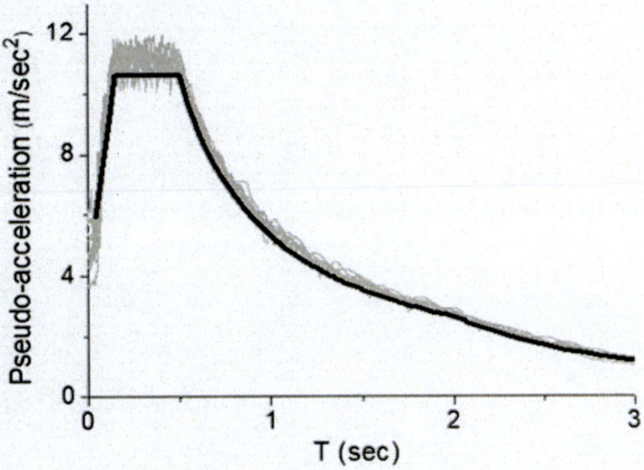

Fig. 5.15 Response spectra of the ground motions used in the design examples compatible to that of EC8 (2004) with soil type B and $PGA = 0.36$ g (after Tzimas 2013, reprinted with permission from UPCE)

FBD method uses a constant value for q. In addition, the FBD method, which employs the equal deformation rule, overestimates the IDR_{max} and the $u_{r,max}$ while the HFD provides values for them very close to the ones from NLTH analysis. Furthermore, the HFD method predicts well the maximum μ_θ, while the FBD is not capable of providing μ_θ at all. One can finally note that the two methods lead to weights very close to each other, i.e., 1721 kN and 1690 kN, for the HFD and FBD designed structures, respectively.

5.8.2 *Eight-Storey Steel Space MRF with Setbacks*

Consider the eight-storey steel space MRF with setbacks, as shown in Fig. 5.16. The storey heights and bay widths are 3.0 m and 6.0 m, respectively. Dead, live and seismic loads, section types, steel grades and performance levels are the same as in the previous Sect. 5.8.1.

The values of the indices Φ_s and Φ_b of the frame are computed from Eq. (5.8) with $n_s = 8$, $L_1 = 18.0$ m, $L_2 = 12.0$ m, $L_3 = 6.0$ m, $H_1 = 24.0$ m, $H_2 = 15.0$ m and $H_3 = 6.0$ m and found to be 1.25 and 2.05, respectively. Thus, the parameter β can be computed from Eq. (5.10) with $b_1 = 0.18$, $b_2 = 0.36$, $b_3 = 0.85$ and $b_4 = 0.24$ taken from Table 5.5 and found to be equal to 0.49. It has been found in Tzimas et al. (2020) that in most cases, the direction with setbacks (here the x direction) controls the design and for this reason, only equations of that direction are considered here.

The frame is initially designed elastically for the FOE and its resulting dimensions are given in Table 5.11. This table also provides the first three natural periods of the frame with its fundamental period being translational and equal to $T_1 = 1.47$ s. It was also found that $u_{r,y} = 0.079$ m and $IDR_y = 0.58\%$, which is less than the limit IO value 0.70% of Table 5.1. Thus, the IO performance level design is acceptable.

Table 5.10 NLTH analyses response results and design estimations for the six-storey steel space MRF (TH: NLTH analysis; EST: estimations of HFD or FBD (EC8) methods)

	HFD method						FBD (EC8) method					
	FOE		DBE		MCE		FOE		DBE		MCE	
	TH	EST	TH	EST	TH	EST	TH	EST	TH	EST	TH	EST
IDR (%)	0.69	0.69	2.00	2.25	2.70	3.15	0.66	0.66	1.70	2.20	2.33	3.30
$u_{r,\max}$ (m)	0.093	0.091	0.238	0.248	0.336	0.360	0.089	0.088	0.218	0.293	0.322	0.44
μ_θ	1.00	1.00	3.08	3.14	4.10	4.65	1.00	1.00	2.84	–	4.66	–

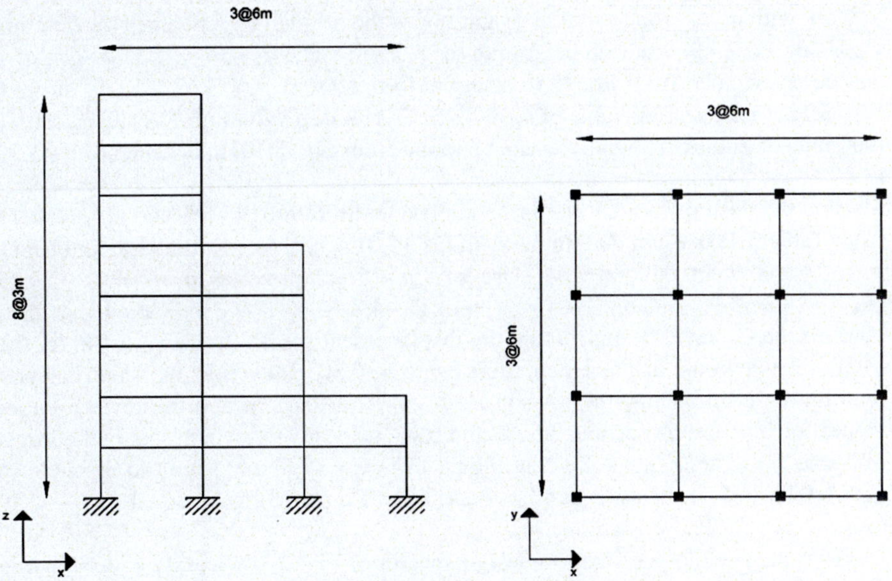

Fig. 5.16 Eight storey steel space MRF with setbacks in elevation and plan view

Table 5.11 Column and beam sections and first three natural periods of the 8-storey steel space MRF with setbacks designed by the HFD and FBD (EC8 2004) methods (indices "i" and "e" stand for internal and external, respectively)

	HFD method FBD (EC8) method					
	IPE		SHS	IPE		SHS
Floor	F_{xe}	F_{ye}	C_e	F_{xi}	F_{yi}	C_i
1	330	300	400 × 16	360	330	400 × 16
2	330	300	400 × 16	400	330	400 × 16
3	330	300	400 × 16	400	330	400 × 16
4	330	300	350 × 16	400	330	350 × 16
5	330	300	350 × 16	360	330	350 × 16
6	330	300	350 × 16	360	–	–
7	300	300	300 × 16	330	–	–
8	300	300	300 × 16	330	–	–

$T_1 = 1.47$ s–$T_2 = 1.37$ s–$T_3 = 1.13$ s

Starting the design from the LS performance level with $IDR_{max} = 2.5\%$ from Table 5.1, one determines from Eq. (5.9) $u_{r,\,\max\,(IDR)} = 0.49 \cdot (3 \cdot 8) \cdot 0.025 = 0.294$ m. Using Eq. (5.12) with $\mu_{\theta,\ max} = 9$ from Table 5.1 for the LS level, one has $\mu_{r,\ \theta} = 1 + 0.79(9 - 1) = 7.32$. Thus, use of Eq. (5.11) leads to $u_{r,\,\max\,(\mu)} = 7.32 \cdot 0.079 = 0.578$ m. Since $u_{r,\,\max\,(d)} =$ min (0.294 m, 0.578 m) = 0.294 m drift controls the LS level design. For a SDOF system with $T = 1.47$ s, the maximum displacement from the displacement design spectrum of Fig. 5.14a is found to be 0.197 m and hence the maximum roof displacement for the 8-storey frame considered is $0.197 \cdot 1.4 = 0.276$ m with the multiplier 1.4 coming from ASCE 41-13 (2014). It is observed that the displacement 0.276 m is lower than the $u_{r,\,\max\,(IDR)} = 0.294$ m meaning that one has to work with $u_{r,\ \max\ (IDR)} = 0.276$ m and revise the original target values for *IDR* and μ_θ. Using Eq. (5.9), one can determine the revised value of $IDR = 2.35\%$. The $\mu_{r,\ d}$ can be determined from Eq. (5.6) as equal to 0.276/0.079 = 3.49 leading through Eq. (5.12) to the revised value of $\mu_\theta = 4.15$. The revised values of *IDR* and μ_θ satisfy the limits of Table 5.1. Finally, q can be found from Eq. (5.13) to be equal to 1 + 1.39 $(3.49 - 1)^{1.05}1.25^{-0.06} = 4.60$.

Starting the design from CP performance level with $IDR_{max} = 5\%$ and $\mu_{\theta,\ max} = 11$ from Table 5.1, one can determine from Eq. (5.9) $u_{r,\ \max\ (IDR)} = 0.588$ m and hence $\mu_{r,\ IDR} = 0.058/0.079 = 7.44$. Use of Eq. (5.12) provides $\mu_{r,\ \theta} = 1 + 0.79$ $(11 - 1) = 8.90$ and thus $\mu_{r,\ d} =$ min (7.44,8.90) = 7.44 indicating that drift controls the CP level design. Using the displacement spectrum of Fig. 5.14a for the MCE, the maximum displacement turns out to be $0.30 \cdot 1.4 = 0.42$ m, which is lower than the 0.588 m. Thus, the revised $u_{r,\ \max\ (IDR)} = 0.42$ m and the revised target values for *IDR* and μ_θ become 3.57% and 6.46, respectively, which satisfy the limits of Table 5.1. The $\mu_{r,\ d}$ can be determined from Eq. (5.6) and found to be equal to 0.42/0.079 = 5.31. Thus, q can be found from Eq. (5.13) to be equal to 1 + 1.39 $(5.31 - 1)^{1.05}1.25^{-0.06} = 7.36$.

From the pseudo-acceleration design spectra of Fig. 5.14b and $T_1 = 1.47$ s, one can determine the design base shears V for the three performance levels IO, LS and CP as follows:

$$V_{IO}/M = S_\alpha^{IO}/q_{IO} = 0.90/1.00 = 0.90\,\text{m/s}^2$$

$$V_{LS}/M = S_\alpha^{LS}/q_{LS} = 3.70/4.60 = 0.80\,\text{m/s}^2$$

$$V_{CP}/M = S_\alpha^{CP}/q_{CP} = 5.40/7.36 = 0.73\,\text{m/s}^2$$

where M is the mass of the framed building.

The above results clearly indicate that the IO performance level controls the design and thus the member dimensions provided in Table 5.11 are the final dimension of the present MRF with setbacks according to the HFD method.

The FBD seismic design method (EC8 2004; FEMA 445 2006) is also used here with a behavior factor $q = 6.5 \times 0.8 = 5.2$. The LS performance level controls the frame design. The cross-sectional dimensions and the first three natural periods of the designed frame are given in Table 5.11. The frame behaves elastically at the IO level and experiences $u_{r,y} = 0.079$ m and $IDR_{max} = 0.58\%$, $u_{r,max} = 0.316$ m and $IDR_{max} = 2.32\%$ and $u_{r,max} = 0.474$ m and $IDR_{max} = 3.48\%$, under the FOE, DBE and MCE respectively.

Table 5.12 provides NLTH analyses response results together with those estimated by the HFD and the FBD methods and serves the purpose of comparing the response results of these two methods. It is observed that the accuracy of the HFD response results is higher than that of the FBD results, as in the previous example. In this example both methods provide designs with the same total weigh of steel equal to 916 kN.

Table 5.12 NLTH analyses response results and design estimations for the eight-storey steel space MRF with setbacks (TH=NLTH analysis results; EST = estimations of HFD or FBD (EC8) methods)

	HFD method						FBD (EC8) method					
	FOE		DBE		MCE		FOE		DBE		MCE	
	TH	EST	TH	EST	TH	EST	TH	EST	TH	EST	TH	EST
IDR (%)	0.60	0.58	2.01	2.07	3.02	2.96	0.60	0.58	2.01	2.32	3.02	3.48
$u_{r,\max}$ (m)	0.086	0.079	0.278	0.246	0.418	0.350	0.086	0.079	0.278	0.316	0.418	0.474
μ_θ	1.00	1.00	3.08	3.66	4.39	5.33	1.00	1.00	3.08	–	4.39	–

5.9 Seismic Design Examples for Steel Plane Frames

This section serves to illustrate the HDF seismic design method and demonstrate its merits by means of two representative examples dealing with one regular plane MRF and one regular plane CBF (with X-braces), as shown in Fig. 5.17. Additional examples dealing with regular plane MRFs, regular plane CBFs (with chevron braces) and irregular plane MRFs (with setbacks or mass irregularities in height) can be found in Tzimas et al. (2013).

The dead (G) and live (Q) loads are assumed to be $G = 4.5$ kN/m^2 and $Q = 3.5$ kN/m^2, while self-weight of the steel members is taken into account during the analysis. The design load combinations are according to EC8 (2004) and EC3 (2009), i.e., the gravity load combination $1.35G + 1.5Q$ and the seismic loading combination $G + 0.3Q \pm E$, where E stands for seismic loads. The steel grade for beams, columns and braces is S275, while the design *PGA* is 0.35 g and soil type is B. The frames are designed using the HFD seismic design method. Modal response spectrum analysis is performed with the aid of SAP 2000 (2010) software and in accordance with the capacity and ductility rules of EC8 (2004). The three performance levels IO, LS and CP and their associated earthquakes FOE, DBE and MCE of Sect. 5.8 are also considered here. Table 5.1, which provides limit values for performance metrics is also used here.

5.9.1 Five-Storey Regular Steel Plane MRF

Here the HFD seismic design method is applied to the design of the five-storey three-bay regular plane MRF of Fig. 5.17a. The frame is initially designed elastically for the FOE, resulting in "HEB400-IPE400, HEB400-IPE500, HEB360-IPE400, HEB360-IPE400, HEB340-IPE300" sections for the five stories of the frame from

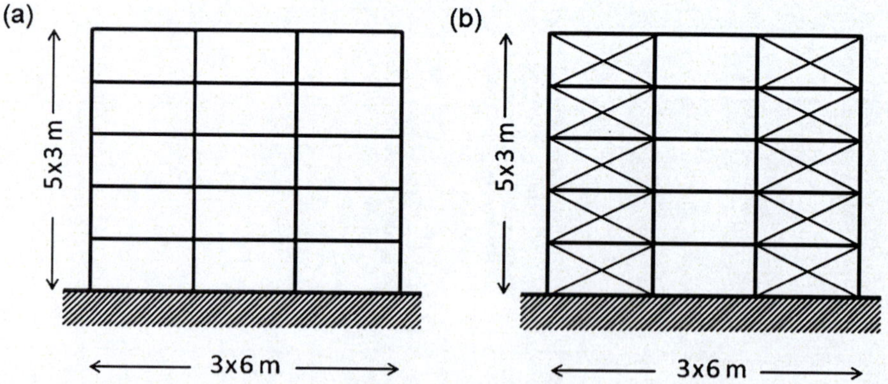

Fig. 5.17 Regular plane steel frames (**a**) MRF; (**b**) CBF (X-braced)

the first to the fifth one, respectively. Section profiles HEB and IPE are used for columns and beams, respectively. The roof displacement at first yielding under the FOE is determined to be $u_{r,\ y} = 0.10m$. The values of the stiffness and strength parameters of the frame are found from Eq. (5.7) to be $\rho = 0.2$ and $\alpha = 2.4$, respectively, while the fundamental period of the designed structure is found to be equal to 1.74 s.

The target values of the *IDR* and μ_θ for the LS performance level are found from Table 5.1 to be 2.5% and 7.0, respectively. Parameter β is computed from Eq. (5.19) as equal to 0.73. Use of Eq. (5.18) results in $u_{r,\ \max\ (IDR)} = 0.73 \cdot 2.5\,\% \cdot (5 \cdot 3) = 0.27$ m, while use of Eq. (5.21) leads to $\mu_{r,\ \theta} = 1 + 1.35 \cdot (7 - 1) \cdot 0.86 \cdot 2.40^{0.43} \cdot 5^{-0.31} = 6.6$. Thus, from Eq. (5.20) $u_{r,\ \max\ (\mu)} = 0.1 \cdot 6.6 = 0.66$ m. It is then found that $u_{r,\ \max\ (d)} = \min\ (0.27{,}0.66) = 0.27$ m and thus, drift controls the LS level design. From Fig. 5.14a, one obtains for $T = 1.74$ s $u_{r,\ \max} = 0.225$ m for the SDOF system and hence $0.225 \cdot 1.4 = 0.315$ m for the five-storey frame. Thus, since 0.315 m > 0.270 m, $u_{r,\ \max\ (d)} = 0.27$ m and there is no need to revise the target values of the *IDR* and μ_θ. Hence $\mu_{r,\ d} = 0.27/0.10 = 2.7$ and using Eq. (5.22) one can find $q = 1 + 1.39(2.7 - 1) = 3.4$. The DBE design spectrum ordinates of Fig. 5.14b are divided by 3.4 and the design yields "HEB450-IPE400, HEB450-IPE450, HEB400-IPE450, HEB400-IPE400, HEB360-IPE360" sections for the frame. The new values of the input variables are $u_{r,\ y} = 0.095m$, $\rho = 0.22$, $\alpha = 3$ leading to a new value of $q = 3.5$ which, being very close to the previous value of 3.4, does not change the previously obtained sections. It is observed that using the values $q = 3.5$ and $\alpha = 3$ in Eqs. (5.21) and (5.22), there results $\mu_\theta = 2.41$. This frame remains elastic under the IO level since it has larger sections than those of the frame initially designed for that level. Elastic analysis for the IO level results in $IDR = 0.65\%$ and $u_{r,\ y} = 0.095$ m. It is observed that 0.65 % < 0.70%, which is the limit value of Table 5.1 for the IO level and 0.095 < 0.10 m as it should be.

The same initial values ($u_{r,\ y} = 0.10$ m, $\rho = 0.2$, $\alpha = 2.4$, $T = 1.74$ s and $\beta = 0.73$) previously used for the LS design level are also used here for designing the frame for the CP level under the MCE. The target values of IDR_{max} and $\mu_{\theta,\ max}$ for the CP level are found from Table 5.1 to be 5% and 9, respectively. The first round of calculations proves again that drift controls the design. Figure 5.14a for the MCE and a SDOF system with $T = 1.74$ s provides $u_{r,\ max} = 0.340$ m. Thus, the $u_{r,\ max}$ for the five-storey frame becomes $0.340 \cdot 1.4 = 0.476$ m. On the other hand, from Eqs. (5.19) and (5.18) with $\beta = 0.73$, one obtains $u_{r,\ \max\ (IDR)} = 0.73 \cdot (3 \cdot 5) \cdot 0.05 = 0.548$ m, which is higher than 0.476 m. This implies that the target IDR_{max} and $\mu_{\theta,\ max}$ values need to be revised, using as $u_{r,\ \max\ (IDR)} = 0.476$ m. Employing Eq. (5.18) one determines the demand $IDR_{max} = 4.3\%$, which is less than the demand value of 5.0% of Table 5.1. Furthermore, for $\mu_{r,\ d} = 0.476/0.10 = 4.76$, Eq. (5.22) yields $q = 6.22$.

From the pseudo-acceleration design spectra of Fig. 5.14b and $T_1 = 1.47$ s, one can determine the design base shears V for the three performance levels IO, LS and CP as follows:

$$V_{IO}/M = S_{\alpha}^{IO}/q_{IO} = 0.95/1.00 = 0.95\,\mathrm{m/s^2}$$

$$V_{LS}/M = S_{\alpha}^{LS}/q_{LS} = 3.50/3.50 = 1.00\,\mathrm{m/s^2}$$

$$V_{CP}/M = S_{\alpha}^{CP}/q_{CP} = 5.20/6.22 = 0.84\,\mathrm{m/s^2}$$

where M is the mass of the frame. The above results clearly indicate that the LS performance level controls the design and thus the member sections "HEB450-IPE450, HEB400-IPE450, HEB400-IPE400, HEB400-IPE400, HEB360-IPE360" are the final ones according to the HFD method.

The FBD method of EC8 (2004) is also used for the design of the frame considered in this example. The method starts with a strength checking at the LS performance level with $q = 6.5$ and leads to the "HEB240-IPE300, HEB240-IPE300, HEB220-IPE270, HEB220-IPE270, HEB200-IPE240" section selection. Then, a deformation checking is conducted for the IO performance level. Use of the equal-displacement rule provides a value of $IDR = 1.23\%$, which is larger than the 0.7% drift limit of Table 5.1. After a number of iterations, the section selection "HEB400-IPE400, HEB400-IPE500, HEB360-IPE400, HEB360-IPE400, HEB340-IPE300" leads to a response with $u_{r,\,\max} = 0.10$ m, $IDR = 0.7\%$ and $\mu_\theta = 1.0$, which satisfies the limits of Table 5.1. Thus, the IO performance level controls the design. The above two frame designs by the HFD and FBD (EC8) methods are compared on the basis of their response results obtained by NLTH analyses involving 10 artificial ground motions compatible with the EC8 (2004) elastic response spectra and soil type B conducted with the Ruaumoko (Carr 2005) computer program.

Table 5.13 provides the mean values of various maximum response quantities obtained by NLTH analyses together with the corresponding estimated values obtained by the HFD and FBD (EC8) methods. It is observed that both methods (with the exception of one *IDR* value provided by the FBD (EC8)) provide deformation values less than the limiting ones in Table 5.1 and that the estimated values by the HFD method are closer to the NLTH analyses values than those of the FBD (EC8) method, which employs the equal displacement rule.

In particular, the FBD (EC8) usually overestimates maximum roof displacements and underestimates *IDR*s. It is worth mentioning that the *IDR* value for the DBE is estimated by the FBD (EC8) to be 2.45%, which is less than the limit value of 2.50% of Table 5.1. However, this *IDR* restriction is not really satisfied because the true *IDR* value is 3.00% > 2.50% implying unsafe design by the FBD (EC8) method. Finally, one can also observe that the FBD (EC8) method cannot estimate μ_θ values like the HFD method. On the other hand, the resulting weights of the two designs are rather close with the weight of the HFD design to be higher than that of the FBD (EC8) design.

Table 5.13 Deformational response results for the 5-storey plane steel MRF designed by the HFD and FBD (EC8) methods obtained by NLTH (TH) analyses and comparison with estimations (EST) by those two design methods

	HFD						FBD (EC8)					
	FOE		DBE		MCE		FOE		DBE		MCE	
	TH	EST	TH	EST	TH	EST	TH	EST	TH	EST	TH	EST
IDR (%)	0.63	0.65	2.4	2.5	3.75	3.65	0.67	0.70	3.00	2.45	4.20	3.7
u_{rmax} (m)	0.110	0.090	0.260	0.270	0.430	0.410	0.120	0.100	0.310	0.350	0.460	0.530
μ_θ	1.00	1.00	2.70	2.40	3.60	3.70	1.00	1.00	2.80	–	3.80	–

5.9.2 Five-Storey Regular Steel Plane X-Braced Frame

Here the HFD seismic design method is applied to the design of the five-storey steel plane X-braced frame of Fig. 5.17b. The seismic action is associated with a $PGA_{DBE} = 0.35$ g. Beam cross sections were selected on the basis of the gravity load combination and were found to be IPE270 for all the stories of the frame.

The frame is initially designed elastically for the FOE resulting in the "HEB300-TUBOD244.5x5.4, HEB300-TUBOD244.5x5.4, HEB280-TUBOD219.1x5, HE280-TUBOD219.1x5, HEB260-TUBOD193.7x4.5" sections for the five stories of the frame, from the first to the fifth one, respectively. Section profiles HEB and TUBO (CHS) are used for columns and diagonal braces, respectively. Parameters $\overline{\lambda}$ and a of the frame are computed from Eq. (5.25) and found to be equal to 1.02 and 0.025, respectively. The natural period of the frame $T_1 = 0.70$ s, while the characteristic period $T_c = 0.50$ s. It is also found that under the FOE $u_{r,\,y} = 0.037$ m. The target values of the IDR_{max} and μ_{cb} for the CP performance level are found from Table 5.1 equal to 2.0% and 10.86, respectively. Use of Eq. (5.23) yields $\beta = 0.59$ and thus, one can calculate from Eq. (5.18) $u_{r,\,\max\,(IDR)} = 0.59 \cdot 2.0\,\% \cdot (5 \cdot 3) = 0.177$ m. On the other hand, one can prove that Eq. (5.20) provides a larger value of the roof displacement and therefore, drift controls the CP level design. The u_{max} of a SDOF system with natural period equal to $T_1 = 0.70$ s is found from Fig. 5.14a for the MCE to be equal to 0.14 m. Thus, the $u_{r,\,\max}$ of the frame will be $1.4 \cdot 0.14 = 0.196$ m > 0.177 m. This implies that the target values of the IDR_{max} and μ_{θ} do not have to be revised and that $\mu_{r,\,d} = 0.177/0.037 = 4.78$. Use of Eq. (5.28) provides the required behavior factor $q = 6.3$. The MCE design spectrum of Fig. 5.14b is reduced by this factor and the design yields "HEB260-TUBOD219.1x5, HEB260-TUBOD219.1x5, HEB240-TUBOD219.1x5, HE240-TUBOD193.7x4.5, HEB220-TUBOD168.3x4" sections for the frame. The new values of the input variables are $u_{r,\,y} = 0.035m$, $\overline{\lambda} = 1.02$ and a $= 0.013$ with the last two ones computed from Eq. (5.25). A second round of calculations provides a value of $q = 6.5$. This value is very close to the previous one and hence the previously found sections do not change. This finalizes the design for the CP performance level. Similar calculation for the IO and LS yielded lighter frames than the one for the CP performance level. The frame experiences an $IDR = 2.0\%$ and $\mu_{cb} = 4.8$ (computed from Eq. (5.26) for known $\mu_{r,\,\theta}$) under the MCE, both in agreement with the limits of Table 5.1. The q factor of the frame under the DBE can be computed as $(PGA_{DBE}/PGA_{MCE}) \cdot q_{MCE} = (0.35/0.53) \cdot 6.5 = 4.29$. Use of this q value can lead to the response values $u_{r,\,max} = 0.088$ m, $IDR_{max} = 1.1\%$ and $\mu_{cb} = 2.22$, which satisfy the limit values of Table 5.1 for the LS level. Similarly, for the FOE, the q factor can be computed as $(PGA_{FOE}/PGA_{MCE}) \cdot q_{MCE} = (0.10/0.53) \cdot 6.5 = 1.23$. The value of this q factor being very close to 1.0 indicates an almost elastic behavior. Thus, an elastic analysis under the elastic FOE spectrum can lead to response values $u_{r,\,max} = 0.036$ m, $IDR_{max} = 0.29\%$ and $\mu_{cb} = 1.0$, which satisfy the limit values of Table 5.1 for the IO level.

The FBD method of EC8 (2004) is also used for the design of the frame considered in this example. The method starts with a strength checking at the LS performance level with $q = 4$, which leads to the same sections as those obtained by the employment of the HFD method. Deformation checking at the IO performance level reveals that the frame satisfies drift requirements and thus strength demands control the design. It is found that the frame designed by the FBD (EC8) method experiences $u_{r,\ max} = 3.5 \cdot 0.036 = 0.126m$ and $IDR_{max} = 3.5 \cdot 0.029 = 1.1\%$ under the DBE and $u_{r,\ max} = 5.3 \cdot 0.036 = 0.19m$ and $IDR_{max} = 5.3 \cdot 0.0029 = 1.54\%$ under the MCE.

The above two frame designs by the HFD and FBD (EC8) methods are compared on the basis of their response results obtained by NLTH analyses as in the previous example conducted with the Ruaumoko (Carr 2005) computer program. Table 5.14 provides the mean values of various maximum response quantities obtained by NLTH analyses together with the corresponding estimated values obtained by the HFD and FBD (EC8) methods. It is observed that both methods provide deformation values less than the limiting ones of Table 5.1, apart from those *IDR* values associated with the MCE, which are slightly higher than the limiting ones. In general, the HFD estimations are slightly better than the FBD ones because the FBD employs the equal displacement rule. Furthermore, the FBD cannot estimate member ductilities.

5.10 Conclusions

On the basis of the preceding developments and application results concerning the hybrid force/displacement (HFD) seismic design method for steel building structures, the following general conclusions can be stated:

1. The HFD method combines the advantages of the well-known FBD and DBD seismic design methods. Thus, it controls both nonstructural and structural deformation through interstorey drift ratio (*IDR*) and local ductility, respectively. This is accomplished by using a behavior factor q, which depends on these deformation metrics and the three basic performance levels of immediate occupancy (IO), life safety (LS) and collapse prevention (CP). Explicit expressions for q and some key deformation quantities are obtained through extensive parametric studies involving many frames under many seismic motions. The present version of the method can be applied to both plane and space moment resisting frames regular and irregular in plan view, elevation and mass distribution as well as to plane X-braced frames.
2. The main characteristics of the HFD design method are: (i) treats drift demands as input variables for the initiation of the design process as the DBD method and in addition treats ductility demands; (ii) does not use a substitute single-degree-of-freedom system, which reduces modeling accuracy, as it is the case with the DBD method; (iii) makes use of the conventional elastic response pseudo-acceleration

Table 5.14 Deformational response results for the 5-storey plane steel CBF designed by the HFD and FBD (EC8) methods obtained by NLTH (TH) analyses and comparison with estimations (EST) by those two design methods

	HFD						FBD (EC8)					
	FOE		DBE		MCE		FOE		DBE		MCE	
	TH	EST	TH	EST	TH	EST	TH	EST	TH	EST	TH	EST
IDR (%)	0.27	0.29	1.20	1.10	2.20	2.00	0.27	0.29	1.20	1.00	2.20	1.54
u_{rmax} (m)	0.038	0.036	0.090	0.090	0.170	0.160	0.038	0.036	0.090	0.130	0.170	0.190
μ_{cb}	1.20	1.00	2.10	2.22	5.00	4.80	1.20	1.00	2.10	–	5.00	–

spectrum analysis in conjunction with a behavior factor q; (iv) it is a truly performance-based seismic design method unlike the FBD of EC8 method, which considers only two performance levels.

3. The main advantage of the HFD design method over the FBD of EC8 is that, as revealed by nonlinear time history (NLTH) analyses, the former method estimates inelastic deformation with high accuracy and not in an approximate manner (using the equal displacement rule) as it is the case with the latter method. Thus, underestimation of *IDR* can falsely imply satisfaction of its limit value, while in reality this limit value has been exceeded. Another advantage of the HFD design method over the FBD method of EC8 is the ability to identify the performance level which controls the design and to provide accurate predictions not only for the maximum values of *IDR* but also for maximum values of local rotational ductility and roof displacement. Resulting total weights of the designed frames by the HFD design method and the FBD method of EC8 were found to be very close with the former method to lead, in general, to heavier sections.
4. For space regular MRFs, accidental eccentricity is taken into account. However, this eccentricity slightly affects the expression for the behavior factor q in terms of the design ductility. HFD design equations for plane regular MRFs can also be used for corresponding space regular MRFs with small eccentricities. However, for medium to high eccentricities and relatively low values of the stiffness parameter ρ, only equations for space MRFs are recommended. In general, space regular MRFs require higher ductility demands and lower behavior factors than the corresponding plane ones.
5. The torsional response component in the irregular in plan view MRFs considered here is smaller than the one in regular frames with a 5% accidental eccentricity. In contrast, the torsional response component of the MRFs with setbacks considered here can be more than two times the one of the corresponding regular MRFs with 5% accidental eccentricity. Frames with mass discontinuity in elevation have higher ductility demands and satisfy the various performance levels for a lower seismic intensity than the corresponding regular frames.

References

ASCE 41–13 (2014) Seismic evaluation and retrofit of existing buildings. American Society of Civil Engineers, Reston, VI

ASCE/SEI 41-17 (2017) Seismic evaluation and retrofit of existing buildings. American Society of Civil Engineers, Reston, VI

ASCE/SEI 7-10 (2010) Minimum design loads and associated criteria for buildings and other structures. American Society of Civil Engineers, Reston, VI

ASCE/SEI 7-16 (2017) Minimum design loads and associated criteria for buildings and other structures. American Society of Civil Engineers, Reston, VI

Bazeos N (2009) Comparison of three seismic design methods for plane steel frames. Soil Dyn Earthq Eng 29:553–562

Carr AJ (2005) Ruaumoko 2D and 3D: programs for inelastic dynamic analysis. Theory and user guide to associated programs. Department of Civil Engineering, University of Canterbury, Christchurch

Chopra AK (2007) Dynamics of structures: theory and applications in earthquake engineering, 3rd edn. Prentice-Hall, Upper Saddle River, NJ

De Luca F, Iervolino I, Cosenza E (2009) Unscaled, scaled, adjusted and artificial spectral matching accelerograms: displacement and energy-based assessment. In: Proceedings of 13th Congress of the Italian National Association of Earthquake Engineering (ANIDIS), Bologna, Italy

Dimopoulos AI, Bazeos N, Beskos DE (2012) Seismic yield displacements of plane moment resisting and X-braced steel frames. Soil Dyn Earthq Eng 41:128–140

EC3 (2009) Eurocode 3, Design of steel structures – Part 1–1: general rules and rules for buildings, EN 1993-1-1. European Committee for Standardization (CEN), Brussels

EC8 (2004) Eurocode 8, Design of structures for earthquake resistance, Part 1: general rules, seismic actions and rules for buildings, EN 1998-1-1. European Committee for Standardization (CEN), Brussels

FEMA 273 (1997) NEHRP guidelines for the seismic rehabilitation of buildings. Federal Emergency Management Agency, Washington, DC

FEMA 445 (2006) Next-generation performance-based seismic design guidelines. Federal Emergency Management Agency, Washington, DC

Gupta A, Krawinkler H (1999) Seismic demands for performance evaluation of steel moment resisting frame structures. Report no 132. John A Blume Earthquake Engineering Center, Department of Civil Engineering, Stanford University, Stanford, CA

Karabalis DL, Cokkinides GJ, Rizos DC, Mulliken JS (1993) An interactive computer code for generation of artificial earthquake records. In: Khozeimeh K (ed) Computing in civil engineering. American Society of Civil Engineers, New York, pp 1122–1155

Karavasilis TL, Bazeos N, Beskos DE (2006) A hybrid force/displacement seismic design method for plane steel frames. In: Mazzolani FM, Wada A (eds) Proceedings of 5th International Conference on the Behavior of Steel Structures in Seismic Areas (STESSA 2006), Yokohama, Japan. Taylor and Francis, London, pp 39–44

Karavasilis TL, Bazeos N, Beskos DE (2007a) Behavior factor for performance-based seismic design of plane steel moment resisting frames. J Earthq Eng 11:531–559

Karavasilis TL, Bazeos N, Beskos DE (2007b) Estimation of seismic drift and ductility demands in plane regular X-braced steel frames. Earthq Eng Struct Dyn 36:2273–2289

Karavasilis TL, Bazeos N, Beskos DE (2008a) Drift and ductility estimates in regular steel MRF subjected to ordinary ground motions: a design oriented approach. Earthquake Spectra 24:431–451

Karavasilis TL, Bazeos N, Beskos DE (2008b) Seismic response of plane steel MRF with setbacks: estimation of inelastic deformation demands. J Constr Steel Res 64:644–654

Karavasilis TL, Bazeos N, Beskos DE (2009) Estimation of seismic inelastic deformation demands in plane steel MRF with vertical mass irregularities. Eng Struct 30:3265–3275

Karavasilis TL, Makris N, Bazeos N, Beskos DE (2010) Dimensional response analysis of multistorey regular steel MRF subjected to pulse-like earthquake ground motions. J Struct Eng ASCE 136:921–932

MATLAB (2009) The language of technical computing, version 2009a. The Mathworks Inc., Natick, MA

PEER (2009) Pacific Earthquake Engineering Research Center, Strong Ground Motion Database, Berkeley, CA. http://peer.berkeley.edu/

Priestley MJN, Calvi GM, Kowalsky MJ (2007) Displacement-based seismic design of structures. IUSS Press, Pavia

SAP 2000 (2010) Structural analysis program 2000, static and dynamic finite element analysis of structures, Version 14. Computers and Structures Inc, Berkeley, CA

Stamatopoulos H, Bazeos N (2011) Seismic inelastic response and ductility estimation of steel planar chevron-braced frames. In: Boudouvis AG, Stavroulakis GE (eds) Proceedings of 7th GRACM International Congress on Computational Mechanics, Athens, Greece

Tzimas AS (2013) A new hybrid force/displacement method for seismic design of space steel structures. Ph.D. Thesis, Department of Civil Engineering, University of Patras, Patras, Greece (in Greek)

Tzimas AS, Karavasilis TL, Bazeos N, Beskos DE (2013) A hybrid force-displacement seismic design method for steel building frames. Eng Struct 56:1452–1463

Tzimas AS, Karavasilis TL, Bazeos N, Beskos DE (2017) Extension of the hybrid force/displacement (HFD) seismic design method to 3D steel moment-resisting frames. Eng Struct 147:486–504

Tzimas AS, Skalomenos KA, Beskos DE (2020) A hybrid seismic design method for steel irregular space moment resisting frames. Journal of Earthquake Engineering, Online. https://doi.org/10.1080/13632469.2020.1733140

Vamvatsikos D, Cornell CA (2002) Incremental dynamic analysis. Earthq Eng Struct Dyn 31:491–514

Zotos PC, Bazeos N (2009) Estimation of seismic response in planar X-braced multi-storey steel frames. In: Papadrakakis M, Lagaros ND, Fragiadakis M (eds) Proceedings of Computational Methods in Structural Dynamics and Earthquake Engineering (COMPDYN 2009), Rhodes, Greece

Chapter 6
Ductility-Based Plastic Design

Abstract A seismic design method for plane steel moment resisting frames is presented. The method designs moment frames with desired ductility and failure mode. The failure mode is of the global type, which ensures a global ductility supply and energy dissipation capacity of these frames. Use is made of first order and second order plastic analysis. Beam sections are first designed to resist vertical loads and the column sections are then determined on the basis of limit analysis for the global collapse mechanism of the frame under vertical load and lateral distributed load applied statically. The distributed load is of the inverted triangle type and can be associated with the collapse prevention performance level. Using second order plastic analysis, one is able to determine during the design the plastic rotation capacity of beams and beam-columns. The whole design procedure uses simplified analytical expressions and thus, one can complete the design by hand. Numerical examples associated with the seismic design of plane steel moment resisting frames are presented and discussed.

Keywords Plastic design method · Seismic ductility design · Plane steel moment resisting frames · Limit analysis · Global analysis · Global collapse mechanism · Lateral static seismic load

6.1 Introduction

Even though current seismic design codes recognize the philosophy of multi performance level design and accept that a structure should resist minor earthquakes almost without damage, moderate earthquakes with repairable damage and major earthquakes without collapse, in reality they consider only a strength checking at the life safety (LS) performance level under the design earthquake and a displacement checking at the immediate occupancy (IO) performance level under the frequent earthquake. The strength checking is based on static elastic analysis with inertia forces obtained from the elastic design spectrum by dividing its ordinated by the behavior (or strength reduction) factor q in order to take into account seismic energy

G. A. Papagiannopoulos et al., *Seismic Design Methods for Steel Building Structures*, Geotechnical, Geological and Earthquake Engineering 51,
https://doi.org/10.1007/978-3-030-80687-3_6

dissipation and ductility. The displacement checking is based on the equal displacement rule. Both the factor q and the equal displacement rule are approximate. The so designed structure is assumed to automatically satisfy the requirements of the collapse prevention (CP) performance level provided that the suggested by the codes detailing provisions are satisfied.

However, it has been observed, that the above procedure does not always lead to the desired failure mode and level of ductility. Inelastic behavior of framed structures is associated with the number and location of the developed plastic hinges and hence with the type of collapse mechanism. Partial collapse mechanisms, like the soft storey mechanism, are characterized by low energy of dissipation and ductility and demonstrate their inability to exploit all the available ductility capacity of the structure and its potential for energy dissipation. Seismic codes like EC8 (1994, 2004) have recognized the above facts and have introduced the concept of "capacity design" in conjunction with that of dissipative and non-dissipative parts in a framed structure, as it has been described in Chaps. 1 and 3. The selection of the dissipative parts where inelastic deformation and hence plastic hinges will develop, has to be done in such a way as the resulting collapse mechanism to exploit the maximum of dissipation of energy and ductility capacity of the structure. For the case of plane steel moment resisting frames (MRFs) this is the global collapse mechanism associated with the rule of "weak beams-strong columns". This "member hierarchy criterion" is expressed in EC8 (1994) by the requirement that the sum of flexural strength of the columns at a joint is greater than the sum of the flexural strength of the beams framing to that joint. An improved criterion of the member hierarchy type was also proposed by Lee (1996) stating that the flexural strength of each column should be greater than the sum of the flexural strength of the beams framing to the joint multiplied by the factor $0.75\beta = 0.75 \cdot 1.1 = 0.825$.

The above member hierarchy criteria may be sufficient to ensure that a soft storey collapse mechanism will not develop, but they cannot secure that failure will occur in a global collapse mechanism. Apparently, Mazzolani and Piluso (1995, 1997) were the first to propose a seismic design method for plane steel MRFs capable of assuring the development of a global collapse mechanism. Their method is based on the kinematic theorem of plastic collapse and the study of $3n_s$ kinematically admissible collapse mechanisms (including the global one), where n_s is the number of stories. Selecting the beam sections on the basis of the known vertical load, the plastic section moduli of the columns are determined iteratively so that the kinematic multiplier of the global mechanism for the corresponding lateral (seismic) forces is less than those of the remaining $3n_s - 1$ mechanisms. This ensures, according to the upper bound theorem, that the true collapse mechanism is the global one, Using second order plastic analysis, Mazzolani and Piluso (1995, 1997) were also able to determine during the design process the plastic rotation capacity of beams and/or beam-to-column connections.

Ghersi et al. (1999) were able to greatly reduce the complexity of the member dimensioning part of the Mazzolani and Piluso (1995, 1997) method on the basis of some approximations introduced during the application of limit analysis. Selecting the beam sections of the basis of the known vertical load and the sections of the columns of the first storey with the aid of an empirical expression, the collapse multiplier for the global collapse mode can be easily obtained and from there all the flexural strengths of the remaining columns specified. From equilibrium considerations, shear and axial forces of the frame can be easily determined and thus the sections of the columns can be selected on the basis of the combined moment-axial force action. A comparison of design methods conducted by Ghersi et al. (1999) on the basis of a plane steel MRF with four bays and six stories revealed that for the same beam sections, the total weight of columns was minimum for the EC8 (1994) design, while it was +7%, +33% and + 55% of that for the Lee (1996), Ghersi et al. (1999) and the Mazzolani and Piluso (1997) designs, respectively. However, only the last two designs experienced a global collapse mechanism with the designs of EC8 (1994) and Lee (1996) experiencing partial collapse mechanisms.

Utilizing the information in the works of Ghersi et al. (1999) and Mazzolani and Piluso (1995, 1997) and combining it with the work of Gioncu and Petcu (1997, 2001) on the ultimate rotation capacity of beams and beam-columns under static and seismic loads, Gioncu and Mazzolani (2002), in their book on seismic ductility of steel structures, were able to develop a comprehensive methodology for ductility design of plane steel MRFs. This method is a multi-level design one and is characterized by its ability to be easily applied using hand calculations. At this point one should note that the name "Ductility-based plastic design method" given to the method of this chapter comes from the combination of the names "Plastic design" in Mazzolani and Piluso (1997) and "Ductiltiy design" in Gioncu and Mazzolani (2002). The theoretical foundations of the method are presented in some detail and two numerical examples serve to illustrate it and demonstrate its simplicity, efficiency and reliability.

6.2 Theoretical Foundations of the Method

This section briefly describes the plastic design methods of Ghersi et al. (1999) and Mazzolani and Piluso (1997) in the framework of the comprehensive methodology of Gioncu and Mazzolani (2002) for ductility design of plane steel MRFs.

6.2.1 Multi-Level Seismic Design

Performance-based seismic design of structures introduces a framework for multi-level criteria design methods (Fajfar and Krawinkler 1997). According to Gioncu (2000), a multi-level design method should consider three levels of design: the serviceability limit state (SLS) for minor frequent earthquakes (20 years return period) associated with no damage, the damageability limit state (DLS) for occasional moderate to strong earthquakes (475 year return period) associated with repairable damage in nonstructural parts and the ultimate or survivability limit

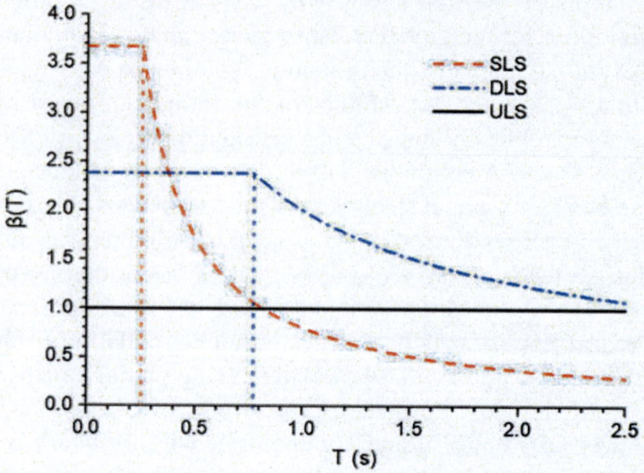

Fig. 6.1 Simplified design spectra for interplate earthquakes and three limit states as described by Eqs. (6.4)–(6.6)

state (ULS) for rare very strong earthquakes (970 years return period) associated with significant structural damage without collapse. The above three performance or limit states correspond more or less to the well-known IO, LS and CP performance levels of FEMA 356 (2000).

The peak ground acceleration (*PGA*) for the above three levels can be determined in terms of the return periods. Thus, according to Lee et al. (2000)

$$a/a_d = (p_r/p_{rd})^{0.28} \tag{6.1}$$

where a_d and p_{rd} are the *PGA* and the return period in years for the DLS (LS), respectively, and a and p_r are the corresponding quantities for any of the other two limit states. Thus, one can express the *PGA*'s a_s and a_u for the SLS and ULS, respectively, in terms of a_d in the forms

$$a_s = 0.412a_d \tag{6.2}$$

$$a_u = 1.22a_d \tag{6.3}$$

Equations (6.2) and (6.3) have slightly different forms (0.50 instead of 0.412 and 1.50 instead of 1.22) in other documents (FEMA 356 2000).

Simplified elastic design acceleration spectra (assuming T_B of EC8 (2004) equal to zero) for the above three limit states are presented in Fig. 6.1 for interplate earthquakes (strong and moderate earthquakes according to EC8 1994) and read for $S = 1$ and $\eta = 1$ as

$$\beta_s(T) = 0.8/T^{1.1} \quad \beta \leq 3.67 \tag{6.4}$$

$$\beta_d(T) = 2.0/T^{0.67} \quad \beta \leq 2.38 \tag{6.5}$$

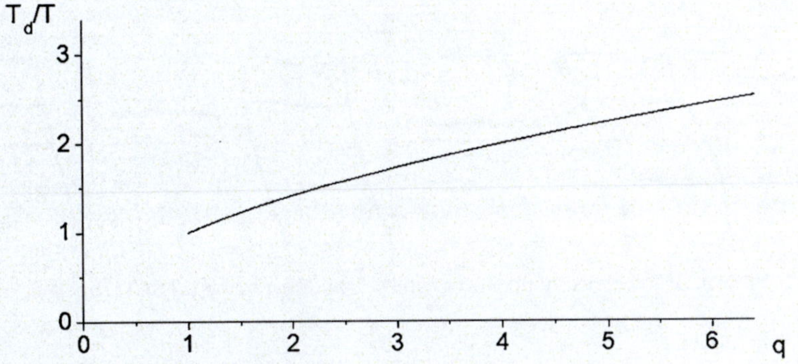

Fig. 6.2 Elongation of structural period of vibration due to deterioration

$$\beta_u(T) = 1.0 \tag{6.6}$$

where β is the ground acceleration over *its own a*/*g*, T is the structural period, S the soil coefficient and η the damping correction factor, while the subscripts s, d and u denote the SLS (IO), DLS (LS) and ULS (CP), respectively. Thus, the base shear seismic forces F_b resulting from spectra for different limit states (Fig. 6.1) take the form

$$F_{bs} = (a_s/g)\beta_s(T)W \tag{6.7}$$

$$F_{bd} = (a_d/g)\beta_d(T_d)W/q \tag{6.8}$$

$$F_{bu} = (a_u/g)\beta_u(T)W \tag{6.9}$$

where W is the structural weight, q is the behavior factor and T_d is the inelastic structural period, which increases over its elastic one T with increasing inelastic deformation (number of plastic hinges) according to the approximate relation $T_d/T = \sqrt{\mu} \cong \sqrt{q}$ (Priestley 1997), as shown in Fig. 6.2. This relation is based on the assumption that the displacement ductility μ equals to q for $T > 0.55$ s, which is usually the case for steel structures. Boushaba and Plumier (1988) studied the gradual increase of inelastic deformation of a five-storey two-bay plane steel moment resisting frame for increasing values of q and observed that even for $q = 6$, the structure is not a plastic collapse mechanism meaning that a high q factor does not assure the development of such a mechanism.

An important aspect in seismic design is the determination of the base shear force distribution along the height of the frames. Figure 6.3 shows such distributions for the three limit cases considered here. For the SLS where the behavior is elastic and the influence of higher mode effects is important, an inverse parabolic distribution seems reasonable (Fig. 6.3a). For the DLS due to the increasing of vibration and decreasing of the higher mode effects, a parabolic distribution seems appropriate (Fig. 6.3b). For the ULS for which the frame has become a collapse mechanism, an inverted triangular distribution is recommended (Fig. 6.3c).

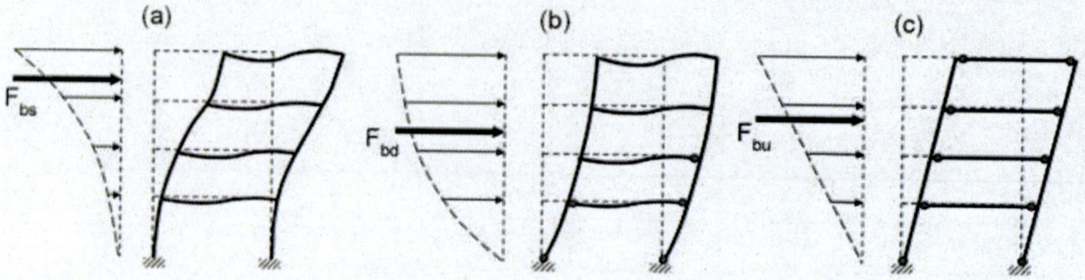

Fig. 6.3 Base shear force distributions along the height of the frame (**a**) SLS; (**b**) DLS; (**c**) ULS

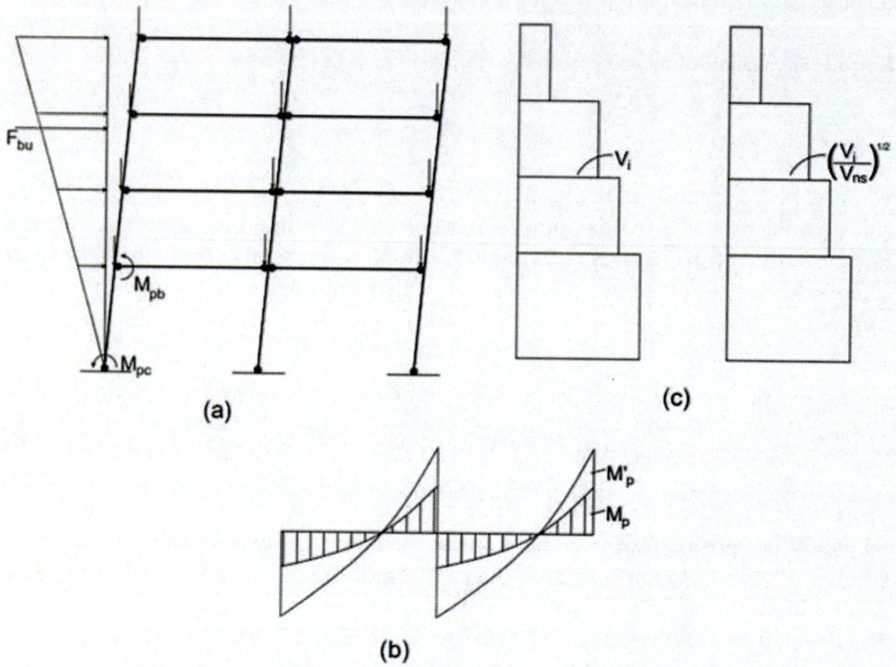

Fig. 6.4 Global mechanism of plane frame: (**a**) external forces and internal moments; (**b**) nominal and actual plastic moments of beams; (**c**) distribution of seismic shear forces V_i and $(V_i/V_{ns})^{0.5}$

6.2.2 Simplified Plastic Design

For a preliminary seismic design with the goal of creating a frame with strong columns and weak beams that can fail in a global collapse mechanism (Fig. 6.4a), one has first to determine the full plastic moment capacity of beams and then design the columns to be stronger than the beams. The cross section selection of the beams is based on the bending moment M_w of beams due to static vertical loads amplified by the factor m_h due to shear forces coming from the lateral (seismic) loads. Thus, the design nominal plastic moment for beams M_{pb} is computed as (Fig. 6.4b, c)

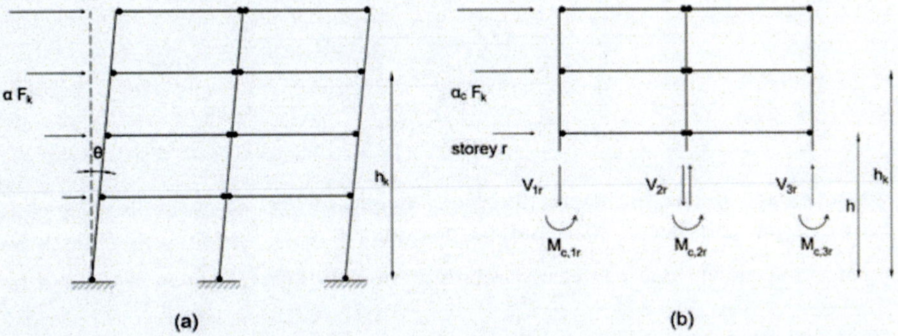

Fig. 6.5 Global collapse mechanism under lateral load: (**a**) equilibrium of the whole frame; (**b**) equilibrium of upper part of the frame

$$M_{pb} = m_h M_w \;, \;\; m_h = (V_i/V_{ns})^{0.5} \tag{6.10}$$

and the section modulus W_{pl} is determined from

$$M_{pb} = W_{pl} f_y \tag{6.11}$$

where V_i and V_{ns} are the storey shear forces corresponding to storey levels i and ns (roof level) (Leelataviwat et al. 1999) and f_y is the nominal yield strength of steel. The actual plastic moment for beams M'_{pb} is then obtained from (Fig. 6.4b)

$$M'_{pb} = m_y M_{pb} \;, \;\; m_y = f_{ya}/f_y \tag{6.12}$$

where f_{ya} is the actual yield strength for steel including overstrength and usually being taken as 1.1 (EC3 1992).

Among the methods described in the introduction of this chapter for capacity seismic design, i.e., the one of EC8 (1994) based on the hierarchy criterion, the one of Mazzolani and Piluso (1995, 1997) based on the global collapse mechanism method and the simplified one of Ghersi et al. (1999) based also on the global collapse mechanism, the last one of those three methods is adopted here due to its simplicity. Consider a plane steel frame with n_s storeys, n_c columns and n_b bays. Indices k, i and j stand for storey, column and bay, respectively, while $M_{pc,\,i1}$ and $M_{pb,\,jk}$ denote plastic moments of the ith column of the first storey and plastic moments of beams of the jth bay and kth storey, respectively.

On the basis of Fig. 6.5a depicting the global collapse mechanism under vertical and lateral (seismic) load, one can equate the energy dissipated by plastic hinges to the work of lateral forces and solve for the collapse multiplier a_c in the form (Ghersi et al. 1999)

$$a_c = \frac{\sum_{i=1}^{n_c} M_{c,i1} + \sum_{k=1}^{n_s} \sum_{j=1}^{n_b} 2M_{b,jk}}{\sum_{k=1}^{n_s} F_k h_k} \tag{6.13}$$

where F_k and h_k are the lateral force and height of the kth storey respectively. Provided that the lateral load can be determined from Eq. (6.9) and the term $\sum_{i=1}^{n_c} M_{c,i1}$ is known, one can easily obtain a_c from Eq. (6.13), since $M_{b,\ jk}$ can be computed from Eqs. (6.10)–(6.12). The term $\sum_{i=1}^{n_c} M_{c,i1}$ can be evaluated from the empirical relation

$$\sum_{i=1}^{n_c} M_{pc,i1} = (0.7 + 0.15n_s) \sum_{j=1}^{n_b} 2M_{pb,j1} \tag{6.14}$$

which has been obtained by Ghersi et al. (1999) through extensive numerical parametric studies involving pushover analyses of more than 100 frames. Consideration of the equilibrium of the upper part of the frame (Fig. 6.5b) results in the relation

$$\sum_{i=1}^{n_c} M_{c,ir} = a_c \sum_{k=r} F_k (h_k - h) - \sum_{k=r}^{n_s} \sum_{j=1}^{n_b} 2M_{b,jk} \tag{6.15}$$

where $h = h_r$ and $h = h_{r-1}$ correspond to the top and bottom section of columns. It is apparent that with known collapse multiplier a_c obtained by using Eqs. (6.13) and (6.14), one can determine from (6.15) the term $\sum_{i=1}^{n_c} M_{c,ir}$ but not every column plastic moment. However, pushover analysis have shown that for columns of the same section in every storey, bending moments in the ULS are very close for all columns. Thus, one can use the average value

$$M_{ac,ir} = \frac{1}{n_c} \sum_{i=1}^{n_c} M_{c,ir} \tag{6.16}$$

for all columns in every storey and thus determine all moment columns. The same assumption is also used when calculating the individual $M_{pc,\ i1}$ from Eq. (6.14). It should be noted that when calculating column moments in terms of beam moments through Eqs. (6.14) and (6.15), the beam moments are amplified by the overstrength factor $m_y = 1.1$ defined by Eq. (6.12b).

Finally, axial forces N_{rj} in columns can be computed from the relation

$$N_{rj} = N_w \pm \sum_{k=r}^{n_s} \frac{2M_{b,kj}}{L_j} \mp \sum_{k=r}^{n_s} \frac{2M_{b,kj+1}}{L_{j+1}} \tag{6.17}$$

where $N_w = wL_j/2$ is the axial force due to vertical distributed loads w. It should be noted that the two summation terms give zero for the case of internal columns. With known bending moments and axial forces in the columns, their appropriate cross sections can be selected.

In conclusion, the above design method is very simple and easy to use for section selection but it cannot be used for determining deformations, exactly because it employs first order limit analysis.

6.2.3 Required Plastic Rotation Capacity

The previously described simplified plastic design method of Ghersi et al. (1999) cannot determine deformation at the ULS or CP level associated with the global collapse mechanism exactly because it is a first order theory. The plastic design method of Mazzolani and Piluso (1995, 1997) is based on second order plastic analysis and thus can determine deformation and especially the required plastic rotation capacity at the ultimate state associated with the global mechanism of collapse. The following presentation from the Gioncu and Mazzolani (2002) book is based on the work of Mazzolani and Piluso (1995, 1997) and some simplifications of the authors of the book on available rotation capacity of plane steel MRFs.

Consider a plane steel MRF under constant vertical load W and earthquake lateral load of an inverted triangle type of distribution at tis ultimate limit state associated with the global collapse mechanism, as shown in Fig. 6.6. The total vertical load W is equal to the sum of the vertical loads w_k at every storey level k, while the lateral force

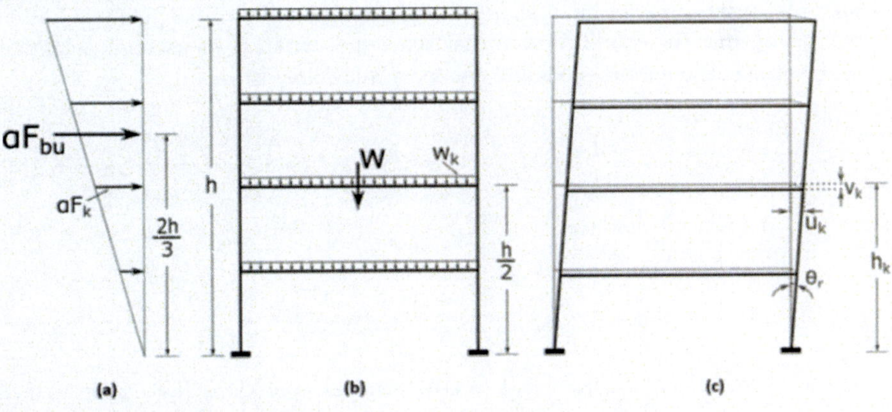

Fig. 6.6 Global plastic mechanism behavior: (**a**) horizontal loads; (**b**) vertical loads; (**c**) plastic mechanism deformations u_k, v_k, θ_r

F_k at level k is expressed in terms of the seismic base shear force at the ultimate limit state F_{bu}, as given by Eq. (6.9), in the form

$$aF_k = \frac{w_k h_k}{\sum_{k=1}^{n_s} w_k h_k} aF_{bu} \tag{6.18}$$

The horizontal and vertical resultant forces αF_{bu} and W are at distances $2\,h/3$ and $h/2$, respectively, from the base of the frame (Fig. 6.6a, b).

During deformation, all plastic hinges in beams and column bases experience a rotation θ_r. Thus, the horizontal and vertical displacements u_k and v_k, respectively, of storey k can be expressed as (Fig. 6.6c)

$$u_k = h_k \sin\theta_r, v_k = h_k(1 - \cos\theta_r) \tag{6.19}$$

or by taking into account that for small θ_r one has $\sin\theta_r \cong \theta_r$ and $\cos\theta_r \cong 1 - 0.5\theta_r^2$, as

$$u_k = h_k\theta_r \ , \ \ v_k = 0.5h_k\theta_r^2 \tag{6.20}$$

The displacement components of the horizontal resultant force are

$$u_F = (2/3)h\theta_r \ , \ \ v_F = (2/3)h0.5\theta_r^2 \tag{6.21}$$

and those of the vertical resultant force are

$$u_W = 0.5h\theta_r \ , \ \ v_W = 0.5h0.5\theta_r^2 \tag{6.22}$$

In order to determine the plastic rotation θ_r of the frame at the ultimate limit state of Fig. 6.6c, use is made of the principle of the minimum of the total potential of the frame stating that for equilibrium at that state, the variation of its total potential should be zero. The internal potential energy of the frame is

$$U = \left(\sum_{i=1}^{n_c} M_{pc,i1} + \sum_{k=1}^{n_s} \sum_{j=1}^{n_b} 2M_{pb,jk} \right) \theta_r \tag{6.23}$$

and its external potential is

$$V = (2/3)aF_{bu}h\theta_r + 0.5Wh0.5\theta_r^2 \tag{6.24}$$

The total potential of the frame is $P = U + V$ and the condition $\delta P = 0$ leads to the relation

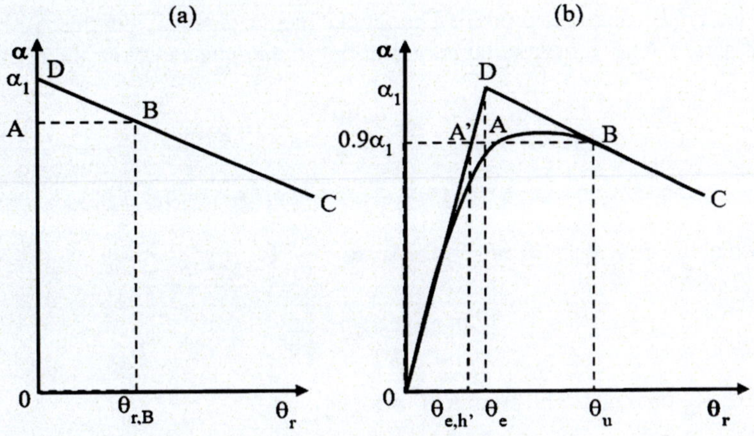

Fig. 6.7 Determination of rotation capacity: (**a**) ultimate rigid plastic rotation; (**b**) ultimate elastoplastic rotation

$$a = a_1 - a_2\theta_r \tag{6.25}$$

where

$$a_1 = (3/2)\left(\sum_{i=1}^{n_c} M_{pc,i1} + \sum_{k=1}^{n_s}\sum_{j=1}^{n_b} 2M_{pb,jk}\right)/F_{bu}h \tag{6.26}$$

$$a_2 = (3/4)W/F_{bu} \tag{6.27}$$

with $a_1 < 1.0$ in order to assure that a plastic mechanism is formed at the ultimate limit state. Eq. (6.25) represents a straight line with negative slope in the coordinate system (a, θ_r) as shown in Fig. 6.7a.

The above results have been obtained on the basis of second order rigid plasticity theory for which the total rotation θ of the frame in its global collapse mode is equal to the plastic rotation θ_r. For an ideal elasto-plastic behavior, the total rotation θ is equal to

$$\theta = \theta_e + \theta_r \tag{6.28}$$

where θ_e is the elastic part of the rotation (Fig. 6.7b). If $\theta_1 = \delta_1/h$ is the rotation due to the seismic force F_{bu} (with $a = 1$),

$$\theta_e = a_1\theta_1 \tag{6.29}$$

This rotation θ_1 can be determined approximately by considering the frame to be a single-degree-of-freedom system with its weight W at the roof level where the lateral load F_i is applied to produce a horizontal displacement δ_i. For this system with natural period T the fundamental period of the frame, one can write the relations

$$F_1 = K\delta_1, K = \left(\tfrac{2\pi}{T}\right)^2\left(\frac{W}{g}\right), g = 9.81\text{m/s}^2$$
$$\delta_1 = \theta_1 h \ , \ F_1 = (3/2)F_{bu} \tag{6.30}$$

from which an expression for θ_1 in the form

$$\theta_1 \cong 0.375T^2\frac{F_{bu}}{hW} \tag{6.31}$$

can be easily obtained. The empirical formula

$$T = 0.0724h^{0.8} \tag{6.32}$$

of Goel and Chopra (1997) for steel building frames in a slightly modified form to be valid in s and m for T and h, respectively, is now used in Eq. (6.31) in conjunction with Eqs. (6.6) and (6.9) to obtain

$$\theta_1 = 0.159 \cdot 10^{-2}h^{0.6}(a_u/g) \tag{6.33}$$

Thus, the elastic rotation θ_e can be computed from Eq. (6.29) with a_1 and θ_1 obtained from Eqs. (6.26) and (6.33), respectively.

As one passes from the rigid plastic behavior (Fig. 6.7a) to the elastoplastic one (Fig. 6.7b), the curve DBC becomes $ODBC$. However, in the elastoplastic case of Fig. 6.7b, the real behavior curve is the continuous one $OABC$, which can be approximately replaced by the tri-linear one $OABC$, where the plateau AB is defined for $a = 0.9a_1$. Since the required ultimate rotation is the one at point B, one has that

$$\theta_{rB} = 0.1(a_1/a_2) \ , \ \theta_{eA} = 0.9\theta_e \tag{6.34}$$

Thus, the required ultimate elastoplastic rotation is

$$\theta_{rr} = 0.1\left(\frac{a_1}{a_2} + \theta_e\right) \tag{6.35}$$

with the ratio a_1/a_2, given by Eqs. (6.26) and (6.27), being the predominant part and θ_e being small. The kinematic ductility μ_θ is defined as

$$\mu_\theta = \frac{\theta_{rr}}{\theta_{eA}} = 0.111\left(1 + \frac{a_1}{a_2\theta_e}\right) \tag{6.36}$$

Table 6.1 Coefficient γ_a for plastic rotation accumulation

a_1	Beams	Columns
0.2	3.5	4.0
0.4	2.5	3.0
0.6	1.5	2.0

At this point one should stress that the above results concerning the required rotation capacity have been obtained on the basis of a static monotonic approach without taking into account the influence of seismic actions.

On the basis of the discussion in Gioncu and Mazzolani (2002), the influence of seismic actions on the required ductility or rotation consists of the following factors:

1. the duration of strong ground motions and the number of strong pulses, which can have an important influence on seismic damage effects
2. the strain rate, which produces an increase of yield stress and thus an increase of the plastic moment and the required rotation. Thus the θ_{rr} of Eq. (6.35) increases and becomes

$$\theta_{rrsr} = \theta_{rr}\left(f_{ysr}/f_y \right) \tag{6.37}$$

where the subscript *sr* stands for strain rate. The ratio f_{ysr}/f_y can be about 1.30–1.50

3. the plastic rotation accumulation, which increases the required rotation θ_{rr} of Eq. (6.35) in accordance with the relation

$$\theta_{rrac} = \theta_{rr}\gamma_a \tag{6.38}$$

where the subscript *ac* stands for accumulation and γ_a is given for beams and columns as a function of a_1 by Table 6.1.

6.2.4 Available Plastic Rotation Capacity

The demonstration of the plastic rotation capacity at the member level is done in two stages as in the case of the required plastic rotation of the previous section: under static and monotonic actions and under seismic actions. The presentation is based on the Gioncu and Mazzolani (2002) book, which follows the work of Gioncu and Petcu (1997, 2001) utilizing local plastic mechanism of collapse for various types of beam and beam-column sections for the construction of simple approximate analytical expressions under the static monotonic and cyclic (seismic) loading. Due to space limitations, only simple I-section beams and beam-columns are discussed here. A complete presentation of all types of section cases can be found in the Gioncu and Mazzolani (2002) book, which also contains a computer code (DUCTROT-M) for easy and fast computations. A more detailed presentation of

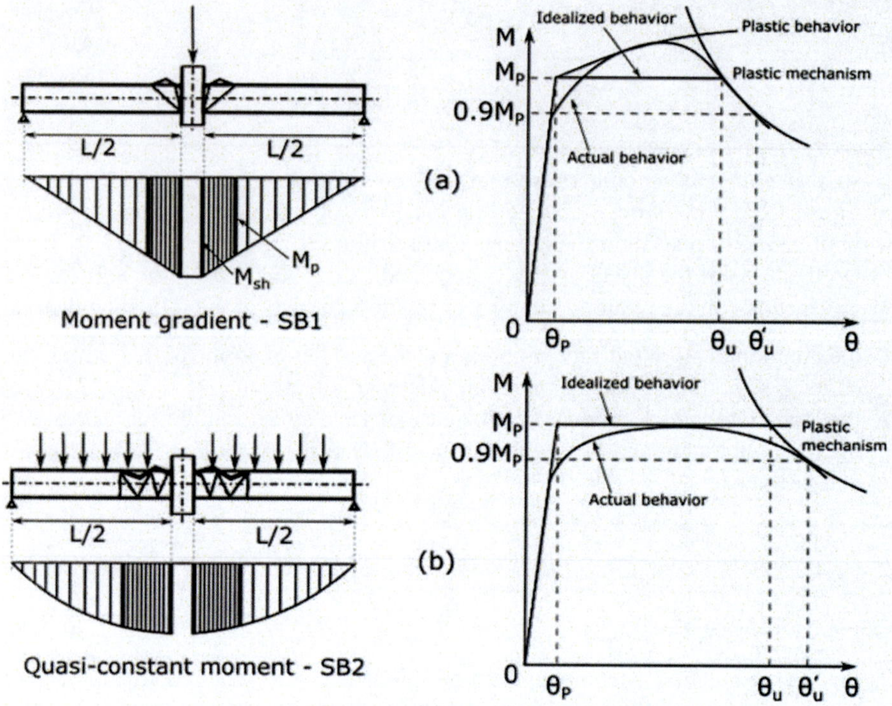

Fig. 6.8 Local plastic mechanism patterns: (**a**) moment gradient and its moment-rotation diagram; (**b**) quasi-constant moment and its moment-rotation diagram

DUCTROT-M program can be found in Gioncu et al. (2012) and Anastasiadis et al. (2012).

The moment M versus rotation θ curve depends on the moment variation along the length of the beam. There are two such moment variations, the gradient moment and the quasi-constant moment, for the concentrated load in the middle of the beam's span and the uniformly distributed load on that span, respectively. Figure 6.8a, b presents the $M - \theta$ curves for the local plastic mechanisms in standard beam (SB) types 1 and 2 for the two corresponding moment gradient and quasi-constant moment cases with the subscripts p and u denoting plastic and ultimate, respectively. The ductility is a function of the rotation capacity θ_u defined at the intersection of the lowering post-buckling $M - \theta$ curve with the theoretical plastic moment M_p, as shown in Fig. 6.8c, d. In practice, use of the value $0.9M_u$ is recommended instead of M_p, as shown in these figures.

The I-section type is the most widely used steel section for beams and beam-columns. For steel building frames, where plastic or compact sections are used, the two local collapse mechanisms considered are the in-plane and the out-of-plane ones. These mechanisms are used for determining the ultimate rotation capacity of beams and beam-columns.

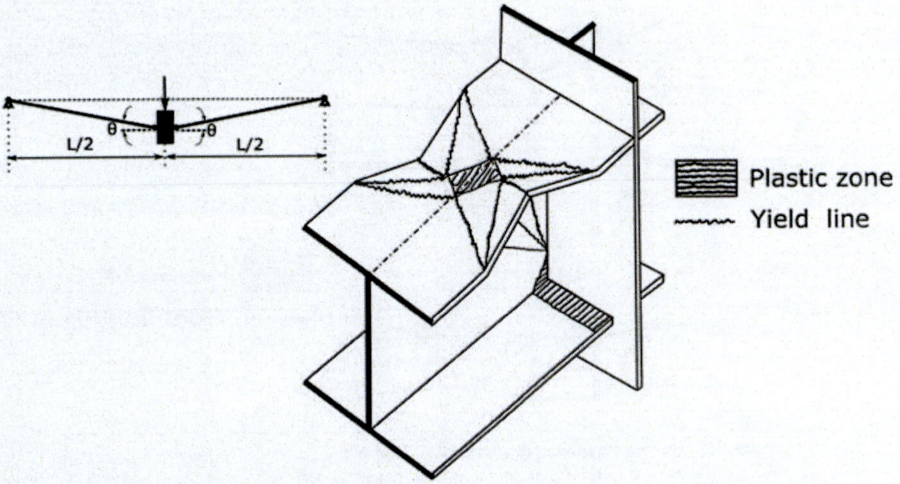

Fig. 6.9 In-plane plastic mechanism including compression flange plastic mechanism, web plastic mechanism and tension flange plastic mechanism

Figure 6.9 depicts the in-plane plastic mechanism, which consists of the compression flange mechanism, the web plastic mechanism and the tension flange plastic mechanism. Using rigid plastic theory including the concepts of plastic hinge and yield line and employing the principle of the vanishing of the variation of the total potential, one is able to explicitly describe the $M - \theta$ mechanism curve in the form (Gioncu and Mazzolani 2002)

$$M/M_p = a_{1M} + a_{2M}(1/\theta)^{0.5} \tag{6.39}$$

for the moment gradient case and

$$M/M_p = a_{1M} + 2a_{2M}(2/\theta)^{0.5} \tag{6.40}$$

for the quasi-constant moment case and express the ultimate plastic rotation θ_r in the form

$$\theta_r = [a_{2M}/(1 - a_{1M})]^2 \tag{6.41}$$

for the moment gradient case and

$$\theta_r = (1/8)[a_{2M}/(1 - a_{1M})]^2 \tag{6.42}$$

for the quasi-constant moment case.

In the above, a_{1M} and a_{2M} are given as complicated explicit expressions of the geometrical details of the I-section type, the yielding strength of steel and three

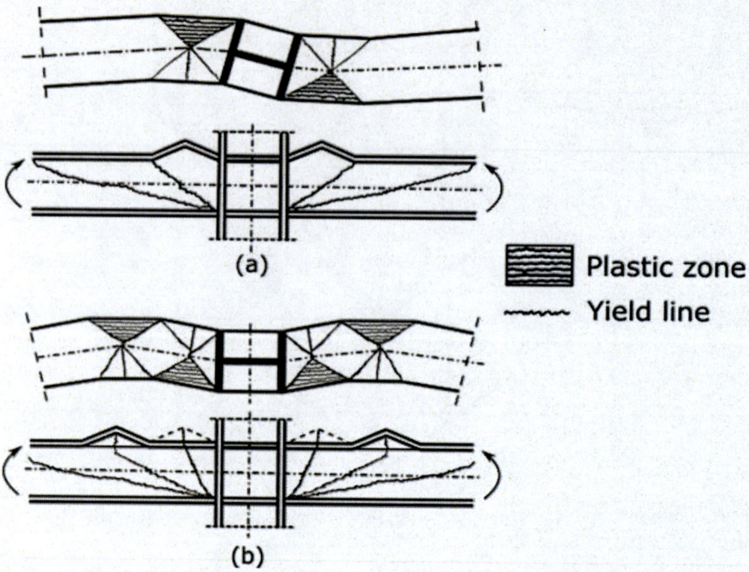

Fig. 6.10 Out-of-plane plastic mechanisms: (**a**) S-shaped plastic mechanism; (**b**) M-shaped plastic mechanism

plastic mechanism parameters. More details can be found in Gioncu and Mazzolani (2002). It should be noted that the above θ_r values can be also obtained from the intersection points of the $M-\theta$ curves with the lines corresponding to M_p or $0.9M_p$.

Figure 6.10 shows the two types of the out-of-plane local collapse mechanisms: the S-shaped plastic mechanism characterized by an asymmetrical shape around the plane of the mid-span stiffener (Fig. 6.10a) and the M-shaped plastic mechanism characterized by a symmetrical shape around the plane of beam mid-span (Fig. 6.10b). Using rigid plastic theory including the concepts of plastic hinge and yield line and employing the principle of vanishing of the total potential, one is able to explicitly describe the $M-\theta$ curve for the S-shaped plastic mechanism in the form

$$M/M_p = a'_{1M} + a'_{2M}(1/\theta)^{0.5} + a'_{3M}(1/\theta)^{0.75} \tag{6.43}$$

where a'_{1M}, a'_{2M} and a'_{3M} are given as complicated explicit expressions of geometrical, material and mechanism parameters, which can be found in Gioncu and Mazzolani (2002). The ultimate rotation can be obtained by the θ coordinate of the intersection point of the above $M-\theta$ curve with the lines corresponding to M_p or $0.9M_p$.

A $M-\theta$ relation of the same form as that of Eq. (6.43) but with different expressions for the coefficients a'_{1M}, a'_{2M} and a'_{3M} holds true for the M-shaped plastic mechanism. Details can be found in Gioncu and Mazzolani (2002).

Closing this discussion on $M-\theta$ curves for local collapse mechanisms, some notes on three particular cases are provided (Gioncu and Mazzolani 2002). The first

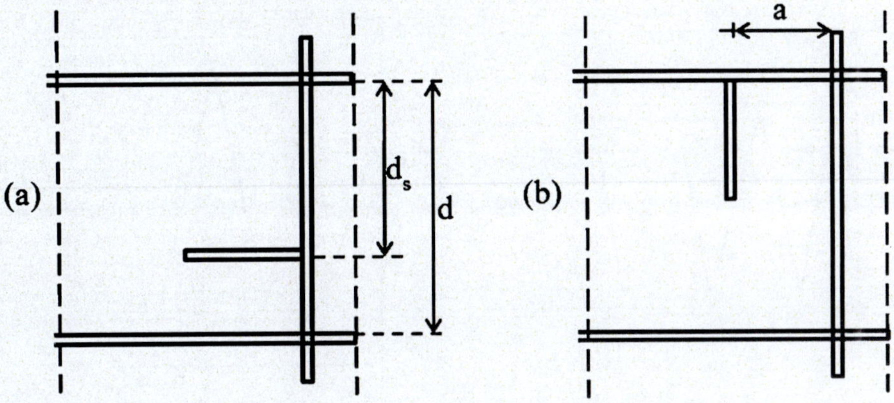

Fig. 6.11 Geometry of web stiffening: (**a**) horizontal stiffeners; (**b**) vertical stiffeners

note has to do with the effect of flange to web junction of rolled I-section type beams on the available rotation capacity. Thus, if θ_r is the rotation capacity without this effect and θ_{rj} with that effect, one has

$$\theta_{rj} = c_j \theta_r \tag{6.44}$$

where

$$c_j = c^2/(c - 0.5t_w - 0.8r)^2 \tag{6.45}$$

with c being half of the flange width b, t_w the thickness of the web and r the radius of flange to web junction of an I-section type. The second note has to do with the effect of web stiffeners on the rotation capacity of beam-columns. With reference to Fig. 6.11, it has been found that rotation capacity θ_r (i) increases for decreasing values of $\delta = d_s/d$ in the range $\delta_h < 0.75$ for horizontal stiffeners (Fig. 6.11a) and (ii) decreases for decreasing values of δ_v = a/b in the range $\delta_v < 1.0$ for vertical stiffeners, where b is the flange width (Fig. 6.11b). Thus, it is recommended to use only horizontal web stiffeners if the goal is to increase rotation capacity and ductility.

The third note has to do with the effect of the axial force on the rotation capacity of beam-columns. In general, the presence of compressive axial force reduces the rotation capacity provided that $n_p = N/N_p > 0.4$, where N stands for the axial force and N_p for its plastic value equal to Af_y with A being the area of the section. It has been observed that in moment resisting steel frames, in general, one has $n_p < 0.4$ and thus, the θ_r determined on the basis of bending (without axial force) can also be used for beam-columns.

Up to now in this section, the load has been assumed to be static monotonic. When the load is seismic, the static response (hence the rotation) has to be modified to take into account seismic effects, like the strain rate influence and the effect of repeated cyclic actions. The strain rate effect is significant for near-fault earthquakes

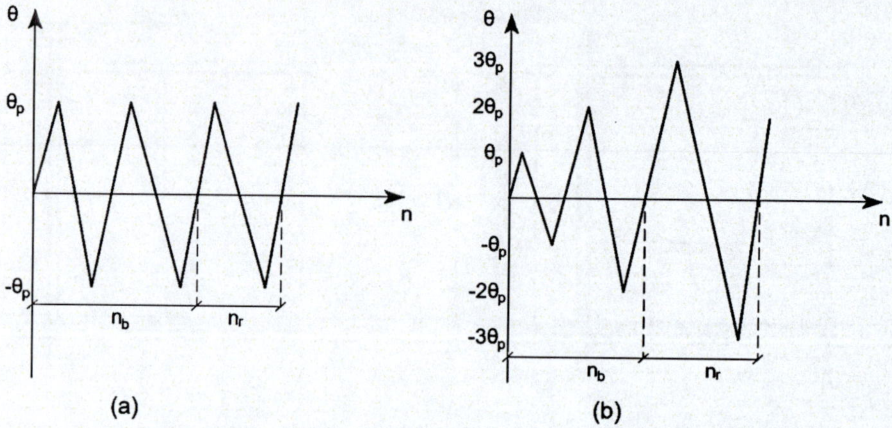

Fig. 6.12 Variation of rotation θ versus number of cycles n: **(a)** constant rotation amplitude; **(b)** increasing rotation amplitude

and consists of increasing the plastic hinge moments over the design ones and of attaining in some sections the fracture moment-rotation capacity. However, since its importance is not significant for far-fault earthquakes considered here, this effect is not further discussed. The interested reader can consult the Gioncu and Mazzolani (2002) book for details.

The effect of repeated cyclic actions on the ductility of structural members, consisting of the reduction of rotation capacity due to the accumulation of plastic deformation and the creation of cracks and section fracture, is significant, especially for the case of intermediate or far-fault earthquakes of interest here. The effect of cyclic actions on the rotation capacity is studied under constant rotation amplitude (Fig. 6.12a) and under increasing rotation amplitude (Fig. 6.12b). Under constant rotation amplitude, the difference in behavior between monotonic and cyclic load starts becoming significant only after plastic buckling at the flanges occurs at a cycle $n_b = \theta_b/\theta_p$. Under increasing rotation amplitude, one observes a more pronounced increase of plastic rotation accumulations and a faster deterioration of member moment capacity as compared to the constant rotation case. If n_r is the number of strong pulses (cyclic amplitudes) after local buckling of flanges

$$n_r = n - n_b \tag{6.46}$$

where n is the total number of strong cycles, the post-critical curve for cyclic actions is

$$M/M_p = a_{1M} + a_{2M}[1/(\theta + \theta_{ar})]^{0.5} \tag{6.47}$$

with θ_{ar} being the accumulated plastic rotation until cycle r (Gioncu and Mazzolani 2002). It turns out that $\theta + \theta_{ar} = n_r\theta_p$ and $\theta + \theta_{ar} = (n_r/2)(n_r + 1)\theta_p$ for the cases of constant and increasing rotation, respectively. The ultimate rotation for cyclic actions θ_{uc} is given in terms of its value for monotonic action θ_u by the relation

$$\theta_{uc} = \theta_u - \theta_{ar} \tag{6.48}$$

Thus, one needs to know the number of pulses producing large plastic deformations and this number varies between 1–2 for intermediate and 3–4 for far-fault earthquakes. Finally, concerning fracture under cyclic actions, one can state that the fracture rotation is $\theta + \theta_{ar}$ with its above expressions for the cases of constant and increasing rotations. However, only for high yield ratios, the number of pulses appears to be realistic ($n < 10$).

6.2.5 *Structural Damage*

Damage represents the degree of deterioration or degradation at material, member or structural level and this is usually quantified by the damage index which varies between the values of 0 (no damage) and 1 (failure) (Powell and Allahabadi 1988). The damage index associated with seismic loading can be defined on the basis of deformation and/or energy of dissipation. Here the book of Gioncu and Mazzolani (2002) is followed and the damage index for structures is defined on the basis of ductility associated with the plastic rotation of plastic hinges and determined at member, storey or structural level.

The member damage index I_{dm} is defined as the ratio of the required plastic rotation θ_r over the available plastic rotation θ_a, i.e.,

$$I_{dm} = \theta_r / \theta_a \tag{6.49}$$

The storey damage index I_{ds} and the global damage index I_{dg} for the whole structure are defined in terms of I_{dm} and have the form

$$I_{ds} = \sum_{1}^{n_b} I_{dm}^2 / \sum_{1}^{n_b} I_{dm} \tag{6.50}$$

$$I_{dg} = \sum_{1}^{n_b} \sum_{1}^{n_s} I_{dm}^2 / \sum_{1}^{n_b} \sum_{1}^{n_s} I_{dm} \tag{6.51}$$

where n_b and n_s denote the number of bays and stories, respectively. In general, the highest damage index is in members and the lowest is the global one. The global damage index gives a general qualitative picture of the damage but provides no information about the amount and location of damage as do the damage indices for members and storeys.

In order to be able to appreciate the degree of damage due to earthquakes, a correspondence between member damage levels and seismic limit states has to be established. Thus, using the rotation capacity θ as a measure of damage, one can establish the following damage levels (Gioncu and Mazzolani 2002):

1. No damage for $\theta < \theta_p$
2. Minor damage for $\theta_p < \theta < 1.5\theta_p$
3. Easy repairable damage for $1.5\theta_p < \theta < \theta_b$
4. Repairable damage for $\theta_b < \theta < \theta_u$
5. Irrepairable damage for $\theta > \theta_u$
6. Extensive damage (partial or complete collapse)

At the level of global damage the following I_{dg} values can be used:

1. No damage for $I_{dg} < 0.05$
2. Minor damage for $I_{dg} < 0.15$
3. Repairable damage for $I_{dg} < 0.50$
4. Collapse prevention for $I_{dg} < 0.80$
5. Near collapse for $I_{dg} > 0.80$
6. Structural collapse for $I_{dg} = 1.00$

6.3 Numerical Examples

This section presents two numerical examples in order to illustrate the ductility-based plastic design method of this chapter and point out its advantages. The first example illustrates the proposed method presented in this chapter as a combination of the methods of Mazzolani and Piluso (1997), Ghersi et al. (1999) and material provided in the book of Gioncu and Mazzolani (2002). The second example illustrates the method of Ghersi et al. (1999) because of its simplicity. As one can see, both methods can be worked out by hand calculations.

6.3.1 Seismic Design of a Six-Storey Two-Bay MRF

In this section, a comprehensive numerical example taken from the book of Gioncu and Mazzolani (2002) is presented in some detail. Consider a six-storey two-bay plane steel moment resisting frame (MRF) with the same height of 3.0 m in all stories and the same span of 6.0 m in both bays, as shown in Fig. 6.13. The vertical dead G and live Q loads at every floor are 35.5 kN/m and 15.0 kN/m, respectively, with the dead load to include the weight of external and internal walls. Thus, the seismic load combination consists of the seismic lateral load (base shear F_b and the vertical load w_k at each floor k equal to $w_k = G + 0.3Q = 35.50 + 0.3 \cdot 15.0 = 40.0$ kN/m. The total weight W of the frame is then $W = 40 \cdot (6 + 6) \cdot 6 = 2880.00$ kN. The

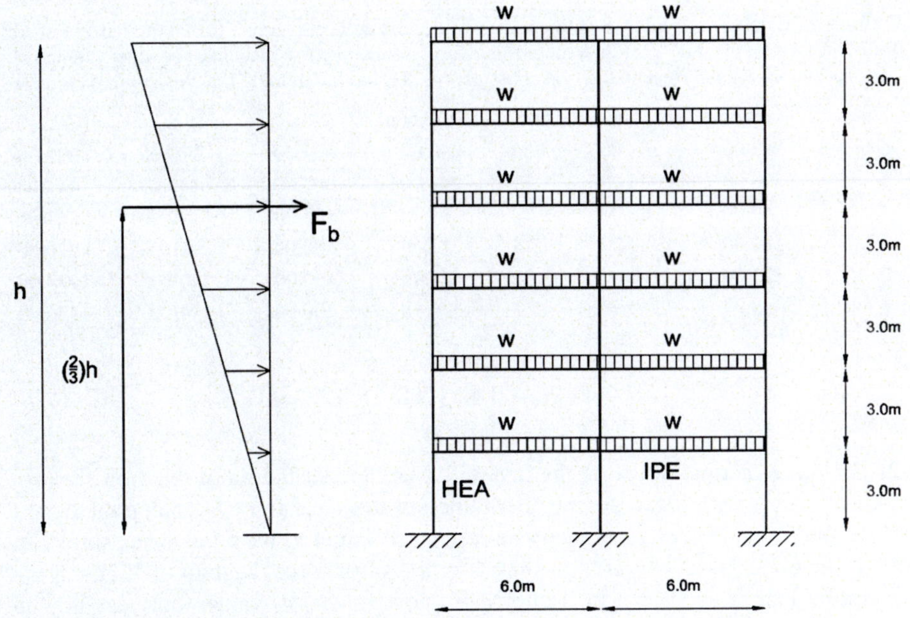

Fig. 6.13 Geometry and loading of the plane steel moment resisting frame considered

seismic load comes from seismic motion with PGA=0.36 g for the DLS (or LS) with soil type C and soil coefficient $S = 1.35$ (EC8 1994). The beam and column profiles are assumed to be of the IPE and HEA types, respectively with steel grade S235 for beams and S275 for columns. The joints are made of welded connections in order to have enough overstrength so as to be considered as non-dissipative areas. The fundamental period of vibration of the frame is approximately determined to be from Eq. (6.32) equal to $T = 0.73$ s.

The design of this frame, in accordance with Sect. 6.2, involves the following steps:

1. Seismic actions: the seismic accelerations a_s, a_d and a_u for the three limit states SLS, DLS and ULS are given with the aid of Eqs. (6.2) and (6.3) as

$$a_s = 0.412 \cdot 0.36\ \text{g} = 0.15\ \text{g} \tag{6.52}$$

$$a_d = 1.00 \cdot 0.36\ \text{g} = 0.36\ \text{g} \tag{6.53}$$

$$a_u = 1.22 \cdot 0.36\ \text{g} = 0.44\ \text{g} \tag{6.54}$$

The global collapse mechanism, which is associated with the ULS, controls the design with a base shear force F_{bu} obtained by combining Eqs. (6.6), (6.9 and (6.54) as

Table 6.2 External and internal actions: (a) distribution of lateral force; (b) distribution of shear force; (c) ratio $(V_k/V_{ns})^{0.5}$; (d) required bending moments; (e) available bending moments; (f) distribution of axial force in internal columns; (g) distribution of axial force in external columns

Storey	(a) kN	(b) kN	(c)	(d) kNm	(e) kNm	(f) kN	(g) kN
6	363.00	363.00	1.00	180.00	307.00	240.00	222.00
5	303.00	666.00	1.35	243.00	307.00	480.00	445.00
4	242.00	908.00	1.58	284.40	307.00	720.00	667.00
3	181.00	1089.00	1.73	311.40	400.00	960.00	920.00
2	121.00	1210.00	1.83	329.40	400.00	1200.00	1173.00
1	60.00	1270.00	1.87	336.60	400.00	1440.00	1426.00

$$F_{bu} = 0.44\ \mathrm{W} = 0.44 \cdot 2880 = 1270.00\ \mathrm{kN} \tag{6.55}$$

2. Section selection: adopting the inverted triangle distribution of the base shear in accordance with Fig. 6.3c, the distribution of the shear force F_{bu} along the height of the frame (forces F_k at every storey k) are found to have values as shown in Table 6.2a. These lateral forces result in the shear force diagram of V_k along the frame height as shown in Table 6.2b. Next to the V_k values one can find in Table 6.2c the amplification factors $m_h = (V_k/V_{ns})^{0.5}$ for the bending moments as shown in the second part of Eq. (6.10). Table 6.2d provides the bending beam moments as defined in the first part of Eq. (6.10) with $M_w = wl^2/8$ for the present two-bay frame. The bending moments of the top storey beams are $m_h M_w = 1.0 \cdot 40 \cdot 6^2/8 = 180$ kNm and together with those of the remaining stories can be found in Table 6.2d. On the basis of the required bending moments for beams, a section selection with IPE400 for the first three top stories and IPE450 for the three lower stories is done, for which the available moments $1307 \cdot 10^{-6} \cdot 235 \cdot 10^3 = 307.14$ kNm and $1702 \cdot 10^{-6} \cdot 235 \cdot 10^3 = 400.00$ kNm, respectively, as shown in Table 6.2e.

With known values of available bending moments, one is now able to compute axial forces in columns by using Eq. (6.17) with $N_w = wl/2$, where $w = G + 0.3Q = 40$ kN/m. Thus, Table 6.2f, g is completed.

The plastic moment of the first storey columns is obtained from Eq. (6.14) and reads $M_{pc,1} = \left(\sum_{i=1}^{3} M_{pc,i1}\right)/3 = (1/3) \cdot (0.7 + 0.15 \cdot 6) \cdot 2 \cdot 2 \cdot 400.00 \cdot 1.1 =$ 938.67 kNm with the factor 1.1 being the overstrength m_y of beam plastic moments. Now it is possible to compute the collapse multiplier a_c from Eq. (6.13) and then to determine all the plastic moments of the remaining columns. However, since the axial force and bending moment attain maximum values at the first storey, only the column section selection at that storey will be presented. Thus, selecting a HEA550 column section, a checking in bending moment 938.67 kNm and axial force 1440.00 kN results in a full plastic moment capacity $M_{pc,\,1} = 4622 \cdot 10^{-6} \cdot 275 \cdot 10^3 = 1271.05$ kNm and a reduced one due to the axial force N of the form (Gioncu and Mazzolani 2002) $M_{Npc,\,1} = 1.11 M_{pc,\,1}[1 - (N/$

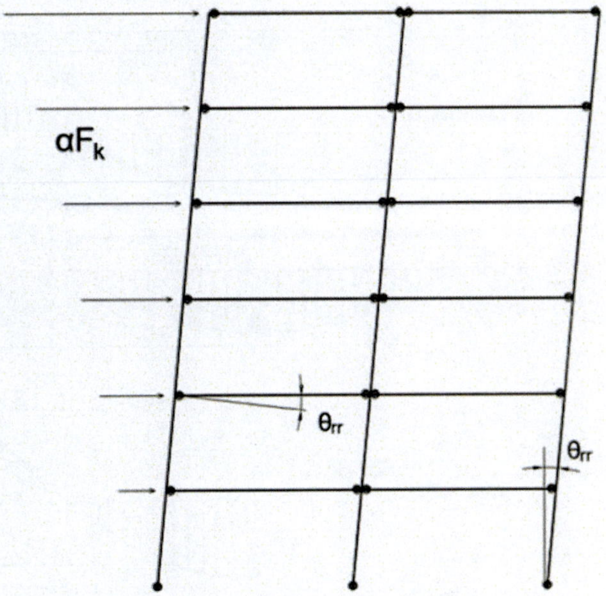

Fig. 6.14 Required rotation under static and monotonic action

N_p)] = 1.11 · 1271.05 · [1 – (1440/5830)] = 1066 kNm. If one follows EC3 (1992), the result is 1140 kNm. It is observed that both 1066 and 1140 kNm are higher than 938.67 kNm.

3. Required plastic rotation: it is first determined for static monotonic load. Using Eqs. (6.26) and (6.27), one finds a_1 and a_2 as $a_1 = (3/2)(3 \cdot 1066 + 2 \cdot 3 \cdot 400 + 2 \cdot 3 \cdot 307)$ $1.1/(1270 \cdot 18) = 0.537 < 1.0$, $a_2 = (3/4)(2880/1270) = 1.70$. From Eq. (6.33), θ_1 can be computed as $\theta_1 = 0.159 \cdot 10^{-2} \cdot 18^{0.6} \cdot 0.44 = 0.00396$ and hence θ_e can be found from Eq. (6.29) to be $\theta_e = 0.537 \cdot 0.00396 = 0.00213$ rad. Thus, the required plastic rotation θ_{rr} from Eq. (6.35) is $\theta_{rr} = 0.1[(0.537/1.70) + 0.00213] = 0.0318$ rad. Figure 6.14 shows the required rotation θ_{rr} under static monotonic action. Finally, in order to take into account the effect of seismic motion on θ_{rr}, use is made of Eq. (6.38) and Table 6.1. From the Table 6.1 and $a_1 = 0.537$ one obtains $\gamma_a = 1.963$ for beams and 2.463 for columns. Thus, using Eq. (6.38) one obtains the required plastic rotations as $\theta_{rrac} = 1.963 \cdot 0.0318 = 0.0624$ rad for beams and $\theta_{rrac} = 2.463 \cdot 0.0318 = 0.0783$ rad for columns.

4. Available plastic rotation: in order to determine it for the three types of sections found here (IPE400 and IPE450 for beams and HEA550 for columns), a two steps approach involving static monotonic conditions and seismic conditions is followed. The presentation is very brief and one should consult the Gioncu and Mazzolani (2002) book for details. The structure is first divided in standard beams (SB) and the available rotation capacity of them is determined under static monotonic load. Five standard beams are considered for IPE400 and IPE450

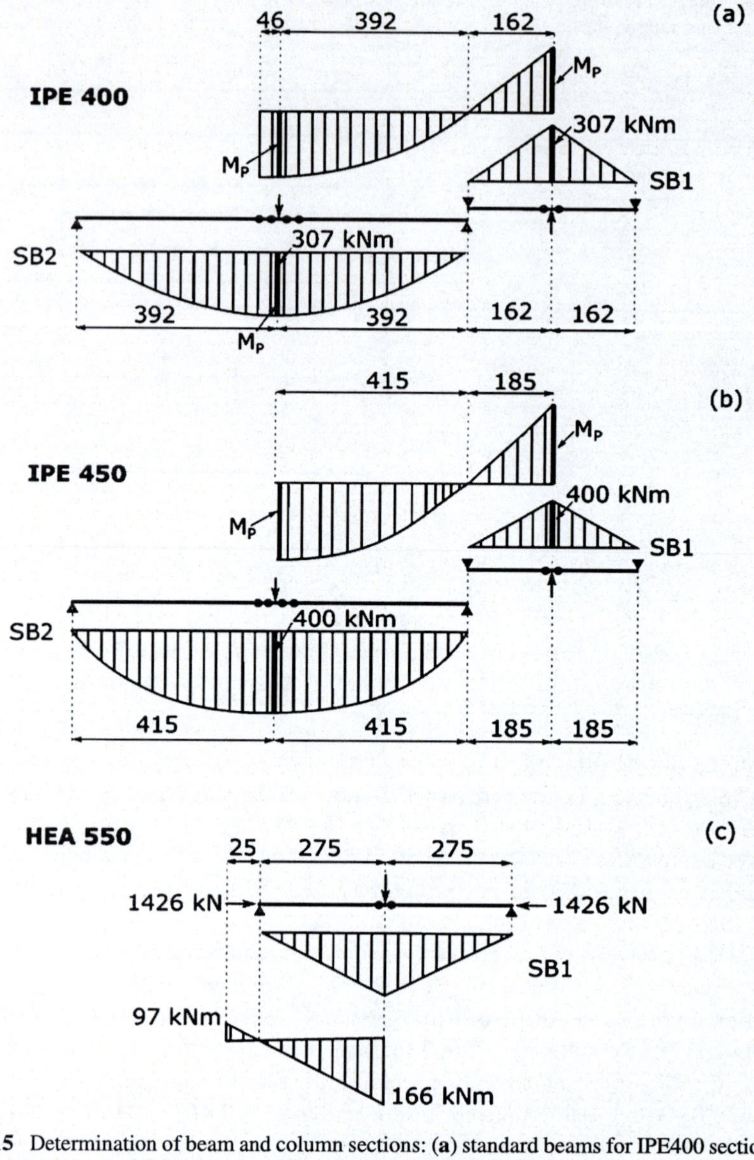

Fig. 6.15 Determination of beam and column sections: (**a**) standard beams for IPE400 sections; (**b**) standard beams for IPE450 sections; (**c**) standard beam for HEA550 sections

beams (Fig. 6.15b, c) and for HEA550 column (Fig. 6.15d). SB1 type is considered for moment gradient variation and SB2 type for quasi-constant moment for beams. SB1 type is used for columns. The moment-rotation curves for the three sections in the case of moment gradient are depicted in Fig. 6.16 using the method of local plastic mechanism. The $0.9M_p$ ductility criterion is used for the computation of the ultimate plastic rotation. The results are as follows:

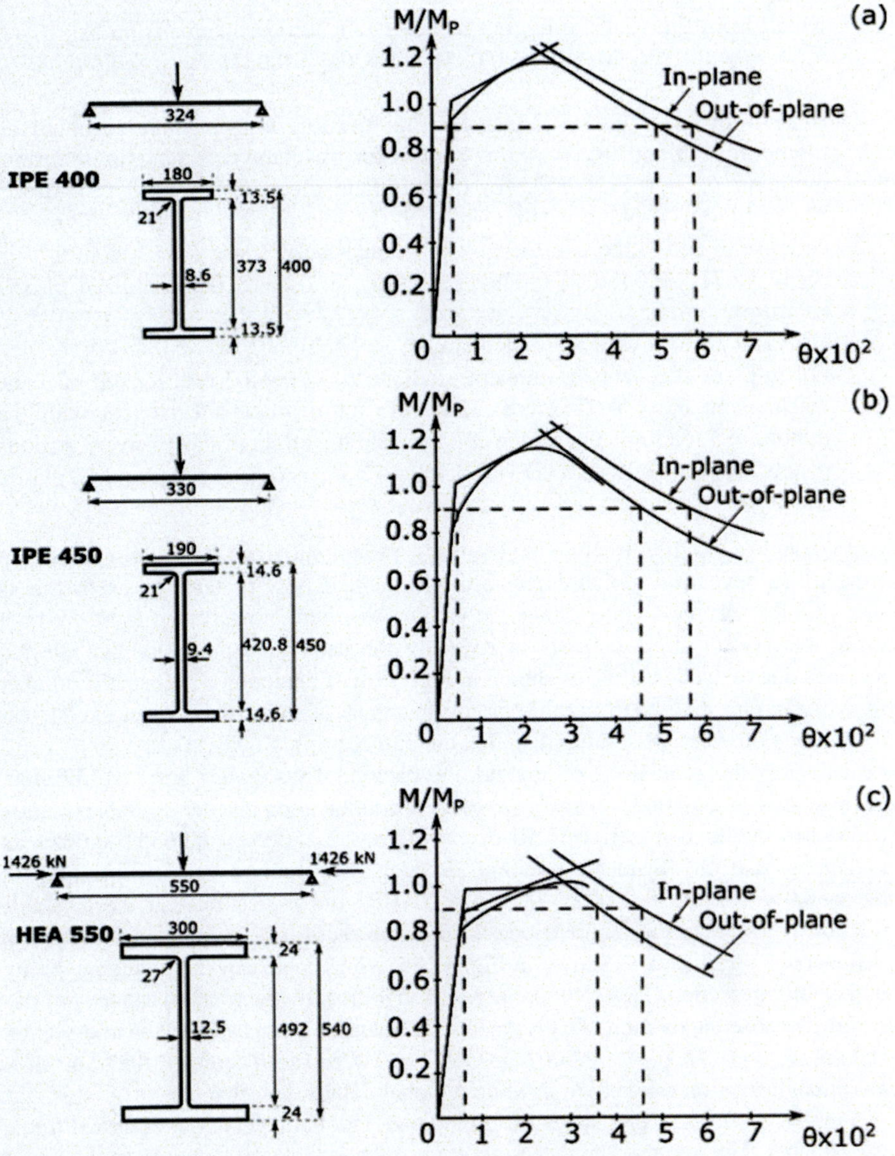

Fig. 6.16 Available rotation capacity: **(a)** IPE400 profiles; **(b)** IPE450 profiles; **(c)** HEA550 profiles

(i) IPE400 profiles with S235 steel (Fig. 6.16a). For in-plane mechanism, moment gradient: $\theta_r = 0.0582$ rad; for in-plane mechanism, constant moment: $\theta_r = 0.1239$ rad; for out-of-plane mechanism: $\theta_r = 0.0499$ rad. Thus, the out-of-plane mechanism with the minimum θ_r controls the profile collapse. The effect of flange-web junctions is taken into account through the coefficient c_j, which from Eq. (6.45) turns out to be $c_j = 1.706$ and hence from Eq. (6.44) $\theta_{rj} = c_j\theta_r = 1.706 \cdot 0.0499 = 0.0851$ rad $> \theta_{rr} = 0.0318$ rad (*required plastic rotation*).

(ii) IPE450 profiles with S235 steel (Fig. 6.16b). For in-plane mechanism, moment gradient: $\theta_r = 0.0570$ rad; for in-plane mechanism, constant moment: $\theta_r = 0.1170$ rad; for out-of-plane mechanism: $\theta_r = 0.0460 rad$. Thus, the out-of-plane mechanism with the minimum θ_r controls the profile collapse and the effect of flange-web junctions results in $\theta_{rj} = 1.671 \cdot 0.0460 = 0.0769$ rad $> \theta_{rr} = 0.0318$ rad (*required plastic rotation*).

(iii) HEA550 profiles with S275 steel and $n_p = N/N_{pl} = 0.247 < 0.4$ (Fig. 6.16c). For in-plane mechanism, moment gradient: $\theta_r = 0.0462$ rad; for out-of-plane mechanism: $\theta_r = 0.0357$ rad. Thus, the out-of-plane mechanism with the minimum θ_r controls the profile collapse and the effect of flange-web junctions results in $\theta_{rj} = 1.508 \cdot 0.0357 = 0.0538$ rad $> \theta_{rr} = 0.0318$ rad (required plastic rotation).

In conclusion, the structure under static monotonic load has enough rotation capacity to develop a global collapse mechanism. It now remains to see If this is also true for seismic actions. Since the earthquakes here are assumed to be far-fault ones, their mean effect consists of reducing the rotation capacity of the selected sections due to cyclic action, as shown in Fig. 6.16d–f depicting θ_r versus the number of cycles n. One can see from those figures that there is no rotation capacity reduction for the first 6 cycles. Assuming then that the number of big cycles is 8, only 8–6 = 2 cycles provide reduction of rotation capacity. Thus, one has: (i) IPE400: $\theta_{rc} = 0.0851 - 2 \cdot 0.0051 = 0.0749 > 0.0624 rad$ indicating that the available rotation exceeds the required one; (ii) IPE450: $\theta_{rc} = 0.0769 - 2 \cdot 0.0052 = 0.0665 > 0.0624 rad$ indicating that the available rotation exceeds the required one; (iii) HEA550: $\theta_{rc} = 0.0538 - 2 \cdot 0.0062 = 0.0414 < 0.0783$ rad indicating that the available rotation is less than the required one. In order to rectify this, a pair of welded web stiffeners is introduced, as shown in Fig. 6.17a, which increases the rotation capacity of the column section. This solution increases significantly the rotation capacity of this section because the rotation of a plastic load mechanism occurs around the mid-section and thus $\theta_{rc} = 0.2051 - 2 \cdot 0.0062 = 0.1927 > 0.0783$ rad. Therefore, the frame now has enough rotation capacity to develop a global collapse mechanism.

Finally, the seismic damage at the member and global level of the designed frame is evaluated. The member damage indices for the three types of sections of the frame are computed on the basis of Eq. (6.49) and read:

For the IPE400: $I_{dm} = 0.0624/0.0749 = 0.833$

For the IPE450: $I_{dm} = 0.0624/0.0665 = 0.938$

For the HEA550: $I_{dm} = 0.0783/0.1927 = 0.406$

The global damage index is computed on the basis of Eq. (6.51) and reads $I_{dg} = 0.798$. These high damage values indicate that, even though the frame is

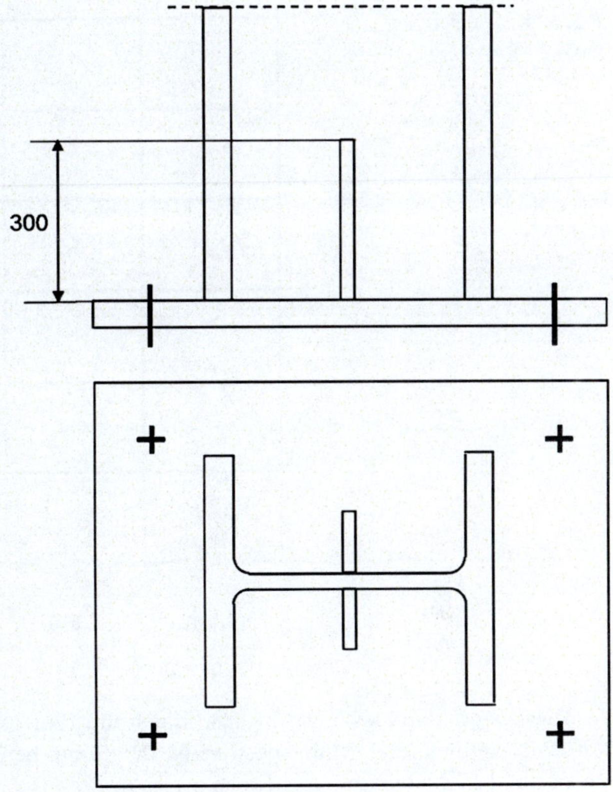

Fig. 6.17 Strengthening of columns at the base by adding a pair of web stiffeners

prevented from collapse, the maximum considered earthquake produces extremely high damage.

6.3.2 *Seismic Design of a Six-Storey Three-Bay MRF*

In this section, the simplified plastic design method of Ghersi et al. (1999), which has been presented in Sect. 6.2 as part of the method of Gioncu and Mazzolani (2002), is illustrated by means of an example in order to demonstrate that the method can be also successfully used separately for the seismic design of plane steel MRFs. However, this method, even though is very simple, cannot determine and check deformation.

Consider the plane steel six-storey, three-bay MRF of Fig. 6.18 subjected to dead load $G = 15.00$ kN/m and live load $Q = 10.00$ kN/m as well as to lateral seismic load with a peak ground acceleration $a_g = 0.35$ g and soil type C. This frame has been designed by Ghersi et al. (1999) following their method as described by Eqs. (6.13)–(6.17).

Fig. 6.18 Geometry of plane steel six-storey, three-bay MRF

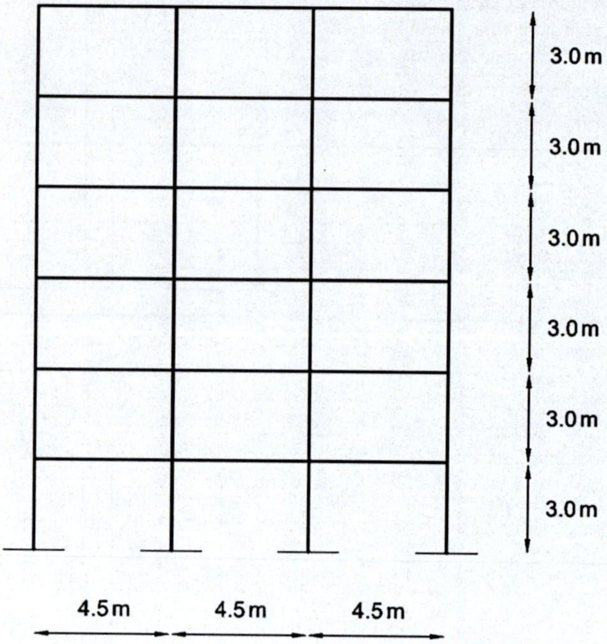

The design starts with the beams, which are dimensioned on the basis of the vertically applied load. More specifically, the design bending moment for beams is determined as the maximum between the values of $M_1 = (1.35G + 1.50Q)L^2/10$ for the vertical load combination and $M_2 = (G + 0.3Q)L^2/4$ for the vertical load of the seismic combination, where $L = 4.50$ m is the bay length, in accordance with EC8 (1994). These values turn out to be $M_1 = 71.40$ kNm and $M_2 = 91.10$ kNm. Selecting an IPE300 cross-section for beams and assuming a S275 grade for steel, one can calculate (EC3 1992) the $M_{b.\ Rd} = W_{pl}f_y/1.1 = 628 \cdot 10^{-6} \cdot 275 \cdot 10^6 \cdot 10^{-3}/1.1 = 157$ kNm > 91.10 kNm. Thus, a section of IPE300 is assigned to all beams in the frame.

Use of Eq. (6.14) with $n_s = 6$ and $M_{pb} = 157$ kNm yields a design bending moment at the column base equal to $M_{pc,1} = (1/4) \cdot \sum_{i=1}^{n_c} M_{pc,i1} = (1/4) \cdot (0.7 + 0.15 \cdot 6) \cdot (2 \cdot 3 \cdot 157) = 376.80$ kNm. Columns are also subjected to axial compressive load due to the $G + 0.3Q = 15 + 0.3 \cdot 10 = 18$ kN/m vertical load of the seismic combination. Use of Eq. (6.17) permits one to determine axial forces for interior and exterior columns. For example, for the sixth storey $N_{ext} = (18 \cdot 4.5)/2 + (2 \cdot 157)/4.5 = 110$ kN, $N_{int} = (18 \cdot 4.5)/2 + (18 \cdot 4.5)/2 = 81$ kN and for the first storey $N_{ext} = 6 \cdot 110 = 662$ kN, $N_{int} = 6 \cdot 81 = 486$ kN. Results for the axial forces in all stories can be found in Table 6.3. With known design moment and axial force for the column of the first storey ($M_{pc,\ 1} = 376.80$ kNm, $maxN_1 = N_{ext,\ 1} = 662$ kN) one can select a HEB300 of steel grade S275 section for which $M_{c.\ Rd} = 1869 \cdot 10^{-6} \cdot 275 \cdot 10^6 \cdot 10^{-3}/1.1 = 467.25$ kNm. This is reduced due to the axial force (EC3 1992) to the value of $M_{c.\ N.\ Rd} = 422$ kNm.

Table 6.3 Design axial forces in columns and lateral forces F_k along the height of the frame

Storey	N_{ext} (kN)	N_{int} (kN)	F_k	h_k	$F_k h_k$
6	110	81	0.286	18	5.143
5	221	162	0.238	15	3.571
4	331	243	0.190	12	2.286
3	441	324	0.143	9	1.286
2	551	405	0.095	6	0.571
1	662	486	0.048	3	0.143

Table 6.4 Computational sequence for column moments M_{ct} and M_{cb}

Storey	$a_c F_k$	$a_c F_k h_k$	$M_{b,\ ik}$	FKT	FMT	M_{ct}	FKB	FMB	M_{cb}
6	161.3	2903.7	942.0	0.0	−942.0	−235.5	484.0	−458.0	−114.5
5	134.4	2016.5	1884.0	484.0	−1400.0	−350.0	1371.2	−512.8	−128.2
4	107.5	1290.5	2826.0	1371.2	−1454.8	−363.7	2581.1	−244.9	−61.2
3	80.7	725.9	3768.0	2581.1	−1186.9	−296.7	4033.0	265.0	66.2
2	53.8	322.6	4710.0	4033.0	−677.0	−169.3	5646.2	936.2	234.0
1	26.9	80.7	5652.0	5646.2	−5.8	−1.5	7340.0	1688.0	422.0

The design bending moments for the columns of stories other than the first can be determined by using first Eq. (6.13) to evaluate a_c and then Eq. (6.15) to compute these moments. More specifically, the numerator of the right hand side of Eq. (6.13) is equal to 4 · 422 + 2 · 3 · 6 · 157 = 7340 kNm, while its denominator $\sum_{i=1}^{n_s} F_i h_i$ is computed as follows: on the assumption of a linear distribution with height (inverted triangular) of the seismic lateral forces F_k and in conjunction with a normalization of them so that their sum to be equal to 1, one can complete the last three columns of Table 6.3 and determine the sum of $F_k h_k$ as equal to 13.00. Thus, from Eq. (6.13), one can find $a_c = 7340/13 = 564.61$.

The design bending moments at the top (M_{ct}) and bottom (M_{cb}) of the columns of the stories of the frame are computed on the basis of Eq. (6.13) and the whole computational sequence is recorded in Table 6.4. Construction of the first three columns of this table is self-explanatory. The fourth column labeled as $M_{b,\ ik}$ represents the term $\sum_{k=r}^{n_s} \sum_{i=1}^{n_c} 2M_{b,ik}$ of the right hand side of Eq. (6.13) and, e.g., its first two values are computed as $1 \cdot (2 \cdot 3 \cdot 157) = 942$ and $2 \cdot (2 \cdot 3 \cdot 157) = 2 \cdot 942 = 1884$. The fifth column of Table 6.4 labeled as FKT represents the term $a_c \sum_{k=r}^{n_s} F_k (h_k - h_r)$ of the right hand side of Eq. (6.13), while the sixth column of Table 6.4 labeled as FMT represents the term $\sum_{j=1}^{n_c} M_{c,jr}$ of the left hand side of Eq. (6.13). The seventh column of Table 6.4 provides the values of the design bending moments M_{ct} at the top of one column in the storey and obtained as 1/4 of the FMT values. The remaining three columns of Table 6.4 concerning the M_{cb} moments are obtained in an analogous manner. Finally, a column section selection is done for the columns of the frame (the same for all the columns of a storey) consisting of HEB300 section for the first storey, HEB240 for the sixth storey and HEB280 for the intermediate stories. The strength of

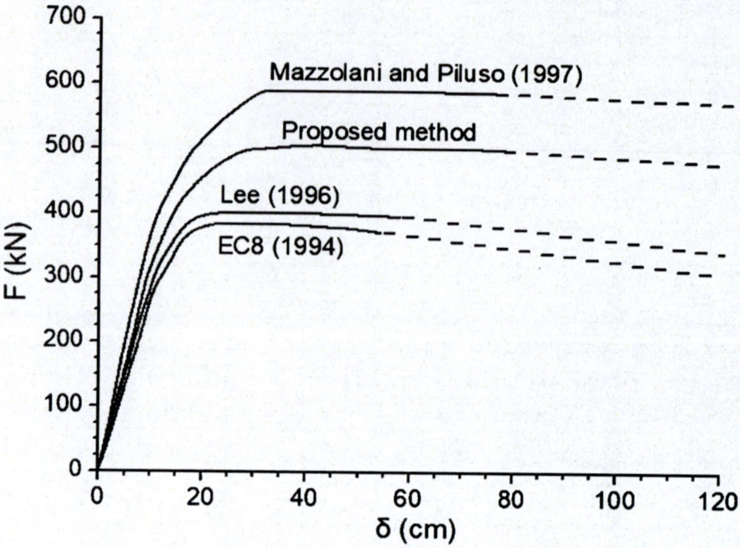

Fig. 6.19 Load-displacement curve of the frame designed according to the proposed and other methods

Table 6.5 Column sections of four design methods (expressions like 220/240 refer to external/internal column sections)

Storey	Present method	EC8 (1994) method	Lee (1996) method	Mazzolani and Piluso (1997) method
6	HEB 240	HEB 220/240	HEB 220/260	HEB 260
5	HEB 280	HEB 220/240	HEB 220/260	HEB 300
4	HEB 280	HEB 220/240	HEB 220/260	HEB 300
3	HEB 280	HEB 220/240	HEB 220/260	HEB 300
2	HEB 280	HEB 220/240	HEB 220/260	HEB 360
1	HEB 300	HEB 220/240	HEB 220/260	HEB 360

these sections is checked in bending moment and axial force in accordance with the EC3 (1992) provisions.

The frame of this example (Fig. 6.18) has been also designed in accordance with the EC8 (1994) with $q = 5$ and the design methods proposed by Lee (1996) and Mazzolani and Piluso (1997). For an easy design comparison, the beam cross-sections have been selected to be the same with those of the present method. This was possible because the maximum beam bending moment in those methods of 98.30 kNm, observed for the EC8 (1994) case, was close to the maximum value of 91.10 kNm of the present method and especially because the selected cross-section of IPE300 has a moment of resistance of 157.00 kNm, far higher than any of those maximum moments. Thus, the four design methods under comparison differ only with respect to their column cross-sections, as shown in Table 6.5.

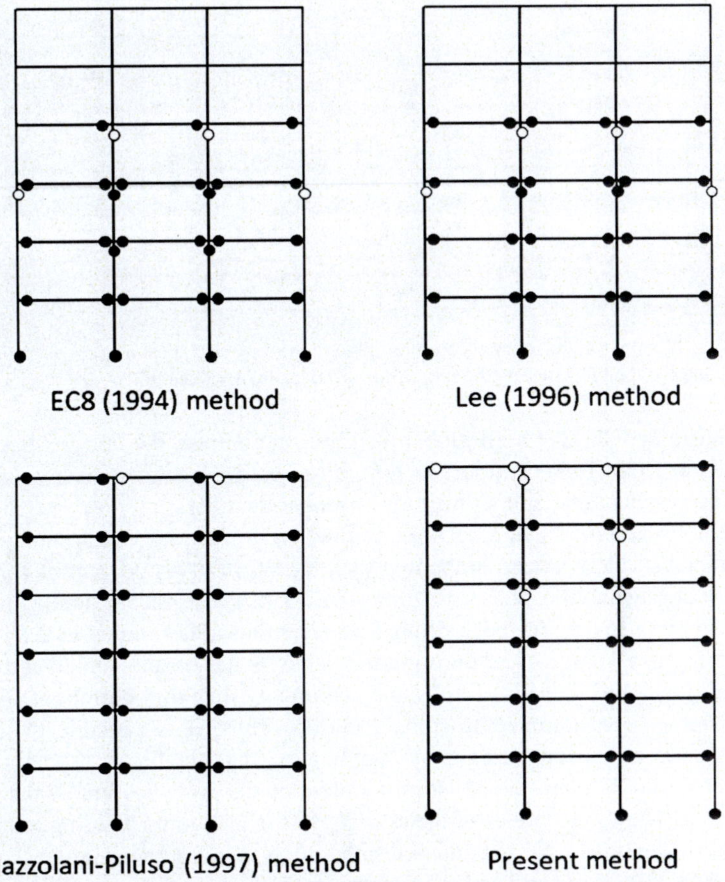

Fig. 6.20 Plastic hinge patterns of the frame designed by four different methods

On the basis of Table 6.5, one can easily observe that the minimum total steel weight of columns is 55 kN and corresponds to the EC8 (1994) design. The corresponding weights of the Lee (1996) and Mazzolani and Piluso (1997) designs are more by 7% and 58% with respect to the EC8 (1994) design, while the one by the present method is more by 33%.

An assessment of the seismic performance of all four designs was done by inelastic static (pushover) and dynamic analyses. Figure 6.19 depicts the force F versus roof displacement δ diagram, while Fig. 6.20 the plastic hinge patterns of the frame for all four design methods as obtained by pushover analyses. In Fig. 6.19 one can see the global strength and ductility of the four designs. In Fig. 6.21 where the white circles denote plastic rotation of the base cross section of the columns greater than the ultimate rotation (assumed to be 0.04) and correspond to the dashed lines of Fig. 6.19, the collapse mechanism is also apparent. One can observe that the present method and the Mazzolani and Piluso (1997) method lead to a global collapse

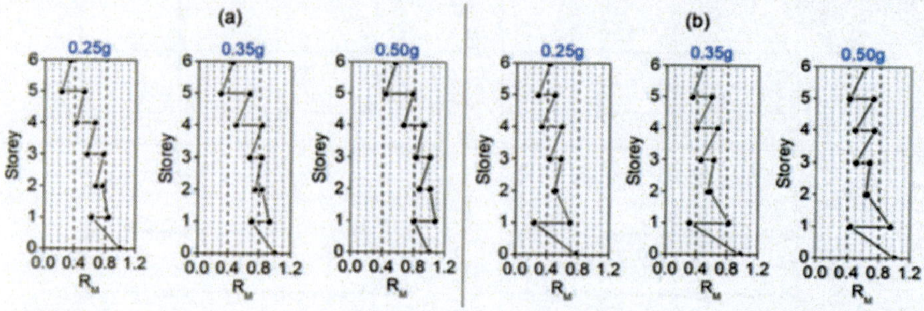

Fig. 6.21 R_M values in the internal columns of the frame designed according to: (**a**) the EC8 (1994) method and (**b**) the proposed method, both for different *PGA* values

mechanism and show a high global ductility. In contrast, the designs based on the methods of EC8 (1994) and Lee (1996) fail in partial mechanisms involving 3 and 5 stories, respectively, and show a lower global ductility.

The four designs have also been studied by inelastic dynamic analyses using 10 artificial accelerograms compatible with the EC8 (1994) provisions for soil type C and a strong motion phase duration of 22.5 s. These accelerograms have been scaled in order to achieve peak ground accelerations (*PGA*) equal to 0.25 g, 0.35 g and 0.50 g. In all cases, ideal elastoplastic behavior has been assumed at the ends of the beams and the bases of the first storey columns, while the columns at all the other stories have been assumed to remain elastic. Thus, it is possible to assess the effectiveness of the four designs by simply checking whether or not the maximum moment is smaller than the resisting moment of the cross-section of the columns. Figure 6.21 shows the values of the ratio R_M of the maximum bending moment over the resisting moment for the internal columns of the frame designs according to: (a) the EC8 (1994) method and (b) the present method for different *PGA* values. It is observed that columns at the bottom stories of the frame designed by the EC8 (1994) method cannot remain elastic for high *PGA* values, while the present method gives acceptable bending moments. It has been found that the other two methods are in agreement with the present method.

6.4 Conclusions

The preceding developments serve to reach the following conclusions:

1. A seismic design method for plane steel moment resisting frames capable of achieving the desired ductility and failure mode has been presented. The failure mode is of the global type and thus use is made of the full ductility capacity of the structure.
2. Beam sections are first determined to resist vertical loads, while column sections are then selected with the aid of limit plastic analysis of the global collapse

mechanism of the frame under vertical load and lateral distributed seismic load applied statically.
3. The required and the available rotation capacity of the beams and beam-columns of the frame are determined analytically, both in two phases: first assuming the load to be static and monotonic and second by considering the cyclic nature of the seismic load.
4. All the expressions used here are in analytical form. Of course, this requires some approximations during both the internal force and moment determination for section selection and the computation of required and available rotation ductility in order to check if sufficient ductility has been assured by the section selection.
5. A comparison of this design method against the design method of EC8 on the basis of a plane steel MRF has revealed that on the assumption of selecting the same beam sections, the columns coming from this design are by about 30% heavier than those of EC8 design. However, a EC8 design cannot always assure a global collapse mechanism, which is always the case with the present design method.

References

Anastasiadis A, Mosoarca M, Gioncu V (2012) Prediction of available rotation capacity and ductility of wide-flange beams. Part 2: applications. J Constr Steel Res 69:176–191

Boushaba B, Plumier A (1988) Relation entre la ductilite locale et le facteur de comportement sismique de structures en acier. Constr Metallique 2:59–70

EC3 (1992) Eurocode 3, Design of steel structures – Part 1–1: general rules and rules for buildings, ENV 1993-1-1. European Committee for Standardization (CEN), Brussels

EC8 (1994) Eurocode 8, Design of structures for earthquake resistance, Part 1: general rules, seismic actions and rules for buildings. European Committee for Standardization (CEN), Brussels

EC8 (2004) Eurocode 8, Design of structures for earthquake resistance, Part 1: general rules, seismic actions and rules for buildings, EN 1998-1-1. European Committee for Standardization (CEN), Brussels

Fajfar P, Krawinkler H (eds) (1997) Seismic design methodologies for the next generation of codes. Taylor and Francis, Boca Raton, FL

FEMA 356 (2000) Prestandard and commentary for the seismic rehabilitation of buildings. Federal Emergency Management Agency, Washington, DC

Ghersi A, Marino E, Neri F (1999) A simple procedure to design steel frames to fail in global mode. In: Dubina D, Ivanyi M (eds) Stability and Ductility of Steel Structures (SDSS' 99). Elsevier Science, London, pp 377–384

Gioncu V (2000) Design criteria for seismic resistant steel structures. In: Mazzolani FM, Gioncu V (eds) Seismic resistant steel structures. CISM courses. Springer, Wien, pp 19–99

Gioncu V, Mazzolani F (2002) Ductility of seismic resistant steel structures. Spon Press, London

Gioncu V, Petcu D (1997) Available rotation capacity of wide-flange beams and beam-columns, Part 1: theoretical approaches; Part 2: experimental and numerical tests. J Constr Steel Res 43:161–217. 219-244

Gioncu V, Petcu D (2001) Ductility of steel members. Part 1: Static and monotonic ductility; Part 2: Seismic ductility, (manuscript contained in Gioncu V, Mazzolani FM (2002) book)

Gioncu V, Mosoarca M, Anastasiadis A (2012) Prediction of available rotation capacity and ductility of wide-flange beams: Part 1: DUCTROT-M computer program. J Constr Steel Res 69:8–19

Goel RK, Chopra AK (1997) Period formulas for moment resisting frame buildings. J Struct Eng ASCE 123:1454–1461

Lee HS (1996) Revised rule for concept of strong-column weak-girder design. J Struct Eng ASCE 122:359–364

Lee LH, Lee HH, Han SW (2000) Method of selecting design earthquake ground motions for tall buildings. Struct Design Tall Spec Build 9:201–213

Leelataviwat S, Goel SC, Stojadinovic B (1999) Towards performance-based seismic design of structures. Earthquake Spectra 15:435–461

Mazzolani FM, Piluso V (1995) A new method to design steel frames failing in global mode including P-Δ effects. In: Mazzolani FM, Gioncu V (eds) Behavior of Steel Structures in Seismic Areas (STESSA 94). E & FN Spon, London, pp 303–309

Mazzolani FM, Piluso V (1997) Plastic design of seismic resistant steel frames. Earthq Eng Struct Dyn 26:167–191

Powell GH, Allahabadi R (1988) Seismic damage prediction by deterministic methods: concepts and procedures. Earthq Eng Struct Dyn 16:719–734

Priestley MJN (1997) Displacement-based seismic assessment of reinforced concrete buildings. J Earthq Eng 1:157–192

Chapter 7
Energy-Based Plastic Design

Abstract A compete methodology for the seismic design of plane steel frames is presented. It has been successfully used for moment resisting frames, eccentrically braced frames, concentrically braced frames, special truss moment frames and buckling restrained braced frames. In a performance-based design framework, the method uses pre-selected target drift and yield mechanisms as main performance limit states, which are directly related to the amount and distribution of damage, respectively. The design base shear for a specific seismic level is obtained by solving a balance of energy equation established by equating the work done by the structure as it is pushed up to the target drift to the seismic energy input on an equivalent to the structure elastic-plastic single-degree-of-freedom system. With a known design base shear, plastic design is used for determining member forces and dimensioning members and detailing connections. Since the base shear has been computed on the basis of the design target drift and yield mechanisms, this design method is accomplished in one step without iterations in contrast to code design methods requiring a two steps approach involving strength and deformation checking with iterations. The method is illustrated by four application examples covering plane steel moment resisting frames, eccentrically braced frames and concentrically braced frames, which also serve the purpose of demonstrating its advantages.

Keywords Plastic design method · Energy-based method · Global collapse mechanism · Performance-based seismic design · Plane steel frames · Moment resisting frames · Braced frames

7.1 Introduction

Current seismic design codes are based on elastic analysis and inelastic structural behavior under strong earthquakes is taken into account indirectly through the behavior (or strength reduction) factor q (or R). Thus, the design base shear force of a structure is obtained by combining elastic modal synthesis with elastic design spectra reduced by q (or R) to account for inelasticity. From the design base shear,

G. A. Papagiannopoulos et al., *Seismic Design Methods for Steel Building Structures*, Geotechnical, Geological and Earthquake Engineering 51,
https://doi.org/10.1007/978-3-030-80687-3_7

the member design forces are obtained by elastic structural analysis and the appropriate section selection is done through a strength checking. Finally, the deformation checking of the so dimensioned structure is performed iteratively until satisfaction of specific requirements. This iterative procedure usually leads to heavier sections than the original ones selected during the strength checking. During the section selection care is taken to observe the strong column-weak beam capacity design rule in order to avoid premature failure before the structure has been able to utilize its ductility capacity. Unfortunately, the inelastic behavior of a code-designed structure during an earthquake may result in a rather undesirable and uncontrollable response or collapse.

What is needed is the development of seismic design methods capable of controlling various design factors, such as design lateral forces, member strength hierarchy, desirable yield mechanism, structural and non-structural deformation etc. for various hazard levels in a performance-based seismic design framework. Methods of this kind that take into account one or more of these design factors have been presented in various chapters of this book starting with the displacement-based design method (Chap. 4), the hybrid force/displacement-based design method (Chap. 5) and the ductility-based plastic design method (Chap. 6). The method presented in this chapter uses preselected target drift and yield mechanism as the main performance objectives. The design base shear for a specific seismic level is obtained by solving a balance of energy equation between the work done by the structure moving to the target drift and the seismic input energy on an equivalent single-degree-of-freedom (SDOF) elastic-plastic system. Use of plastic design and the known base shear leads to member dimensioning and connection detailing. The whole procedure requires one step (strength checking), since the second step (deformation checking) is automatically satisfied from the start. The appropriate yielding mechanism is also automatically satisfied from the start.

The above briefly described method, which is the subject matter of this chapter, has been originally developed by Leelataviwat et al. (1999) for plane steel moment resisting frames (MRFs) and sequentially improved by Leelataviwat et al. (2002), Chao et al. (2007) and Lee et al. (2004), to include a better distribution of the lateral seismic forces and an explicit expression for the energy modification factor. The method was also extended to the seismic design of plane steel eccentrically braced frames (EBFs) by Chao and Goel (2006), special truss moment frames (STMFs) by Chao and Goel (2008), concentrically-braced frames (CBFs) by Bayat et al. (2010), buckling-restrained braced frames (BRBFs) by Sahoo and Chao (2010) and buckling-restrained knee-braced truss moment frames by Wongpakdee et al. (2014). Finally, one should mention an overview paper by Goel et al. (2010) that has examples only for MRF and especially the comprehensive book by Goel and Chao (2008), which has extensive theory and application examples pertaining to all kinds of plane steel frames mentioned previously.

At this point one should observe that the name of the method started as 'performance-based seismic design', changed to 'energy-based seismic design using yield mechanism and target drift' or 'performance-based seismic design using target drift and yield mechanism' and eventually became 'performance-based plastic design'.

The name given to the method in this book is 'energy-based plastic design' for two reasons: first for uniformity reasons the term 'performance-based' was taken out since almost all the methods considered here are 'performance-based' and second for differentiating it from the similar plastic design type of method considered in the previous chapter. Thus, the term 'energy-based' was added in front of the term 'plastic design' common to both methods. Indeed, the two methods are of the plastic design type and use the yielding mechanism as pre-selected design factor. However, the present method uses in addition the drift as a preselected design factor and employs a new lateral force distribution and the energy modification factor for a more accurate treatment. One should also point out that, while the method of Chap. 6 determines design moments for beams of MRFs directly on the basis of gravity loads, the present method determines those moments indirectly on the basis of equations coming from plastic analysis. Both methods determine axial forces in columns and the design moments at column bases in similar ways. Finally, the present method has been successfully extended and applied to a number of other plane steel frames (EBFs, CBFs etc.) in addition to the MRFs.

In the following sections, the theoretical foundations of the method will be presented followed by numerical examples for illustrating the method and demonstrating its merits.

7.2 Theoretical Foundations for MRFs

According to the present method, the design base shear force is determined for the given seismic level and selected target drift and yield mechanism by solving an energy-based algebraic equation (Goel et al. 2010). This equation is obtained on the basis of the energy balance equation

$$E_e + E_p = \gamma\left(\frac{1}{2}MS_v^2\right) \tag{7.1}$$

where E_e and E_p are the elastic and plastic energies, respectively, required to move the structure up to the target drift, S_v is the design spectral elastic pseudo-velocity, M is the total structural mass and γ is the energy modification factor, required for Eq. (7.1) to be valid (Fig. 7.1a). Having in mind the MRF structure of Fig. 7.1b considered here and using the expressions

$$E_e = \frac{1}{2}\left(\frac{W}{g}\right)\left(\frac{T}{2\pi}\frac{V_y}{W}g\right)^2 \tag{7.2}$$

$$E_p = V_y\left(\sum_{i=1}^{N}\lambda_i h_i\right)\theta_p \tag{7.3}$$

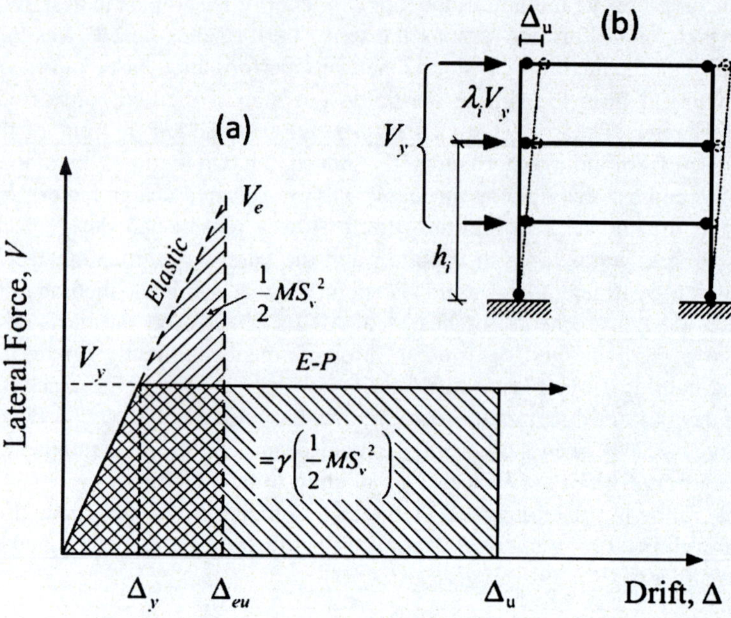

Fig. 7.1 Energy balance and MRF under lateral forces (after Goel and Chao 2008, reprinted with permission from ICC)

$$S_v = S_a\left(\frac{T}{2\pi}\right) \tag{7.4}$$

where W is the total structural weight, g is the gravity acceleration, V_y is the design base shear, θ_p is the target plastic drift ratio, h_i is the height of storey i from the base, λ_i is the base shear multiplier at storey level i, T is the fundamental period of the structure and S_a is the spectral pseudo-acceleration (usually given in the form ag with a being a dimensionless number), Eq. (7.1) takes the form

$$\left(\frac{V_y}{W}\right)^2 + \left(h^* \frac{8\pi^2\theta_p}{T^2 g}\right)\left(\frac{V_y}{W}\right) - \gamma(S_a/g)^2 = 0 \tag{7.5}$$

Solving Eq. (7.5) for its positive real root, the base shear can be obtained in the form

$$\frac{V_y}{W} = \frac{-\mathrm{a} + \sqrt{\mathrm{a}^2 + 4\gamma(S_a/g)^2}}{2} \tag{7.6}$$

where a is a dimensionless parameter given as

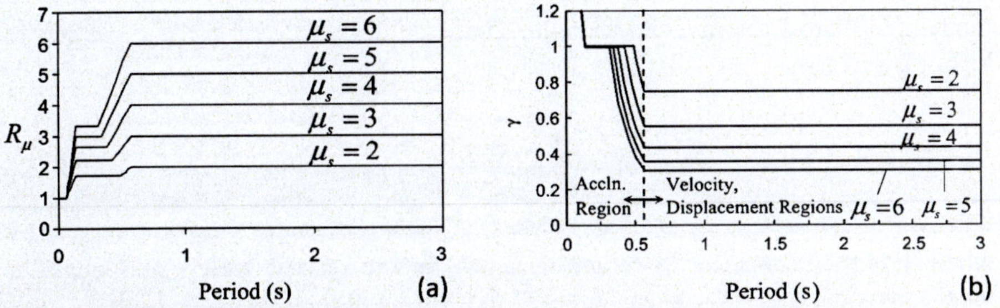

Fig. 7.2 Relations among R_μ, μ_s, T and γ for an elastoplastic SDOF: (**a**) idealized inelastic spectra; (**b**) energy modification factor γ in terms of μ_s and T (after Goel and Chao 2008, reprinted with permission from ICC)

$$a = h^* \frac{8\pi^2\theta_p}{T^2 g} \tag{7.7}$$

with

$$h^* = \sum_{i=1}^{N} \lambda_i h_i \tag{7.8}$$

and the fundamental period T usually determined approximately at this stage by the simple UBC (1994) formula $T = 0.035h_n^{0.75}$ where h_n is the total height of the frame in feet. The energy modification factor γ can be obtained from the expression

$$\gamma = \frac{2\mu_s - 1}{R_\mu^2} \tag{7.9}$$

for an elastoplastic SDOF, where μ_s is the structural ductility and R_μ is the ductility reduction factor given in terms of μ_s and T. Adopting the $R_\mu = f(\mu_s, T)$ simple relations of Newmark and Hall (1982) depicted in Fig. 7.2a, one can evaluate R_μ from there and use Eq. (7.9) to determine γ. Alternatively, one can eliminate R_μ between Eq. (7.9) and the relations depicted in Fig. 7.2a and obtain expressions providing γ in terms of μ_s and T in pictorial form as shown in Fig. 7.2b.

It should be noted here that Eq. (7.6) is based on the assumptions of an ideal elastoplastic behavior and full hysteretic loops for the structure. This implies that Eq. (7.6) is applicable to ductile steel MRFs considered here, to ductile steel EBFs considered in the next section as well as to ductile steel STMFs and BRBFs. For structures exhibiting pinched hysteretic loops like CBFs with buckling type of braces some modification of Eq. (7.6) is required as described in subsequent sections.

The design lateral force distribution along the height of an n-storey building frame (Fig. 7.1b) presented in codes, follows the shape of the elastic fundamental mode, i.e., that of the inverted triangle. Thus, the design lateral force F_i^* at storey level i is given as

$$F_i^* = \left(\frac{w_i h_i}{\sum_{j=1}^{n} w_j h_j}\right) V_y \tag{7.10}$$

where V is the base shear force and w_i and h_i denote seismic weight and height at level i from the base. The shear force V_i^* at level i will then be of the form

$$V_i^* = \left(\frac{\sum_{j=i}^{n} w_j h_j}{\sum_{j=1}^{n} w_j h_j}\right) V_y \tag{7.11}$$

Goel and Chao (2008) have found, through extensive inelastic dynamic analyses on building frames, that the distribution of maximum storey shears along the height is quite different than the elastic first mode one of the codes and can be described by the empirical expression

$$V_i = \left(\frac{\sum_{j=i}^{n} w_j h_j}{\sum_{j=1}^{n} w_j h_j}\right)^{0.75T^{-0.2}} V_y \tag{7.12}$$

where T is the fundamental period of the structure. Defining the ratio β_i as

$$\beta_i = \frac{V_i}{V_n} = \left(\frac{\sum_{j=i}^{n} w_j h_j}{w_n h_n}\right)^{0.75T^{-0.2}} \tag{7.13}$$

with V_n computed from Eq. (7.12) for $i = n$ ($\beta_{n+1} = 0$), one is able to write for the lateral force F_i at storey level i

$$F_i = (\beta_i - \beta_{i+1}) \left(\frac{w_n h_n}{\sum_{j=1}^{n} w_j h_j}\right)^{0.75T^{-0.2}} V_y \tag{7.14}$$

Thus, on the basis of Eqs. (7.10) and (7.14), the base shear multiplier λ_i has the form

$$\lambda_i = \left(\frac{w_i h_i}{\sum_{j=1}^{n} w_j h_j}\right) \tag{7.15}$$

for the elastic first mode shape case and

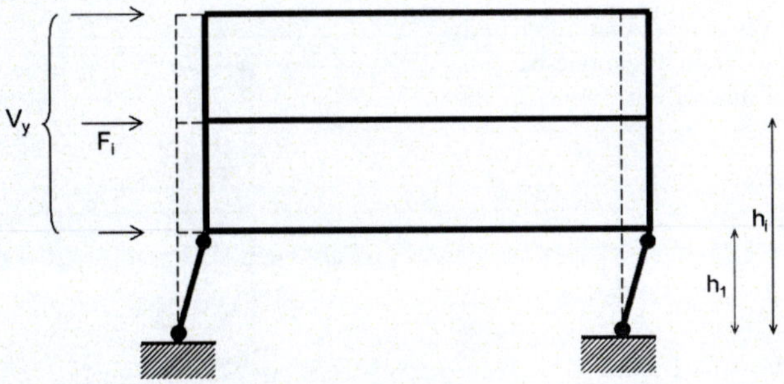

Fig. 7.3 One bay frame with first soft-storey mechanism

$$\lambda_i = (\beta_i - \beta_{i+1})\left(\frac{w_n h_n}{\sum_{j=1}^{n} w_j h_j}\right) \tag{7.16}$$

for the inelastic empirical shape case, respectively.

Following Fig. 7.1b depicting the desired yielding mechanism and equating the plastic work done at the hinges of the frame to the work done by the lateral seismic forces, one receives

$$\sum_{i=1}^{n} 2\beta_i M_{pbr} + 2M_{pc} = \sum_{i=1}^{n} F_i h_i = \left(\sum_{i=1}^{n} \lambda_i h_i\right) V_y = h^* V_y \tag{7.17}$$

where M_{pbr} is a reference plastic moment for beams, M_{pc} is the plastic moment at the column bases for the one bay frame considered here and β_i is an amplification factor expressing the effect of shear forces on beam moments found to be equal to $(V_i/V_n)^{1/2}$ (Leelataviwat et al. 1999). For a frame with m bays and hence $m + 1$ columns, the left hand side of Eq. (7.17) has to be replaced by the expression $\sum_{j=1}^{m}\sum_{i=1}^{n} 2\beta_i M_{pbrj} + \sum_{j=1}^{m+1} M_{pcj}$, with $M_{pbrj} = M_{pbr}$ and $M_{pcj} = M_{pc}$. The M_{pc} of Eq. (7.17) can be easily obtained by considering the first-soft-storey mechanism of Fig. 7.3 and writing down the balance of work in the form

$$4M_{pc} = V_y h_1 \tag{7.18}$$

In order to prevent the above mechanism to occur, one should consider the overstrength factor ψ and write the solution of Eq. (7.18) in the form

$$M_{pc} = \psi V_y h_1/4 \tag{7.19}$$

where ψ= 1.10–1.50. For a frame with m bays one should replace V_y in Eq. (7.19) by V_y/m. From the above, one can easily see that with known V_y, M_{pc} can be computed from Eq. (7.19) and then M_{pbr} can be determined from Eq. (7.17). The beam moments M_{pbi} can be finally determined from $M_{pbi} \geq (\beta_i/\varphi)M_{pbr}$ with $\varphi = 0.9$

Fig. 7.4 Column free body diagram in the one-bay frame of Fig. 7.1

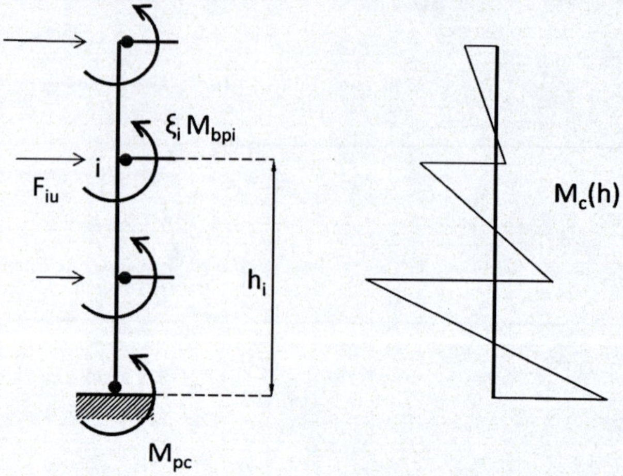

(AISC 1994) and the beam sections selected using $Z_b f_y$ with Z_b being the plastic modulus and f_y the nominal yield strength of steel used for beams. With known design moments for beams, their section selection can be done with the aid of a steel structures code (e.g. AISC 1994).

The final step has to do with the design of columns, which has to comply with the strong column-weak beam rule of capacity design. In order to ensure this, it is assumed that all plastic hinges in the beams are fully strain-hardened when the structural drift is at the target ultimate level. Thus, the beam nominal plastic moments M_{pbi} are amplified by an overstrength factor ξ_i and the design moment distribution in the columns along the height $M_c(h)$ is determined by employing equilibrium at a column free body diagram subjected to beam moments $\xi_i M_{pbi}$, the moment M_{pc} at the column base and the updated values F_{iu} of the lateral forces due to this beam moment amplification, as shown in Fig. 7.4. The overstrength factor ξ_i, which is defined as the product of the overstrength due to the difference between nominal and actual yield strengths of steel and the overstrength due to strain hardening, can be taken as 1.10 (Leelataviwat et al. 1999). On account of Eqs. (7.10) or (7.14), one has

$$F_{iu} = \lambda_i V_u \tag{7.20}$$

where V_u is the updated base shear. Then the equilibrium of forces for a column like the one in Fig. 7.4 reads

$$\sum\nolimits_{i=1}^{n} F_{iu} h_i = M_{pc} + \sum\nolimits_{i=1}^{n} \xi_i M_{pbi} \tag{7.21}$$

Substitution of Eq. (7.20) into Eq. (7.21) and solution of the resulting equation for V_u yields

$$V_u = (1/h^*)\left(M_{pc} + \sum\nolimits_{i=1}^{n} \xi_i M_{pbi}\right) \tag{7.22}$$

where h^* is given by Eq. (7.8). Thus, in view of Eq. (7.22), Eq. (7.20) takes the form

$$F_{iu} = (\lambda_i/h^*)\left(M_{pc} + \sum\nolimits_{i=1}^{n} \xi_i M_{pbi}\right) \tag{7.23}$$

Then, the variation of the column design moment $M_c(h)$ along the height h with the column treated as a cantilever can be expressed as

$$M_c(h) = \sum\nolimits_{i=1}^{n} \delta_i \xi_i M_{pbi} - \sum\nolimits_{i=1}^{n} \delta_i F_{iu}(h_i - h) \tag{7.24}$$

where $\delta_i = 1$ for $h < h_i$ and $\delta_i = 0$ for $h > h_i$ and M_{pbi}, M_{pc} and F_{iu} are given by Eqs. (7.17), (7.19) and (7.23), respectively. In addition, the variation of the column axial force $P_c(h)$ along the height h of the column can be expressed as

$$P_c(h) = \sum\nolimits_{i=1}^{n} \delta_i \left(\frac{2\xi_i M_{pbi}}{L}\right) + P_{cg}(h) \tag{7.25}$$

where L is the length of the beams and $P_{cg}(h)$ is the axial force due to gravity loads at height h. With known values of $M_c(h)$ and $P_c(h)$, one can dimension the column considering it to be under a bending moment and axial force combination. At this point one should observe that Eqs. (7.24) and (7.25) are for columns in an equivalent one-bay frame. For the multi-bay frame, the internal forces of the exterior columns remain the same as those in the one-bay frame, while the internal forces of the interior columns can be approximately taken as follows: their moment as twice that of the one-bay frame and their axial seismic force as zero. A final note has to do with the consideration of P-Δ effects. Column moments computed from Eq. (7.24) are amplified by using factors, like B_1 and B_2 of the AISC (2005a) in order to take P-Δ effects into account.

The two flowcharts in Figs. 7.5 and 7.6 taken from the book of Goel and Chao (2008) describe the basic steps of the method pertaining to the computation of the design base shear and the lateral force distribution (Fig. 7.5) and to its application to MRF including the design of elements (Fig. 7.6).

Closing this section, one could briefly discuss three general issues, which are related not only to steel MRFs considered here but to other steel frames, such as EBFs or CBFs considered in the following sections. The first issue has to do with the determination of the yield drift ratio θ_y (%) needed in order to compute the plastic drift ratio θ_p (%) from the relation $\theta_p = \theta_u - \theta_y$, where θ_u (%) is the assumed target drift ratio. With known θ_p, one can easily determine the base shear ratio V/W from Eqs. (7.6) and (7.7). Table 7.1 provides approximate values of yield drift ratios θ_y for steel MRFs, EBFs, CBFs and STMFs taken from the book of Goel and Chao (2008). These values are very useful for design purposes.

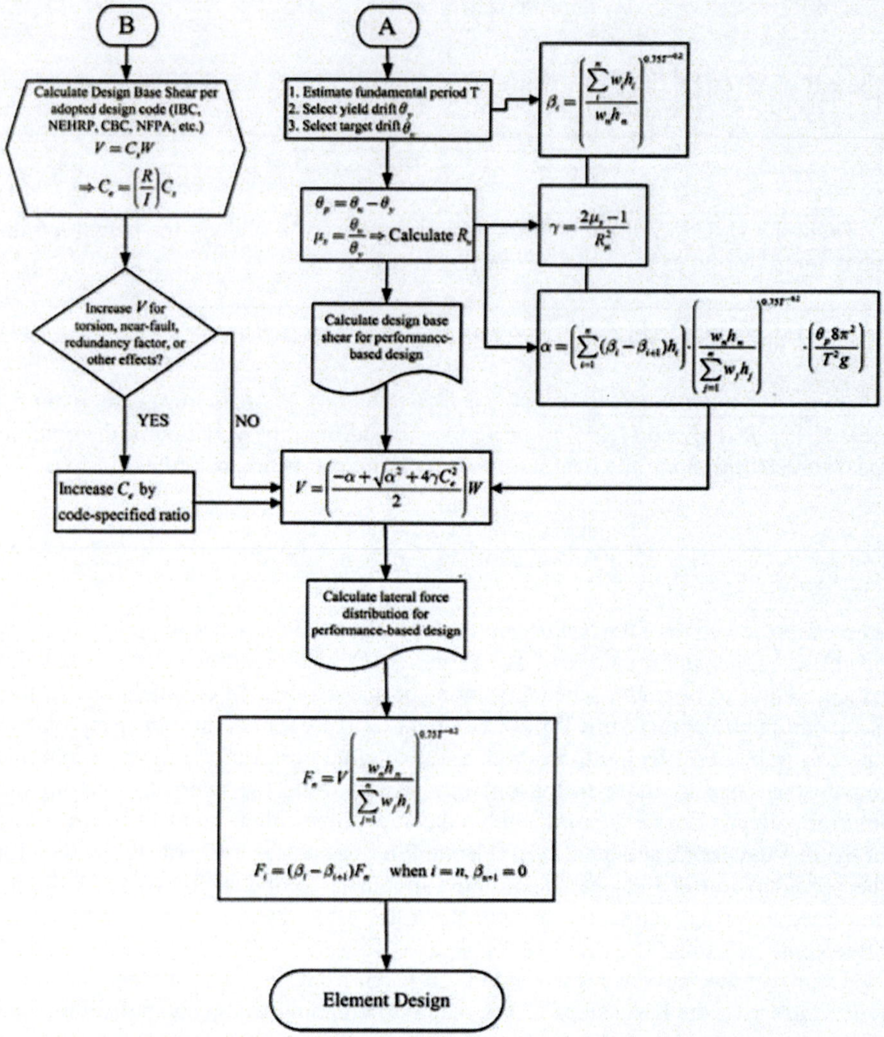

Fig. 7.5 Flowchart for determining design base shear and lateral force distribution with $C_e = S_a$ being normalized with respect to g (after Chao and Goel 2006, reprinted with permission from AISC)

The second issue concerns the determination of the fundamental period T of various types of steel frames needed for the computation of the base shear ratio V/W from Eqs. (7.6) and (7.7). A good approximation of T for preliminary design purposes is offered from the very simple expressions of the form $T = T(h)$, where h is the total height of the steel frame. Table 7.2 presents such expressions for various types of steel frames taken from ASCE/SEI 7-10 (2010) apart from that for CBFs which was taken from Tremblay (2005).

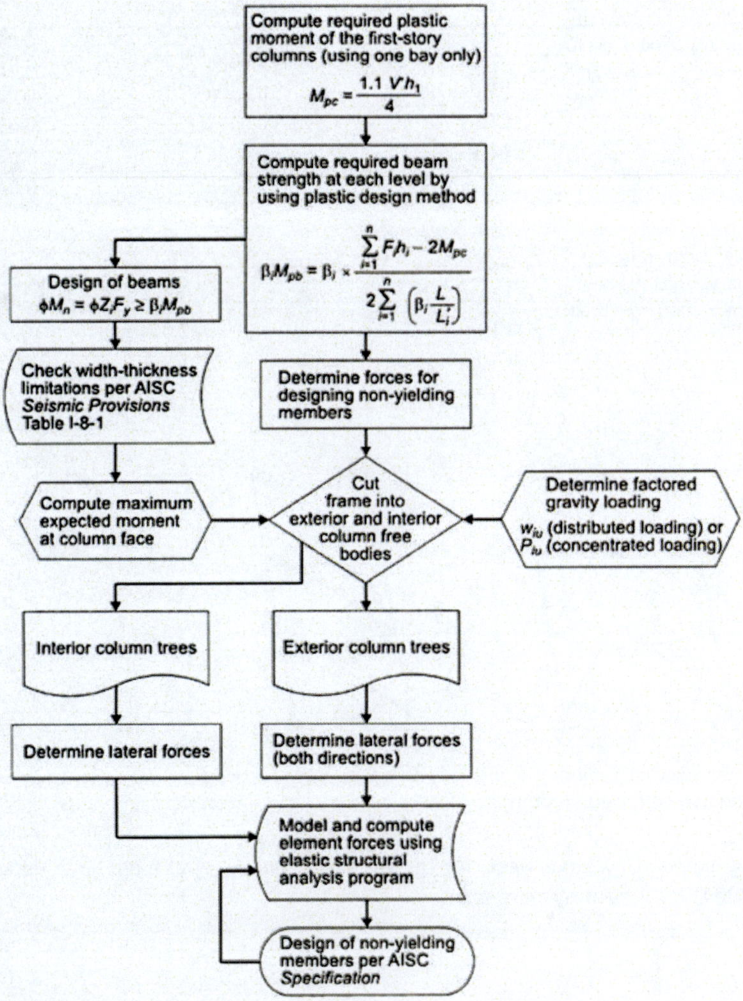

Fig. 7.6 Flowchart for moment frames including element design (after Goel and Chao 2008, reprinted with permission from ICC)

Table 7.1 Yield drift ratios for various types of steel frames

Frame type	Yield drift ratio θ_y
MRF	1.00%
EBF	0.50%
CBF	0.30%
STMF	0.75%

The third issue has to do with the three performance levels associated with the method, namely the IO, LS and CP corresponding to FOE, DBE and MCE. $IDR(\theta_u)$ limit values for various types of frames are given in Table 7.3 (Goel and Chao 2008;

Table 7.2 Expressions of fundamental period T (s) for various types of steel frames

Frame type	T (h in ft)	T (h in m)
MRFs	$0.028h^{0.80}$	$0.0724h^{0.80}$
EBFs	$0.030h^{0.75}$	$0.0731h^{0.75}$
CBFs	$0.0076h$	$0.025h$
BRBFs	$0.030h^{0.75}$	$0.0731h^{0.75}$
All others	$0.020h^{0.75}$	$0.0488h^{0.75}$

Table 7.3 Limit $IDR(\theta_u)$ values for various types of steel frames and performance levels

Frame type	IO (FOE)	LS (DBE)	CP (MCE)
MRFs	1.0%	2.0%	3.0%
EBFs	1.0%	2.0%	3.0%
CBFs	1.0%	1.5%	2.5%
STMFs	1.0%	2.0%	3.0%

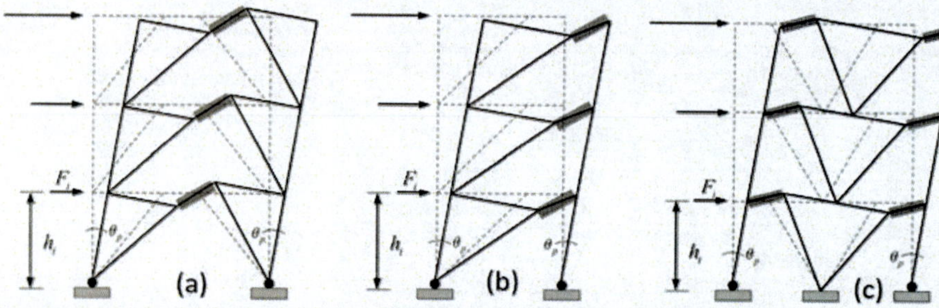

Fig. 7.7 Preselected yield mechanism of EBFs with various configurations (after Chao and Goel 2006, reprinted with permission from AISC)

Bayat et al. 2010). Values of θ_u for the IO performance level have been taken from EC8 (2004) by assuming bare frames.

7.3 Theoretical Foundations for EBFs

The present section is based on the work of Chao and Goel (2006). Consider the one- or two- bay multistory plane steel eccentrically braced frames (EBFs) of Fig. 7.7 with short seismic links shown with very thick line segments. The frames are under vertical load w_i (including dead and live load) at every storey i and lateral seismic forces F_i at every storey level i. It is assumed that all inelastic deformations are restricted in the seismic links exhibiting shear yielding. In Fig. 7.7, one can also see the desired global yield mechanisms for the frames consisting of shear yielding in the links plus flexural plastic hinges at the column bases. For all these three configurations of EBFs in Fig. 7.7, the design base shear ratio V/W, where V is the design base shear and W the total frame weight, is given again by Eqs. (7.6)–(7.9) developed for MRFs with θ_p being the global inelastic drift ratio equal to the difference between the

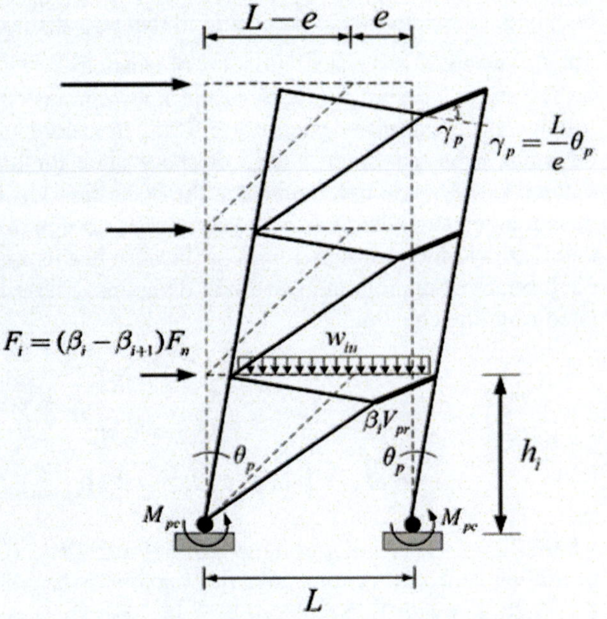

Fig. 7.8 One-bay EBF with preselected yield mechanism (after Chao and Goel 2006, reprinted with permission from AISC)

preselected target drift ratio θ_u and the yield drift ratio θ_y usually assumed to be 0.5% for EBFs.

On the basis of the one-bay EBF of Fig. 7.8 with its preselected yield mechanism, one can calculate the required strength of its shear links. From Fig. 7.8, the link plastic rotation angle $\gamma_p = (L/e)\theta_p$, where L is the bay length of the frame and e is the link length. Then, the required shear strength of shear links $\beta_i V_{pr}$ for the EBF of Fig. 7.8 at any level i is of the form (Chao and Goel 2006)

$$\beta_i V_{pr} = \left(\beta_i / L \sum\nolimits_{i=1}^{n} \beta_i\right)\left(\sum\nolimits_{i=1}^{n} F_i h_i + \frac{1}{2} L(L-e) \sum_{i=1}^{n} w_{iu} - 2M_{pc}\right) \tag{7.26}$$

where h_i is the height of the storey i from the base, F_i are the lateral forces at storey i (valid only for one-bay), w_{iu} is the uniformly distributed vertical load at storey i equal to $1.2G + 0.5Q$ (AISC 2005b), V_{pr} is the required link shear strength at the top level, M_{pc} is the required plastic moment at the column bases and β_i is the shear distribution ratio of Eq. (7.13). Moment M_{pc} is evaluated from Eq. (7.19) with V_y being the base shear for one-bay braced frame or V_y/m for a frame with m braced bays. The design of shear links is then based on the equation (AISC 2005a)

$$\varphi V_n = 0.9 V_p = 0.9\left[0.6 f_y \left(d_b - 2t_f\right) t_w\right] \geq \beta_i V_{pr} \tag{7.27}$$

where f_y is the nominal yield strength of steel and d_b, t_f and t_w denote the total height, flange thickness and web thickness of the I section. In addition, for avoiding local buckling, the width to thickness ratio in flanges satisfies $b/t_f \leq 0.38\sqrt{E/f_y}$ (AISC 2005b), where E is the steel modulus of elasticity.

The seismic design of braces, beams and columns outside the shear links is done following the capacity design rules. Thus, in accordance with AISC (2005b), one determines the maximum values of shear force for links V_u, link moment at the column side M_C and link moment at the brace side M_B, taking into account material overstrength, strain hardening and shear resistance in the link flanges, and then uses them to determine design forces in braces, beams and columns of the basis of equilibrium of appropriate free-body diagrams. These maximum values are calculated from the equations

$$V_u = 1.25 R_y V_p \tag{7.28}$$

$$M_C = R_y M_p \tag{7.29}$$

$$M_B = \left[e\left(1.25 R_y V_p\right) - R_y M_p\right] \geq 0.75 R_y M_p \tag{7.30}$$

where $R_y = f_y$ expected/f_y nominal and $e \leq 1.6M_p/V_p$. It should be noted that for the beam segment outside the link, the maximum link shear V_u is calculated from Eq. (7.28) but with 1.25 replaced by 1.10. The whole design procedure is described by the flowcharts of Figs. 7.5 and 7.9.

7.4 Theoretical Foundations for CBFs

The present section is based on the work of Bayat et al. (2010). Consider a one-bay three storey plane steel concentrically braced frame (CBF) of the chevron type, as shown in Fig. 7.10, as a representative CBF. This frame is under the vertical loads (on its horizontal beams) consisting of dead and live loads and lateral (seismic) loads F_i at every storey level i. The frame is pushed by the lateral forces to its target drift limit state, as shown in Fig. 7.10. Dissipative parts in this frame are the braces, which deform inelastically in the form of yielding and buckling. Thus, its design yield mechanism shown in Fig. 7.10 is the result of inelastic deformation in the braces and plastic hinges at the column bases.

The design base shear expression for steel MRFs, EBFs or BRBFs was derived in the two previous sections of this chapter on the assumption of elastoplastic hysteretic behavior of these frames. However, bracing buckling in a CBF results in a pinched hysteresis loop with an area A_1, which is smaller than the area A_2 for an elastoplastic hysteresis loop of a BRBF of the same geometry, as shown in Fig. 7.11 describing the corresponding force-displacement relationships. The factor $\eta = A_1/A_2 < 1$, defined as the ratio of the corresponding dissipation energies, can be used in the balance of energy Eq. (7.1) to multiply its left hand side ($E_e + E_p$). This finally leads to a modified base shear expression (7.6) with γ being replaced by γ/η. An approximate value for η can be taken between 0.35 and 0.50. Another approximate but more

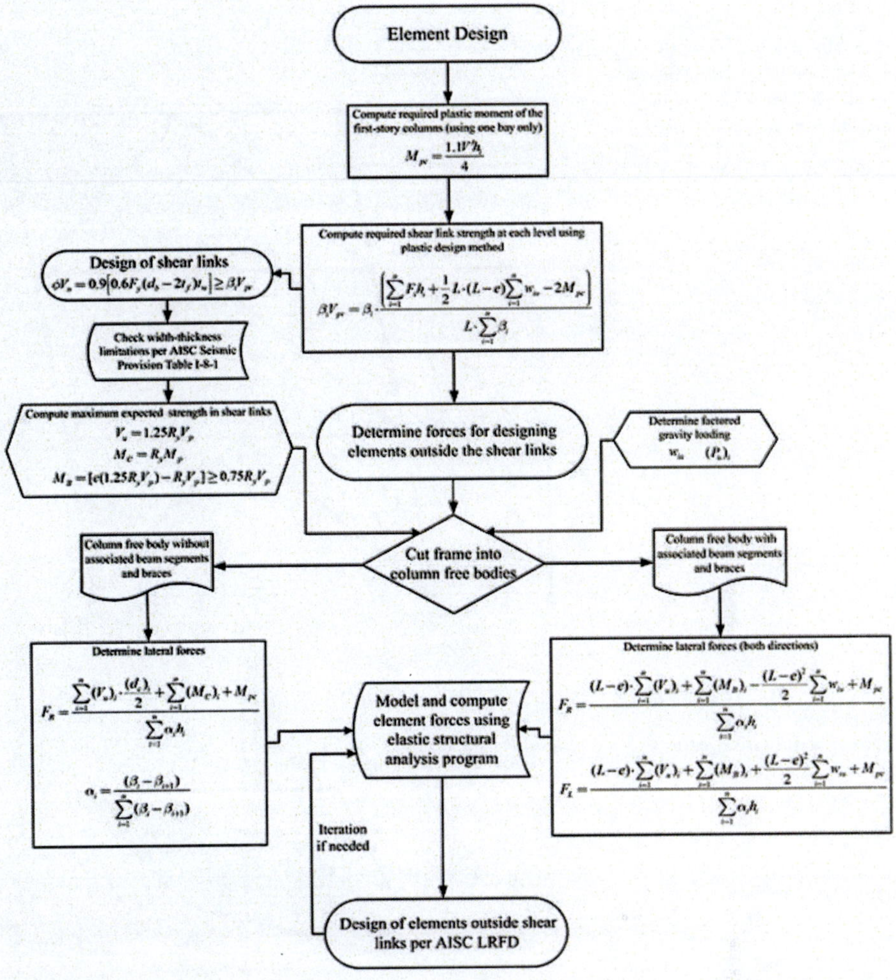

Fig. 7.9 Flowchart for EBFs including element design (after Chao and Goel 2006, reprinted with permission from AISC)

accurate way to quantify this pinching effect is to use in Eq. (7.6) for the base shear expression a modified target drift equal to θ_p/C_2, where C_2 is a factor mainly depending on the fundamental period of the frame T and receiving the value of 1 for $T \geq 1s$ and larger than 1 for $T < 1s$. This factor has been obtained by extensive parametric studies on the peak seismic displacement response of SDOF degrading hysteretic systems (FEMA 356 2000). Bayat et al. (2010) have proposed a similar factor (called new factor λ), which takes into account in addition to pinching, P-Δ effects. Figure 7.12 shows the variation of this new λ factor with period T together with the factor C_2 defined for the IO, LS and CP performance levels. The new λ factor is equal to 1.10 for $T \geq 1s$ and increases linearly for $T < 1$ s till the value of

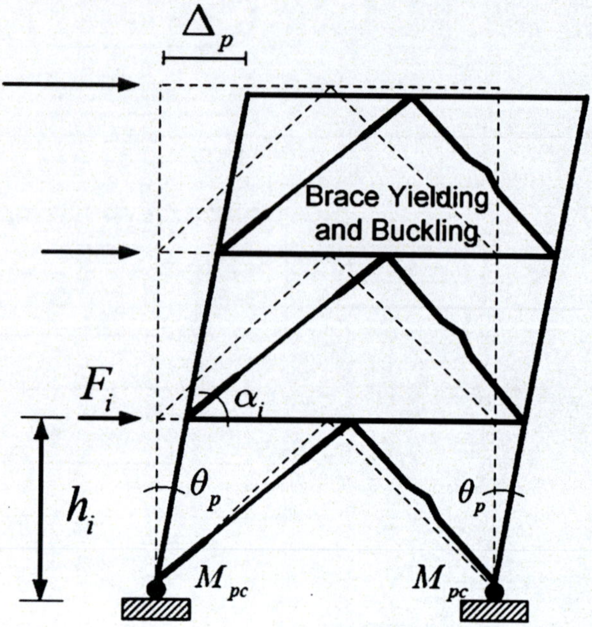

Fig. 7.10 Target yield mechanism on CBF with chevron bracing (after Bayat et al. 2010, reprinted with permission from UMCEE)

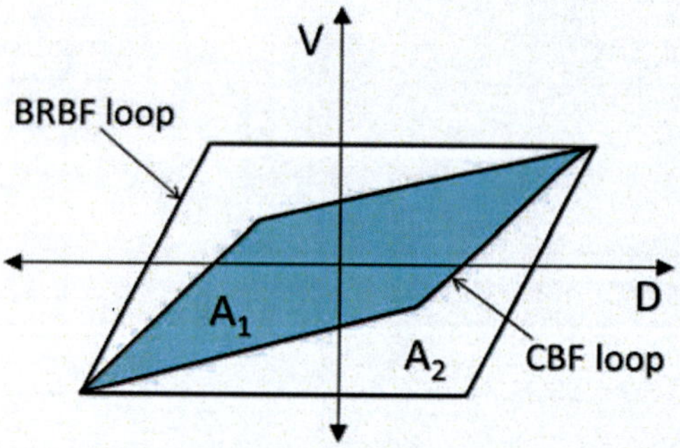

Fig. 7.11 Hysteresis loops for BRBFs and CBFs and corresponding dissipation areas A_2 and A_1, respectively

1.25 for $T = 0$. It is observed that the new λ factor is closer to C_2 for the CP level when $T = 0.50 - 0.75$ s and approaches C_2 for the LS level and longer T.

Once the base shear force has been computed by using Eqs. (7.6)–(7.8) with θ_p replaced by θ_p/λ, the seismic design member forces can be determined and the member design can be performed. The design of bracing members involves the checking of three criteria: the strength, the fracture and the compactness criterion. According to the strength criterion, the braces are designed in order to resist the total design storey shear (columns do not participate) at their ultimate state (plastic design) of tension yielding and post-buckling. Thus

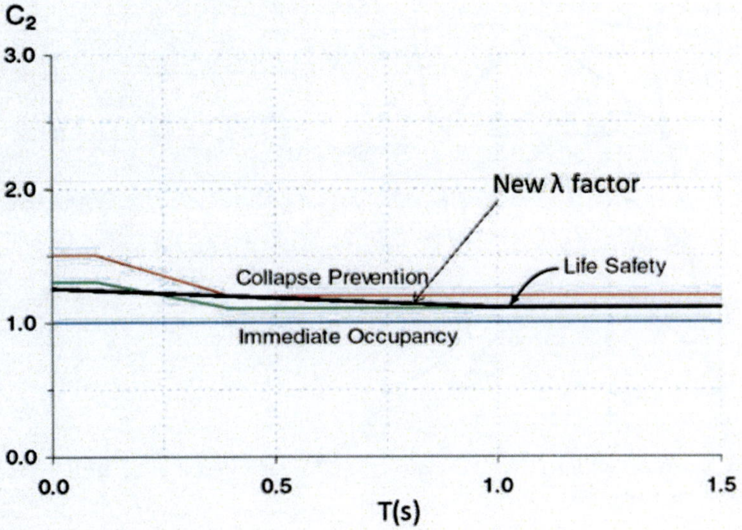

Fig. 7.12 Comparison of the new λ-factor with C_2 values in FEMA 356 (2000) for site class B and framing type 1

$$V_i \leq 0.9\, cos\, a_i \left(P_y + 0.5P_{cr}\right)_i \tag{7.31}$$

where V_i is the shear at storey level i for an equivalent one-bay CBF, P_y is the nominal axial tensile strength of braces, P_{cr} is the nominal axial compressive strength of the braces and a_i is the angle of braces with the horizontal directions (Fig. 7.10) with $P_y = f_y A_g$ and A_g being the gross sectional area. The axial buckling stress f_{cr} is given as

$$\begin{aligned} f_{cr} &\leq \left(0.658^{\,f_y/f_e}\right) f_y \quad for \ f_e \geq 0.44 f_y \\ f_{cr} &= 0.877 f_e \quad for \ f_e < 0.44 f_y \end{aligned} \tag{7.32}$$

where $f_e = \pi^2 E/(KL/r)^2$. The post-buckling strength of $0.5P_{cr}$ refers to braces buckling in-plane. For braces buckling out-of-plane the post-buckling strength becomes $0.3P_{cr}$. The effective length factor k is taken as 0.5 and 0.85 for in-plane (K_x) and out-of-plane (K_y) directions, respectively. If $K_xL/r_x > K_yL/r_y$ in-plane buckling is ensured, where L is the brace length and r the radius of gyration of the cross-section.

According to the fracture criterion, used for HSS (hollow structural sections) braces, the brace fracture life N_f is computed from the empirical relationship

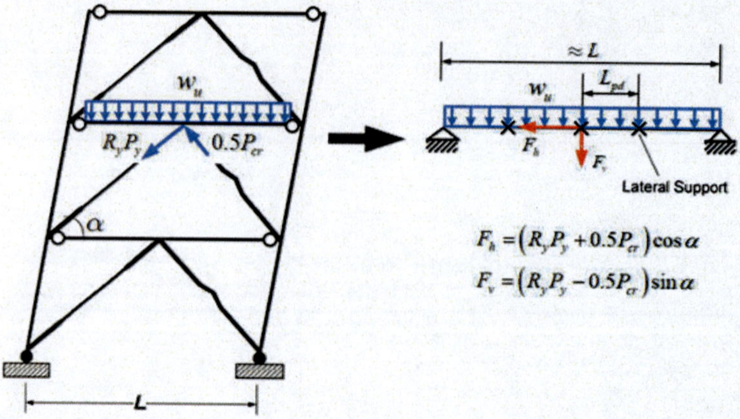

Fig. 7.13 Beam design forces for a chevron-type CBF (after Bayat et al. 2010, reprinted with permission from UMCEE)

$$N_f = \begin{cases} 262(b/d)(KL/r)/[(b-2t)/t]^2 & for\ KL/r > 60 \\ 262(b/d)60/[(b-2t)/t]^2 & for\ KL/r \le 60 \end{cases} \quad (7.33)$$

where d and b are the gross section depth and width of the section, respectively ($b \geq d$), t is the wall thickness and KL/r is the slenderness ratio. A minimum value of $N_f = 100$ is suggested for HSS braces. Finally, the compactness criterion for the brace section (b/t or $h/t_w \leq 0.64\sqrt{E/f_y}$) is automatically satisfied for $N_f \geq 100$. The design of non-yielding (non-dissipative) members (beams and columns) is done in accordance with the capacity design rule. The beams intersected by the braces should be designed on the assumption that braces do not support gravity loads. The beams should be designed to support vertical and horizontal unbalanced forces coming from the difference between tension and compression brace forces after buckling, as shown in Fig. 7.13 with F_h being the horizontal unbalanced force and R_y the ratio of the expected over the nominal yield strength. Beams are assumed to be pinned because shear splices are used at the ends and designed as beam-columns due to the presence of large axial forces.

The columns are designed only for axial forces because very small or zero moments are transferred into them due to beam shear splices. Axial forces come mainly from gravity loads and vertical components of brace forces. In the pre-buckling limit state, the design axial force in an exterior column is (Fig. 7.14a)

$$P_u = (P_{tr})_i + (P_b)_i + (P_{cr}sina)_{i+1} \quad (7.34)$$

where $(P_{tr})_i$ is the tributary gravity ($1.2G + 0.5Q$) load from the transverse direction at storey level i, $(P_b)_i = (1/2)w_uL$ is the reaction of the beam at level

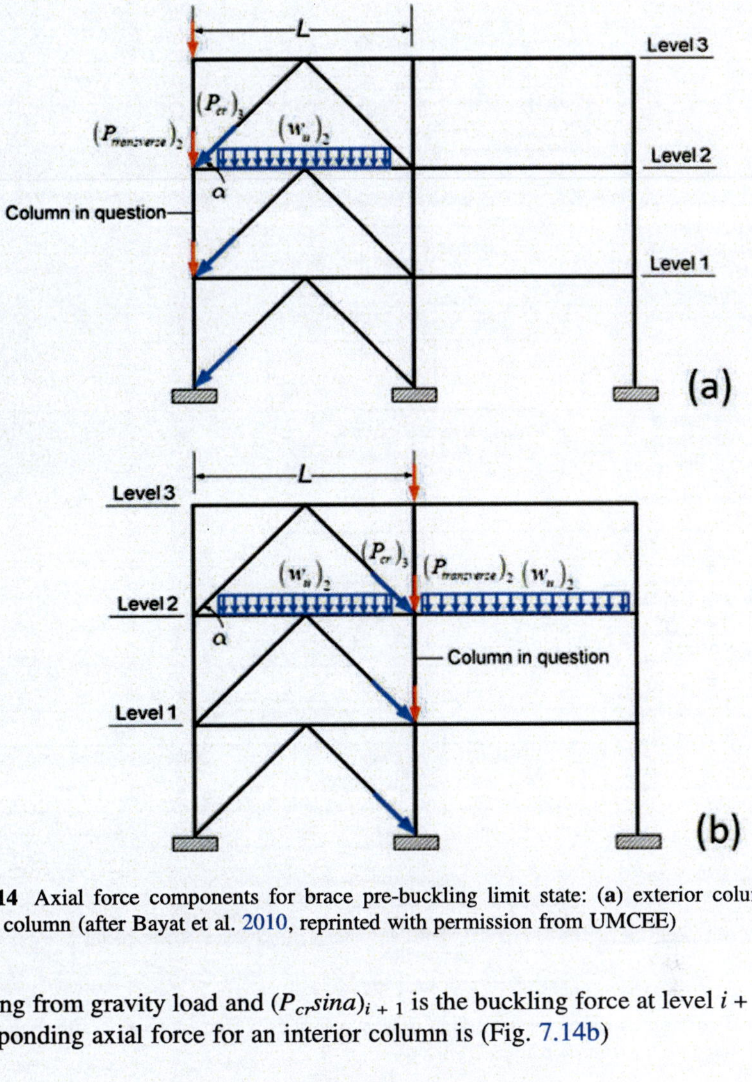

Fig. 7.14 Axial force components for brace pre-buckling limit state: (**a**) exterior column; (**b**) interior column (after Bayat et al. 2010, reprinted with permission from UMCEE)

i coming from gravity load and $(P_{cr}sina)_{i+1}$ is the buckling force at level $i + 1$. The corresponding axial force for an interior column is (Fig. 7.14b)

$$P_u = (P_{tr})_i + \sum (P_b)_i + (P_{cr}sina)_{i+1} \tag{7.35}$$

In the post-buckling limit state, the design axial force is given again by Eqs. (7.34) and (7.35) but with the addition of the extra term $(1/2)F_v$ in their right hand sides representing the unbalanced vertical force F_v as expressed in Fig. 7.13. Thus, the design axial force P_u for a column is the governing one coming from the pre-buckling or post-buckling limit state. Column design is finally done by using Eq. (7.32) with $K = 1$ (AISC 2005a). According to Bayat et al. (2010), the column section compactness criterion of $b/t \leq 0.30\sqrt{E/f_y}$ (AISC 2005b) is reserved only for the first level columns where plastic hinges may occur at their bases, while for

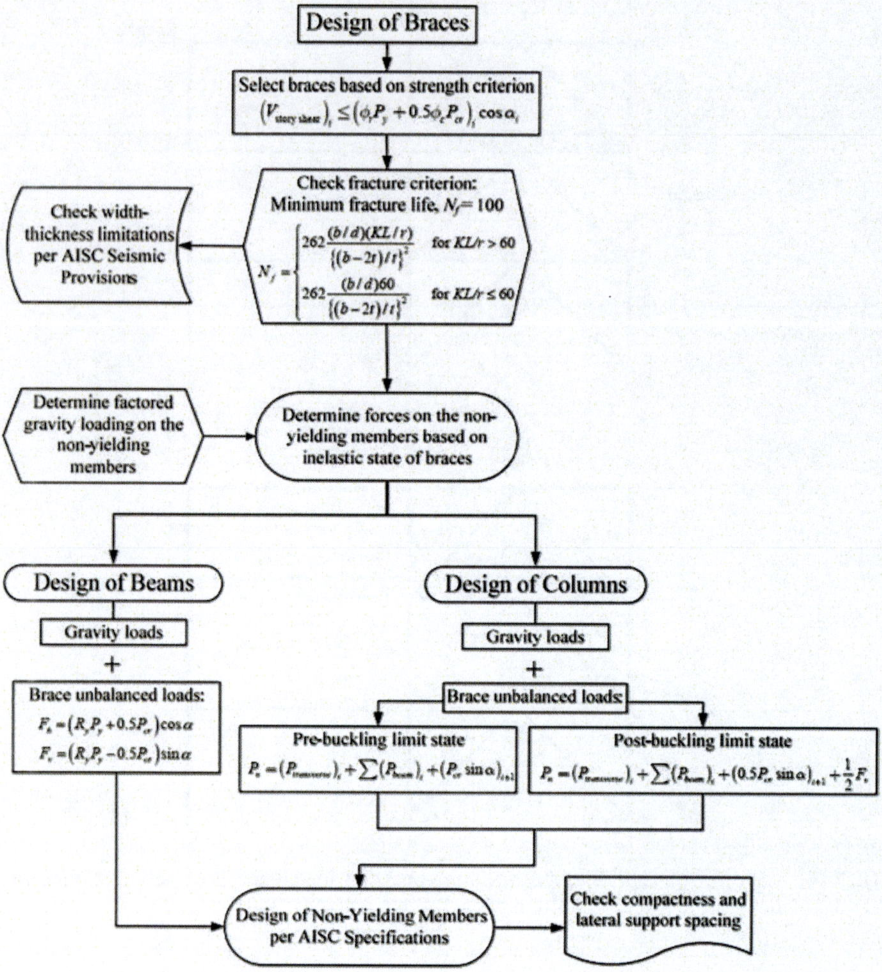

Fig. 7.15 Performance-based plastic design flowchart for CBF: Element design (after Bayat et al. 2010, reprinted with permission from UMCEE)

columns at all other levels the less restrictive $b/t \leq 0.38\sqrt{E/f_y}$ criterion can be used.

Flowcharts in Figs. 7.5 and 7.15 describe the whole design procedure for which more details can be found in Bayat et al. (2010).

7.5 Numerical Examples

This section presents four numerical examples to illustrate the energy-based plastic design method of this chapter and demonstrate its advantages.

7.5.1 *Seismic Design of a MRF*

Consider a five-storey, three-bay steel MRF, which was a part of the lateral load resisting system of a building significantly damaged during the 1994 Northridge earthquake. This frame was designed by the UBC (1994) and the proposed method and the two designs compared on the basis of pushover and NLTH analyses (Leelataviwat 1998; Leelataviwat et al. 1999). The frame has a bay span of 25 ft, a height of 15 ft for the first floor and heights of 14 ft for the remaining four floors. The frame resists seismic forces coming from a weight $W = 2850.55$ kips distributed to the floors as 668.75, 525.90, 525.90, 525.90 and 604.10 kips from the first to the fifth floor. Figure 7.16, taken from Leelataviwat et al. (1999), shows the original frame design as well as the redesigned one by the present method.

The design of the frame by the present method assumes a target drift ratio of $\theta_u = 2\%$ under the DBE (or the LS performance level) and since the yielding drift ratio is usually taken for MRFs as $\theta_y = 1\%$, the plastic drift (or plastic rotation) is equal to $\theta_p = \theta_u - \theta_y = 1\%$. The fundamental period of the frame on account of the formula $T = 0.035h_n^{0.75}$ with $h_n = 71$ ft turns out to be $T = 0.86$ s. Thus, from the pseudo-acceleration design spectrum of UBC (1994) for the DBE with $Z = 0.4$ (seismic zone 4), $I = 1.0$ (standard occupancy), $S = 1.5$ (soil type S3) and $R = 8$ (for ductile steel MRF), one has for $T = 0.86$ s that $S_a = 0.83$ g. Using Eqs. (7.7) and

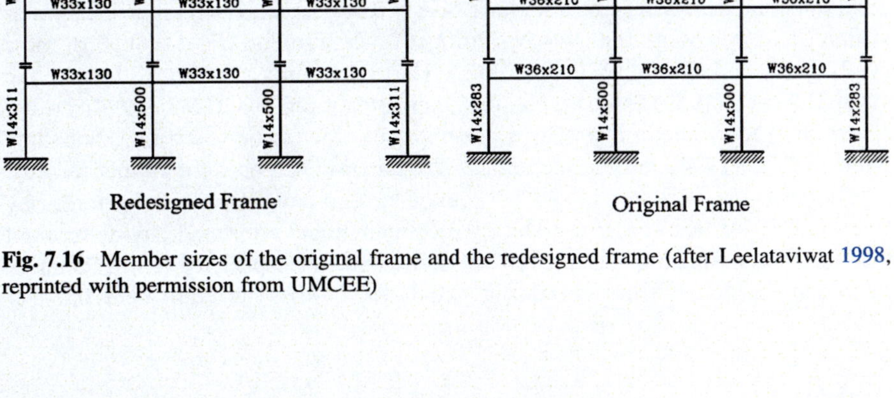

Fig. 7.16 Member sizes of the original frame and the redesigned frame (after Leelataviwat 1998, reprinted with permission from UMCEE)

(7.8) and assuming for simplicity $\gamma = 1$ and a seismic shear force distribution of the inverted triangle type (Eq. 7.15), one can compute $a = 1.73$. Thus, Eq. (7.6) can yield $V_y/W = 0.333$ leading to a base shear force $V_y = 0.333 \cdot 2850.55 = 949.23$ kips. At this point and before proceeding any further, it is interesting to check, which of the two performance levels corresponding to the DBE and the MCE controls the design. For the MCE one can assume a target drift ratio $\theta_u = 3\%$ implying $\theta_p = 2\%$ and determine $S_a = 1.5 \cdot 0.83$ g $= 1.245$ g. Thus, using Eqs. (7.6)–(7.8), one can determine for $\gamma = 1$ that $V_y/W = 0.40$ meaning that the MCE controls the design. However, this is not really true as one can easily find by repeating the computations for the realistic case of $\gamma \neq 1$. Indeed, for the DBE case one has $\mu_s = 2$, $R_\mu = 2$ and hence from Eq. (7.9) $\gamma = 0.75$ and for the MCE case $\mu_s = 3$, $R_\mu = 3$ and hence from Eq. (7.9) $\gamma = 0.55$. Thus, using again Eqs. (7.6)–(7.8) with the above values of γ, one receives V_y/W equal to 0.26 and 0.23 for the DBE and MCE cases, respectively, implying that the DBE really controls the design.

In any case, on the basis of the $V_y/W = 0.333$ ratio for the DBE with $\gamma = 1$ and the simple inverted triangle type of shear force distribution, the design proceeds as described in Sect. 7.2 by assuming steel grade with $f_y = 50$ ksi for all members and an overstrength factor equal to 1.05 for all beams. The section selection for beams and columns is shown in Fig. 7.16 together with that of the original frame done on the basis of the UBC (1994) with a V_y/W ratio equal to 0.09. Thus, with a $V_y/W = 0.333$ for the redesigned frame, i.e., 3.7 times that of the original frame, the design base shear and hence the member design forces of the redesigned frame are larger than those of the original frame. However, the total steel weight of 154.6 kips for the redesigned frame is almost equal to the 153.2 kips weight of the original frame. This is because in most frames designed by codes, the design is controlled by drift rather than strength. By looking at the member sizes of Fig. 7.16, one can observe that the redesigned frame has larger column sizes and smaller beam sizes than the corresponding ones of the original frame, as a result of enforcing the strong column-weak beam rule in the preselected mechanism of the design by the present method.

In order to further compare the two designs of Fig. 7.16, nonlinear static (push-over) and dynamic analyses of the original and redesigned frames were performed by the nonlinear finite element program SNAP-2DX (Rai et al. 1996). Lateral load distribution in the pushover analysis was the one of UBC (1994) for design lateral forces. NLTH analyses were conducted with the aid of four earthquake motions: the 1940 Imperial Valley/El Centro station ($PGA = 0.32$ g), the 1994 Northridge/Sylmar station ($PGA = 0.84$ g), the 1994 Northridge/Newhall station ($PGA = 0.59$ g) and a synthetic ground motion ($PGA = 1.00$ g). The real records were scaled to have the same Housner (1959) intensity as that computed from the UBC (1994) design spectrum. The synthetic record was generated in order to be spectrum compatible with the UBC (1994) design spectrum. A 20 s duration was used for all four records.

The analyses were conducted on equivalent one-bay frames of the three-bay frames of Fig. 7.16 in order to reduce the computational effort and having in mind that this approximation has proven to be acceptable, especially if one is more interested in global behavior (Basha and Goel 1994). The equivalent one-bay

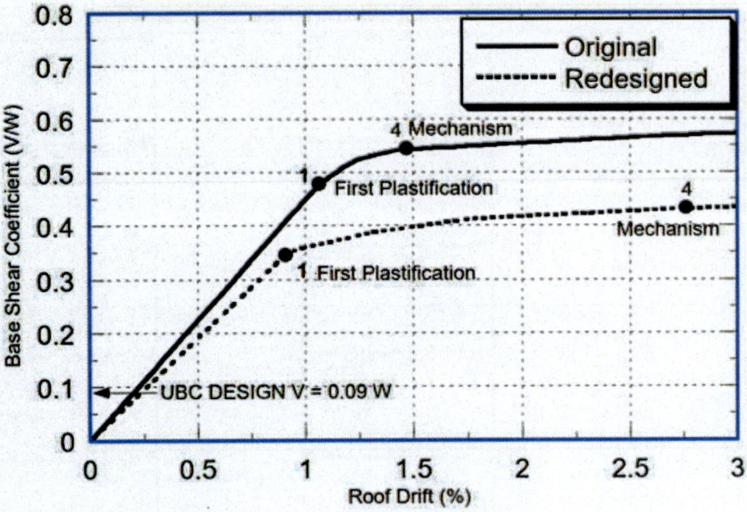

Fig. 7.17 Base shear versus roof drift response of the original and the redesigned frames (after Leelataviwat 1998, reprinted with permission from UMCEE)

frame has the same moment of inertia, area, modulus of elasticity and yield moment in beams as those in the three-bay frame. The corresponding properties in columns of the one-bay frame are equal to one-sixth of the sum of those in the three-bay frame. The equivalent one-bay frame has one-third of the mass of the three-bay frame. The floor masses were lumped at the beam-to-column joints, while viscous damping was assumed to be 2% of the critical one and proportional only to the mass matrix.

Figure 7.17 shows the base shear ratio V_y/W versus the roof drift ratio (%) as obtained from pushover analysis, while Fig. 7.18 depicts the sequence of plastic hinge formation in association with Fig. 7.17 for both the original and the redesigned equivalent one-bay frames. It is observed in Fig. 7.17 that, while the original frame satisfies the UBC (1994) drift ratio limit, the redesigned one slightly exceeds it. Both frames exhibit significant overstrength above the UBC (1994) design force level, six and four times for the original and the redesigned frame, respectively. It is also observed in Fig. 7.18 that as far as the original frame is concerned, the first plastic hinges were formed at the column bases and the final yield mechanism was a soft first storey mechanism indicating not a good behavior. On the contrary, the redesigned frame exhibited an expected behavior by developing all plastic hinges in beams and the column bases resulting in a global mechanism of collapse as intended. Furthermore, plastic hinges at column bases were formed last and not first as in the original frame.

Some results from the dynamic analyses are presented in Figs. 7.19 and 7.20. In Fig. 7.19 one can clearly see that the maximum storey drift ratios of the redesigned frame are smaller than those of the original frame and slightly exceed (for only two earthquakes) the limit value of 2%. On the contrary, those of the original frame clearly exceed the limit value of 2% reaching the value of 3% for the case of the

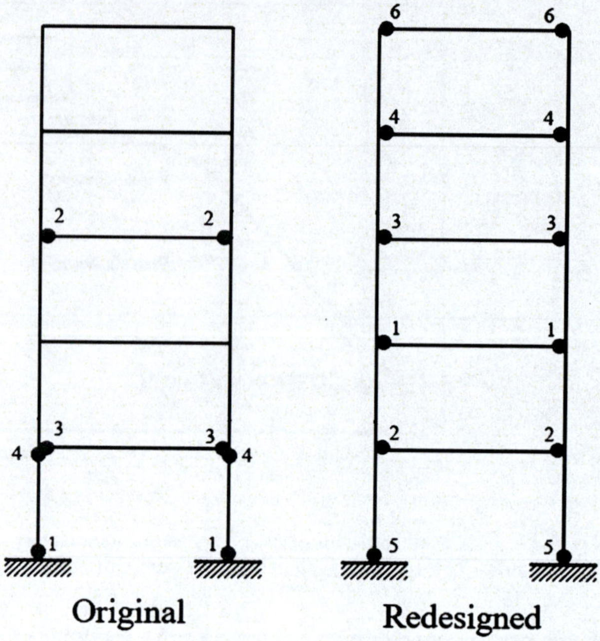

Fig. 7.18 Sequences of inelastic activity under increasing lateral forces (after Leelataviwat 1998, reprinted with permission from UMCEE)

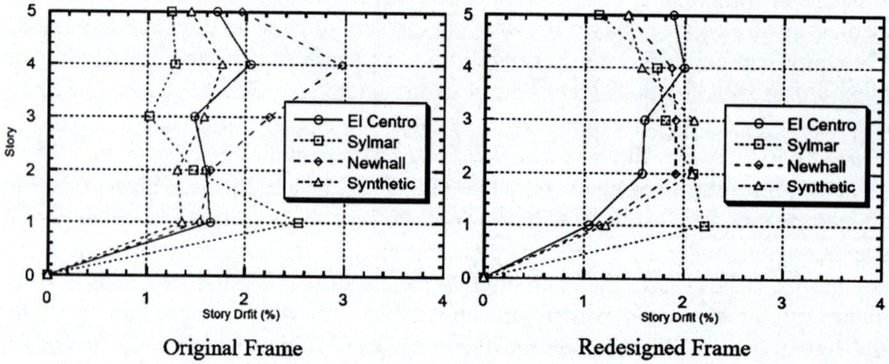

Fig. 7.19 Maximum story drifts of the original and the redesigned frames (after Leelataviwat 1998, reprinted with permission from UMCEE)

Sylmar earthquake motion. An important observation is also that, even though maximum drifts of the redesigned frame had exceeded the 2% limit for the static case, they do not do that for the more reliable dynamic case.

Figure 7.20 shows the plastic hinge pattern and the rotational ductility demands (ratio of the maximum end rotation to that at the elastic limit) for the redesigned frame for all four earthquake motions. In all cases global collapse mechanisms are observed. Analogous to Fig. 7.20 results for the original frame that can be found in Leelataviwat et al. (1999) clearly show soft-storey mechanisms and many plastic

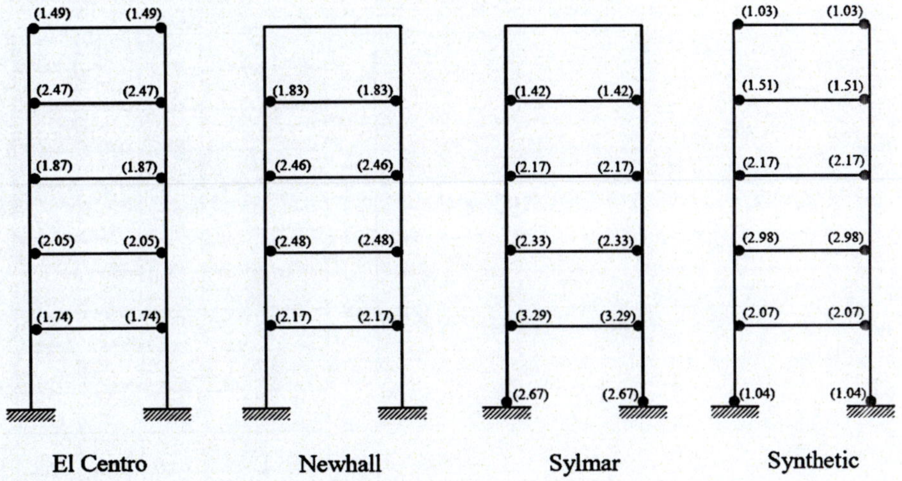

Fig. 7.20 Location of inelastic activity under the four selected earthquakes (after Leelataviwat 1998, reprinted with permission from UMCEE)

hinges in columns as in the static (pushover) case. Another observation is that the rotational demands at the base of the redesigned frame were much smaller than those of the original frame.

7.5.2 Seismic Design of Two MRFs

Consider the two plane steel MRFs of Fig. 7.21 taken from Lee et al. (2004) that were designed by the present method and studied with respect to their seismic performance in that reference. The geometry of these frames as well as their member section sizes obtained on the basis of the AISC (1994) specifications are shown in Fig. 7.21. These frames are a 9-storey 5-bay frame (Fig. 7.21a) and a 20-storey 5-bay frame (Fig. 7.21b). The frames were designed on the assumption of a 2% target drift and a 50 ksi steel yield strength. Design parameters pertaining to the present method are given in Table 7.4 also taken from Lee et al. (2004). In that table S_a/g is the normalized design elastic pseudo-acceleration according to UBC (1994) for the DBE.

The seismic performance of these frames was studied by non-linear analyses under static and dynamic conditions with the aid of the computer program SNAP-2DX (Rai et al. 1996). Strain hardening and viscous damping of 2% were assumed for all members.

Figure 7.22 depicts the base shear versus the roof drift ratio for the two frames as obtained from pushover analysis and demonstrates that the yield drift ratio and the design base shear of the frames are very close to the values assumed in the design (Table 7.4). Figure 7.23 shows the profiles of the maximum storey drift ratios of the

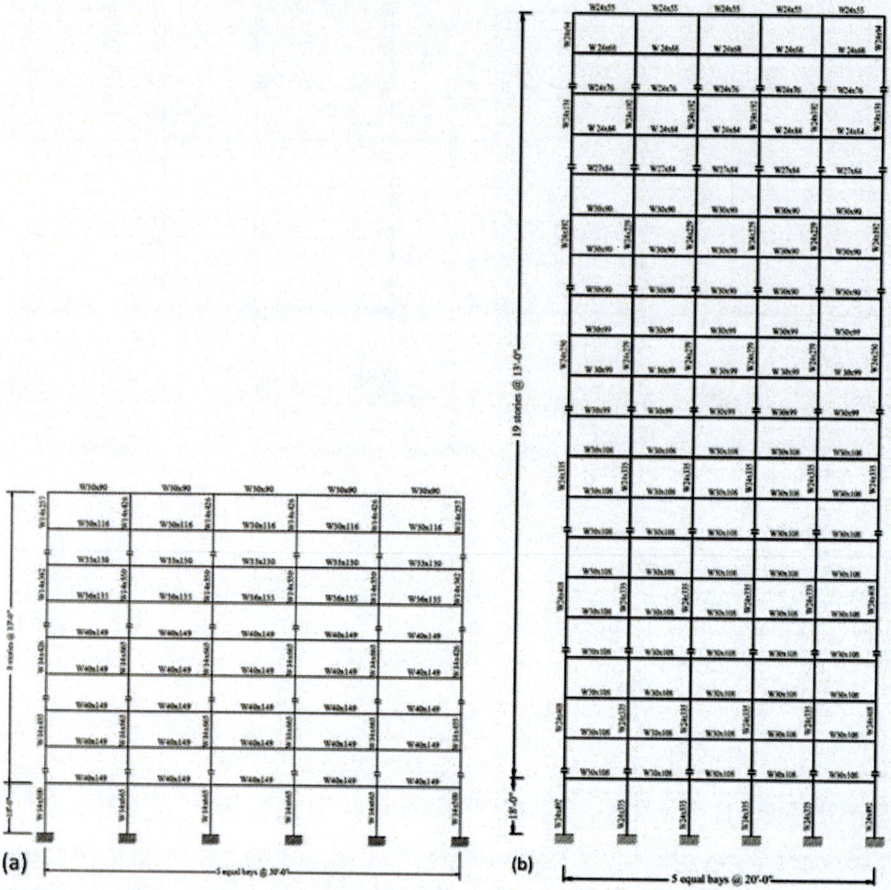

Fig. 7.21 Member sizes of the 9 and 20-storey frames designed for 2% target drift (after Lee et al. 2004, reprinted with permission from CAEE)

Table 7.4 Design parameters (2% target drift limit) of the two structures

Number of stories	Period (s)	S_a/g	Assumed θ_y	θ_p	γ	α	V/W
9	1.285	0.635	0.01	0.01	0.750	1.505	0.179
20	2.299	0.431	0.0075	0.0125	0.609	1.272	0.083

frames due to the four earthquake motions described in the previous example. One can observe that the storey drift ratio do not generally exceed the 2% limit. The slight exceedance observed in two stories of the 9-storey frame does not occur for all earthquakes. Mean values of the storey drift ratio at those two stories are smaller than the 2% limit value, even though mean values are acceptable for a larger than four number of earthquakes.

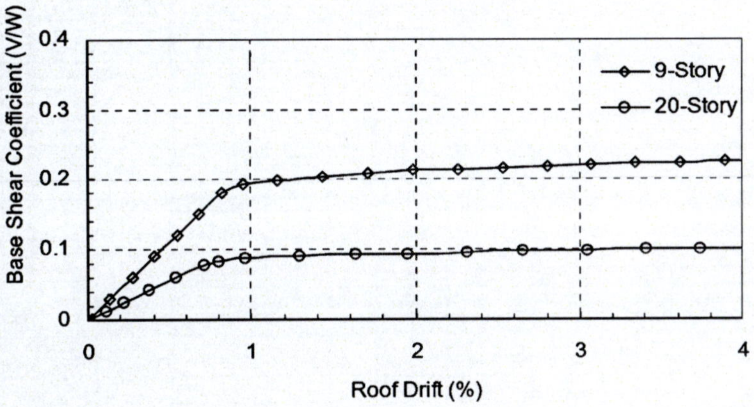

Fig. 7.22 Base shear versus roof drift responses from nonlinear pushover analysis (after Lee et al. 2004, reprinted with permission from CAEE)

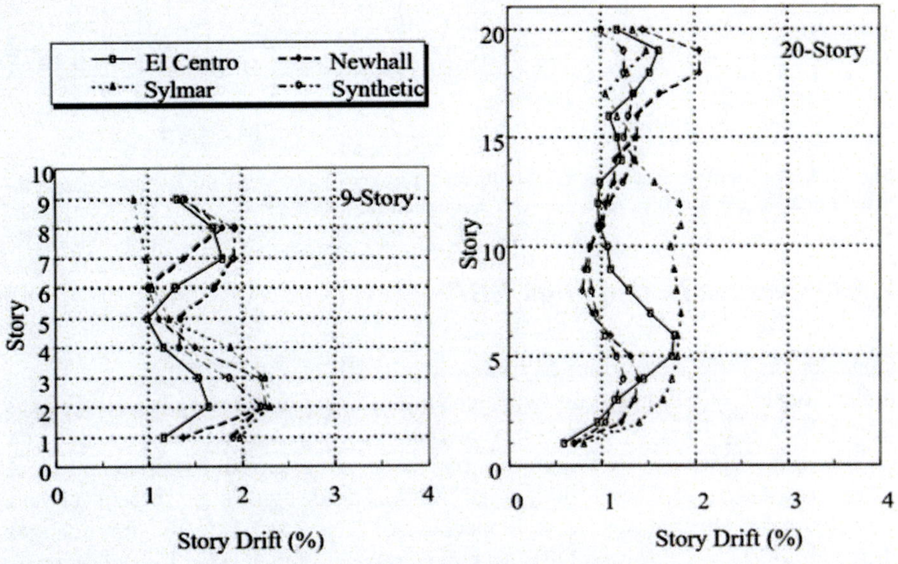

Fig. 7.23 Maximum storey drifts of the frames due to selected earthquake records (after Lee et al. 2004, reprinted with permission from CAEE)

Finally, Fig. 7.24 presents the plastic hinge patterns and the rotational ductility demands for the two frames under the four earthquake motions considered here. One can observe that plastic hinges are formed only in beams and the column bases leading to global type of collapse mechanisms. It was also observed that plastic hinges at the column bases were formed after those in beams in most cases and the rotational ductility demands at the column bases are much smaller than those in the beams.

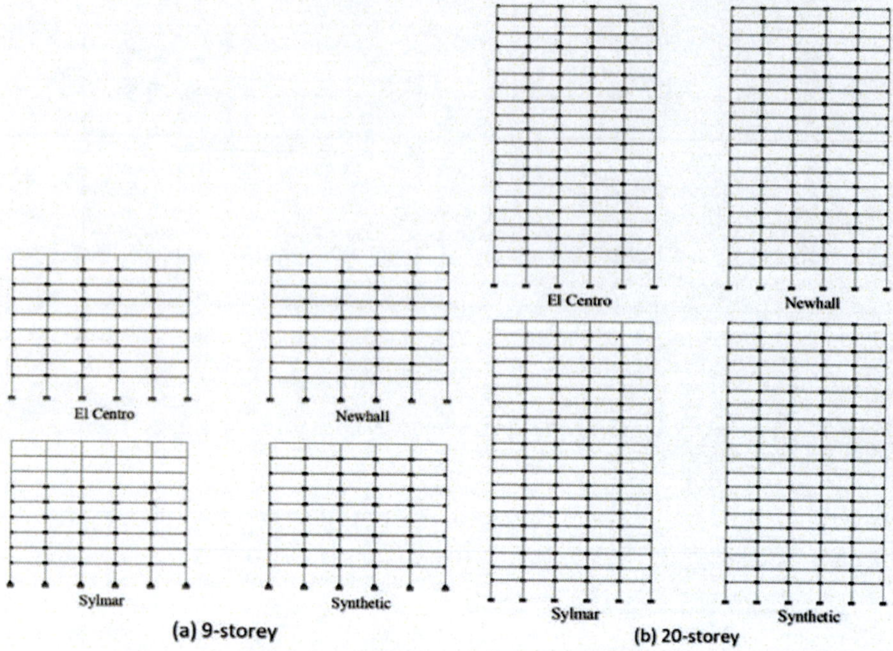

Fig. 7.24 Location of inelastic activity in 9 and 20-storey frames under the four selected earthquakes (after Lee et al. 2004, reprinted with permission from CAEE)

7.5.3 Seismic Design of an EBF

Consider the three-storey four-bay plane steel EBF of Fig. 7.25, part of a three-dimensional steel building with two such frames along its shorter plan view direction, seismically designed by the conventional approach of IBC (2000) by Richards (2004) (Fig. 7.25a) and by the present plastic design approach by Chao and Goel (2006) and Goel and Chao (2008) (Fig. 7.25b). The design by the present method was controlled by a preselected yield mechanism (Fig. 7.7b) and a maximum target drift ratio of $\theta_u = 2\%$ for the DBE. The design parameters needed for the determination of the V/W ratio as obtained by Goel and Chao (2008) read as follows: $T = 0.418$ s, $S_a = 1.587$ g, $\theta_y = 0.5\%$, $\theta_p = \theta_u - \theta_y = 1.5\%$, $\mu_s = \theta_u/\theta_y = 4$, $R_\mu = 2.93$, $\gamma = 0.814$ and a $= 6.583$. Thus, $V/W = 0.297$. Since the total structural weight is 4630 kips, the base seismic shear for one frame is $V = 0.297 \cdot (4630/2) = 0.297 \cdot 2315 = 687$ kips and $687/2 = 343.50$ kips for one-bay braced frame. The floor weights for one frame are 757.50, 757.50 and 800.00 kips for the first, second and third floor, respectively, while the lateral force distribution according to Eq. (7.16) consists of the forces 100.50, 207.50 and 379.00 kips for the first, second and third floor, respectively, with their sum being equal to the base shear of 687.00 kips as it should.

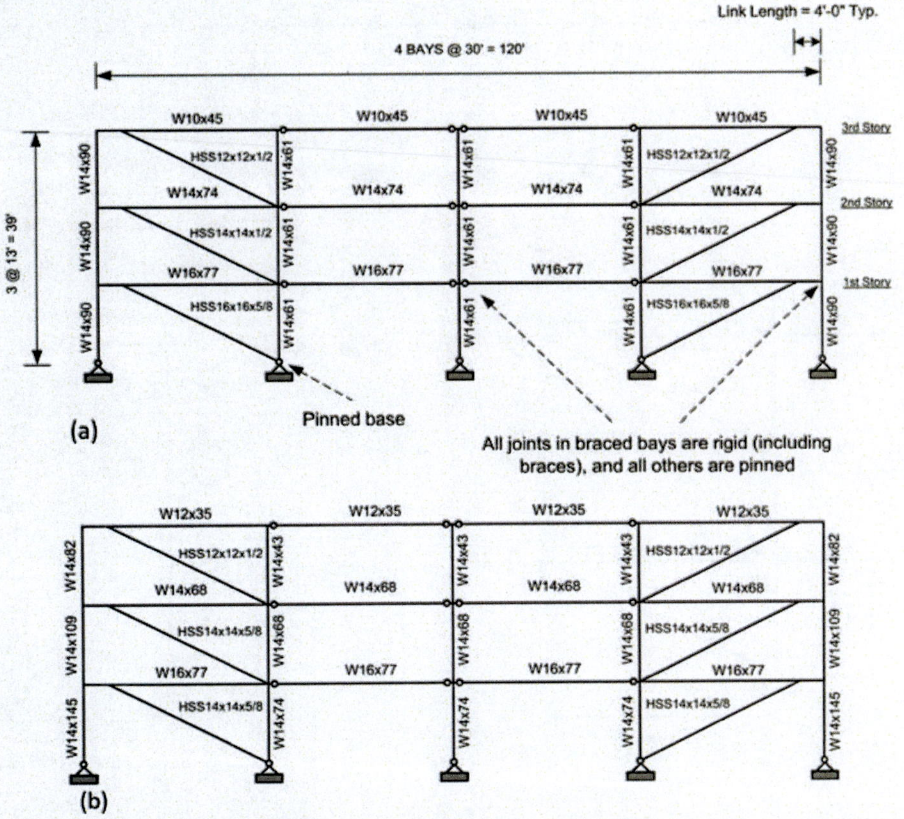

Fig. 7.25 Member sections of three-storey EBFs designed based on (**a**) IBC (2000) approach and (**b**) proposed design approach (after Chao and Goel 2006, reprinted with permission from AISC)

For the link design, the values of the floor lateral forces for one-bay braced frame reading 50.25, 103.75 and 189.50 kips for the first, second and third, respectively, are used. Thus, using Eq. (7.26) one can determined the values 164, 140 and 90 kips for $\beta_i V_{pr}$ and select sections W16 × 77, W14 × 68 and W12 × 35 for the link beams of first, second and third floor, respectively, The corresponding to these sections φV_n values from Eq. (7.27) are 184, 141 and 93 kips satisfying the relation $\varphi V_n \geq \beta_i V_{pr}$. With known sections for the links, one can determine for the first, second and third floor, the values of $V_p = 204.60$, 156.30 and 103.20 kips, $M_p = 633.30$, 479.20 and 213.30 kip-ft, $V_u = 281.30$, 214.90 and 141.90 kips, $M_C = 696.70$, 527.10 and 234.70 kip-ft and $M_B = 522.50$, 395.30 and 250.20 kip-ft, respectively.

The member design procedure by the present method is pictorially described in Fig. 7.26. The exterior columns are designed for vertical loads only, while the interior columns are designed on the basis of capacity design. The free body

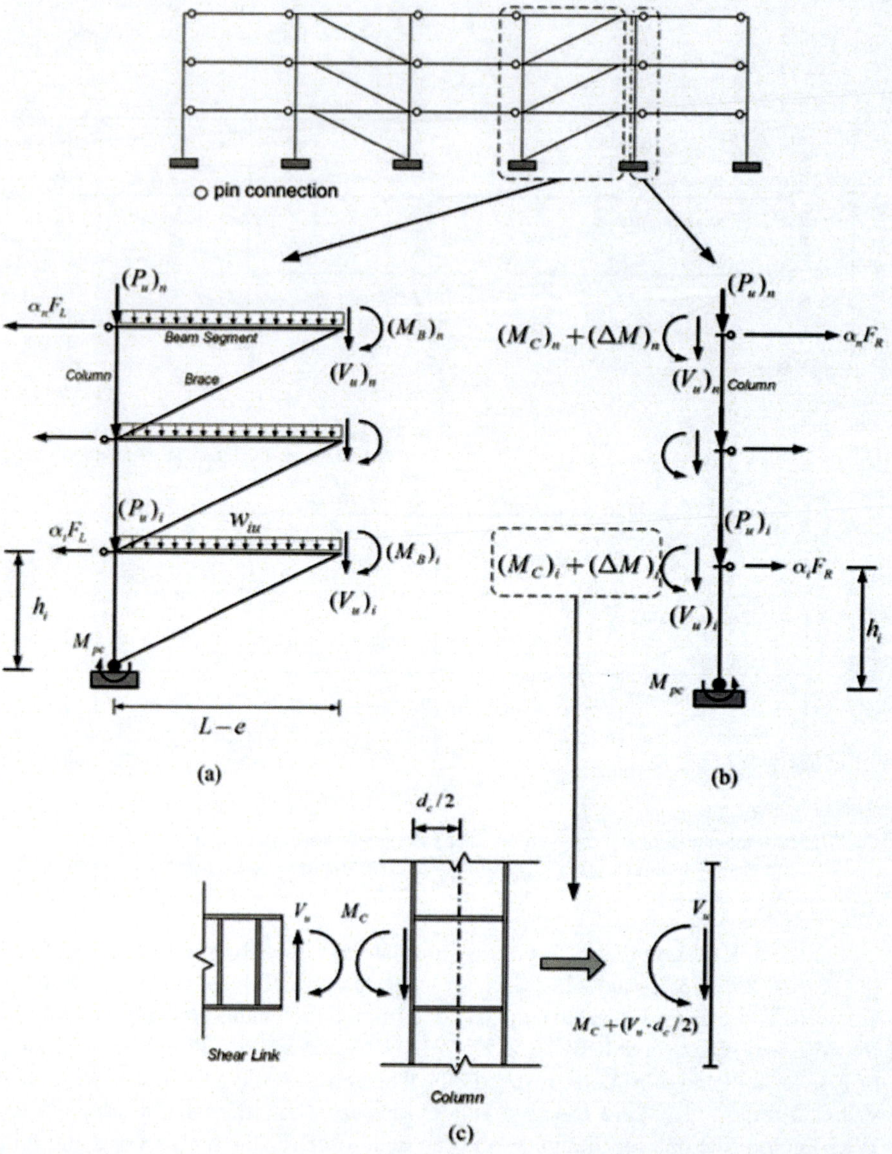

Fig. 7.26 Free body diagram of interior columns and associated beam segments and braces: **(a)** lateral forces acting toward left; **(b)** lateral forces acting toward right; **(c)** illustration showing $(\Delta M)_i$ in the figure (after Chao and Goel 2006, reprinted with permission from AISC)

diagrams for the interior columns with and without braces and beam segments are shown in Fig. 7.26a, b, respectively. For the interior columns of Fig. 7.26a, the sum of the applied lateral forces F_L acting to the left is obtained from the equilibrium of moments with respect to column bases and reads

$$F_L = \left(1/\sum_{i=1}^{n} a_i h_i\right) \times \left[(L-e)\sum_{i=1}^{n}(V_u)_i + \sum_{i=1}^{n}(M_B)_i + \frac{(L-e)^2}{2}\sum_{i=1}^{n} w_{iu} + M_{pc}\right] \quad (7.36)$$

where

$$a_i = \left(F_i/\sum_{i=1}^{n} F_i\right) = \frac{\beta_i - \beta_{i+1}}{\sum_{i=1}^{n}(\beta_i - \beta_{i+1})} \quad (7.37)$$

with $\beta_{n+1} = 0$ for $i = n$. For the interior columns of Fig. 7.26b, the sum of the applied lateral forces F_R acting to the right is expressed as

$$F_R = \left(1/\sum_{i=1}^{n} a_i h_i\right)\left[\sum_{i=1}^{n}(V_u)_i((d_c)_i/2) + \sum_{i=1}^{n}(M_C)_i + M_{pc}\right] \quad (7.38)$$

where d_c is the depth of the column section.

With factored gravity loads on beam segments w_{iu}= 0.877, 0.957 and 0.957 for first, second and third floor, respectively, one can determine the corresponding column design forces $(P_u)_i$= 43.10, 43.10 and 39.50 kips and the updated lateral forces a_iF_L= 87.40, 180.80 and 329.90 kips for the case with beam segments and braces. With $(\Delta M)_i = (V_u)_i \cdot (d_c/2)$= 164.10, 125.40 and 82.80 kips-ft for the first, second and third floor, respectively, one can determine the corresponding column design forces $(P_u)_i$= 38.10, 38.10 and 39.50 kips and the updated lateral forces a_iF_R= 6.80, 14.10 and 25.70 kips, where $d_c/2 = 7$ in has been taken assuming W14 for columns.

After the above lateral forces F_L and F_R have been computed via Eqs. (7.36) and (7.38), the required strength of beam segments, braces and columns can be easily obtained by an elastic analysis on the basis of the free body diagram of Fig. 7.26. The terms a_iF_R, a_iF_L, $(V_u)_i$, $(M_B)_i$, $(M_C)_i$, w_{iu} and $(P_u)_i$ denote applied loads, with $(P_u)_i$ being axial forces in columns due to vertical loads. Finally, beam segments, braces and columns are dimensioned in accordance with AISC (2005a) specifications.

A comparison of the two designs on the basis of Fig. 7.25 indicates that the IBC (2000) design has heavier link sections and lighter column sections than those of the present design. This leads to different performances under seismic loads, as shown in response results obtained with the aid of NLTH analyses. Figure 7.27 depicts the inelastic activity of the two frame designs for two seismic events (Imperial Valley 1940, El Centro and Loma Prieta 1989, Gilroy) selected to have an occurrence of 10% in 50 years corresponding to the DBE. Inelasticity was limited to shear links and column bases in the frame designed by the present method, as expected. On the contrary, inelasticity is observed not only in links and column bases but also in the exterior columns, which is undesirable. Figure 7.28 shows the maximum interstorey drift ratios for both frame designs due to two seismic events (Imperial Valley, 1979, Array 5 and Northridge, 1994, Rinaldi) selected to be of 10% in 50 years occurrence.

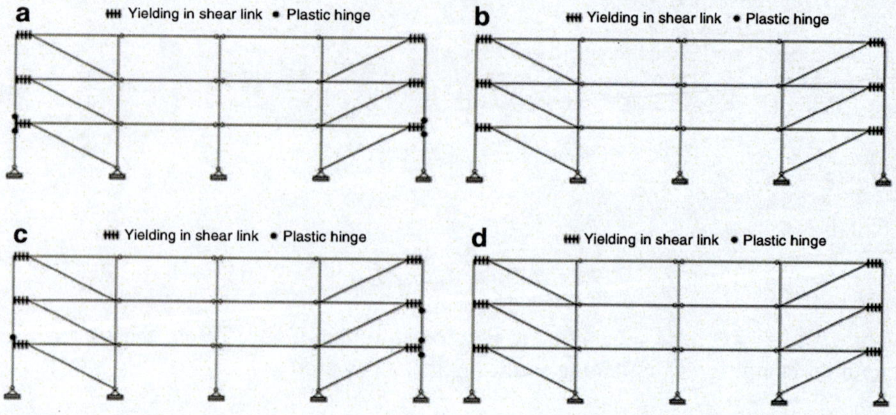

Fig. 7.27 Inelastic activity in three-storey: (**a**) EBF designed by IBC and (**b**) EBF designed by proposed method during the Imperial Valley, 1940, El Centro event; (**c**) EBF designed by IBC and (**d**) EBF designed by the proposed method during the Loma Prieta, 1989, Gilroy event (after Chao and Goel 2006, reprinted with permission from AISC)

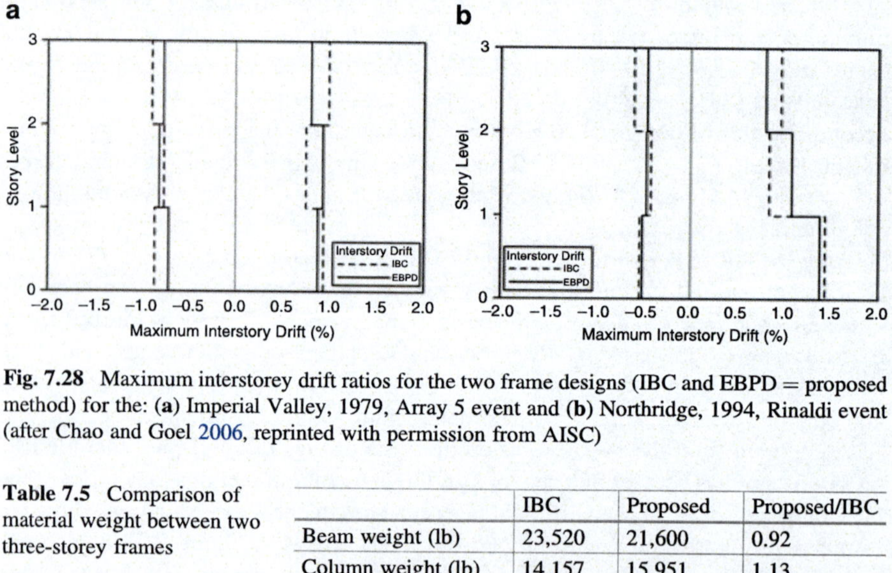

Fig. 7.28 Maximum interstorey drift ratios for the two frame designs (IBC and EBPD = proposed method) for the: (**a**) Imperial Valley, 1979, Array 5 event and (**b**) Northridge, 1994, Rinaldi event (after Chao and Goel 2006, reprinted with permission from AISC)

Table 7.5 Comparison of material weight between two three-storey frames

	IBC	Proposed	Proposed/IBC
Beam weight (lb)	23,520	21,600	0.92
Column weight (lb)	14,157	15,951	1.13
Brace weight (lb)	17,005	17,202	1.01
Total weight (lb)	54,682	54,753	1.00

In both designs these ratios do not exceed the preselected target drift value of 2%. Finally, in Table 7.5 one can see a comparison of the material weight between the two designs. It is observed that, even though the total weight is almost the same in the two designs, beam weight is smaller than column weight in the present design in agreement with the capacity design, while the opposite is true in the IBC design. This clearly shows the advantages of the present method, in agreement with the results of Fig. 7.27.

Up to now all the results for the three-storey frame of Fig. 7.25 had to do with the DBE associated with a 2% assumed target drift. In a true performance-based design approach one should also investigate the case of the MCE associated with a 3% assumed target drift and check if this performance level leads to a higher or lower base shear value than the one of the DBE in order to see which one of the two controls the design.

7.5.4 Seismic Design of a CBF

Consider the chevron type three-storey two-bay CBF of Fig. 7.29, part of a three-dimensional building with four such frames to resist seismic forces along the short plan view direction, originally designed by Sabelli (2000) and re-designed by Bayat et al. (2010) using the present method. The frame's distributed gravity loads $w_1 = 1.13$ kip/ft and $w_2 = 0.95$ kip/ft as well as the axial gravity loads on columns $L_1 = 26.17$ kips, $L_2 = 37.56$ kips, $L_3 = 22.4$ kips and $L_4 = 32.88$ kips coming form transverse beams and the exterior wall and shown in Fig. 7.29 come from the 1.2G + 0.5Q combination of dead (G) and live (Q) load. The frame considered here resists 1/4 of the total structural weight of 6503 kips. Thus, one has here $W = 1625.75$ kips.

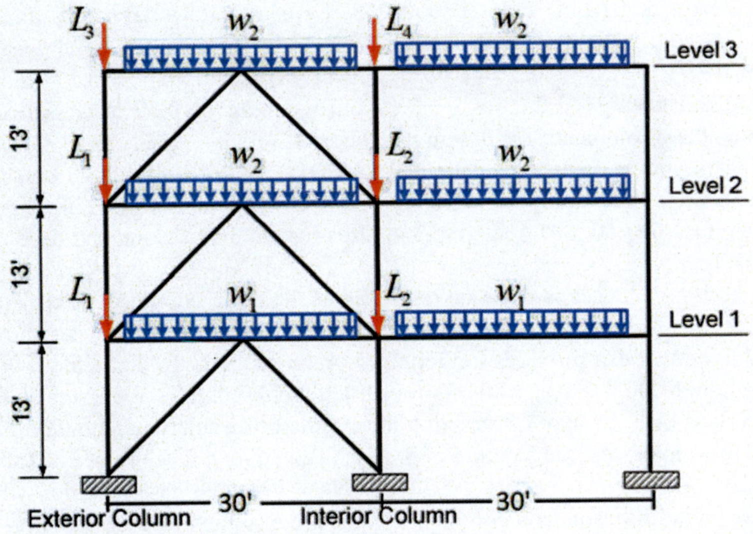

Fig. 7.29 Geometry and gravity load of 3-storey CBF (after Bayat et al. 2010, reprinted with permission from UMCEE)

Table 7.6 Required brace strength and selected sections for the 3-storey CBF

Floor	a	V_i (kips)	$(V_{storey\ shear})_i/0.9\cos a_i$ (kips)	Brace section	Brace nominal strength $(P_y + 0.5P_{cr})_i$
3	41°	423.4	623	2HSS4–1/2 x 4–1/2 x 3/8	683
2	41°	667.4	981	2HSS5 x 5 x 1/2	996
1	41°	787.5	1158	2HSS6 x 6 x 1/2	1266

Table 7.7 Brace fracture life calculation for the 3-storey CBF

Floor	Brace length L (in)	r_x (in)	r_y (in)	$\frac{K_xL}{r_x}$	$\frac{K_yL}{r_y}$	$(b - 2t)/t$	$0.64\sqrt{E/f_y}$	N_f
3	238.2	1.67	3.06	71.3	66.2	10.0	16.1	187
2	238.2	1.82	3.35	63.6	60.5	8.0	16.1	268
1	238.2	2.23	3.99	53.5	50.7	10.0	16.1	157

Table 7.8 Nominal axial strength of the braces selected for the 3-storey CBF

Floor	Brace cross sectional area (in^2)	F_e (ksi)	$0.44F_y$ (ksi)	F_{cr} (ksi)	$0.5P_{cr}$ (kips)	P_y (kips)
3	10.96	56.3	20.2	32.68	179.1	504.2
2	13.76	66.6	20.2	34.54	271.5	723.0
1	19.48	100.1	20.2	37.95	369.6	896.1

The design parameters needed for determining the design base shear of the frame for the DBE with target drift ratio $\theta_u = 1.25\%$ have as follows (Bayat et al. 2010): $S_a = 1.392$ g, $T = 0.30$ s from Table 7.2, $\theta_y = 0.3\%$ from Table 7.1 leading to $\theta_p = \theta_u - \theta_y = 1.25 - 0.3 = 0.95\%$, $\mu_s = \theta_u/\theta_y = 1.25/0.3 = 4.17$ leading to $R_\mu = 2.71$ and hence $\gamma = 1.0$. Thus, a = 7.52 and assuming $\eta = 0.50$, one can determine the $V/W = 0.484$ and hence the design base shear $V = 0.484 \cdot 1625.75 = 787.50$ kips with a distribution along the height equal to 120.00, 244.00 and 423.40 kips for the first, second and third storey, respectively of one CBF frame of Fig. 7.29. The storey shears are then 787.50, 667.40 and 423.40 kips for the first, second and third storey, respectively.

The design of braces is done on the basis of the three criteria of Sect. 7.4. The brace sections are built-up double tubes (HSS with $f_y = 46$ ksi) as shown in Table 7.6, which also provides the required brace strength. In that table a denotes the brace angle (Fig. 7.10), while its fourth and sixth columns are associated with Eq. (7.31). Tables 7.7 and 7.8 describe the results of calculations pertaining to the fracture life and brace axial strength of braces, respectively. One can see in Table 7.7 that $K_xL/r_x > K_yL/r_y$, $(b - 2t)/t < 16.1$ and $N_f > 100$. The design of non-yielding members includes beams and columns. Beams are designed as beam-columns with $K = 1.0$ because in addition to bending moments are also subjected to the horizontal components of the axial forces of braces. Table 7.9 provides the design parameters for the beams, while Table 7.10 their selected cross-sections for the three stories of the frame as well as their lateral support spacing. The value of $R_y = 1.4$ (AISC

Table 7.9 Design parameters for beams of the 3-storey CBF

Floor	Uniformly distributed gravity loading w_u (kips/ft)	R_yP_y (kips)	$0.5P_{cr}$ (kips)	F_h (kips)	F_v (kips)	P_u (kips)	M_u (kips-ft)
3	0.95	703.9	179.1	669	345	334.5	2689
2	1.13	1013.0	271.5	972	487	486.0	3723
1	1.13	1254.5	369.6	1227	580	613.5	4433

Table 7.10 Determination of lateral support spacing for beams of the 3-storey CBF

Floor	Selected beam section	r_y (in)	M_1 (kips-ft)	M_2 (kips-ft)	L_{pd} (ft)	L_b (ft)
3	W40 x 183	2.49	1838	2689	8.19	5
2	W40 x 235	2.54	2530	3723	8.39	5
1	W40 x 278	2.52	3013	4433	8.32	5

2005b), whereas the values of F_h and F_v in Table 7.9 are computed on the basis of the equations of Fig. 7.13. The spacing of lateral supports L_b in Table 7.10 is taken as 5.0 ft in all stories, which is less than the values of $L_{pd} = [0.12 + 0.076(M_1/M_2)](E/f_y)$ r_y (AISC 2005b) with the moments M_1 and M_2 computed at 5 ft from mid-span and at mid-span, respectively.

Tables 7.11 and 7.12 summarize the required strength checks for exterior and interior columns of the 3-storey CBF in accordance with Eqs. (7.34) and (7.35) for interior and exterior columns, respectively in conjunction with the notes concerning the extra term $(1/2 \cdot F_v)$. In both cases, as it turns out, the post-buckling limit state governs the design with P_u (cumulative) being 1124 and 1205 kips, respectively for exterior and interior columns. Using the maximum axial force of 1205 kips, one can select a column section W12 × 120 (with axial strength equal to 1325 kips) for all columns in all stories. Figure 7.30a shows the final member sections for the 3-storey CBF, while Fig. 7.30b the corresponding ones as obtained on the basis of the FEMA-302 (1997) design provisions with $R = 6$. A comparison of the two above designs results in 36,838 lbs of steel for the CBF of Fig. 7.30a and 25,033 lbs for the CBF of Fig. 7.30b. Even though the design by the present method is heavier than the design by the FEMA 302 (1997) method, its seismic performance is superior to that of the latter method as is shown in the following paragraphs dealing with the results of NLTH analyses.

The two frame designs of Fig. 7.30 were seismically analyzed with the aid of the SNAP-2DX (Rai et al. 1996) software by modelling all beams and columns as beam-column elements and beam-to-column connections as follows: moment resisting connections due to gusset plates at the first and second levels of the CBF of Fig. 7.30b and pinned connections at all three levels of the CBF of Fig. 7.30a due to beam splices.

Figures 7.31 and 7.32 present the plastic hinge formation and maximum plastic rotations due to the Imperial Valley, 1979, Array 5 (LA02) ground motion (10% in 50 years) for the 3-storey CBF as designed by the FEMA 302 (1997) (Fig. 7.31) and the present method (Fig. 7.32). The heavy damage inflicted on the CBF of FEMA

Table 7.11 Design parameters for exterior columns of the 3-storey CBF

Exterior columns									
			Pre-buckling limit state			Post-buckling limit state			
Floor	$P_{trasnverse}$ (kips)	P_{beam} (kips)	$P_{cr}sina$ (kips)	P_u (total kips)	P_u (cumulative kips)	$0.5P_{cr}sina$ (kips)	$1/2 \cdot F_v$	P_u (total kips)	P_u (cumulative kips)
3	22.24	14.25	0	36	36	0	173	209	209
2	26.17	16.95	235	278	314	117	243	404	613
1	26.17	16.95	356	399	713	178	290	511	1124

Table 7.12 Design parameters for interior columns of the 3-storey CBF

Interior columns								
			Pre-buckling limit state			Post-buckling limit state			
Floor	$P_{trasnverse}$ (kips)	P_{beam} (kips)	$P_{cr}sina$ (kips)	P_u (total kips)	P_u (cumulative kips)	$0.5P_{cr}sina$ (kips)	$1/2 \cdot F_v$	P_u (total kips)	P_u (cumulative kips)
3	32.88	28.5	0	61	61	0	173	234	234
2	37.56	33.9	235	306	367	117	243	432	666
1	37.56	33.9	356	427	795	178	290	539	1205

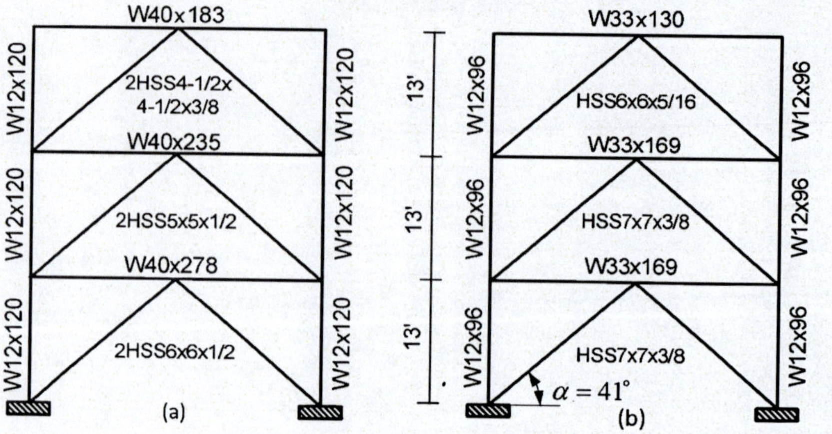

Fig. 7.30 Member sections for the 3-storey CBF designed by (**a**) the present method and (**b**) the FEMA 302 (1997) provisions (after Bayat et al. 2010, reprinted with permission from UMCEE)

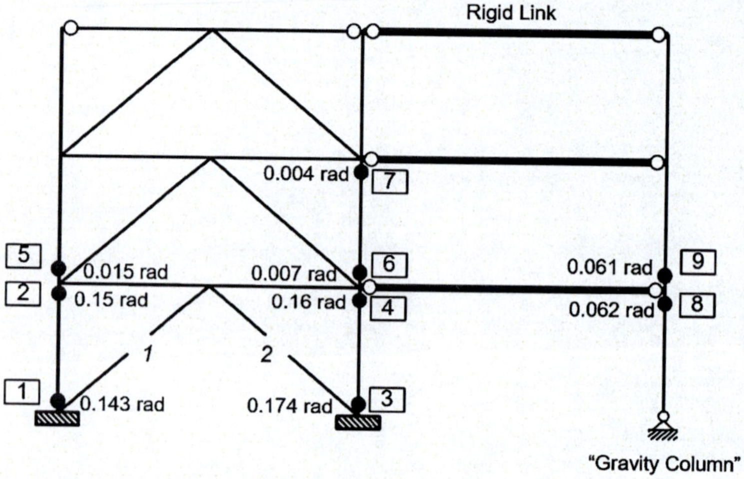

Fig. 7.31 Plastic hinge formation and maximum plastic hinge rotations of the 3-storey CBF designed by FEMA 302 (1997) under the imperial Valley, 1979, Array 5 (LA02) ground motion (10% in 50 years) (after Bayat et al. 2010, reprinted with permission from UMCEE)

302 (1997) with many plastic hinges leading eventually to collapse after 20 s is apparent. The absence of plastic hinges in the CBF of the present method is also apparent clearly demonstrating the superior seismic performance of the design by the present method. Similar performance of the two designs has been observed under many other ground motions not only of the 10% in 50 years type but also of the 2% in 50 years type. One can look at Bayat et al. (2010) for more details.

Concerning the drift ratio responses, mean maximum values of drift ratios for the two CBF designs under 11 earthquake motions of the 10% in 50 years type reveal

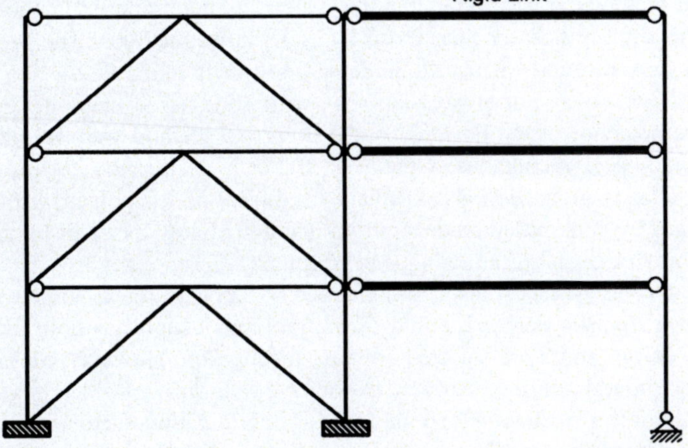

Fig. 7.32 Absence of plastic hinges in 3-storey CBF designed by present method under the LA02 ground motion (10% in 50 years) (after Bayat et al. 2010, reprinted with permission from UMCEE)

that the FEMA 302 (1997) design experiences a large drift of about 3.25% at the first floor due to brace failure and hinges at columns, while the design by the present method shows a more uniform drift distribution along the height of the CBF with values within the design target drift ratio of 1.25%.

7.6 Conclusions

This section enumerates the conclusions derived from all the previous material presented in this chapter, which read as follows:

1. A performance-based seismic design method for various types of plane steel frames, such as, moment resisting and eccentrically or concentrically braced ones, has been presented. The method uses pre-selected target drift and yield mechanisms as main performance limit states and accomplishes the design in one step and without iterations.
2. The design base shear for a specific seismic level is determined by solving an energy equation that equates the work done by the frame as it is pushed up to the target drift to the seismic energy input on an equivalent to the frame single-degree-of-freedom elastoplastic system. This base shear in conjunction with plastic design is used for member dimensioning of the frame. This is the reason for calling the method energy-based plastic design method.
3. The proposed method has been successfully applied here to plane steel moment resisting frames, eccentrically braced frames and concentrically braced frames. In all cases the designs by the proposed method were compared against designs by

code-based methods. It was found that the designs by the proposed method were found to be slightly heavier than those by code-based methods. The proposed method always resulted in maximum interstorey drift ratios below the target values, while this was not always the case with code-based methods. Finally, the seismic behavior of the designs by the proposed method were found to be superior to those by code-based methods. Indeed, designs by the proposed method resulted in global collapse mechanisms and no plastic hinges in columns, while designs by code-based methods in premature collapse mechanisms including soft-storey ones and plastic hinges in columns.

4. The present energy-based plastic design method appears to be similar to the ductility-based plastic design method of the previous chapter as both methods use plastic design and a pre-selected collapse mechanism. However, the present method uses in addition the drift as a pre-selected design factor and employs a new lateral force distribution and the energy modification factor for a more accurate treatment. In addition, the present method has been extended and applied to many other types of plane steel frames in addition to moment resisting ones.

References

AISC (1994) Load and resistance factor design, 2nd edn. American Institute of Steel Construction, Chicago, IL

AISC (2005a) Specification for structural steel buildings. American Institute of Steel Construction, Chicago, IL

AISC (2005b) Seismic provisions for structural steel buildings. American Institute of Steel Construction, Chicago, IL

ASCE/SEI 7-10 (2010) Minimum design loads and associated criteria for buildings and other structures. American Society of Civil Engineers, Reston, VI

Basha H, Goel SC (1994) Seismic resistant truss moment frames with ductile Vierendeel segment Report UMCEE 94-29. Department of Civil and Environmental Engineering, University of Michigan, Ann Arbor MI

Bayat MR, Goel SC, Chao SH (2010) Performance-based plastic design of earthquake resistant concentrically braced steel frames, Research report UMCEE 10-02. Department of Civil and Environmental Engineering, The University of Michigan, Ann Arbor MI

Chao SH, Goel SC (2006) Performance-based seismic design of eccentrically braced frames using target drift and yield mechanism as performance criteria. Eng J AISC, Third Quarter 43:173–200

Chao SH, Goel SC (2008) Performance-based plastic design of special truss moment frames. Eng J AISC, Second Quarter 45:127–150

Chao SH, Goel SC, Lee SS (2007) A seismic design lateral force distribution based on inelastic state of structures. Earthquake Spectra 23:547–569

EC8 (2004) Eurocode 8, Design of structures for earthquake resistance, Part 1: general rules, seismic actions and rules for buildings, EN 1998-1-1. European Committee for Standardization (CEN), Brussels

FEMA 302 (1997) NEHRP recommended provisions for seismic regulations for new buildings and other structures. Federal Emergency Management Agency, Washington, DC

FEMA 356 (2000) Prestandard and commentary for the seismic rehabilitation of buildings. Federal Emergency Management Agency, Washington, DC

Goel SC, Chao SH (2008) Performance-based plastic design: earthquake resistant steel structures. International Code Council, Washington DC

Goel SC, Liao WC, Bayat MR, Chao SH (2010) Performance-based plastic design (PBPD) method for earthquake-resistant structures: an overview. Struct Des Tall Spec Build 19:115–137

Housner GW (1959) Behavior of structures during earthquakes. J Eng Mech Div ASCE 85:109–130

IBC (2000) International building code. International Code Council, Washington, DC

Lee SS, Goel SC, Chao SH (2004) Performance-based seismic design of steel moment frames using target drift and yield mechanism. In: Proceedings of the 13th World Conference on Earthquake Engineering, Vancouver, BC, Canada, Paper No. 266

Leelataviwat S (1998) Drift and yield mechanism based seismic design and upgrading of steel moment frames. Ph.D. Thesis, Department of Civil and Environmental Engineering, University of Michigan, Ann Arbor, MI

Leelataviwat S, Goel SC, Stojadinovic B (1999) Towards performance-based seismic design of structures. Earthquake Spectra 15:435–461

Leelataviwat S, Goel SC, Stojadinovic B (2002) Energy-based seismic design of structures using yield mechanism and target drift. J Struct Eng ASCE 128:1046–1054

Newmark NM, Hall WJ (1982) Earthquake spectra and design, Engineering monographs on earthquake criteria, structural design and strong motion records, vol 3. University of California, Berkeley, CA, Earthquake Engineering Research Institute

Rai DC, Goel SC, Firmansjah J (1996) SNAP-2DX: A general purpose computer program for nonlinear structural analysis, Report EERC 96-21. Department of Civil and Environmental Engineering, University of Michigan, Ann Arbor, MI

Richards PW (2004) Cyclic stability and capacity design of steel eccentrically braced frames. Ph.D. Thesis, Department of Structural Engineering, University of California, San Diego, CA

Sabelli R (2000) Research on improving the design and analysis of earthquake resistant steel braced frames. FEMA/EERI Report. Federal Emergency Management Agency, Washington DC

Sahoo DR, Chao SH (2010) Performance-based plastic design method for buckling-restrained braced frames. Eng Struct 32:2950–2958

Tremblay R (2005) Fundamental periods of vibration of braced steel frames for seismic design. Earthquake Spectra 21:833–860

UBC (1994) Uniform building code. International conference of building officials, Whittier, CA

Wongpakdee N, Leelataviwat S, Goel SC, Liao WC (2014) Performance-based design and collapse evaluation of buckling-restrained knee braced truss moment frames. Eng Struct 60:23–31

Chapter 8
Design Using Modal Damping Ratios

Abstract A performance-based seismic design method for plane steel moment resisting and braced framed structures is presented. It is a force-based design method employing equivalent viscous modal damping ratios ξ_k instead of the behavior (or strength reduction) factor q (or R) to account for inelastic energy of dissipation. These modal damping ratios ξ_k are defined for an equivalent linear multi-degree-of-freedom structure to the original non-linear multi-degree-of-freedom structure, which has the same mass and elastic stiffness as the non-linear one. In addition, they are functions of the structural period, the target inter-storey drift ratio and member plastic rotation and the soil type. Empirical expressions of these ξ_k for the first few significant modes are obtained through extensive parametric studies involving non-linear time-history analyses of many frames under many far-fault earthquakes and different deformation targets. These ξ_k are used for seismic design through an elastic acceleration design spectrum with high amounts of damping. The presented method, which is illustrated by numerical examples, is more rational and provides results of higher accuracy in one step (strength checking) than code-based design methods requiring two steps (strength and deformation checkings).

Keywords Equivalent linear structure · Equivalent modal damping ratios · Seismic design method · Absolute acceleration design spectra · High amounts of damping · Performance-based design

8.1 Introduction

Modern seismic design codes, such as EC8 (2004), employ the force-based design method (FBD), which uses forces as the main design parameters and designs inelastically deforming structures by linear analysis involving an elastic acceleration design spectrum with ordinates divided by the behavior (strength reduction) factor q (or R) in order to take inelasticity into account. Behavior factor q is given as a single constant value for different types of structures, without consideration of their

G. A. Papagiannopoulos et al., *Seismic Design Methods for Steel Building Structures*, Geotechnical, Geological and Earthquake Engineering 51,
https://doi.org/10.1007/978-3-030-80687-3_8

dynamic and deformational characteristics. An extensive treatment of the FBD method in the framework of the EC8 (2004) has been given in Chap. 3.

As discussed in Chap. 1, performance-based design (PBD) approach (SEAOC 1995; Bozorgnia and Bertero 2004) introduced a general design framework for seismic structural design with a comprehensive damage control. Thus, the structure is designed for a number of performance levels, each defined by a pair of seismic intensity and damage level. Since damage is directly related to deformation, by controlling deformation metrics such as the inter-storey drift ratio (*IDR*) and the member plastic rotation (θ_p) at every performance level, damage can be also controlled.

Even though EC8 (2004) accepts the performance-based design approach, its use is only partial. Design according to EC8 (2004) is done for two performance levels: the ultimate limit state (ULS) or life safety (LS) performance level under the design basis earthquake (DBE) which involves a strength checking and the damage limit state (DLS) or immediate occupancy (IO) performance level, which involves a displacement checking. However, displacements are calculated by applying the equal displacement rule, which underestimates or overestimates them depending on the fundamental period (Gupta and Krawinkler 2000). From the above, one can observe that code-based FBD methods are characterized by two main approximations: the use of a crude single value of q for all modes and the equal displacement rule for displacement determination. Thus, the need for a more rational, accurate, efficient and truly performance-based FBD method to more effectively control both strength and displacements is apparent.

In this chapter, a performance-based seismic design method for plane steel framed buildings, which combines the FBD method with a rational and efficient way to directly control deformation and hence damage is proposed. The method is based on the original work of Papagiannopoulos and Beskos (2006, 2010) regarding the seismic design of plane steel moment resisting frames (MRFs) under near-fault and long duration earthquakes. This work was improved by Loulelis (2015) and Loulelis et al. (2018) to include MRFs with panel zone effects and material degradation under far-fault motions in order to be compatible with the FBD method of EC8 (2004). The method was also extended by Loulelis (2015), Kalapodis (2017), Kalapodis and Papagiannopoulos (2020) and Kalapodis et al. (2020) to plane steel braced frames (concentrically braced frames-CBFs and eccentrically braced frames-EBFs).

The proposed method is a FBD method using the concept of the equivalent viscous modal damping ratios ξ_k (k is the mode number) to account for inelastic energy dissipation instead of that of the behavior (or strength reduction) factor q (or R). These modal damping ratios ξ_k are defined for the equivalent linear MDOF system to the original nonlinear MDOF structure. This equivalent system has the mass and the elastic stiffness of the nonlinear structure and is characterized only by its equivalent damping ratios and not by equivalent damping and stiffness as it is usually the case in the existing literature. Furthermore, these equivalent modal damping ratios ξ_k are constructed as functions of the periods of the structure, the target non-structural and structural deformation (expressed in terms of *IDR* and θ_p), respectively, for at least three performance levels and soil type. Thus, the proposed FBD method is more rational and leads to more accurate results in one step (only strength checking) than code-based FBD methods requiring two steps (strength and deformation checking). In addition, it is a truly performance-based design method with at least three performance levels. Its only disadvantage is that it requires acceleration design spectra with high amounts of damping. Finally, in comparison with the direct DBD method, the proposed method (i) replaces the nonlinear MDOF system by an equivalent linear MDOF and not a SDOF system that cannot take into account higher mode and P-Δ effects and (ii) employs the acceleration design spectrum and not the unfamiliar to engineers displacement design spectrum.

For every type of plane steel frames (MRFs, CBFs and EBFs) considered here, a large number of them (from 1 to 20 storeys) is analyzed by non-linear time-history (NLTH) analysis under 100 far-fault seismic motions (25 for each A–D soil type of EC8 2004) considering various deformation target pairs (IDR, θ_p). The response databank created by these analyses, is used to derive explicit expressions for equivalent modal damping ratios ξ_k through nonlinear regression analysis. These damping ratios ξ_k are functions of period T, deformation indices (IDR, θ_p) and soil type (A–D according to EC8 2004) and can be used in conjunction with pseudo-acceleration response/design spectra with high values of viscous damping ratios ξ to conduct seismic design in a PBD framework. In other words, the design base shear of the structure can be determined through spectrum analysis using equivalent modal damping ratios ξ_k instead of the behavior (strength reduction) factor q.

Seismic displacements cannot be computed by the proposed method because the equal displacement rule, used by code-based FBD methods, cannot be applied here. However, displacement checking is not required because the modal damping ratios ξ_k, which depend on deformation, have been constructed in order to satisfy deformation limits.

The proposed method is illustrated by performing seismic designs of plane steel MRFs, CBFs and EBFs. These designs are compared against those obtained by using the FBD of EC8 (2004) on the basis of NLTH analyses. It is concluded that, unlike the code-based approach, the proposed method employing equivalent modal damping ratios leads to more accurate and deformation/damage controlled results.

8.2 Theoretical Background of the Method

On the basis of the equivalent linearization concept of Jacobsen (1930), revisited by Papagiannopoulos (2018), as well as that of the energy balance in structures under seismic motion of Housner (1959), this section deals with the construction of an equivalent MDOF linear structure to the original MDOF nonlinear structure. The equivalent structure has the mass and the elastic stiffness of the original structure and modal damping ratios that take into account the effects of all nonlinearities. This implies that a balance (equivalence) between the viscous damping work and that of nonlinearities should be established. This equivalence between nonlinear and damping work is achieved with the aid of a modal damping identification model for linear structures developed by Papagiannopoulos and Beskos (2006).

The matrix equation of motion for a N degrees-of-freedom, linear, viscously damped plane building frame subjected at its base to an earthquake acceleration $\ddot{u}_g = \ddot{u}_g(t)$ is described by Eq. (2.1) which is repeated here for easy reference and reads

$$M\ddot{u} + C\dot{u} + Ku = -MI\ddot{u}_g \tag{8.1}$$

In the above, M, C, K are the mass, damping and stiffness matrices, respectively, of the structure; $\ddot{u}$, $\dot{u}$ and u are the vectors of acceleration, velocity and displacement, respectively, of the structure relative to its base and I is the unit vector. Expressing the solution in the form

$$u = \Phi q \tag{8.2}$$

where Φ is the modal matrix and $q = q(t)$ the modal amplitude vector and assuming that viscous damping is given in terms of modal damping ratios ξ_k $(k = 1, 2, .. N)$, one can uncouple Eq. (8.1) into a system of N second-order differential equations of the form

$$\ddot{q}_j(t) + 2\xi_j\omega_j\dot{q}_j(t) + \omega_j^2 q_j(t) = -\Gamma_j\ddot{u}_g(t) \tag{8.3}$$

where $j = 1, 2, \ldots. N$ is the mode number, ω_j and ξ_j are the undamped natural frequency and modal damping ratio of the jth mode, respectively, Γ_j is the participation factor for the jth mode and overdots indicate differentiation with respect to time t.

Transformation of Eq. (8.3) in the frequency domain by using Fourier transform, leads to the transformed solution

$$\bar{q}_j(\omega) = \frac{-\Gamma_j\bar{\ddot{u}}_g(\omega)}{(\omega_j^2 - \omega^2) + i(2\xi_i\omega_j\omega)} \tag{8.4}$$

where overbars denote transformed quantities, ω is the frequency and $i = \sqrt{-1}$. A transfer function $R(\omega)$ is defined in the frequency domain as the ratio of the absolute roof acceleration of the building over the acceleration at its base as

$$R(\omega) = \frac{\bar{\ddot{U}}_r(\omega)}{\bar{\ddot{u}}_g(\omega)} \tag{8.5}$$

where $\bar{\ddot{U}}_r(\omega) = \bar{\ddot{u}}_r(\omega) + \bar{\ddot{u}}_g(\omega)$ with $\bar{\ddot{u}}_r(\omega)$ being the transformed roof displacement, which can be computed from the Fourier transformed form of Eq. (8.2). Thus, with the aid of Eqs. (8.4), (8.5) can be expressed as

$$R(\omega) = 1 + \sum_j \frac{\varphi_{rj}\Gamma_j\omega^2}{(\omega_j^2 - \omega^2) + i(2\xi_i\omega_j\omega)} \tag{8.6}$$

where φ_{rj} is the roof component of the jth modal shape vector. In order to avoid complex arithmetic, one can introduce the modulus of $R(\omega)$ in the form

$$\begin{aligned}
|R(\omega)|^2 \\
&= 1 + 2\sum_{j=1}^{N} \frac{\varphi_{rj}\Gamma_j\omega^2(\omega_j^2 - \omega^2)}{(\omega_j^2 - \omega^2)^2 + (2\xi_i\omega_j\omega)^2} \\
&+ \sum_{j=1}^{N} \frac{\varphi_{rj}^2\Gamma_j^2\omega^4[(\omega_j^2 - \omega^2)^2 + 4\xi_j^2\omega_j^2\omega^2]}{[(\omega_j^2 - \omega^2)^2 + (2\xi_j\omega_j\omega)^2]^2} \\
&+ 2\sum_{\substack{j=1 \\ j \neq m, m > j}}^{N} \frac{\varphi_{rj}\Gamma_j\varphi_{rm}\Gamma_m\omega^4[(\omega_j^2 - \omega^2)(\omega_m^2 - \omega^2) + 4\xi_j\xi_m\omega_j\omega_m\omega^2]}{[(\omega_j^2 - \omega^2)^2 + (2\xi_j\omega_j\omega)^2][(\omega_m^2 - \omega^2)^2 + (2\xi_m\omega_m\omega)^2]}
\end{aligned} \tag{8.7}$$

Equation (8.7) for $\omega = \omega_k$ represents a system of N nonlinear algebraic equations which, on the assumption that $|R(\omega_k)|$, φ_{rj} (or φ_{rm}), ω_k and Γ_j (or Γ_m) are known, can be solved numerically to obtain the modal damping values ξ_k. This way one can identify modal damping values of a building on the basis of measured values of the roof acceleration due to a known seismic motion, provided that φ_{rj}, ω_k (or ω_m) and Γ_j are known or can be measured. Modal damping values can be found only for the modes that appear in the transfer function and these are the few ones significantly contributing to the response. The roof acceleration is used here for the definition of the transfer function because the roof component of any modal shape vector is always non-zero. The above identification model is valid on the assumption that the structure possesses classical (or proportional) damping. The steel plane frames considered here have a uniform distribution of damping throughout (damping is not localized) and thus they essentially possess classical damping (Chopra 2007).

According to Papagiannopoulos and Beskos (2006, 2010), a linear plane framed structure exhibits a smooth transfer function modulus versus frequency curve with well-defined visible peaks at least for the first few modes significantly contributing to the response. These peaks correspond to the resonant frequencies of the structure. Modal damping ratios can then be calculated by using the resonant frequencies and their corresponding transfer function moduli as well as the participation factors and modal shapes of the structure obtained by modal analysis. The concept of the transfer function, originally defined for linear systems, can also be extended to nonlinear systems. However, in that case, the transfer function modulus loses its smoothness. This lack of smoothness in the transfer function modulus versus frequency curve is characterized by multiple peaks or a jagged (distorted) shape (Papagiannopoulos and Beskos 2010).

It has been proven by McVerry (1980), that this jaggedness is strictly caused here by non-linearities and thus, modal damping cannot be determined uniquely as in the case of linear structures. It is also known (Worden and Tomlinson 2001) that a nonlinear system under a single harmonic force can exhibit a strong nonlinear response that depends not only on the excitation frequency, but also on frequencies considerably apart from that frequency. In case of an earthquake force, which has a rich frequency content, this behavior is amplified and the jagged (distorted) form of the transfer function curve is created. The degree of jaggedness depends on the dynamic properties of the structure, the level of the structural nonlinearity and the characteristics of the seismic motion.

The goal here is to construct a linear structure equivalent to the original nonlinear one. The construction of this equivalent linear structure is based on Jacobsen's (1930) idea of balancing the work of dissipation due to nonlinearities in the nonlinear structure with that due to viscous damping in the linear structure. This balancing is achieved by adding viscous damping in the nonlinear structure, resulting in decreasing changes in its stiffness. The criterion that signifies that this work balancing has been achieved, i.e., that the original nonlinear structure has become an equivalent linear with high damping, is that the jagged (non-smooth) transfer function modulus versus frequency curve of the nonlinear structure has become a smooth one. This smoothness (or monotonicity) criterion is associated with the behavior of the function $|R(\omega)|$ and its derivative $|R'(\omega)| = d|R(\omega)|/d\omega$ (Papagiannopoulos 2008).

Increasing progressively the damping of the nonlinear structure by using Rayleigh (classical) type of viscous damping (Chopra 2007), one is able through NLTH analysis to determine for every value of damping its roof response and then its transfer function in the frequency domain with the aid of Fourier transform. For every value of damping, the smoothness criterion is used to find out how smooth the transfer function versus frequency curve is. At that particular value of damping for which complete smoothness of the curve with clearly visible peaks has been achieved, the original nonlinear structure has become a linear one for which Eq. (8.7) is applicable. For that equivalent linear structure to the original nonlinear one, Eq. (8.7) is evaluated at the first few resonant frequencies significantly contributing to the response (cumulative mass participation greater than 90%) and the resulting system of nonlinear algebraic equations is solved for the corresponding modal damping ratios ξ_k. A mode that does not exhibit a peak in the transfer function curve is considered to be overdamped. Figure 8.1 clearly shows the above described procedure for the case of the 10-storey 3-bay plane frame of Fig. 8.2a under the Cape Mendocino 1992 earthquake accelerogram (Fig. 8.2b). In general, the satisfaction of the smoothness criterion for the transfer function does not occur simultaneously for all modes. When the satisfaction of this criterion is accomplished for one or more modes, the values of $|R(\omega)|$ at the resonant values of ω at these modes are not allowed to alter and their changes in the next cycles of the formation of the $|R(\omega)|$ are ignored. The whole procedure may be termed as per mode piecewise linearization.

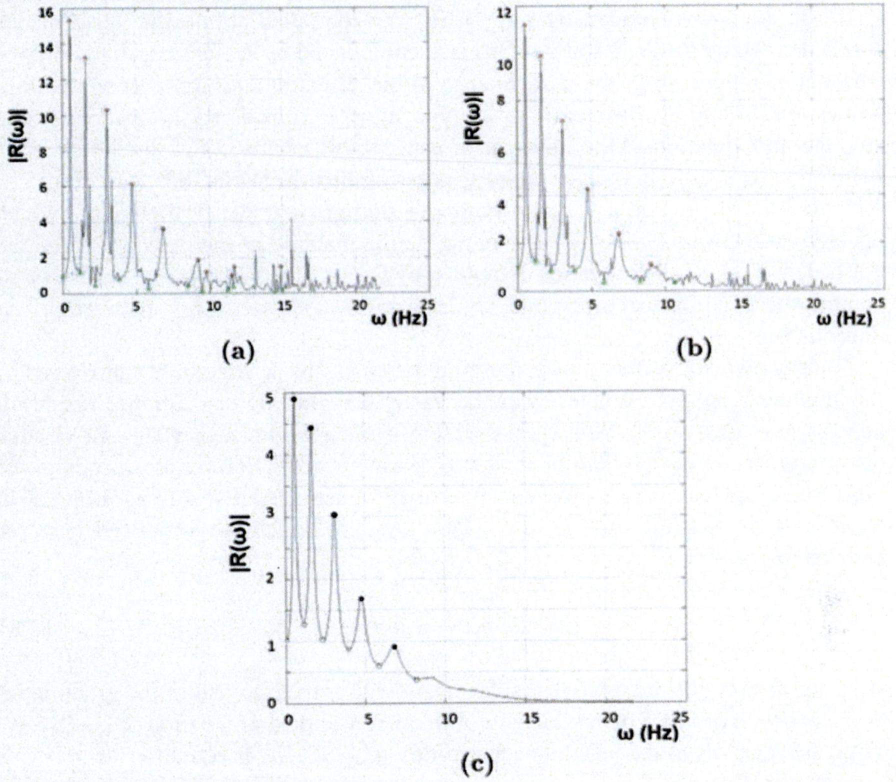

Fig. 8.1 Modulus of transfer function versus frequency for various amounts of damping for the frame of Fig. 8.2: (**a**) 3% damping (distorted shape); (**b**) 15.5% damping (smooth shape of first mode); (**c**) 24.6% damping (smooth shape of first five modes) (after Loulelis 2015, reprinted with permission from UPCE)

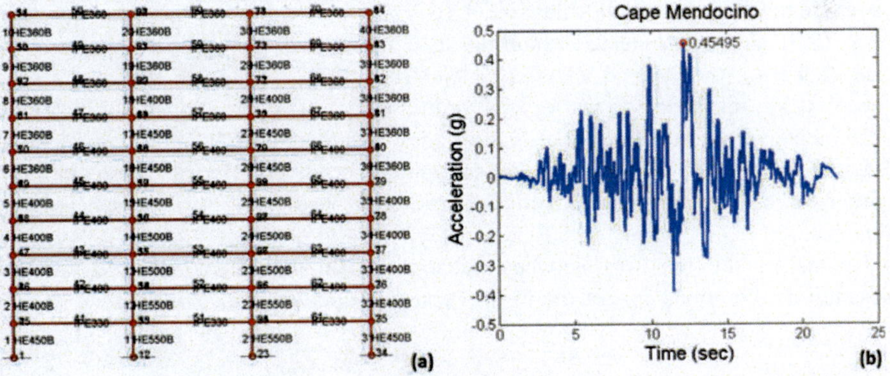

Fig. 8.2 Ten-storey three-bay steel moment resisting frame under the Cape Mendocino accelerogram (after Loulelis 2015, reprinted with permission from UPCE)

These equivalent modal damping ratios are computed at specific performance levels defined by the *IDR* and θ_p. This is accomplished by scaling appropriately the earthquake motion so as the performance of the structure does not exceed the limit values of *IDR* and θ_p. Each earthquake time signal is considered just up to the time step that the violation of the limit performance value occurs. After this time step, a band of zeros is used to replace the rest values of the time signal. Since all nonlinear structures have an initial viscous damping during their elastic behavior, which usually varies from 2% to 5%, depending on the material of the structure, from the equivalent modal damping ratios computed by Eq. (8.7), a 5% damping is subtracted so that these ratios to correspond to dissipation associated only with nonlinear deformation.

With known equivalent modal damping ratios ξ_k, the seismic base shear force of the nonlinear MDOF structure can be easily determined by response spectrum analysis provided an elastic response/design spectrum with high values of viscous damping ($5\% < \xi < 100\%$) is available. Elastic acceleration response spectra for high values of damping ratios can be easily constructed by solving Eq. (2.11) expressing the seismic motion of a SDOF system, which is reproduced here for easy reference as

$$\ddot{u} + 2\xi\omega\dot{u} + \omega^2 u = -\ddot{u}_g \tag{8.8}$$

In the above, u is the relative displacement with respect to that of the ground and $\omega = 2\pi/T$ with ω and T being the natural frequency and natural period, respectively. Thus, from Eq. (8.8) the absolute acceleration $\ddot{u}^a = \ddot{u} + \ddot{u}_g$ is equal to

$$\ddot{u}^a = -2\xi\omega\dot{u} - \omega^2 u \tag{8.9}$$

The maximum value of $\ddot{u}^a$ defines the absolute spectral acceleration S_a^α, which reduces to the spectral pseudo-acceleration S_{a} for zero damping ξ. The latter one is related to the spectral displacement S_d by the relation $S_{\text{a}} = \omega^2 S_d$ (Eq. 2.12). It has been found that for small values of ξ ($\xi < 10\%, T < 0.15s$), the term $2\xi\omega\dot{u}$ in Eq. (8.9) can be neglected resulting in $S_a^\alpha \approx \text{S}_{\text{a}}$ and the practical validity of Eq. (2.12) for those small values of ξ and T (Lin and Chang 2003). Thus, elastic acceleration response spectra for high values of damping are absolute ones. These can be converted to pseudo-acceleration ones with the aid of damping modification factors (Hatzigeorgiou 2010; Papagiannopoulos et al. 2013). This is desirable in order to conform with the existing seismic codes using only pseudo-acceleration spectra.

Consider the relations between spectral absolute accelerations and spectral pseudo-accelerations $S_{\text{a},\ k}$ of the form (Papagiannopoulos et al. 2013)

$$S_{a,k}^\alpha(T_{d,k}, \xi_{eq,k}) = p_1 S_{a,k}(T_k, \xi_{eq,k}) \tag{8.10}$$

where p_1 is a modification factor, $T_{d,k} = T_k/\sqrt{1-\xi_k^2}$ is the damped natural period and k denotes the mode number. One can also introduce the modification factor p_2 through the relation

$$S^{\alpha}_{a,k}(T_{d,k}, \xi_{eq,k}) = p_2 S^{\alpha}_{a,k}(T_k, \xi_{eq,k}) \tag{8.11}$$

Combining Eqs. (8.10) and (8.11), one can obtain the relation

$$S_{a,k}(T_k, \xi_{eq,k}) = (p_2/p_1) S^{\alpha}_{a,k}(T_k, \xi_{eq,k}) \tag{8.12}$$

which makes possible the conversion of an absolute acceleration spectrum with high damping to a pseudo-acceleration with high damping. Figure 8.3a, b displays the pseudo-acceleration spectra with high damping values (5 % $\leq \xi \leq$ 100%), *PGA* = *0.30* g and soil types B and C, respectively. These pseudo-acceleration spectra have been constructed from their corresponding absolute ones using Eq. (8.12). The absolute acceleration spectra for a specific soil type are the mean spectra of 25 ground motions for that soil type compatible to the EC8 (2004) elastic acceleration spectrum with 5% damping and *PGA* = *0.30* g with the spectrum of every one of these motions constructed by solving Eq. (8.8). The pseudo-acceleration design spectra of Fig. 8.3 can be utilized for a *PGA* other than 0.30 g with the aid of the analogy S_a(*PGA*) = S_a(*0.30* g)(*PGA*/*0.30* g). At this point one should observe that construction of elastic acceleration response spectra for high amounts of damping with the aid of the damping correction factor $\eta = \sqrt{10/(5+\xi)}$ of Eq. (3.5) is possible up to a specific damping value because the restriction of $\eta \geq 0.55$ leads to $\xi \leq 28.06\%$, which is a small value compared to the maximum value of $\xi = 100\%$ associated with the spectra of Fig. 8.3. It is demonstrated later in this Chapter that values of ξ up to 100% are needed to design for the LS and CP seismic performance levels. Finally, one should note that analogous pseudo-acceleration spectra with high amounts of damping (up to 100%) can be obtained on the basis of the elastic response spectrum with 5% damping of EC8 (2004).

For a given modal period T_k and equivalent modal damping ratio ξ_k + 0.05, one can easily determine from an acceleration response spectrum the structural acceleration for mode k and from there the seismic base shear force for that mode. The base shear force is finally obtained by using the SRSS rule according to Eq. (2.17). The damping ratio 0.05 corresponds to the elastic damping of the nonlinear structure which had not been considered when evaluating ξ_k values corresponding to the nonlinear work of dissipation.

As it has already been mentioned in Sect. 8.1, deformations, which are needed for checking deformation satisfaction at every performance level, cannot be obtained by the proposed method. However, the use of deformation dependent equivalent damping ratios constructed so as to satisfy deformation limits, makes any displacement determination unnecessary.

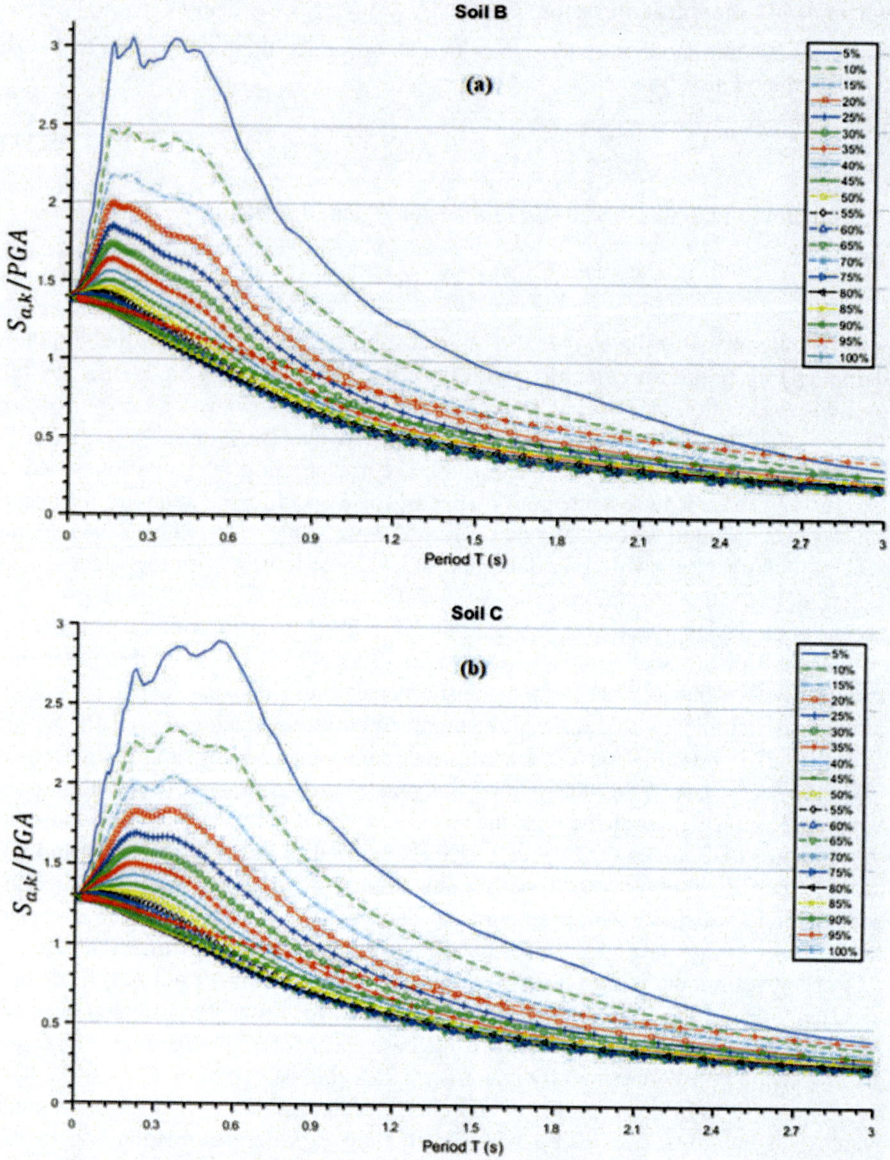

Fig. 8.3 Mean pseudo-acceleration spectra with high damping values ($5\ \% \leq \xi \leq 100\%$) and $PGA = 0.30$ g, derived by Eq. (8.12): (**a**) soil type B; (**b**) soil type C (after Kalapodis 2017, reprinted with permission from UPCE)

8.3 Modal Damping Ratios for Plane Steel MRFs

In this section, explicit expressions for the equivalent modal damping ratios pertaining to plane steel MRFs are derived on the basis of the theoretical background in the previous section and a response databank created by extensive parametric studies involving many frames under many seismic motions.

8.3.1 *Steel Frames Considered*

The steel frames considered here for parametric analyses are 20 plane regular and orthogonal MRFs with storey height equal to 3.0 m and bay width equal to 5.0 m. The number of bays varies from 2 to 6, while the number of stories varies from 2 to 18 in order to have a wide range of periods. The gravity load combination G + 0.3Q is equal to 27.5 kN/m with G and Q being the dead and live load, respectively. The grade of the steel material is S275 MPa. The frames have been designed according to EC3 (2009) and EC8 (2004) provisions with the aid of the computer program SAP 2000 (2010) employing a $PGA = 0.24$ g, soil type B and a strength reduction factor $q = 4$. Section types for beams and columns are assumed to be IPE and HEB, respectively. All connections of steel members are moment resisting ones and column sections have their strong axis perpendicular to the plane of the frame. Table 8.1 lists the 20 designed steel frames with their section dimensions and their fundamental period T_1. In this table expressions of the form 400/450/450/450/400–400 (1–5) mean that (a) from the first to the fifth storey the columns and beams have the same variation at every storey and (b) at every storey one has column sections HEB400 and HEB450 for the first bay, HEB450 and HEB450 for the next two bays and HEB450 and HEB400 for the last bay and that all beams have IPE400 sections. Moreover, expressions of the form 450–330 (1–3) mean that for stories 1 to 3, all columns have HEB450 sections and all beams have IPE330 sections.

8.3.2 *Seismic Motions and Performance Levels*

For the seismic analysis of the 20 plane steel MRFs of Table 8.1, 100 recorded, far-field (ordinary) seismic motions were selected from PEER (2009) and COSMOS (2013) seismic motion databases. These seismic motions, grouped into four categories corresponding to the four soil types A, B, C and D of EC8 (2004), and listed in Tables 8.2–8.5 (Loulelis 2015; Kalapodis 2017) together with their date, record name, direction, station name and *PGA* value. Their selection was done on the basis of the following criteria: (i) they were recorded 20–40 km away from the fault; (ii) they had a moment magnitude 5.2–7.7 and an effective duration 7.0–45.0 s and

Table 8.1 Moment resisting frames considered (after Kalapodis et al. 2021)

Frame No.	Storeys/ bays	Frame sections *HEB (columns) - IPE (beams)*	Period T_1 (s)
1	2/3	360–300(1), 360–270(2)	0.37
2	3/3	400–300(1), 360–270(2–3)	0.61
3	4/3	400–300(1–2), 360–270(3–4)	0.83
4	5/3	400–300(1–2), 360–270(3–5)	1.10
5	6/3	400–300(1–3), 360–270(4–6)	1.33
6	7/3	450–330(1), 400–300(2–4), 360–270(5–7)	1.52
7	7/2	330–330 (1–4) and 260–330 (5–7)	1.29
8	8/3	450–330(1), 400–300(2–4), 360–270(5–8)	1.79
9	9/3	450–330(1–2), 400–300(3–5), 360–270(6–9)	1.98
10	10/3	450–330(1–3), 400–300(4–6), 360–270(7–10)	2.17
11	11/3	450–330(1–4), 400–300(5–6), 360–270(8–11)	2.35
12	12/3	500–360(1), 450–330(2–5), 400–300(6–8), 360–270(9–12)	2.50
13	12/4	340/360/400/360/340–400 (1–4), 320/340/360/340/320–400 (5–8), 300/320/340/320/300–400 (9–12)	1.58
14	13/3	500–360(1–2), 450–330(3–6), 400–300(7–9), 360–270(10–13)	2.66
15	14/3	500–360(1–3), 450–330(4–7), 400–300(8–10), 360–270(11–14)	2.82
16	15/3	550–400(1–2), 500–360(3–5), 450–330(6–8), 400–300(9–11), 360–270(12–15)	2.89
17	15/5	450/450/500/500/450/450–400 (1–4), 400/450/450/450/450/ 400–400 (5–8), 360/400/450/450/400360–400 (9), 340/360/400/ 400/360/340–360 (10–12), 340/360/400/400/360/340–360 (13–15)	2.02
18	16/3	550–400(1–4), 500–360(5–7), 450–330(8–10), 400–300(11–13), 360–270(14–16)	2.93
19	18/6	550/550/600/600/600/550/550–400 (1–6), 500/500/550/550/550/ 500/500–400 (7–12) & 450/450/500/500/500/450/450–360 (13–18)	2.18
20	18/6	550–400 (1–6), 500–400 (7–12), 450–360 (13–18)	2.42

(iii) they had to drive a frame to all performance levels with the minimum scaling factor. The intensity measure employed here was the *PGA* because of its simplicity and popularity. For every structure, each accelerogram was appropriately scaled to cover four seismic performance levels as defined in SEAOC (1999) and FEMA 356 (2000). Use was made of the Incremental Dynamic Analysis (Vamvatsikos and Cornell 2002) with a maximum scale factor of 10 in order to minimize the scaling influence on the seismic structural response (De Luca et al. 2009).

Limit values for the four seismic performance levels case, taken from SEAOC (1999), or FEMA 356 (2000), are provided in Table 8.6. In that table, SP1 or IO stands for Immediate Occupancy, SP2 or DL stands for Damage Limitation, SP3 or LS stands for Life Safety, SP4 or CP stands for Collapse Prevention, *IDR* denotes interstorey drift ratio, θ_p is the member plastic rotation and θ_y is the member rotation at yield. According to FEMA 356 (2000), $\theta_y = W_{pl}L_bf_y/6EI_b$ for beams and $\theta_y = (W_{pl}L_cf_y/6EI_c)(1 - P/P_y)$ for columns, where W_{pl} is the plastic section modulus,

Table 8.2 Recorded far-fault earthquake ground motions corresponding to soil type A (after Loulelis 2015 and Kalapodis 2017, reprinted with permission from UPCE)

No.	Date	Record name	Comp.	Station name	*PGA* (g)
1	1987/10/04	Whittier Narrows	NS	24399 Mt. Wilson—CIT Station	0.158
2	1987/10/04	Whittier Narrows	EW	24399 Mt. Wilson—CIT Station	0.142
3	1999/09/20	Chi-Chi, Taiwan	056-N	HWA056	0.107
4	1999/09/20	Chi-Chi, Taiwan	056-E	HWA056	0.107
5	1987/10/01	Whittier Narrows	NS	24399 Mt. Wilson—CIT Station	0.186
6	1987/10/01	Whittier Narrows	EW	24399 Mt. Wilson—CIT Station	0.123
7	1971/02/09	San Fernando	N069	127 Lake Hughes 9	0.157
8	1971/02/09	San Fernando	N159	127 Lake Hughes 9	0.134
9	1994/01/17	Northridge	NS	90019 San Gabriel—E. Gr. Ave.	0.256
10	1994/01/17	Northridge	EW	90019 San Gabriel—E. Gr. Ave.	0.141
11	1999/09/20	Chi-Chi, Taiwan	NS	TAP103	0.177
12	1999/09/20	Chi-Chi, Taiwan	EW	TAP103	0.122
13	1994/01/17	Northridge	N005	90017 LA—Wonderland Ave.	0.172
14	1994/01/17	Northridge	N175	90017 LA—Wonderland Ave.	0.112
15	1975/06/07	Northern California	N060	1249 Cape Mendocino, Petrolia	0.115
16	1975/06/07	Northern California	N150	1249 Cape Mendocino, Petrolia	0.179
17	1986/07/08	Northern Palm Springs	NS	12206 Silent Valley	0.139
18	1989/10/18	Loma Prieta	N205	58539 San Francisco	0.105
19	1989/10/18	Loma Prieta	NS	47379 Gilroy Array 1	0.473
20	1989/10/18	Loma Prieta	EW	47379 Gilroy Array 1	0.411
21	1999/08/17	Kocaeli, Turkey	NS	Gebze	0.137
22	1999/08/17	Kocaeli, Turkey	EW	Gebze	0.244

(continued)

Table 8.2 (continued)

No.	Date	Record name	Comp.	Station name	*PGA* (g)
23	1992/06/28	Landers	NS	21081 Amboy	0.115
24	1992/06/28	Landers	EW	21081 Amboy	0.146
25	1999/09/20	Chi-Chi, Taiwan	N034	TCU046	0.133

Table 8.3 Recorded far-fault earthquake ground motions corresponding to soil type B (after Loulelis 2015 and Kalapodis 2017, reprinted with permission from UPCE)

No.	Date	Record name	Comp.	Station name	*PGA*(g)
1	1992/04/25	Cape Mendocino	NS	89,509 Eureka	0.154
2	1992/04/25	Cape Mendocino	EW	89,509 Eureka	0.178
3	1980/06/09	Victoria, Mexico	N045	6604 Cerro Prieto	0.621
4	1980/06/09	Victoria, Mexico	N135	6604 Cerro Prieto	0.587
5	1992/04/25	Cape Mendocino	EW	89324 Rio Dell Overpass	0.385
6	1992/04/25	Cape Mendocino	NS	89324 Rio Dell Overpass	0.549
7	1978/08/13	Santa Barbara	N048	283 Santa Barbara Courthouse	0.203
8	1978/08/13	Santa Barbara	N138	283 Santa Barbara Courthouse	0.102
9	1999/09/20	Chi-Chi, Taiwan	NS	TCU095	0.712
10	1999/09/20	Chi-Chi, Taiwan	EW	TCU095	0.378
11	1979/08/06	Coyote Lake	N213	1377 San Juan Bautista	0.108
12	1979/08/06	Coyote Lake	N303	1377 San Juan Bautista	0.107
13	1994/01/17	Northridge	NS	90021 LA—N Westmoreland	0.361
14	1994/01/17	Northridge	EW	90021 LA—N Westmoreland	0.401
15	1986/07/08	Northern Palm Springs	NS	12204 San Jacinto—Soboba	0.239
16	1986/07/08	Northern Palm Springs	EW	12204 San Jacinto—Soboba	0.250
17	1970/09/12	Lytle Creek	N115	290 Wrightwood	0.162
18	1970/09/12	Lytle Creek	N205	290 Wrightwood	0.200
19	1989/10/18	Loma Prieta	NS	58065 Saratoga—Aloha Ave	0.324
20	1989/10/18	Loma Prieta	EW	58065 Saratoga—Aloha Ave	0.512
21	1992/06/28	Landers	NS	22170 Joshua Tree	0.284
22	1992/06/28	Landers	EW	22170 Joshua Tree	0.274
23	1976/09/15	Friuli, Italy	NS	8014 Forgaria Cornino	0.212
24	1976/09/15	Friuli, Italy	EW	8014 Forgaria Cornino	0.260
25	1999/09/20	Chi-Chi, Taiwan	N045	TCU045	0.512

f_y is the steel yield strength, L is the member length, I is the moment of inertia, E is the elastic modulus, P is the axial force and P_y is the axial yield force in the column, while subscripts b and c stand for beam and column, respectively.

The proposed seismic design method is used here as a performance based one of the deterministic type (SEAOC 1999; FEMA 356 2000).

Table 8.4 Recorded far-fault earthquake ground motions corresponding to soil type C (after Loulelis 2015 and Kalapodis 2017, reprinted with permission from UPCE)

No.	Date	Record name	Comp	Station name	*PGA*(g)
1	1999/09/20	Chi-Chi, Taiwan	NS	NST	0.388
2	1999/09/20	Chi-Chi, Taiwan	EW	NST	0.309
3	1983/05/02	Coalinga	EW	36227 Parkfield	0.147
4	1983/05/02	Coalinga	NS	36227 Parkfield	0.131
5	1999/11/12	Duzce, Turkey	NS	Bolu	0.728
6	1999/11/12	Duzce, Turkey	EW	Bolu	0.822
7	1979/10/15	Imperial Valley	N015	6622 Compuertas	0.186
8	1979/10/15	Imperial Valley	N285	6622 Compuertas	0.147
9	1979/10/15	Imperial Valley	N012	6621 Chihuahua	0.270
10	1979/10/15	Imperial Valley	N282	6621 Chihuahua	0.284
11	1999/08/17	Kocaeli, Turkey	NS	Atakoy	0.105
12	1999/08/17	Kocaeli, Turkey	EW	Atakoy	0.164
13	1989/10/18	Loma Prieta	NS	1028 Hollister City Hall	0.247
14	1989/10/18	Loma Prieta	EW	1028 Hollister City Hall	0.215
15	1984/04/24	Morgan Hill	NS	57382 Gilroy Array # 4	0.224
16	1984/04/24	Morgan Hill	EW	57382 Gilroy Array # 4	0.348
17	1994/01/17	Northridge	NS	90057 Canyon Country	0.482
18	1994/01/17	Northridge	EW	90057 Canyon Country	0.410
19	1971/02/09	San Fernando	EW	135 LA – Hollywood	0.210
20	1971/02/09	San Fernando	NS	135 LA – Hollywood	0.174
21	1981/04/26	Westmorland	NS	5169 Westmorland Fire Sta	0.368
22	1981/04/26	Westmorland	EW	5169 Westmorland Fire Sta	0.496
23	1987/11/24	Superstition Hills	NS	01335 El Centro Imp. Co. Cent	0.258
24	1987/11/24	Superstition Hills	EW	01335 El Centro Imp. Co. Cent	0.358
25	1980/01/27	Livermore	EW	57187 San Ramon	0.301

8.3.3 *Frame Modeling and Analysis*

The seismic response of the steel MRFs of Table 8.1 is obtained by NLTH analysis with the aid of the Ruaumoko (Carr 2005) computer program. Diaphragm action is assumed at every floor and P – Δ effects are taken into account by using the "Large-Displacements" option of the Ruaumoko (Carr 2005) software. The inelastic behavior of steel members is simulated by the elastoplastic model with 3% strain hardening concentrated at the plastic hinges at their two ends. Columns are simulated by the steel beam-column model of Fig. 8.4 that takes into account the interaction of bending moment and axial force, where M_0 is equal to the moment capacity of the member $M_{pl.\ Rd} = W_{pl} f_y$ and P_{yc} and P_{yt} are the axial strengths in compression and tension, respectively, both equal to $N_{pl.\ Rd} = Af_y$ with A being the area of the steel column section.

Connections are assumed to be rigid but panel zone effects are not neglected. The panel zone effect is taken into account by the "scissors model" (FEMA 356 2000),

Table 8.5 Recorded far-fault earthquake ground motions corresponding to soil type D (after Loulelis 2015 and Kalapodis 2017, reprinted with permission from UPCE)

No.	Date	Record Name	Comp.	Station Name	*PGA* (g)
1	1981/04/26	Westmorland	N045	5062 Salton Sea Wildlife Red.	0.199
2	1981/04/26	Westmorland	N135	5062 Salton Sea Wildlife Ref.	0.176
3	1987/11/24	Superstition, Hills	N045	5062 Salton Sea Wildlife Refuge	0.119
4	1987/11/24	Superstition, Hills	N135	5062 Salton Sea Wildlife Refuge	0.167
5	1994/0 l/17	Northridge	N064	90011 Montebello—Bluff Rd.	0.128
6	1994/01/17	Northridge	N154	90011 Montebello—Bluff Rd.	0.179
7	1989/10/18	Loma Prieta	NS	58117 Treasure Island	0.159
8	1989/10/18	Loma Prieta	EW	58117 Treasure Island	0.100
9	1999/08/17	Kocaeli, Turkey	NS	Ambarli	0.184
10	1999/08/17	Kocaeli, Turkey	EW	Ambarli	0.249
11	1989/10/18	Loma Prieta	N043	1002 APEEL 2—Redwood City	0.274
12	1989/10/18	Loma Prieta	N133	1002 APEEL 2—Redwood City	0.220
13	1979/10/15	Imperial Valley	N040	5057 El Centro Array 3	0.112
14	1979/10/15	Imperial Valley	Nl30	5057 El Centro Array 3	0.179
15	1999/09/20	Chi-Chi, Taiwan	N041	CHY041	0.639
16	1999/09/20	Chi-Chi, Taiwan	N131	CHY041	0.302
17	1989/10/18	Loma Prieta	N047	1002 APEEL 2—Redwood City	0.274
18	1989/10/18	Loma Prieta	N137	1002 APEEL 2—Redwood City	0.220
19	1999/09/20	Chi-Chi, Taiwan	EW	TAP003	0.126
20	1999/0S/20	Chi-Chi, Taiwan	NS	TAP003	0.106
21	1995/01/16	Kobe	NS	Nishi-Akashi	0.503
22	1995/01/16	Kobe	EW	Nishi-Akashi	0.509
23	1999/09/20	Chi-Chi, Taiwan	N040	TCU040	0.123
24	1999/09/20	Chi-Chi, Taiwan	N130	TCU040	0.149
25	1995/01/16	Kobe	NS	Kakogawa	0.345

Table 8.6 Performance levels and their response limit values for plane steel MRFs

Performance level	*IDR*	θ_p
SP1 or IO	0.7%	0
SP2-or DL	1.5%	θ_y
SP3 or LS	2.5%	3.5 θ_y
SP4 or CP	5.0%	6.5 θ_y

which simulates the shear deformation of the panel zone with the aid of a rotational spring with zero length. This model consists of two nodes having the same coordinates, located at the center of the joint. The rigid end-blocks of the beams are half the joint width and those for the columns half the joint height, as shown in Fig. 2.4b. The nonlinear behavior of the panel zone can be represented by a tri-linear model, as shown in Fig. 8.5. Even though the more complicated panel zone model of Krawinkler (1978) is more accurate than the scissors model, the latter one was selected here as a good compromise between simplicity and accuracy.

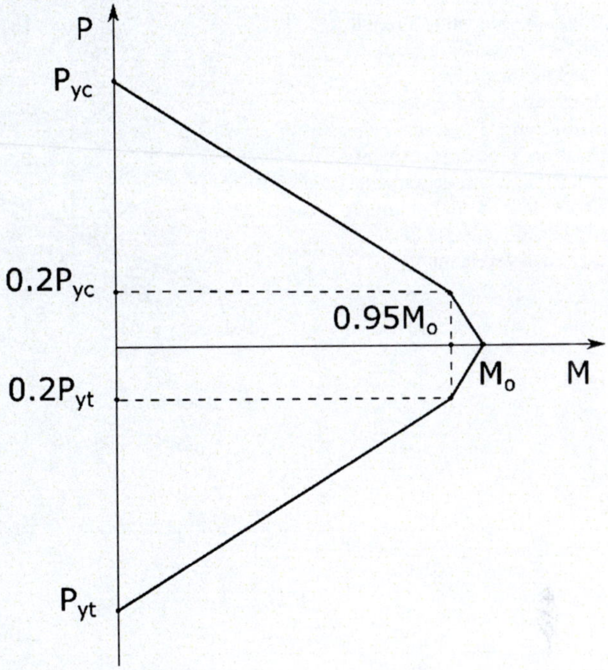

Fig. 8.4 Bending moment-axial force interaction diagram for steel columns

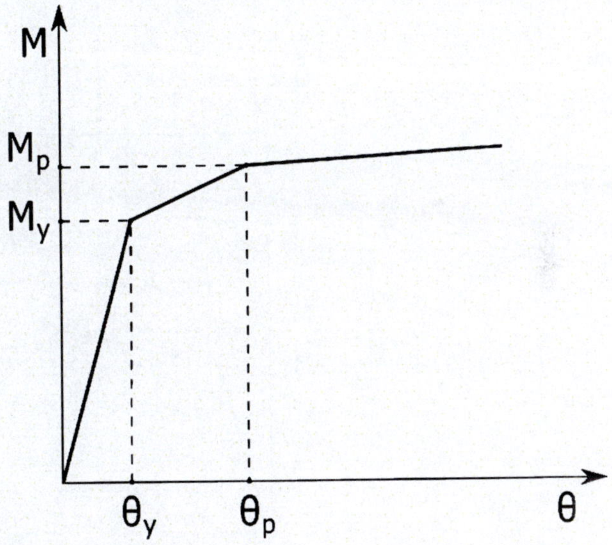

Fig. 8.5 Tri-linear behavior of panel zone according to the scissors model

During seismic loading, structural members experience stiffness and strength degradation or deterioration (Lignos and Krawinkler 2011), as shown in Fig. 8.6a. Here, use is made of the Ruaumoko (Carr 2005) model with a strength degradation that depends on the member ductility, or the number of inelastic cycles. Thus, the

Fig. 8.6 Modeling strength and stiffness degradation of steel members: (**a**) hysteresis rule for moment M (or force F) versus rotation θ (or displacement d); (**b**) strength degradation model in terms of ductilities; (**c**) backbone curve of M versus θ relation

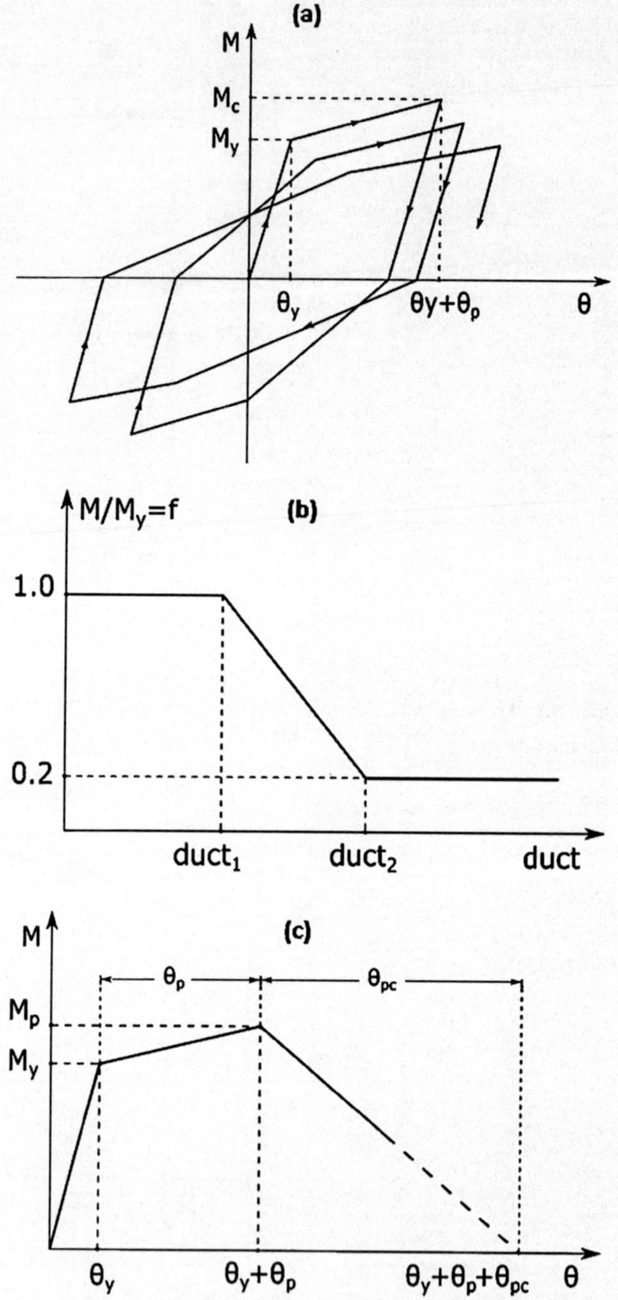

initial strength of the member M_y is multiplied by a deterioration factor f that is a linear function of the ductility (duct) of the form

$$f = \frac{rduct - 1}{duct_2 - duct_1}(duct - duct_1) + 1 \tag{8.13}$$

as shown in Fig. 8.6b. In Eq. 8.13 *rduct* is the residual ductility, while $duct_1$ and $duct_2$ are the values of ductility at which degradation begins and stops, respectively. The $duct_1$ and $duct_2$ parameters can be calculated in terms of the rotations θ_y, θ_p, θ_{pc} of the backbone moment versus rotation curve of a steel member (Fig. 8.5) by the relations

$$duct_1 = \frac{\theta_y + \theta_p}{\theta_y} \tag{8.14}$$

$$duct_2 = \frac{\theta_y + \theta_p + \theta_{pc}}{\theta_y}$$

while the *rduct* can be expressed as

$$rduct = 0.7\kappa \tag{8.15}$$

In the above, θ_y is the yield rotation, θ_p is the plastic rotation, θ_{pc} is the post capping rotation corresponding to a negative tangent stiffness and κ is the ratio of the residual strength M_r to the yield strength M_y. Lignos and Krawinkler (2011) have developed empirical relationships that provide the rotations θ_y, θ_p, θ_{pc} in terms of the geometric and material parameters of steel components on the basis of more than 300 tests available in the literature. These relationships were used here to compute the $duct_1$ and $duct_2$ from Eqs. (8.14). In Eq. (8.15) the factor 0.7 stands for cyclic deterioration effects, while the value of κ can be approximately taken as 0.4.

8.3.4 Design Equations for Modal Damping Ratios

Using the method described in Sect. 8.2 and the results of the databank created by seismically analyzing the frames considered in this section, empirical expressions for the equivalent modal damping ratios ξ_k in terms of the period T for various performance levels and soil types can be obtained. These expressions are derived from the lower bound values of the response databank and provided in explicit form in Tables 8.7, 8.8, 8.9, and 8.10 for the first four modes and soil types A, B, C and D of EC8 (2004), respectively. It should be observed that in Tables 8.7, 8.8, 8.9, and 8.10, the limit *IDR* values defining performance levels are those given in Table 8.6. The modal damping ratios ξ_k of Tables 8.7, 8.8, 8.9, and 8.10 can be applied to the seismic design of a steel MRF against far-fault seismic motions. Similar expressions for ξ_k's derived from MRFs subjected to near-fault and long duration seismic

Table 8.7 Modal damping ratios for MRFs and soil type A (after Kalapodis et al. 2021)

SP level	Mode 1	Mode 2	Mode 3	Mode 4
SP1-IO IDR = 0.7%	$\xi_1 = 0.237T + 1.51$ $(0.37 \leq T \leq 2.93)$	$\xi_2 = -8.0T + 2.3$ $(0.10 \leq T \leq 0.20)$ $\xi_2 = 0.488T + 0.602$ $(0.20 \leq T \leq 1.0)$	$\xi_3 = 0.1T + 0.693$ $(0.07 \leq T \leq 0.57)$	$\xi_4 = -2.174T + 1.35$ $(0.07 \leq T \leq 0.30)$ $\xi_4 = 0.714T + 0.486$ $(0.30 \leq T \leq 0.37)$
SP2-DL IDR = 1.5%	$\xi_1 = -0.51T + 9.50$ $(0.37 \leq T \leq 2.93)$	$\xi_2 = -86.0T + 21.9$ $(0.10 \leq T \leq 0.20)$ $\xi_2 = -1.585T + 5.017$ $(0.20 \leq T \leq 1.0)$	$\xi_3 = -51.36T + 13.47$ $(0.09 \leq T \leq 0.20)$ $\xi_3 = -1.892T + 3.578$ $(0.20 \leq T \leq 0.57)$	$\xi_4 = -71.429T + 20.34$ $(0.15 \leq T \leq 0.24)$ $\xi_4 = -6.385T + 4.73$ $(0.24 \leq T \leq 0.37)$
SP3-LS IDR = 2.5%	$\xi_1 = -0.712T + 36.56$ $(0.37 \leq T \leq 2.93)$	$\xi_2 = -7.831T + 20.488$ $(0.19 \leq T \leq 1.0)$	$\xi_3 = -10.0T + 18.0$ $(0.20 \leq T \leq 0.50)$	$\xi_4 = 100.0$ $(0.07 \leq T \leq 0.37)$
SP4-CP IDR = 5%	$\xi_{1a} = -22.83T + 90.94$ $(0.37 \leq T \leq 0.83)$ $\xi_{1b} = 9.42T + 64.18$ $(0.83 \leq T \leq 2.93)$	$\xi_2 = 100.0$ $(0.10 \leq T \leq 1.0)$	$\xi_3 = 100.0$ $(0.07 \leq T \leq 0.57)$	$\xi_4 = 100.0$ $(0.07 \leq T \leq 0.37)$

Table 8.8 Modal damping ratios for MRFs and soil type B (after Kalapodis et al. 2021)

SP level	Mode 1	Mode 2	Mode 3	Mode 4
SP1-IO IDR = 0.7%	$\xi_1 = -4.0T + 4.98$ $(0.37 \le T \le 0.82)$ $\xi_1 = 0.77T + 1.07$ $(0.82 \le T \le 2.93)$	$\xi_2 = -16.0T + 3.8$ $(0.10 \le T \le 0.20)$ $\xi_2 = 1.583T + 0.283$ $(0.20 \le T \le 1.0)$	$\xi_3 = 0.30T + 0.779$ $(0.07 \le T \le 0.57)$	$\xi_{4a} = -3.91T + 1.774$ $(0.07 \le T \le 0.3)$ $\xi_{4b} = 2.143T\text{-}0.043$ $(0.30 \le T \le 0.37)$
SP2-DL IDR = 1.5%	$\xi_1 = -0.79T + 9.29$ $(0.37 \le T \le 2.93)$	$\xi_2 = -96.0T + 23.9$ $(0.10 \le T \le 0.20)$ $\xi_2 = 0.244T + 4.65$ $(0.20 \le T \le 1.0)$	$\xi_3 = -12.917T + 10.563$ $(0.09 \le T \le 0.57)$	$\xi_4 = -24.3T + 11.69$ $(0.15 \le T \le 0.37)$
SP3-LS IDR = 2.5%	$\xi_1 = -0.395T + 32.85$ $(0.37 \le T \le 2.93)$	$\xi_2 = -18.0T + 27.96$ $(0.20 \le T \le 0.62)$ $\xi_2 = -4.5T + 19.59$ $(0.62 \le T \le 1.0)$	$\xi_3 = -1.54T + 13.18$ $(0.20 \le T \le 0.57)$	$\xi_4 = 100.0$ $(0.07 \le T \le 0.37)$
SP4-CP IDR = 5%	$\xi_1 = -3.07T + 74.14$ $(0.37 \le T \le 2.0)$ $\xi_1 = 22.55T + 22.89$ $(2.0 \le T \le 2.93)$	$\xi_2 = 100.0$ $(0.10 \le T \le 1.0)$	$\xi_3 = 100.0$ $(0.07 \le T \le 0.57)$	$\xi_4 = 100.0$ $(0.07 \le T \le 0.37)$

Table 8.9 Modal damping ratios for MRFs and soil type C (after Kalapodis et al. 2021)

SP level	Mode 1	Mode 2	Mode 3	Mode 4
SP1-IO IDR = 0.7%	$\xi_1 = -1.507T + 3.358$ $(0.37 \leq T \leq 1.10)$ $\xi_1 = 0.833T + 0.783$ $(1.10 \leq T \leq 2.93)$	$\xi_2 = -14.0T + 3.50$ $(0.10 \leq T \leq 0.20)$ $\xi_2 = 0.854T + 0.529$ $(0.2 \leq T \leq 1.0)$	$\xi_3 = 0.166T + 0.785$ $(0.07 \leq T \leq 0.57)$	$\xi_4 = -2.391T + 1.467$ $(0.07 \leq T \leq 0.30)$ $\xi_4 = 2.143T + 0.107$ $(0.30 \leq T \leq 0.37)$
SP2-DL IDR = 1.5%	$\xi_1 = -1.07T + 10.21$ $(0.37 \leq T \leq 2.93)$	$\xi_2 = -85.0T + 24.0$ $(0.10 \leq T \leq 0.20)$ $\xi_2 = -1.829T + 7.37$ $(0.20 \leq T \leq 1.0)$	$\xi_3 = -10.833T + 9.375$ $(0.07 \leq T \leq 0.57)$	$\xi_4 = -30.625T + 14.33$ $(0.21 \leq T \leq 0.37)$
SP3-LS IDR = 2.5%	$\xi_1 = -2.372T + 41.88$ $(0.37 \leq T \leq 2.93)$	$\xi_2 = -1.48T + 13.144$ $(0.37 \leq T \leq 0.62)$	$\xi_3 = 100.0$ $(0.30 \leq T \leq 0.57)$	$\xi_4 = 100.0$ $(0.07 \leq T \leq 0.37)$
SP4-CP IDR = 5%	$\xi_1 = -30.44T + 92.26$ $(0.37 \leq T \leq 0.83)$ $\xi_1 = 9.18T + 59.38$ $(0.83 \leq T \leq 2.93)$	$\xi_2 = 100.0$ $(0.10 \leq T \leq 1.0)$	$\xi_3 = 100.0$ $(0.07 \leq T \leq 0.57)$	$\xi_4 = 100.0$ $(0.07 \leq T \leq 0.37)$

Table 8.10 Modal damping ratios for MRFs and soil type D (after Kalapodis et al. 2021)

SP level	Mode 1	Mode 2	Mode 3	Mode 4
SP1-IO IDR = 0.7%	$\xi_1 = 0.547T + 1.298$ $(0.37 \le T \le 2.93)$	$\xi_2 = -17.0T + 4.10$ $(0.10 \le T \le 0.20)$ $\xi_2 = 0.427T + 0.615$ $(0.20 \le T \le 1.0)$	$\xi_3 = -0.104T + 0.859$ $(0.07 \le T \le 0.57)$	$\xi_4 = -2.667T + 1.687$ $(0.07 \le T \le 0.37)$
SP2-DL IDR = 1.5%	$\xi_1 = -1.641T + 13.11$ $(0.37 \le T \le 2.93)$	$\xi_2 = -0.482T + 6.792$ $(0.20 \le T \le 1.0)$	$\xi_3 = -15.0T + 11.45$ $(0.07 \le T \le 0.57)$	$\xi_4 = -26.54T + 15.22$ $(0.16 \le T \le 0.37)$
SP3-LS IDR = 2.5%	$\xi_1 = -0.313T + 40.384$ $(0.37 \le T \le 2.93)$	$\xi_2 = -9.85T + 23.843$ $(0.37 \le T \le 1.0)$	$\xi_3 = 100.0$ $(0.30 \le T \le 0.57)$	$\xi_4 = 100.0$ $(0.07 \le T \le 0.37)$
SP4-CP IDR = 5%	$\xi_1 = -7.5T + 85.275$ $(0.37 \le T \le 2.93)$	$\xi_2 = 100.0$ $(0.10 \le T \le 1.0)$	$\xi_3 = 100.0$ $(0.07 \le T \le 0.57)$	$\xi_4 = 100.0$ $(0.07 \le T \le 0.37)$

motions have been developed by Papagiannopoulos and Beskos (2010). It should be also noted that the structural modeling used in Papagiannopoulos and Beskos (2010) does not take into account panel zone and material deterioration effects, as it is the case with Kalapodis et al. (2021).

8.4 Modal Damping Ratios for Plane Steel Braced Frames

In this section, explicit expressions for the equivalent modal damping ratios of plane steel braced frames are derived on the basis of the theoretical background in Sect. 8.2 and a response databank created by extensive parametric studies involving many braced frames under many seismic motions.

8.4.1 *Steel Frames and Seismic Motions Considered*

Here plane eccentrically braced frames (EBFs) of chevron type (Fig. 8.7a), and concentrically braced frames (CBFs) of chevron type (Fig. 8.7b) with buckling-restrained braces (BRBs) are considered by following the work of Kalapodis (2017), Kalapodis et al. (2020) and Kalapodis and Papagiannopoulos (2020). Figure 8.8a clearly demonstrates the superior hysteretic behavior of BRBs over conventional braces, while Fig. 8.8b depicts a typical cross section of a BRB. All the aforementioned braced frames have 3 bays with a span of 5.0 m each. Plane eccentrically braced frames (EBFs) with diagonal braces and concentrically braced frames (CBFs) with diagonal BRBs have been also considered but they are not presented in this Chapter. The interested reader may consult Kalapodis (2017) and Loulelis (2015) for these two types of frames.

Column sections are orientated in such a way that their strong axis is perpendicular to the plane of the frame. Beam-column connections are considered as moment resisting ones. Beams are considered as continuously restrained for lateral torsional buckling due to the presence of the concrete floor slab. Concerning the EBFs, use is made of short, intermediate and long seismic links with lengths X = 0.5, 1.0 and 1.5 m, respectively, as shown in Fig. 8.7a. The uniformly distributed vertical load to beams due to the G + 0.3Q loading combination is equal to w = 27.5 kN/m. The grade of steel is S275. The EBFs have been designed according to EC3 (2009) and EC8 (2004) provisions with the aid of SAP 2000 (2010) software, for a $PGA = 0.24$ g, soil type B and a strength reduction factor $q = 4$ corresponding to medium ductility class structures. Frames with BRBs, which are not considered in Eurocodes, are designed by modifying the EC8 (2004) rules for steel chevron braced frames (Bosco et al. 2015). The grade of steel used for beams, columns and BRBs is S275. The buckling-restrained braced frames (BRBFs) have been designed according to EC8 (2004) provisions with the aid of SAP 2000 (2010) software, for a $PGA = 0.24$ g,

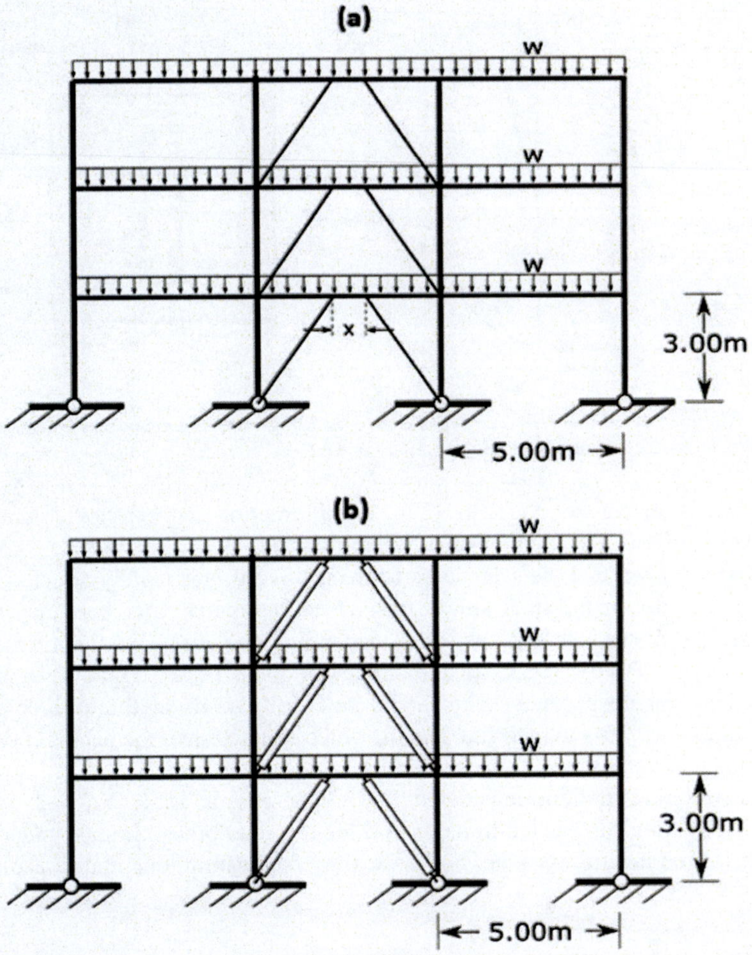

Fig. 8.7 Braced frames of chevron type: (**a**) Eccentrically braced frames; (**b**) Concentrically braced frames with buckling-restrained braces

soil type B and a strength reduction factor $q = 4$ corresponding to medium ductility class structures.

Tables 8.11, 8.12, 8.13, and 8.14 provide the number of stories, the sections (HEB for columns, IPE for beams, CHS or orthogonal core for braces) and the first natural period of the braced frames of Fig. 8.7a, b. In Tables 8.11, 8.12, 8.13, and 8.14, expressions of the form 260–300-193.7 × 4.5(3–4) indicate that for stories 3 and 4, all columns have HEB260 sections, all beams have IPE300 sections and all braces have CHS 193.7x4.5 sections. Moreover, in Table 8.14, expressions of the form 300/360–300–18.9(3–4) mean that for stories 3 and 4, interior columns are HEB360, exterior columns are HEB300, beams are IPE300 and the area of the

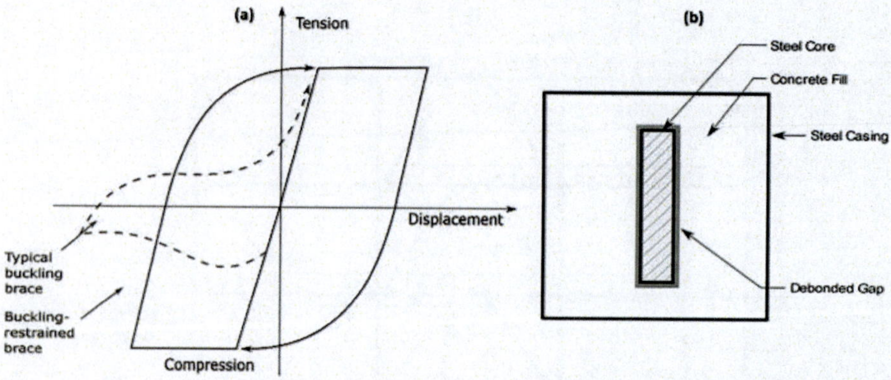

Fig. 8.8 Comparison of the hysteretic behavior of a buckling-restrained brace and a standard brace (**a**); typical cross-section of a buckling-restrained brace (**b**)

rectangular steel core is equal to 18.9 cm^2. If interior and exterior columns are identical (say HEB360), then one can write, e.g., 360–300–18.9(3–4).

It is well known that the seismic performance of EBFs is mainly affected by the inelastic behavior of the steel link beams. These dissipative members may have a shear, flexural or shear/flexural behavior, depending on their geometry and mechanical properties (Popov 1983; Engelhardt and Popov 1989). Thus, even though 84 EBFs are considered here, they should practically be divided into three groups of 28 frames according to their link length, which characterizes the behavior of those link beams and consequently the seismic behavior of the frame.

The earthquake motions considered here are the ones in Tables 8.2, 8.3, 8.4, and 8.5. These motions are scaled in order to drive the steel braced frames under study into the desired level of damage. For the scaling factor limit, one may consult Sect. 8.3.2.

8.4.2 *Modeling of Frames Considered*

The construction of modal damping ratios is accomplished through extensive parametric studies involving NLTH analyses of the frames considered under the seismic motions selected. These analyses are conducted with the aid of the Ruaumoko 2D (Carr 2005) computer program. Geometrical nonlinearities are taken into account by using the "Large-displacement" analysis option. A diaphragm action is also considered at every floor due to the presence of the concrete slab.

Beam members and long and intermediate seismic links are simulated on the basis of the Giberson beam model with two rotational springs (plastic hinges) at its two ends and without taking into account the interaction between moment capacity and axial force. Short seismic links are also simulated by the Giberson beam model with two additional translational springs at both ends in order to simulate the shear behavior of the element (Richards and Uang 2006). Columns are finally simulated

Table 8.11 Eccentrically braced frames of long links (1.5 m): chevron bracing (after Kalapodis 2017, reprinted with permission from UPCE)

Frame	Stories	Frame sections HEB (columns)–IPE (beams)–CHS (brace)	Period (s)
1	2	240–300–152.4 × 4(1-2)	0.336
2	2	200–300–139.7 × 4(1–2)	0.348
3	3	260–330–193.7 × 4.5(1), 240–300–193.7 × 4.5(2), 240–300–168.3 × 4(3)	0.414
4	3	240–300–193.7 × 4.5(1), 240–270–168.3 × 4(2–3)	0.456
5	6	340–360–219.1 × 5(1–2), 300–300–193.7 × 4.5(3–4), 260–270–193.7 × 4.5(5), 260–270–168.3 × 4(6)	0.700
6	6	300–360–219.1 × 5(1–2), 260–300–193.7 × 4.5(3–4), 240–270–168.3 × 4(5–6)	0.720
7	9	450–400–244.5 × 5.4(1), 400–360–244.5 × 5.4(2–3), 340–330–219.1 × 5(4–6), 300–300–193.7 × 4.5(7), 260–270–168.3 × 4(8–9)	0.931
8	9	450–400–219.1 × 5(1), 400–360–219.1 × 5(2–3), 340–330–193.7 × 4.5(4–5), 300–300–193.7 × 4.5(6–7), 260–270–168.3 × 4(8–9)	0.976
9	12	550–450–244.5 × 5.4(1), 500–450–244.5 × 5.4(2–3), 450–400–219.1 × 5(4–5), 400–360–219.1 × 5(6–7), 360–360–193.7 × 4(8–9), 300–300–168.3 × 4(10–12)	1.123
10	12	500–400–219.5 × 5(1), 450–400–219.5 × 5(2–3), 400–360–193.7 × 4.5(4–5), 360–330–193.7 × 4.5(6–7), 300–300–193.7 × 4.5(8–9), 260–300–193.7 × 4.5(10), 260–270–168.3 × 4(11–12)	1.289
11	15	650–450–244.5 × 5.4(1), 600–450–244.5 × 5.4(2), 550–400–244.5 × 5.4(3–4), 500–360–219.1 × 5(5–6), 450–330–193.7 × 4.5(7–8), 400–330–193.7 × 4.5(9–10), 360–300–193.7 × 4.5(11–12), 320–300–168.3 × 4(13), 280–270–168.3 × 4(14–15)	1.520
12	15	600–450–219.1 × 5(1), 550–400–219.1 × 5(2–3), 500–360–193.7 × 4.5(4–5), 450–360–193.7 × 4.5(6–7), 400–330–193.7 × 4.5(8–9), 360–300–193.7 × 4.5(10–11), 320–300–193.7 × 4.5(12–13), 280–270–168.3 × 4(14–15)	1.570
13	17	700–450–244.5 × 5.4(1–2), 650–400–244.5 × 5.4(3–4), 600–400–219.1 × 5(5–6), 500–360–219.1 × 5(7–9), 400–330–193.7 × 4.5(10–12), 360–300–193.7 × 4.5(13–15), 300–270–193.7 × 4.5(16–17)	1.669
14	17	700–450–244.5 × 5.4(1), 600–400–244.5 × 5.4(2–3), 550–400–219 × 5(4-5), 450–360–193.7 × 4.5(6–7), 400–360–193.7 × 4.5(8–9), 360–330–193.7 × 4.5(10–11), 320–360–193.7 × 4.5(12–13), 280–330–168.3 × 4(14–15), 260–270–168.3 × 4(16–17)	1.731

as in the case of MRFs described in Sect. 8.3.3. Concerning BRBs, use is also made of the Giberson beam model with elastic hinges at its two ends. All steel members obey a bilinear hysteretic rule with strain hardening equal to 0.03 for beams and columns and 0.012 for steel braces (FEMA 356 2000).

Even though the Ruaumoko (Carr 2005) program used here can simulate strength and stiffness degradation, as explained in Sect. 8.3.3, no such degradations are

Table 8.12 Eccentrically braced frames of intermediate links (1.0 m): chevron bracing (after Kalapodis 2017, reprinted with permission from UPCE)

Frame	Storey	Frame sections HEB (columns)–IPE (beams)–CHS (brace)	Period (s)
1	2	240–330–193.7 × 4.5(1), 220–330–193.7 × 4.5(2)	0.219
2	2	220–300–168.3 × 4(1), 200–300–168.3 × 4(2)	0.254
3	3	260–360–219.1 × 5.4(1), 240–300–193.7 × 4.5(2–3)	0.344
4	3	220–300–193.7 × 4.5(1), 200–300–168.3 × 4(2–3)	0.418
5	6	340–400–244.5 × 5.4(1–2), 300–360–219.1 × 5(3–4), 280–270–193.7 × 4.5(5–6)	0.593
6	6	300–360–244.5 × 5.4(1–2), 260–330–193.7 × 4.5(3–4), 220–270–168.3 × 4(5–6)	0.659
7	9	360–400–273 × 5.6(1–2), 340–360–244.5 × 5.4(3–4), 320–330–219.1 × 5(5–6), 280–270–193.7 × 4.5(7–9)	0.911
8	9	360–450–244.5 × 5.4(1), 320–400–219.1 × 5(2), 300–360–193.7 × 4.5(3–4), 280–330–193.7 × 4.5(5–6), 240–270–168.3 × 4(7–9)	0.983
9	12	500–450–273.5 × 5.6(1), 450–450–273.5 × 5.6(2), 400–400–244.5 × 5.6(3–4), 360–360–219.1 × 5(5–6), 320–330–193.7 × 4.5(7–8), 280–300–193.7 × 4.5(9–10), 280–270–168.3 × 4(11–12)	1.242
10	12	450–450–244.5 × 5.4(1), 400–400–219.1 × 5(2–3), 360–400–219.1 × 5(4), 340–360–193.7 × 4.5(5–6), 300–330–193.7 × 4.5(7–8), 260–330–168.3 × 4(9–10), 240–300–168.3 × 4(11–12)	1.297
11	15	650–450–273.5 × 5.6(1–2), 500–400–244.5 × 5.4(3–4), 450–360–219.1 × 5(5–6), 400–330–193.7 × 4.5(7–8), 360–300–193.7 × 4.5(9–10), 320–300–168.3 × 4(11–12), 280–270–168.3 × 4(13–15)	1.582
12	15	600–450–244.5 × 5.4(1), 500–450–244.5 × 5.4(2–3), 450–360–219.1 × 5(4–6), 400–360–219.1 × 5(7–8), 360–360–193.7 × 4.5(9–10), 320–360–193.7 × 4.5(11–12), 280–330–168.3 × 4(13–15)	1.673
13	17	700–450–298.5 × 5.9(1–2), 650–400–273.5 × 5.6(3–4), 550–400–244.5 × 5(5–6), 500–360–244.5 × 5(7–8), 450–360–219.1 × 5(9–10), 360–360–193.7 × 4.5(11–12), 300–330–193.7 × 4.5(13–14), 280–300–168.3 × 4(15–17)	1.794
14	17	700–450–273.5 × 5.6(1), 650–450–273.5 × 5.6(2), 600–400–273.5 × 5.6(3–4), 550–400–244.5 × 5(5–6), 450–360–219.1 × 5(7–8), 400–360–193.7 × 4.5(9), 360–360–193.7 × 4.5(10–11), 320–330–193.7 × 4.5(12–13), 280–300–168.3 × 4(14–15), 260–300–168.3 × 4(16–17)	1.873

considered here because of (i) the required very high computational demands and (ii) their not significant effect on the seismic response (NIST 2015; Azad and Topkaya 2017; Loulelis et al. 2018).

For the modeling of EBFs, rigid links are used at member intersections for simulating both member eccentricity from the centroidal axes and the effect of the

Table 8.13 Eccentrically braced frames of short links (0.5 m): chevron bracing (after Kalapodis 2017, reprinted with permission from UPCE)

Frame	Storey	Frame sections HEB (columns)–IPE (beams)–CHS (brace)	Period (s)
1	2	260–300–168.3 × 4(1), 240–300–168.3 × 4(2)	0.256
2	2	240–300–152.4 × 4(1–2)	0.267
3	3	280–330–244.5 × 5.4(1), 260–300–219.1 × 5(2–3)	0.289
4	3	260–300–219.1 × 5 (1), 240–300–193.7 × 4.5(2–3)	0.309
5	6	360–400–244.5 × 5.4(1), 360–360–244.5 × 5.4(2), 320–330–244.5 × 5.4(3), 320–330–219.1 × 5(4), 280–300–219.1 × 5(5–6)	0.535
6	6	320–400–219.1 × 5(1), 320–360–219.1 × 5(2), 280–330–219.1 × 5(3), 280–330–193.7 × 4.5(4), 260–300–193.7 × 4.5(5–6)	0.571
7	9	450–400–244.5 × 5.4(1–2), 400–360–244.5 × 5.4(3–4), 360–330–219.1 × 5(5–6), 320–300–219.1 × 5(7), 320–300–193.7 × 4.5(8), 280–300–193.7 × 4.5(9)	0.769
8	9	400–400–219.1 × 5(1–2), 360–360–219.1 × 5(3–4), 320–330–193.7 × 4.5(5–6), 280–300–193.7 × 4.5(7), 320–300–168.3 × 4(8), 260–300–168.3 × 4(9)	0.814
9	12	550–450–244.5 × 5.4(1–2), 500–400–244.5 × 5.4(3), 500–360–244.5 × 5.4(4), 450–360–244.5 × 5.4(5–6), 400–330–219.1 × 5(7–8), 360–300–219.1 × 5(9–10), 300–300–193.7 × 4.5(11–12)	1.157
10	12	500–450–219.1 × 5(1), 500–400–219.1 × 5(2), 450–400–219.1 × 5(3–4), 400–360–219.1 × 5(5–6), 360–330–193.7 × 4.5 (7–8), 320–300–193.7 × 4.5(9), 320–300–168.3 × 4(10) 280–300–168.3 × 4(11–12)	1.213
11	15	600–500–273.5 × 5.6(1–2), 550–450–244.5 × 5.4(3), 550–400–244.5 × 5.4(4), 500–400–244.5 × 5.4(5–6), 450–360–219.1 × 5(7–8), 400–360–219.1 × 5(9–10), 360–300–193.7 × 4.5(11–12), 320–300–193.7 × 4.5(13–15)	1.472
12	15	550–500–244.5 × 5.4(1), 500–450–219.1 × 5(2–3), 500–400–219.1 × 5(4), 450–400–219.1 × 5(5–6), 400–360–193.7 × 4.5(7–8), 360–360–193.7 × 4.5(9–10), 320–330–168.3 × 4(11), 320–300–168.3 × 4(12–14), 280–300–168.3 × 4(15)	1.532
13	17	700–500–273.5 × 5.6(1–2), 650–450–273.5 × 5.6(3–4), 600–400–244.5 × 5.4(5–6), 550–400–244.5 × 5.4(7), 550–360–219.1 × 5(8), 500–360–219.1 × 5(9), 500–360–219.1 × 5(10), 450–330–219.1 × 5(11–12), 400–330–193.7 × 4.5(13–14), 360–300–193.7 × 4.5(15–16), 300–300–193.7 × 4.5(17)	1.671
14	17	650–500–244.5 × 5.4(1–2), 600–450–244.5 × 5.4(3), 600–450–219.1 × 5(4), 550–400–219.1 × 5(5–6), 500–400–219.1 × 5(7), 500–360–193.7 × 4.5(8), 450–360–193.7 × 4.5(9–10), 400–330–193.7 × 4.5(11–12), 360–330–193.7 × 4.5(13–15), 320–300–168.3 × 4(16), 280–300–168.3 × 4(17)	1.728

Table 8.14 Concentrically braced frames with buckling-restrained braces considered (after Kalapodis 2017, reprinted with permission from UPCE)

Frame	Stories	Frame sections HEB (columns)–IPE (beams)–Steel core (BRB)	Period (s)
1	2	240–300–15.4(1–2)	0.244
2	2	220–300–15.4(1), 220–300–14.0(2)	0.250
3	3	260–360–26.6(1), 260–330–18.9(2), 240–300–15.4(3)	0.288
4	3	240–330–26.6(1), 240–330–18.9(2), 220–300–15.4(3)	0.292
5	6	400–400–40.6(1), 400–400–33.6(2), 320–360–26.6(3–4), 260–360–21.0(5), 260–330–14.0(6)	0.485
6	6	360–360–40.6(1), 360–360–33.6(2), 300–360–26.6(3), 300–330–26.6(4), 240–330–21.0(5), 240–330–14.0(6)	0.502
7	9	550–500–40.6(1–2), 450–500–33.6(3), 400–450–33.6(4), 360–400–26.6(5–6), 300–400–21.0(7–8), 240–400–18.9(9)	0.694
8	9	500–450–40.6(1), 450–450–33.6(2–3), 400–450–26.6(4–5), 300–400–21.0(6–7), 260–400–16.1(8), 220–360–15.4(9)	0.740
9	12	650–550–40.6(1–2), 600–500–33.6(3–4), 500–450–33.6(5–6), 400–400–26.6(7–8), 360–400–21.0(9–10), 300–360–17.5(11–12)	0.966
10	12	600–500–40.6(1–2), 550–500–33.6(3–4), 450–450–26.6(5–6), 360–400–21.0(7–8), 300–400–18.9(9), 260–360–17.5(10–12)	1.030
11	15	800–550–40.6(1–2), 700–500–33.6(3–4), 600–500–26.6(5–6), 500–450–26.6(7–8), 450–400–21.0(9–11), 400–360–18.9(12–13), 320–360–17.5(14), 320–360–16.1(15)	1.246
12	15	700–550–40.6(1–2), 650–500–33.6(3–5), 500–450–26.6(6–8), 400–400–21.0(9–10), 360–360–21.0(11–13), 300–330–17.5 (14–15)	1.294
13	17	800–550–40.6(1–4), 700–500–33.6(5–6), 600–450–33.6(7–8), 500–400–26.6(9–10),450–360–21.0(11–13),400–360–18.9(14–15), 360–330–17.5(16–17)	1.457
14	17	800–550–40.6(1–2), 700–500–33.6(3–4), 600–450–33.6(5–6), 500–400–26.6(7–9),400–400–21.0(10–12),360–360–18.9(13–14), 300–330–17.5(15–16), 260–330–14.0(17)	1.501

gusset plate there, if such a plate exists (Fig. 8.9). Before evaluating the lengths of rigid links, the gusset plates are dimensioned on the basis of capacity design (Hsiao et al. 2012; Okazaki et al. 2013) in order to secure the occurrence of the first yielding of a frame in its dissipative members, i.e., seismic links for the EBFs and braces for the CBFs.

For a designed gusset plate with known side lengths, as shown in Fig. 8.9, the length b of the column rigid segment is equal to the length of the vertical side of the gusset plate, while the length 0.75a of the beam is equal to the length of the horizontal side of the gusset plate times 0.75. In addition, the rigid link length AB on the diagonal is equal to the length of the segment between the work point A and the start of the bracing B (Fig. 8.9). Concerning CBFs with BRBs the braces are assumed to be pin-connected at intersections and without taking the gusset plate influence into account. Thus, in this case, rigid links are related only to the geometric characteristic of the intersected members.

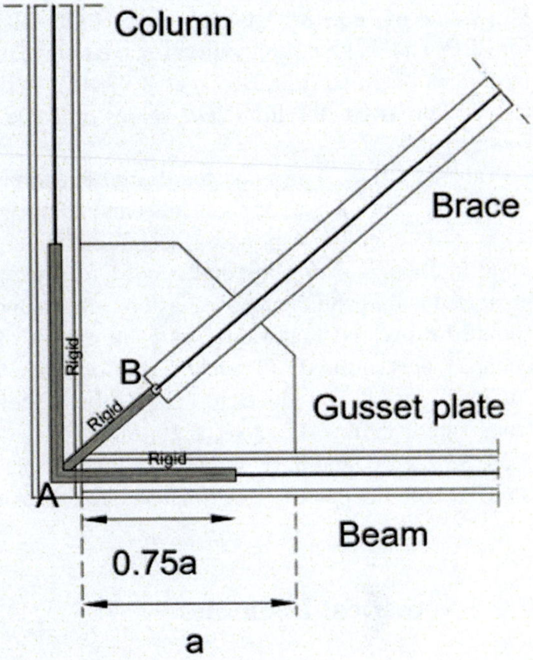

Fig. 8.9 Modeling of gusset plate at beam-column-brace intersection

Table 8.15 Seismic performance (SP) levels for CBFs with BRBs and EBFs (after Kalapodis 2017, reprinted with permission from UPCE)

SP level	*IDR*	μ_δ (brace ductility)	θ_{link} (rad)
SP1-IO	0.4%	1.0	0.0
SP2-DL	1.3%	3.6	Not provided
SP3-LS	2.2%	6.2	0.02 (long links), 0.08 (short links)
SP4-CP	3.2%	8.0	Not provided

Panel zone effect in the modeling of frames has been considered only for joints without gusset plates. For simulating panel zone effect here use is made of the "scissors model", which has been described in Sect. 8.3.3.

8.4.3 Design Equations for Modal Damping Ratios

Following the same procedure as in the case of MRFs, modal damping ratios ξ_k for each EBF and CBF can be computed to create a databank of ξ_k's. Regression analyses on the lowest ξ_k values with period T are then conducted in order to construct simple expressions for ξ_k to be used for seismic design purposes. These design equations correspond to the seismic performance (SP) levels of Table 8.15 and to soil types A-D of EC8 (2004). One should note here that the limit *IDR* and

θ_{link} values of Table 8.15, taken from SEAOC (1999) and EC8 (2004), are valid only for EBFs and that the limit values for axial brace ductility valid for CBFs with BRBs have been obtained from Bosco et al. (2015). In the absence of limit *IDR* values for CBFs with BRBs, the limit *IDR* values of Table 8.15 are also used here for these frames.

Tables 8.16–8.31 provide the design equations of ξ_k for EBFs of chevron type with long, intermediate or short links and CBFs with BRBs of chevron type for soil types A–D. The corresponding design equations for EBFs of diagonal type with long, intermediate or short links and CBFs with BRBs of diagonal type are not presented herein and the reader may consult Kalapodis (2017) and Loulelis (2015). It should be stressed that values of ξ_k in excess of 100% have been found for the collapse prevention (CP) seismic performance level for all EBFs and BRBFs studied. Thus, design equations of ξ_k for the CP performance level have been omitted from Tables 8.16, 8.17, 8.18, 8.19, 8.20, 8.21, 8.22, 8.23, 8.24, 8.25, 8.26, 8.27, 8.28, 8.29, 8.30, and 8.31. In case that one wishes to design an EBF or a BRBF for the CP level, the use of $\xi_k = 100\%$ is obligatory for all *k* modes (Kalapodis 2017).

8.5 Numerical Examples

In this section three numerical examples dealing with plane steel MRFs, CBFs and EBFs considered in this chapter are presented in some detail in order to illustrate the design method using modal damping ratios and demonstrate its merits.

8.5.1 Ten-Storey Three-Bay Plane Steel MRF

Consider a regular orthogonal plane steel MRF with ten stories of 3.0 m height and three bays of 5.0 m span, as shown in Fig. 8.10. HEB and IPE profiles are used for the columns and beams, respectively. The gravity load combination on beams consisting of dead plus 0.30 of live loads is equal to 27.5 kN/m^2. The seismic action is that corresponding to the elastic response spectrum of EC8 (2004) for $PGA = 0.36$ g and soil type B. The steel material is assumed to be of grade S275 for both columns and beams. The frame is seismically designed first by the proposed method using modal damping ratios ξ_k in conjunction with the S_a spectra of Fig. 8.3 (modified for 0.36 g) and then by the method of EC8 (2004). Both designs are finally assessed by using NLTH analyses.

The design by the proposed method is done first for the LS performance level corresponding to the DBE. The initial sections of the frame are assumed to be 340/340/340/340–330 (1–6) and 320/320/320/320–300 (7–10) with the meaning of these numbers as explained in Sect. 8.3.1. For this section selection, the first four natural periods are found to be $T_1 = 2.069$ s, $T_2 = 0.693$ s, $T_3 = 0.376$ s and $T_4 = 0.243$ s. Use of these period values in Table 8.8 provide the corresponding modal damping ratios $\xi_1 = 32.03\%$, $\xi_2 = 16.47\%$, $\xi_3 = 12.60\%$ and $\xi_4 = 100.0\%$.

Table 8.16 Modal damping ratios for EBFs (chevron type) with long links and soil type A (after Kalapodis 2017, reprinted with permission from UPCE)

SP level	Mode 1	Mode 2	Mode 3	Mode 4
SP1-IO IDR = 0.4%	$\xi_1 = 0.29T + 1.10$ $(0.30 \le T \le 1.70)$	$\xi_2 = -8.33T + 2.59$ $(0.10 \le T \le 0.25)$ $\xi_2 = 0.63T + 0.35$ $(0.25 \le T \le 0.60)$	$\xi_3 = -1.93T + 1.07$ $(0.14 \le T \le 0.35)$	$\xi_4 = -2.03T + 0.99$ $(0.09 \le T \le 0.24)$
SP2-DL IDR = 1.3%	$\xi_1 = -37.92\,T + 50.87$ $(0.30 \le T \le 1.0)$ $\xi_1 = -8.77T + 15.58$ $(0.30 \le T \le 1.70)$	$\xi_2 = -24.0T + 18.52$ $(0.35 \le T \le 0.60)$	$\xi_3 = -69.57T + 27.71$ $(0.24 \le T \le 0.35)$	$\xi_4 = 100.0$ $(0.09 \le T \le 0.24)$
SP3-LS IDR = 2.2%	$\xi_1 = -12.23T + 70.16$ $(0.30 \le T \le 1.10)$ $\xi_1 = 9.92\,T + 45.84$ $(1.10 \le T \le 1.70)$	$\xi_2 = 100.0$ $(0.10 \le T \le 0.60)$	$\xi_3 = 100.0$ $(0.14 \le T \le 0.35)$	$\xi_4 = 100.0$ $(0.09 \le T \le 0.24)$

Table 8.17 Modal damping ratios for EBFs (chevron type) with long links and soil type B (after Kalapodis 2017, reprinted with permission from UPCE)

SP level	Mode 1	Mode 2	Mode 3	Mode 4
SP1-IO IDR = 0.4%	$\xi_1 = 0.14T + 1.45$ $(0.30 \leq T \leq 1.70)$	$\xi_2 = -7.75T + 2.55$ $(0.10 \leq T \leq 0.25)$ $\xi_2 = 1.14T + 0.33$ $(0.25 \leq T \leq 0.60)$	$\xi_3 = 0.48T + 0.43$ $(0.14 \leq T \leq 0.35)$	$\xi_4 = -1.35T + 1.13$ $(0.09 \leq T \leq 0.24)$
SP2-DL IDR = 1.3%	$\xi_1 = -15.11T + 38.14$ $(0.30 \leq T \leq 1.70)$	$\xi_2 = -30.56T + 22.39$ $(0.35 \leq T \leq 0.50)$ $\xi_2 = 29.0T - 7.39$ $(0.50 \leq T \leq 0.60)$	$\xi_3 = -104.29T + 38.00$ $(0.23 \leq T \leq 0.30)$ $\xi_3 = 18.0T + 1.31$ $(0.30 \leq T \leq 0.35)$	$\xi_4 = 100.0$ $(0.09 \leq T \leq 0.24)$
SP3-LS IDR = 2.2%	$\xi_1 = -12.23T + 71.16$ $(0.30 \leq T \leq 1.70)$	$\xi_2 = 100.0$ $(0.10 \leq T \leq 0.60)$	$\xi_3 = 100.0$ $(0.14 \leq T \leq 0.35)$	$\xi_4 = 100.0$ $(0.09 \leq T \leq 0.24)$

Table 8.18 Modal damping ratios for EBFs (chevron type) with long links and soil type C (after Kalapodis 2017, reprinted with permission from UPCE)

SP level	Mode 1	Mode 2	Mode 3	Mode 4
SP1-IO IDR = 0.4%	$\xi_1 = 0.58T + 1.0$ $(0.30 \leq T \leq 1.70)$	$\xi_2 = -7.59T + 2.60$ $(0.10 \leq T \leq 0.25)$ $\xi_2 = 1.17T + 0.41$ $(0.25 \leq T \leq 0.60)$	$\xi_3 = 1.94T + 0.22$ $(0.14 \leq T \leq 0.35)$	$\xi_4 = -1.44T + 1.25$ $(0.09 \leq T \leq 0.24)$
SP2-DL IDR = 1.3%	$\xi_1 = -26.46T + 41.64$ $(0.30 \leq T \leq 0.90)$ $\xi_1 = -6.26T + 23.46$ $(0.90 \leq T \leq 1.70)$	$\xi_2 = -16.13T + 14.23$ $(0.35 \leq T \leq 0.60)$	$\xi_3 = -53.45T + 22.60$ $(0.23 \leq T \leq 0.35)$	$\xi_4 = 100.0$ $(0.09 \leq T \leq 0.24)$
SP3-LS IDR = 2.2%	$\xi_1 = -11.43T + 63.89$ $(0.30 \leq T \leq 1.70)$	$\xi_2 = 100.0$ $(0.10 \leq T \leq 0.60)$	$\xi_3 = 100.0$ $(0.14 \leq T \leq 0.35)$	$\xi_4 = 100.0$ $(0.09 \leq T \leq 0.24)$

Table 8.19 Modal damping ratios for EBFs (chevron type) with long links and soil type D (after Kalapodis 2017, reprinted with permission from UPCE)

SP level	Mode 1	Mode 2	Mode 3	Mode 4
SP1-IO IDR = 0.4%	$\xi_1 = 0.14T + 1.45$ $(0.30 \leq T \leq 1.70)$	$\xi_2 = -7.95T + 2.63$ $(0.10 \leq T \leq 0.25)$ $\xi_2 = 0.77\,T + 0.45$ $(0.25 \leq T \leq 0.60)$	$\xi_3 = -1.46T + 1.01$ $(0.14 \leq T \leq 0.35)$	$\xi_4 = -4.05T + 1.49$ $(0.09 \leq T \leq 0.24)$
SP2-DL IDR = 1.3%	$\xi_1 = -15.11T + 36.14$ $(0.30 \leq T \leq 1.00)$	$\xi_2 = -20.48T + 17.77$ $(0.35 \leq T \leq 0.60)$	$\xi_3 = -22.61T + 13.27$ $(0.24 \leq T \leq 0.35)$	$\xi_4 = 100.0$ $(0.09 \leq T \leq 0.24)$
SP3-LS IDR = 2.2%	$\xi_1 = -6.48T + 59.20$ $(0.30 \leq T \leq 1.70)$	$\xi_2 = 100.0$ $(0.10 \leq T \leq 0.60)$	$\xi_3 = 100.0$ $(0.14 \leq T \leq 0.35)$	$\xi_4 = 100.0$ $(0.09 \leq T \leq 0.24)$

Table 8.20 Modal damping ratios for EBFs (chevron type) with intremediate links and soil type A (after Kalapodis 2017, reprinted with permission from UPCE)

SP level	Mode 1	Mode 2	Mode 3	Mode 4
SP1-IO IDR = 0.4%	$\xi_1 = -1.90\ T + 2.89$ $(0.20 \leq T \leq 0.60)$ $\xi_1 = 0.50\ T + 1.45$ $(0.60 \leq T \leq 1.90)$	$\xi_2 = -13.06\ T + 4.36$ $(0.08 \leq T \leq 0.25)$ $\xi_2 = 1.10$ $(0.25 \leq T \leq 0.60)$	$\xi_3 = -0.46\ T + 1.16$ $(0.12 \leq T \leq 0.35)$	$\xi_4 = -11.92\ T + 3.47$ $(0.09 \leq T \leq 0.22)$ $\xi_4 = -2.50\ T + 1.40$ $(0.22 \leq T \leq 0.24)$
SP2-DL IDR = 1.3%	$\xi_1 = 1.79\ T + 31.64$ $(0.20 \leq T \leq 1.35)$ $\xi_1 = 13.79\ T + 15.44$ $(1.35 \leq T \leq 1.90)$	$\xi_2 = 4.29\ T + 9.71$ $(0.30 \leq T \leq 0.65)$	$\xi_3 = 3.12\ T + 6.06$ $(0.19 < T < 0.35)$	$\xi_4 = 100.0$ $(0.09 \leq T \leq 0.24)$
SP3-LS IDR = 2.2%	$\xi_1 = -2.35\ T + 71.47$ $(0.20 \leq T \leq 1.90)$	$\xi_2 = 100.0$ $(0.08 \leq T \leq 0.60)$	$\xi_3 = 100.0$ $(0.12 < T < 0.35)$	$\xi_4 = 100.0$ $(0.09 \leq T \leq 0.24)$

Table 8.21 Modal damping ratios for EBFs (chevron type) with intermediate links and soil type B (after Kalapodis 2017, reprinted with permission from UPCE)

SP level	Mode 1	Mode 2	Mode 3	Mode 4
SP1-IO IDR = 0.4%	$\xi_1 = -0.26T + 2.50$ $(0.20 \leq T \leq 1.90)$	$\xi_2 = -15.12T + 4.78$ $(0.08 \leq T \leq 0.25)$ $\xi_2 = -0.62T + 1.16$ $(0.25 \leq T \leq 0.64)$	$\xi_3 = -12.35T + 3.44$ $(0.13 \leq T \leq 0.22)$ $\xi_3 = -1.59T + 1.07$ $(0.22 \leq T \leq 0.34)$	$\xi_4 = -33.77T + 6.27$ $(0.09 \leq T \leq 0.16)$ $\xi_4 = 0.91T + 0.72$ $(0.16 \leq T \leq 0.24)$
SP2-DL IDR = 1.3%	$\xi_1 = -28.75T + 43.75$ $(0.20 \leq T \leq 0.60)$ $\xi_1 = 3.46T + 24.42$ $(0.60 \leq T \leq 1.90)$	$\xi_2 = -3.03T + 9.97$ $(0.32 \leq T \leq 0.64)$	$\xi_3 = -10.0T + 10.50$ $(0.25 \leq T \leq 0.34)$	$\xi_4 = 100.0$ $(0.09 \leq T \leq 0.24)$
SP3-LS IDR = 2.2%	$\xi_1 = -30.95T + 81.19$ $(0.20 \leq T \leq 1.90)$	$\xi_2 = 100.0$ $(0.08 \leq T \leq 0.64)$	$\xi_3 = 100.0$ $(0.13 \leq T \leq 0.34)$	$\xi_4 = 100.0$ $(0.09 \leq T \leq 0.24)$

Table 8.22 Modal damping ratios for EBFs (chevron type) with intermediate links and soil type C (after Kalapodis 2017, reprinted with permission from UPCE)

SP level	Mode 1	Mode 2	Mode 3	Mode 4
SP1-IO IDR = 0.4%	$\xi_1 = 0.18T + 1.87$ $(0.20 \leq T \leq 1.90)$	$\xi_2 = -24.29T + 6.34$ $(0.08 \leq T \leq 0.22)$ $\xi_2 = -0.48T + 1.10$ $(0.22 \leq T \leq 0.65)$	$\xi_3 = -15.39T + 3.92$ $(0.13 \leq T \leq 0.19)$ $\xi_3 = -1.87T + 1.36$ $(0.19 \leq T \leq 0.35)$	$\xi_4 = -19.15T + 3.98$ $(0.09 \leq T \leq 0.13)$ $\xi_4 = -6.67T + 2.36$ $(0.13 \leq T \leq 0.24)$
SP2-DL IDR = 1.3%	$\xi_1 = -9.33T + 37.86$ $(0.20 \leq T \leq 0.90)$ $\xi_1 = 29.44$ $(0.90 \leq T \leq 1.90)$	$\xi_2 = 3.03T + 8.03$ $(0.32 \leq T \leq 0.65)$	$\xi_3 = -9.38T + 9.78$ $(0.19 \leq T \leq 0.35)$	$\xi_4 = 100.0$ $(0.09 \leq T \leq 0.24)$
SP3-LS IDR = 2.2%	$\xi_1 = -16.0T + 70.10$ $(0.20 \leq T \leq 0.95)$ $\xi_1 = 14.30T + 41.33$ $(0.95 \leq T \leq 1.90)$	$\xi_2 = 100.0$ $(0.08 \leq T \leq 0.65)$	$\xi_3 = 100.0$ $(0.12 \leq T \leq 0.35)$	$\xi_4 = 100.0$ $(0.09 \leq T \leq 0.24)$

Table 8.23 Modal damping ratios for EBFs (chevron type) with intermediate links and soil type D (after Kalapodis 2017, reprinted with permission from UPCE)

SP level	Mode 1	Mode 2	Mode 3	Mode 4
SP1-IO IDR = 0.4%	$\xi_1 = -5.75T + 5.25$ $(0.20 \le T \le 0.60)$ $\xi_1 = -0.08T + 1.85$ $(0.60 \le T \le 1.90)$	$\xi_2 = -28.57T + 7.39$ $(0.08 \le T \le 0.22)$ $\xi_2 = -0.70T + 1.25$ $(0.22 \le T \le 0.65)$	$\xi_3 = -1.82T + 1.34$ $(0.13 \le T \le 0.35)$	$\xi_4 = -2.67T + 1.34$ $(0.09 \le T \le 0.24)$
SP2-DL IDR = 1.3%	$\xi_1 = -10.67T + 40.13$ $(0.20 \le T \le 0.90)$ $\xi_1 = 2.11T + 28.63$ $(0.90 \le T \le 1.90)$	$\xi_2 = -5.0T + 14.75$ $(0.35 \le T \le 0.65)$	$\xi_3 = -4.54\,T + 11.09$ $(0.24 \le T \le 0.35)$	$\xi_4 = 100.0$ $(0.09 \le T \le 0.24)$
SP3-LS IDR = 2.2%	$\xi_1 = -18.67T + 74.73$ $(0.20 \le T \le 0.90)$ $\xi_1 = 2.11T + 56.3$ $(0.90 \le T \le 1.90)$	$\xi_2 = 100.0$ $(0.08 \le T \le 0.65)$	$\xi_3 = 100.0$ $(0.13 \le T \le 0.35)$	$\xi_4 = 100.0$ $(0.09 \le T \le 0.24)$

Table 8.24 Modal damping ratios for EBFs (chevron type) with short links and soil type A (after Kalapodis 2017, reprinted with permission from UPCE)

SP level	Mode 1	Mode 2	Mode 3	Mode 4
SP1-IO IDR = 0.4%	$\xi_1 = 1.80$ $(0.22 \le T \le 1.70)$	$\xi_2 = -56.11T + 12.0$ $(0.09 \le T \le 0.20)$ $\xi_2 = 0.14T + 0.75$ $(0.20 \le T \le 0.55)$	$\xi_3 = -241.67T + 32.60$ $(0.11 \le T \le 0.13)$ $\xi_3 = -3.68T + 1.66$ $(0.13 \le T \le 0.30)$	$\xi_4 = -1.43T + 0.95$ $(0.12 \le T \le 0.21)$
SP2-DL IDR = 1.3%	$\xi_1 = -21.82T + 35.46$ $(0.22 \le T \le 0.80)$ $\xi_1 = -5.53T + 22.42$ $(0.80 \le T \le 1.70)$	$\xi_2 = -16.11T + 13.56$ $(0.36 \le T \le 0.55)$	$\xi_3 = -33.33T + 15.0$ $(0.20 \le T \le 0.30)$	$\xi_4 = 100.0$ $(0.12 \le T \le 0.21)$
SP3-LS IDR = 2.2%	$\xi_1 = 30.36T + 50.71$ $(0.22 \le T \le 0.80)$ $\xi_1 = -38.89T + 106.11$ $(0.80 \le T \le 1.70)$	$\xi_2 = 100.0$ $(0.09 \le T \le 0.55)$	$\xi_3 = 100.0$ $(0.11 \le T \le 0.30)$	$\xi_4 = 100.0$ $(0.12 \le T \le 0.21)$

Table 8.25 Modal damping ratios for EBFs (chevron type) with short links and soil type B (after Kalapodis 2017, reprinted with permission from UPCE)

SP level	Mode 1	Mode 2	Mode 3	Mode 4
SP1-IO IDR = 0.4%	$\xi_1 = -0.14T + 1.67$ $(0.22 \leq T \leq 1.70)$	$\xi_2 = -43.30T + 9.47$ $(0.09 \leq T \leq 0.20)$ $\xi_2 = 0.81$ $(0.20 \leq T \leq 0.55)$	$\xi_3 = -60.0T + 10.30$ $(0.11 \leq T \leq 0.16)$ $\xi_3 = -0.71T + 0.81$ $(0.16 \leq T \leq 0.30)$	$\xi_4 = -2.22T + 1.17$ $(0.12 \leq T \leq 0.21)$
SP2-DL IDR = 1.3%	$\xi_1 = -26.16T + 41.28$ $(0.22 \leq T \leq 1.10)$ $\xi_1 = -0.83T + 13.43$ $(1.10 \leq T \leq 1.70)$	$\xi_2 = -30.56T + 21.30$ $(0.36 \leq T \leq 0.55)$	$\xi_3 = -25.0T + 14.0$ $(0.20 \leq T \leq 0.30)$	$\xi_4 = 100.0$ $(0.12 \leq T \leq 0.21)$
SP3-LS IDR = 2.2%	$\xi_1 = -13.70T + 66.88$ $(0.22 \leq T \leq 1.70)$	$\xi_2 = 100.0$ $(0.09 \leq T \leq 0.55)$	$\xi_3 = 100.0$ $(0.11 \leq T \leq 0.30)$	$\xi_4 = 100.0$ $(0.12 \leq T \leq 0.21)$

Table 8.26 Modal damping ratios for EBFs (chevron type) with short links and soil type C (after Kalapodis 2017, reprinted with permission from UPCE)

SP level	Mode 1	Mode 2	Mode 3	Mode 4
SP1-IO IDR = 0.4%	$\xi_1 = -3.77T + 5.0$ $(0.22 \le T \le 0.78)$ $\xi_1 = 0.11T + 1.97$ $(0.78 \le T \le 1.70)$	$\xi_2 = -34.29T + 8.74$ $(0.09 \le T \le 0.22)$ $\xi_2 – 0.61\ T + 1.33$ $(0.22 \le T \le 0.55)$	$\xi_3 = -69.17T + 11.92$ $(0.10 \le T \le 0.16)$ $\xi_3 = -1.07T + 1.01$ $(0.16 \le T \le 0.30)$	$\xi_4 = -2.86T + 1.35$ $(0.12 \le T \le 0.21)$
SP2-DL IDR = 1.3%	$\xi_1 = -24.18T + 37.60$ $(0.22 \le T \le 1.10)$ $\xi_1 = 1.82T + 8.99$ $(1.10 \le T \le 1.70)$	$\xi_2 = -22.22T + 17.21$ $(0.30 \le T \le 0.55)$	$\xi_3 = -44.0T + 19.20$ $(0.20 \le T \le 0.30)$	$\xi_4 = 100.0$ $(0.12 \le T \le 0.21)$
SP3-LS IDR = 2.2%	$\xi_1 = -50.0T + 79.0$ $(0.22 \le T \le 0.48)$ $\xi_1 = -10.03T + 59.79$ $(0.48 \le T \le 1.70)$	$\xi_2 = -33.33T + 34.33$ $(0.45 \le T \le 0.55)$	$\xi_3 = 100.0$ $(0.11 \le T \le 0.30)$	$\xi_4 = 100.0$ $(0.12 \le T \le 0.21)$

Table 8.27 Modal damping ratios for EBFs (chevron type) with short links and soil type D (after Kalapodis 2017, reprinted with permission from UPCE)

SP level	Mode 1	Mode 2	Mode 3	Mode 4
SP1-IO IDR = 0.4%	$\xi_1 = 0.14T + 1.67$ $(0.22 \leq T \leq 1.70)$	$\xi_2 = -33.13T + 9.48$ $(0.09 \leq T \leq 0.25)$ $\xi_2 = -1.33T + 1.53$ $(0.25 \leq T \leq 0.55)$	$\xi_3 = -0.50T + 0.85$ $(0.11 \leq T \leq 0.30)$	$\xi_4 = -2.86T + 1.35$ $(0.12 \leq T \leq 0.21)$
SP2-DL IDR = 1.3%	$\xi_1 = -18.61T + 32.47$ $(0.22 \leq T \leq 1.10)$ $\xi_1 = 1.67T + 10.17$ $(1.10 \leq T \leq 1.70)$	$\xi_2 = -18.57T + 15.21$ $(0.20 \leq T \leq 0.55)$	$\xi_3 = -20.0T + 11.0$ $(0.20 \leq T \leq 0.30)$	$\xi_4 = 100.0$ $(0.12 \leq T \leq 0.21)$
SP3-LS IDR = 2.2%	$\xi_1 = -55.56T + 83.33$ $(0.22 \leq T \leq 0.60)$ $\xi_1 = -9.09T + 55.46$ $(0.60 \leq T \leq 1.70)$	$\xi_2 = -5.56T + 20.06$ $(0.45 \leq T \leq 0.55)$	$\xi_3 = 100.0$ $(0.11 \leq T \leq 0.30)$	$\xi_4 = 100.0$ $(0.12 \leq T \leq 0.21)$

Table 8.28 Modal damping ratios for CBFs with BRBs (chevron type) and soil type A (after Kalapodis 2017, reprinted with permission from UPCE)

SP level	Mode 1	Mode 2	Mode 3	Mode 4
SP1-IO IDR = 0.4%	$\xi_1 = -4.62T + 3.31$ $(0.22 \leq T \leq 0.45)$ $\xi_1 = 0.24T + 1.12$ $(0.45 \leq T \leq 1.50)$	$\xi_2 = -6.15T + 2.18$ $(0.11 \leq T \leq 0.24)$ $\xi_2 = -0.20T + 0.75$ $(0.24 \leq T \leq 0.48)$	$\xi_3 = -2.65T + 1.42$ $(0.10 \leq T \leq 0.27)$	$\xi_4 = -5.41T + 1.70$ $(0.11 \leq T \leq 0.19)$
SP2-DL IDR = 1.3%	$\xi_1 = -46.05T + 59.05$ $(0.22 \leq T \leq 1.00)$ $\xi_1 = -2.0T + 15.0$ $(1.00 \leq T \leq 1.50)$	$\xi_2 = -17.31T + 14.15$ $(0.25 \leq T \leq 0.48)$	$\xi_3 = -33.60T + 13.7$ $(0.14 \leq T \leq 0.27)$	$\xi_4 = 100.0$ $(0.11 \leq T \leq 0.19)$
SP3-LS IDR = 2.2%	$\xi_1 = -7.58T + 78.82$ $(0.22 \leq T \leq 0.90)$ $\xi_1 = -45.0T + 112.50$ $(0.90 \leq T \leq 1.50)$	$\xi_2 = 100.0$ $(0.11 \leq T \leq 0.48)$	$\xi_3 = 100.0$ $(0.10 \leq T \leq 0.27)$	$\xi_4 = 100.0$ $(0.11 \leq T \leq 0.19)$

Table 8.29 Modal damping ratios for CBFs with BRBs (chevron type) and soil type B (after Kalapodis 2017, reprinted with permission from UPCE)

SP level	Mode 1	Mode 2	Mode 3	Mode 4
SP1-IO IDR = 0.4%	$\xi_1 = -3.33T + 2.77$ $(0.22 \leq T \leq 0.50)$ $\xi_1 = 0.40T + 0.90$ $(0.50 \leq T \leq 1.50)$	$\xi_2 = -16.67T + 3.83$ $(0.11 \leq T \leq 0.17)$ $\xi_2 = -1.27T + 1.22$ $(0.17 \leq T \leq 0.48)$	$\xi_3 = -4.60T + 1.57$ $(0.10 \leq T \leq 0.27)$	$\xi_4 = -6.51T + 2.03$ $(0.11 \leq T \leq 0.19)$
SP2-DL IDR = 1.3%	$\xi_1 = -38.16T + 51.16$ $(0.22 \leq T \leq 1.00)$ $\xi_1 = 9.61T + 3.39$ $(1.00 \leq T \leq 1.50)$	$\xi_2 = -54.29T + 28.03$ $(0.25 \leq T \leq 0.42)$ $\xi_2 = 57.33T - 18.84$ $(0.42 \leq T \leq 0.48)$	$\xi_3 = -5.44T + 9.97$ $(0.18 \leq T \leq 0.27)$	$\xi_4 = 100.0$ $(0.11 \leq T \leq 0.19)$
SP3-LS IDR = 2.2%	$\xi_1 = 28.57T + 61.14$ $(0.22 \leq T \leq 0.50)$ $\xi_1 = -26.53T + 88.69$ $(0.50 \leq T \leq 1.50)$	$\xi_2 = 100.0$ $(0.11 \leq T \leq 0.48)$	$\xi_3 = 100.0$ $(0.10 \leq T \leq 0.27)$	$\xi_4 = 100.0$ $(0.11 \leq T \leq 0.19)$

Table 8.30 Modal damping ratios for CBFs with BRBs (chevron type) and soil type C (after Kalapodis 2017, reprinted with permission from UPCE)

SP level	Mode 1	Mode 2	Mode 3	Mode 4
SP1-IO IDR = 0.4%	$\xi_1 = -9.58T + 5.60$ $(0.21 \le T \le 0.48)$ $\xi_1 = 0.39T + 0.81$ $(0.48 \le T \le 1.50)$	$\xi_2 = -33.33T + 7.0$ $(0.11 \le T \le 0.18)$ $\xi_2 = -1.30T + 1.24$ $(0.18 \le T \le 0.48)$	$\xi_3 = -3.70T + 1.48$ $(0.10 \le T \le 0.27)$	$\xi_4 = -8.87T + 2.13$ $(0.11 \le T \le 0.19)$
SP2-DL IDR = 1.3%	$\xi_1 = -50.0T + 65.0$ $(0.22 \le T \le 1.00)$ $\xi_1 = -1.82T + 16.82$ $(1.00 \le T \le 1.50)$	$\xi_2 = -53.14T + 27.05$ $(0.25 \le T \le 0.42)$ $\xi_2 = 30.99T - 8.29$ $(0.42 \le T \le 0.48)$	$\xi_3 = -7.66T + 9.37$ $(0.18 \le T \le 0.23)$ $\xi_3 = 15.38T + 3.34$ $(0.23 \le T \le 0.27)$	$\xi_4 = 100.0$ $(0.11 \le T \le 0.19)$
SP3-LS IDR = 2.2%	$\xi_1 = -31.25T + 84.50$ $(0.22 \le T \le 1.20)$ $\xi_1 = 2.86T + 43.57$ $(1.20 \le T \le 1.50)$	$\xi_2 = 100.0$ $(0.11 \le T \le 0.48)$	$\xi_3 = 100.0$ $(0.10 \le T \le 0.27)$	$\xi_4 = 100.0$ $(0.11 \le T \le 0.19)$

Table 8.31 Modal damping ratios for CBFs with BRBs (chevron type) and soil type D (after Kalapodis 2017, reprinted with permission from UPCE)

SP level	Mode 1	Mode 2	Mode 3	Mode 4
SP1-IO IDR = 0.4%	$\xi_1 = 1.50$ $(0.22 \leq T \leq 1.50)$	$\xi_2 = -24.23T + 6.26$ $(0.11 \leq T \leq 0.24)$ $\xi_2 = 1.01T + 0.21$ $(0.24 \leq T \leq 0.48)$	$\xi_3 = -37.77T + 5.72$ $(0.10 \leq T \leq 0.14)$ $\xi_3 = 3.42T - 0.05$ $(0.14 \leq T \leq 0.27)$	$\xi_4 = -9.86T + 2.28$ $(0.11 \leq T \leq 0.19)$
SP2-DL IDR = 1.3%	$\xi_1 = -47.37T + 61.37$ $(0.22 \leq T \leq 1.00)$ $\xi_1 = 3.64T + 10.36$ $(1.00 \leq T \leq 1.50)$	$\xi_2 = -82.14T + 37.71$ $(0.25 \leq T \leq 0.38)$ $\xi_2 = 26.17T - 3.44$ $(0.38 \leq T \leq 0.48)$	$\xi_3 = -111.11T + 29.94$ $(0.18 \leq T \leq 0.22)$ $\xi_3 = 66.0T - 9.02$ $(0.22 \leq T \leq 0.27)$	$\xi_4 = 100.0$ $(0.11 \leq T \leq 0.19)$
SP3-LS IDR = 2.2%	$\xi_1 = 9.68T + 74.68$ $(0.22 \leq T \leq 0.48)$ $\xi_1 = -39.0T + 98.04$ $(0.48 \leq T \leq 1.50)$	$\xi_2 = 100.0$ $(0.11 \leq T \leq 0.48)$	$\xi_3 = 100.0$ $(0.10 \leq T \leq 0.27)$	$\xi_4 = 100.0$ $(0.11 \leq T \leq 0.19)$

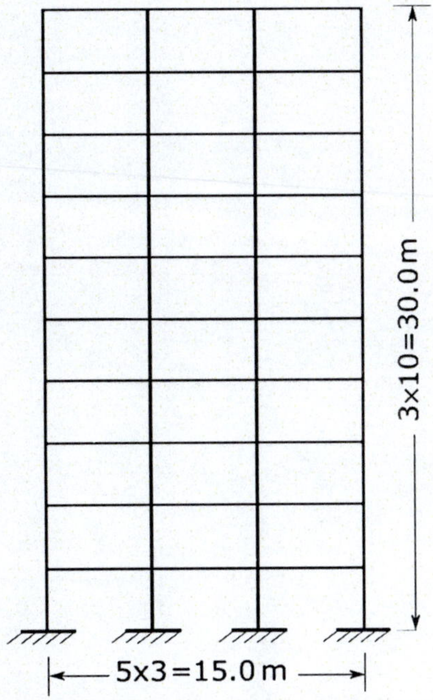

Fig. 8.10 Geometry of the plane steel MRF considered in example of Sect. 8.5.1

From the mean pseudo-acceleration spectra of Fig. 8.3a for the above values of ξ and T, one obtains $S_{a1} = 0.122$ g, $S_{a2} = 0.514$ g, $S_{a3} = 0.796$ g and $S_{a4} = 0.460$ g. These spectral acceleration values lead to a design base shear of 487.31 kN with the aid of the SAP 2000 (2016) computer program. It should be noted that since SAP 2000 (2016) works with only one value of modal damping, a modified design spectrum is created in the program with ordinates the four previously computed S_a values corresponding to the four natural periods, as shown in Fig. 8.11.

Distributing this base shear value of 487.31 kN along the height of the frame and performing a strength checking that takes into account the capacity design rules, one can arrive at the section selection of 360/400/400/360–360 (1–3), 360/400/400/360–330 (4), 340/400/400/340–330 (5–7) and 340/360/360/340–330 (8–10). For these sections, the first four natural periods and corresponding modal damping ratios are $T_1 = 1.858$ s and $\xi_1 = 32.12\%$, $T_2 = 0.600$ s and $\xi_2 = 17.16\%$, $T_3 = 0.333$ s and $\xi_3 = 12.67\%$, $T_4 = 0.214$ s and $\xi_4 = 100.0\%$, respectively. For these values of T and ξ, one obtains from Fig. 8.3a, $S_{a1} = 0.137$ g, $S_{a2} = 0.583$ g, $S_{a3} = 0.788$ g and $S_{a4} = 0.468$ g, respectively, and hence with the aid of SAP 2000 (2016) a base shear equal to 542.02 kN. For this new base shear value, the strength checking is repeated and is found to be satisfied for the sections of the previous step indicating that these sections are the final ones for the LS performance level. No displacement checking is required for the LS level because displacement requirements are automatically satisfied by using the displacement dependent modal damping ratios ξ_k.

Next the section selection for the LS level is checked according to the proposed method for strength satisfaction for the IO and CP performance levels corresponding

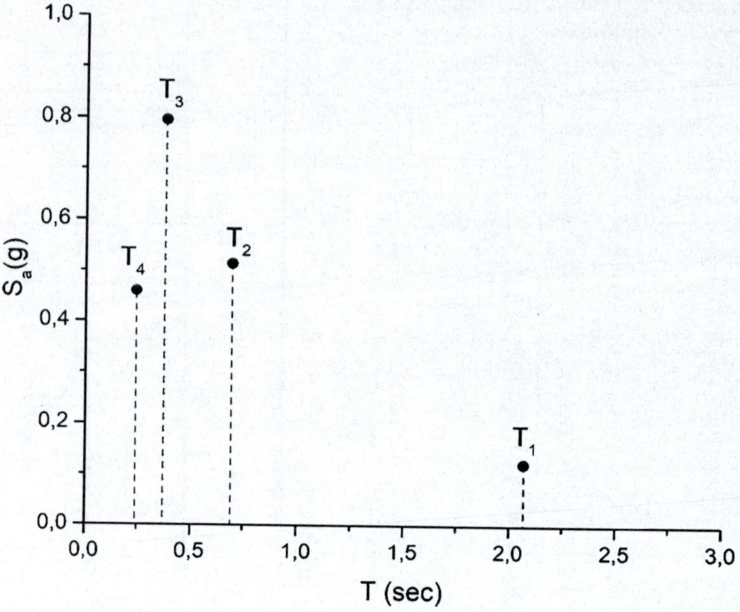

Fig. 8.11 Modified response spectrum

to the FOE and MCE, respectively. The S_a spectra for the IO and CP performance levels are derived by multiplying the modified for 0.36 g ordinates of Fig. 8.3a that correspond to DBE, by 0.5 and 1.5, respectively. On the basis of the aforementioned first four natural periods, the modal damping ratios and corresponding S_a values for the IO performance level are $\xi_1 = 1.95\%$ and $S_{a1} = 0.145$ g, $\xi_2 = 0.89\%$ and $S_{a2} = 0.448$ g, $\xi_3 = 0.73\%$ and $S_{a3} = 0.525$ g, $\xi_4 = 0.88\%$ and $S_{a4} = 0.540$ g, whereas those for the CP performance level are $\xi_1 = 81.68\%$ and $S_{a1} = 0.197$ g, $\xi_2 = 100.0\%$ and $S_{a2} = 0.572$ g, $\xi_3 = 100.0\%$ and $S_{a3} = 0.661$ g, $\xi_4 = 100.0\%$ and $S_{a4} = 0.702$ g. It should be noted that since the modal damping ratios for the IO level are below 5%, one makes use of the 5%-damped, modified for 0.36 g and multiplied by 0.5, spectrum ordinates of Fig. 8.3a. Application of the proposed method provides base shears 523.08 and 701.24 kN for the IO and CP levels, respectively. Taking into account that strength checking of all selected sections is satisfied for the IO and CP levels, indicates that among the three levels, i.e., IO, LS and CP, the CP level, associated with the highest base shear, controls the design.

Nonlinear dynamic analyses with the aid of SAP 2000 (2016) of the three designs (for IO, LS and CP levels) by the proposed method are performed by employing 10 seismic motions compatible with the elastic design spectra for the IO, LS and CP levels, obtained on the basis of the modified for 0.36 g S_a elastic spectrum with 5% damping of Fig. 8.3a that corresponds to the LS level. The values of this spectrum are multiplied by 0.5 and 1.5 in order to account for the IO and CP levels, respectively. The mean values of maximum base shear (defined as that at first yielding for the LS and CP levels and as the maximum one for the IO level) and maximum *IDR* are 524.16 kN and 0.54%, 603.55 kN and 1.79%, 798.66 kN and 2.91%, respectively for the IO, LS and CP levels. It is observed that in all cases, the obtained from non-linear dynamic analyses *IDR* values do not exceed the limits of 0.7%, 2.5% and 5.0% for the IO, LS and CP levels of Table 8.6, as expected because of the displacement dependence of the employed modal damping ratios ξ_k. However, these values of *IDR* are significantly lower than the limit ones indicating a conservative design. This can be attributed to the fact that the used values of ξ_k have been constructed as lower bounds and thus lead to higher design seismic forces and hence to stiffer frames and smaller drifts. Nevertheless, the proposed method even though results in conservative design, satisfies all three seismic performance levels constituting thereby a truly performance-based seismic design method.

A comparison of the frame designed by the proposed method is finally done against that designed by the method of EC8 (2004) for the case of a design spectrum defined for $PGA = 0.36$ g and soil type B. Selecting an initial value of $q = 6.5$ and performing the necessary strength and damage limit state (limit *IDR* of 0.7%) checkings according to Chap. 3, one needs two iterations to finally end up with sections 360/360/360/360–360 (1–4) and 340/340/340/340–330 (5–10). It should be noted that iterations are only with respect to the displacement/damage checking and that for the finally selected sections one can determine that there corresponds a value of $q = 4.0$. The above sections being heavier than the initial ones coming from using $q = 6.5$ and the fact that they correspond to $q < 6.5$ indicates that the damage limitation state controls the design.

A nonlinear dynamic analysis of the EC8 (2004) design for the previously considered 10 seismic motions now compatible with the elastic design spectrum with 5% damping of EC8 (2004) for $PGA = 0.36$ g and soil type B, provides a base shear (at first yielding) of 401.24 kN and *IDR* of 2.07%. Thus, the *IDR* limit of 2.5% is not exceeded for the LS level as in the case of the design by the proposed method. However, the design by the EC8 method is lighter than the one by the proposed method as this is evident by looking at the respective section selections and the *IDR* values.

8.5.2 Five-Storey Three-Bay Plane Steel EBF

Consider a five-storey three-bay plane steel chevron EBF with long links of the type shown in Fig. 8.7a. The length of each bay is 5.0 m and the height of each storey is 3.0 m. HEB, IPE and CHS sections made of S275 grade steel are used for columns, beams and braces, respectively. The gravity load combination is G + 0.3Q = 27.5 kN/m with G and Q being the dead and live loads, respectively. The seismic action is that corresponding to the elastic response spectrum of EC8 (2004) with $PGA = 0.36$ g and soil type B. The EBF is seismically designed first by the proposed method using modal damping ratios ξ_k in conjunction with the S_a spectra of Fig. 8.3 and then by the method of EC8 (2004) with $q = 4.0$ (Table 3.6).

The design by the proposed method is done here only for the LS performance level associated with a target $IDR = 2.2\%$ (Table 8.15), even though such a design may not be the controlling one in a performance-based design framework with three or four performance levels. The initial sections of the frame are assumed to be 260–300–193.7 × 4.5 (1–2) and 240–270–168.3 × 4 (3–5) with the meaning of these numbers as explained in Sect. 8.4.1. For this section selection, the first four natural periods are found to be $T_1 = 0.655$ s, $T_2 = 0.230$ s, $T_3 = 0.136$ s and $T_4 = 0.097$ s. Use of these period values in Table 8.16 provides the corresponding modal damping ratios $\xi_1 = 63.15\%$, $\xi_2 = 100.0\%$, $\xi_3 = 100.0\%$ and $\xi_4 = 100.0\%$. From the mean pseudo-acceleration spectra of Fig. 8.3a for the above values of ξ and T, one obtains $S_{a1} = 0.304$ g, $S_{a2} = 0.464$ g, $S_{a3} = 0.490$ g and $S_{a4} = 0.487$ g. These spectral acceleration values lead to a design base shear of 542.88 kN with the aid of the SAP 2000 (2016) computer program. It is recalled that since SAP 2000 (2016) works with only one value of modal damping, a modified design spectrum is created in SAP 2000 (2016) with ordinates the four previously computed S_a values corresponding to the four natural periods.

Distributing this base shear value of 542.88 kN along the height of the frame and performing a strength checking that takes into account the capacity design rules, one can arrive at the section selection of 280–330–193.7 × 4.5 (1–2), 260–300–193.7 × 4.5 (3), 260–300–168.3 × 4.0 (4–5). For these sections, the first four natural periods and corresponding modal damping ratios are $T_1 = 0.596$ s and $\xi_1 = 63.87\%$, $T_2 = 0.211$ s and $\xi_2 = 100.0\%$, $T_3 = 0.122$ s and $\xi_3 = 100.0\%$, $T_4 = 0.088$ s and $\xi_4 = 100.0\%$, respectively. For these values of T and ξ, one obtains from Fig. 8.3a, $S_{a1} = 0.324$ g, $S_{a2} = 0.468$ g, $S_{a3} = 0.483$ g and $S_{a4} = 0.490$ g, respectively, and hence with the aid of SAP 2000 (2016) a base shear equal to 582.71 kN. For this new base shear value, the strength checking is repeated and found to be satisfied, indicating that the sections and the base shear obtained in the previous step are the final ones for the LS performance level. No displacement checking is required for the LS level because displacement requirements are automatically satisfied by using the displacement dependent modal damping ratios ξ_k.

Nonlinear dynamic analyses with the aid of SAP 2000 (2016) of the designed EBF by the proposed method are performed by employing 10 seismic motions compatible with the elastic design spectrum obtained on the basis of the modified for 0.36 g S_a elastic spectrum with 5% damping of Fig. 8.3a that corresponds to the LS level. The mean value of maximum base shear (defined at first yielding), maximum IDR and θ_{link} (link rotation) were found to be 623.02 kN, 1.59% and 0.0212 rad, respectively. It is observed that the obtained from non-linear dynamic analyses maximum IDR value does not exceed the limit of 2.2% for the LS level of Table 8.6, as expected, because of the displacement dependence of the employed modal damping ratios ξ_k. However, this value of IDR is significantly lower than the limit one indicating a conservative design. This can be attributed to the fact that the used values of ξ_k have been constructed as lower bounds and thus lead to higher design seismic forces and hence to a stiffer EBF and smaller drifts.

A comparison of the frame designed by the proposed method is finally done against that designed by the method of EC8 (2004) for the case of a design spectrum defined for $PGA = 0.36$ g and soil type B. Selecting an initial value of $q = 4.0$ and performing the necessary strength and damage limit state (limit *IDR* of 0.7%) checkings according to Chap. 3, one needs two iterations to finally end up with sections 280–330–219.1 × 5.0 (1–2), 260–330–219.1 × 5.0 (3), 260–300–168.3 × 4.0 (4–5). It should be noted that iterations are only with respect to the displacement/damage checking and that for the finally selected sections one can determine that there corresponds a value of $q = 2.5$. The above sections being heavier than the initial ones coming from using $q = 4.0$ and the fact that they correspond to $q < 4.0$ indicates that the damage limitation state controls the design.

Nonlinear dynamic analyses of the EC8 (2004) design for the previously considered 10 seismic motions now compatible to the elastic design spectrum with 5% damping of EC8 (2004) for $PGA = 0.36$ g and soil type B, provide the following mean value for maximum base shear (defined at first yielding), maximum *IDR* and θ_{link} (link rotation), i.e., 821.24 kN, 1.57% and 0.0204 rad. Thus, the *IDR* limit of 2.2% is not exceeded for the LS level as in the case of the design by the proposed method. The designs by the proposed and the EC8 (2004) methods are almost the same as this is evident by looking at the respective section selections and the *IDR* values. However, the EC8 (2004) design also satisfies the damage limit state (IO) for which the design by the proposed method has not been checked. It should be also noted that for both methods the mean θ_{link} value is marginally higher than the value of 0.02 rad permitted for the targeted *IDR* level according to Table 8.15.

8.5.3 Five-Storey Three-Bay Plane Steel CBF

The proposed method using modal damping ratios ξ_k in conjunction with the S_a spectra of Fig. 8.3 is now employed to seismically design a five-storey and three-bay CBF with chevron type BRBs of the type shown in Fig. 8.7b. The gravity load combination is G + 0.3Q = 27.5 kN/m with G and Q being the dead and live loads, respectively. The length of each bay is 5.0 m and the height of each storey is 3.0 m. HEB and IPE are used for columns and beams, whereas the section of the BRB is the one shown in Fig. 8.8b. All steel sections are of S275 grade. Dimensioning of the BRBs follows the work of Bosco et al. (2015).

The design by the proposed method is done here only for the LS performance level associated with a target $IDR = 2.2\%$ (Table 8.15), even though such a design may not be the controlling one in a performance-based design framework with three performance levels. The initial sections of the frame are assumed to be 300–300–27 (1–2) and 300–270–21 (3–5) with the meaning of these numbers as explained in Sect. 8.4.1. For this section selection, one computes by the proposed method a design base shear of 550.11 kN. Distributing this base shear along the height of the frame and performing a strength checking that takes into account the capacity design rules, one can arrive at the section selection of 320–330–27 (1–2), 300–300–21 (3) and 300–300–17 (4–5). For these sections, the first four natural periods and corresponding modal damping ratios are $T_1 = 0.60$ s and $\xi_1 = 72.77\%$, $T_2 = 0.209$ s and $\xi_2 = 100.0\%$, $T_3 = 0.120$ s and $\xi_3 = 100.0\%$, $T_4 = 0.082$ s and $\xi_4 = 100.0\%$, respectively. For these values of T and ξ, one obtains from Fig. 8.3a, $S_{a1} = 0.317$ g, $S_{a2} = 0.468$ g, $S_{a3} = 0.483$ g and $S_{a4} = 0.490$ g, respectively, and hence with the aid of SAP 2000 (2016) a base shear equal to 566.32 kN. For this new base shear value, the strength checking is repeated and found to be satisfied, indicating that the sections and the base shear obtained in the previous step are the final ones for the LS performance level. No displacement checking is required for the LS level because displacement requirements are automatically satisfied by using the displacement dependent modal damping ratios ξ_k.

Nonlinear dynamic analyses with the aid of SAP 2000 (2016) of the designed frame by the proposed method are performed by employing 10 seismic motions compatible with the elastic design spectrum obtained on the basis of the modified for 0.36 g S_a elastic spectrum with 5% damping of Fig. 8.3a that corresponds to the LS level. The mean values of maximum base shear (defined at first yielding), maximum *IDR* and μ_δ (brace ductility) were found to be 641.53 kN, 1.47% and 3.22, respectively. It is observed that the obtained from non-linear dynamic analyses maximum *IDR* value does not exceed the limit of 2.2% for the LS level of Table 8.6, as expected, because of the displacement dependence of the employed modal damping ratios ξ_k. However, this value of *IDR* is significantly lower than the limit one, indicating a conservative design. This can be attributed to the fact that the used values of ξ_k have been constructed as lower bounds and thus lead to higher design seismic forces and hence to a stiffer frame and smaller drifts.

A comparison of the frame designed by the proposed method is finally done against that designed by the method of EC8 (2004) for the case of a design spectrum defined for $PGA = 0.36$ g and soil type B. Selecting an initial value of $q = 4.0$ and performing the necessary strength and damage limit state (limit *IDR* of 0.7%) checkings according to Chap. 3, one needs two iterations to finally end up with sections 320–330–24 (1–2), 300–300–24 (3) and 300–300–21 (4–5). It should be noted that iterations are only with respect to the displacement/damage checking and that for the finally selected sections one can determine that there corresponds a value of $q = 3.5$. The above sections being heavier than the initial ones coming from using $q = 4.0$ and the fact that they correspond to $q < 4.0$ indicates that the damage limitation state controls the design.

Nonlinear dynamic analyses of the EC8 (2004) design for the previously considered 10 seismic motions now compatible with the elastic design spectrum with 5% damping of EC8 (2004) for $PGA = 0.36$ g and soil type B, provide as mean values for maximum base shear (defined at first yielding), maximum *IDR* and μ_δ (brace ductility), the values 661.11 kN, 1.44% and 3.18, respectively. Thus, the *IDR* limit of 2.2% is not exceeded for the LS level as in the case of the design by the proposed method. The design by the proposed and the EC8 (2004) methods are almost the same as this is evident by looking at the respective section selections and the *IDR* values. However, the EC8 (2004) design also satisfies the damage limit state (IO) for which the design by the proposed method has not been checked.

8.6 Conclusions

On the basis of the preceding developments, the following conclusions can be stated:

1. A seismic design method that determines the design base shear of plane steel framed structures through spectrum analysis and modal synthesis by using equivalent viscous modal damping ratios ξ_k and design acceleration spectra with high amounts of damping has been presented.
2. Explicit empirical expressions providing ξ_k's as functions of period, interstorey drift ratio and plastic rotation or ductility for the first few significant modes and for various performance levels and soil types have been derived from a very large response databank employing nonlinear regression analysis. This databank is the result of extensive parametric studies on many plane steel frames under many seismic motions done with the aid of nonlinear dynamic analyses.
3. The plane steel frames considered here include moment resisting frames, concentrically braced frames with buckling-restrained braces of chevron type and eccentrically braced frames with long, intermediate and short seismic links in a chevron configuration. The seismic motions are of the far-fault type and cover all four soil types A, B, C and D according to EC8.
4. The proposed seismic design method is illustrated by performing seismic designs of almost all kinds of frames considered here, which are compared against those done by the EC8 method and are validated using nonlinear dynamic analyses. Unlike the conventional code-based approach that considers a single behavior (or strength reduction) factor for all modes, the proposed approach employing equivalent modal damping ratios leads to more accurate and deformation/damage controlled results in only one step (strength checking) since deformation requirements are automatically satisfied.
5. It has been found on the basis of the presented examples that the proposed method, which is a truly performance-based seismic design method, produces heavier member sections than the ones coming from the EC8 method when it is used as a performance-based one with three performance levels, or almost the same sections with those of the EC8 method when it is used as a method associated with only the life safety performance level.

References

Azad SK, Topkaya C (2017) A review of research on steel eccentrically braced frames. J Constr Steel Res 128:53–73

Bosco M, Marino EM, Rossi PP (2015) Design of steel frames equipped with BRBs in the framework of Eurocode 8. J Constr Steel Res 113:43–57

Bozorgnia Y, Bertero VV (eds) (2004) Earthquake engineering: from engineering seismology to performance-based engineering. CRC Press, Boca Raton, FL

Carr AJ (2005) Ruaumoko 2D and 3D: programs for inelastic dynamic analysis. Theory and user guide to associated programs. Department of Civil Engineering, University of Canterbury, Christchurch

Chopra AK (2007) Dynamics of structures: theory and applications in earthquake engineering, 3rd edn. Prentice-Hall, Upper Saddle River, NJ

COSMOS (2013) Consortium of organizations for strong motion observation systems, San Francisco. http://www.cosmos-eq.org/

De Luca F, Iervolino I, Cosenza E (2009) Unscaled, scaled, adjusted and artificial spectral matching accelerograms: displacement and energy-based assessment. In: Proceedings of 13th Congress of the Italian National Association of Earthquake Engineering (ANIDIS), Bologna, Italy

EC3 (2009) Eurocode 3, Design of steel structures – Part 1–1: general rules and rules for buildings, EN 1993-1-1. European Committee for Standardization (CEN), Brussels

EC8 (2004) Eurocode 8, Design of structures for earthquake resistance, Part 1: general rules, seismic actions and rules for buildings, EN 1998-1-1. European Committee for Standardization (CEN), Brussels

Engelhardt MD, Popov EP (1989) On design of eccentrically braced frames. Earthquake Spectra 5:495–511

FEMA 356 (2000) Prestandard and commentary for the seismic rehabilitation of buildings. Federal Emergency Management Agency, Washington, DC

Gupta A, Krawinkler H (2000) Estimation of seismic drift demands for frame structures. Earthq Eng Struct Dyn 29:1287–1305

Hatzigeorgiou GD (2010) Damping modification factors for SDOF systems subjected to near-fault, far-fault and artificial earthquakes. Earthq Eng Struct Dyn 39:1239–1258

Housner GW (1959) Behavior of structures during earthquakes. J Eng Mech Div ASCE 85:109–130

Hsiao PC, Lehman DE, Roeder CW (2012) Improved analytical model for special concentrically braced frames. J Constr Steel Res 73:80–94

Jacobsen LS (1930) Steady forced vibration as influenced by damping. Trans ASME 52:169–181

Kalapodis NA (2017) Seismic design of plane steel braced frames with the use of three new methods. PhD Thesis, Department of Civil Engineering, University of Patras, Patras, Greece (in Greek)

Kalapodis NA, Papagiannopoulos GA (2020) Seismic design of plane steel braced frames using equivalent modal damping ratios. Soil Dyn Earthq Eng 129:105947

Kalapodis NA, Papagiannopoulos GA, Beskos DE (2020) A comparison of three performance-based seismic design methods for plane steel braced frames. Earthq Struct 18:27–44

Kalapodis NA, Muho EV, Beskos DE (2021) Seismic design of plane steel MRFs, EBFs and CBFs by improved direct displacement-based design method. Soil Dyn Earthq Eng (submitted)

Krawinkler H (1978) Shear in beam-column joints in seismic design of steel frames. AISC Eng J 15:82–91

Lignos DG, Krawinkler H (2011) Deterioration modeling of steel components in support of collapse prediction of steel moment frames under earthquake loading. J Struct Eng ASCE 137:1291–1302

Lin YY, Chang KC (2003) Study on damping reduction factor for buildings under earthquake ground motions. J Struct Eng ASCE 129:206–214

Loulelis DG (2015) Seismic design of planar steel frames with modal strength reduction factors for three performance levels. PhD Thesis, Department of Civil Engineering, University of Patras, Patras, Greece (in Greek)

Loulelis DG, Papagiannopoulos GA, Beskos DE (2018) Modal strength reduction factors for seismic design of steel moment resisting frames. Eng Struct 154:23–37

McVerry GH (1980) Structural identification in the frequency domain from earthquake records. Earthq Eng Struct Dyn 8:161–180

NIST (2015) Seismic design of steel buckling-restrained braced frames: a guide for practicing engineers, GCR 15-917-34, NEHRP seismic design technical brief no. 11. National Institute of Standards and Technology (NIST), Gaithersburg, MD

Okazaki T, Lignos DG, Hikino T, Kajiwara K (2013) Dynamic response of a steel concentrically braced frame. J Struct Eng ASCE 139:515–525

Papagiannopoulos GA (2008) Seismic design of steel frames using equivalent modal damping ratios or modal strength reduction factors. PhD Thesis, Department of Civil Engineering, University of Patras, Patras, Greece (in Greek)

Papagiannopoulos GA (2018) Jacobsen's equivalent damping concept revisited. Soil Dyn Earthq Eng 115:82–89

Papagiannopoulos GA, Beskos DE (2006) On a modal damping identification model of building structures. Arch Appl Mech 76:443–463

Papagiannopoulos GA, Beskos DE (2010) Towards a seismic design method for plane steel frames using equivalent modal damping ratios. Soil Dyn Earthq Eng 30:1106–1118

Papagiannopoulos GA, Hatzigeorgiou GD, Beskos DE (2013) Recovery of spectral absolute acceleration and spectral relative velocity from their pseudo-spectral counterparts. Earthq Struct 4:489–508

PEER (2009) Pacific Earthquake Engineering Research Center, Strong Ground Motion Database, Berkeley, CA. http://peer.berkeley.edu/

Popov EP (1983) Recent research on eccentrically braced frames. Eng Struct 5:3–9

Richards PW, Uang CM (2006) Testing protocol for short links in eccentrically braced frames. J Struct Eng ASCE 132:1183–1191

SAP 2000 (2010) Structural analysis program 2000, static and dynamic finite element analysis of structures, Version 14. Computers and Structures Inc, Berkeley, CA

SAP 2000 (2016) Structural analysis program 2000, static and dynamic finite element analysis of structures, Version 20. Computers and Structures Inc, Berkeley, CA

SEAOC (1995) A framework for performance-based design. Structural Engineers Association of California, Vision 2000 Committee, Sacramento, CA

SEAOC (1999) Recommended lateral force requirements and commentary, 7th edn. Structural Engineers Association of California, Sacramento, CA

Vamvatsikos D, Cornell CA (2002) Incremental dynamic analysis. Earthq Eng Struct Dyn 31:491–514

Worden K, Tomlinson GR (2001) Nonlinearity in structural dynamics: detection, identification and modeling. Institute of Physics Publishing, London

Chapter 9
Design Using Modal Behavior Factors

Abstract A performance based seismic design method for plane steel moment resisting and braced framed structures is described. It is a force-based seismic design method employing different modal (or strength reduction) factors for the first four significant modes of the frame, instead of the same constant behavior factor for all modes as in all current design codes. These modal behavior factors are functions of the modal periods of the structure, different soil types and different performance targets. Thus, the method automatically satisfies deformation demands at all performance levels without requiring deformation checks, as in all current design codes. The method is theoretically based on the construction of the equivalent linear structure to the original nonlinear one and the equivalent modal damping ratios of the previous chapter. The modal behavior factors are determined from the equivalent modal damping ratios with the aid of the modal damping reduction factors. Empirical expressions for the modal behavior factors as functions of period, deformation/damage and soil types for the seismic design of steel plane moment resisting and braced frames are derived. These expressions are appropriately converted to ones which can be used directly in conjunction with code defined elastic pseudo-acceleration design spectra with 5% damping. The proposed method is illustrated with representative numerical examples that demonstrate its advantages over code-based seismic design methods.

Keywords Modal behavior factors · Modal damping reduction factors · Seismic design method · Performance-based design · Pseudo-acceleration design spectra

9.1 Introduction

During the last 50 years or so various seismic design methods for steel structures have been developed. The majority of these methods are forced-based design (FBD) methods with forces being the main design parameters. These methods use elastic response/design spectra and the concept of the behavior (or strength reduction) factor to determine the design base shear through response spectrum analysis.

G. A. Papagiannopoulos et al., *Seismic Design Methods for Steel Building Structures*, Geotechnical, Geological and Earthquake Engineering 51,
https://doi.org/10.1007/978-3-030-80687-3_9

Current seismic design codes, such as the EC8 (2004) or the IBC (2018) are based on the FBD method and use a behavior factor q, which is characterized by the same constant value for all modes and depends only on the type and material of the structure. During the last 30 years or so, more rational expressions of the behavior factor in terms of period and ductility have been also proposed (e.g., Miranda and Bertero 1994; Mazzolani and Piluso 1996; Cuesta et al. 2003; Karavasilis et al. 2007).

In the previous chapter, a FBD method for the seismic design of plane steel framed buildings based on the concept of the equivalent modal damping ratios has been presented and its advantages demonstrated. The method uses equivalent modal damping ratios in conjunction with an elastic absolute acceleration design spectrum with high values of damping to determine the seismic base shear of the structure. The method has been developed for plane steel MRFs by Papagiannopoulos and Beskos (2010), Loulelis et al. (2018) and Kalapodis et al. (2021) and for plane steel CBFs and EBFs by Loulelis (2015), Kalapodis et al. (2020) and Kalapodis and Papagiannopoulos (2020). In this method, equivalent modal damping ratios play the role of the behavior factor in the sense that both take into account nonlinearities (material and geometric) in the framework of an elastic analysis with the aid of the response/design elastic spectra. However, engineers are not familiar with the concept of equivalent modal damping ratios and even more with absolute acceleration design spectra.

In this chapter, a FBD method based on the more familiar to engineers concepts of modal behavior factors and pseudo-acceleration design spectra is presented. This method has originally been developed by Papagiannopoulos and Beskos (2011) for plane steel MRFs and improved by Loulelis et al. (2018) so as to include more refined modeling (steel deterioration and panel zone effects) and expressions for modal behavior factors appropriate for using them in conjunction with pseudo-acceleration design spectra. Furthermore, far-fault motions were considered in Loulelis et al. (2018) instead of near-fault and long duration motions considered in Papagiannopoulos and Beskos (2011). Finally, many more structures under many more excitations were used in the former work than in the latter one for the construction of the modal behavior expressions in order to increase their reliability. Loulelis (2015), Kalapodis (2017), Kalapodis et al. (2018, 2020) and Kalapodis and Papagiannopoulos (2020) extended the method based on modal behavior factors to the design of various types of plane steel CBFs and EBFs.

At this point one should observe that use of different values of behavior factor q for different natural periods in a response spectrum analysis is a more rational approach than the universally employed one using the same value of q for all modes (Priestley 2003; Papagiannopoulos 2018). The proposed method succeeds in using different values of q for different natural periods through the modal behavior factors q_k. The present chapter describes the above method based on modal behavior factors as applied to various types of plane steel frames, illustrates its applications by means of numerical examples and demonstrates its advantages over the conventional code-based FBD of EC8 (2004).

9.2 Derivation of Modal Behavior Factors

9.2.1 Theoretical Background

Consider the equivalent linear multi-degree-of-freedom (MDOF) system to the original nonlinear MDOF system as defined in Sect. 8.2. For this equivalent linear system modal damping ratios ξ_k (k is the mode number) have been defined and explicitly determined in terms of period, deformation/damage, soil type and performance level for various types of plane steel frames. Because these modal damping ratios correspond to high amounts of damping, the concept of modal damping reduction factors $B_{d,\ k}$ is important. These modal damping reduction factors $B_{d,\ k}$ are absolute for high amounts of damping and are defined using absolute acceleration spectral values as

$$B_{d,k} = S_{a,k}(T_k, \xi_k)/S_{a,k}(T_k, \xi_{5\%}) \tag{9.1}$$

where ξ_k is the damping ratio, $S_{a,\ k}(T_k, \xi_{5\%})$ the absolute maximum acceleration of the structure with 5% damping (typical damping in frames and spectra) and $S_{a,\ k}(T_k, \xi_k)$ the absolute maximum acceleration of the structure with other than 5% damping. Of course, for small amounts of damping (less than 10%) and low natural periods (less than 0.15 s), these $B_{d,\ k}$ practically coincide with those defined for pseudo-acceleration spectra. The contribution of the kth mode to the seismic design force can be given as $M_k^* S_{a,k}(T_k, \xi_{eq,k})$ where M_k^* is the effective modal mass for the kth mode. The absolute behavior (strength reduction) factor for the kth mode can be expressed as the ratio of the modal contribution in the elastic base shear over the modal contribution in the base shear at first yield (Papagiannopoulos and Beskos 2011), i.e.,

$$\begin{aligned} q_k &= V_{el,k}/V_{y,k} = M_k^* S_{a,k}(T_k, \xi_{5\%})/M_k^* S_{a,k}(T_k, \xi_{eq,k}) \\ &= S_{a,k}(T_k, \xi_{5\%})/S_{a,k}(T_k, \xi_{eq,k}) = 1/B_{d,k} \end{aligned} \tag{9.2}$$

It is observed that the above definition of q_k is the same as that for q in EC8 (2004) and that q_k can be obtained as the inverse of $B_{d,\ k}$ and hence as function of ξ_k. Having those absolute behavior factors q_k as well as absolute response spectra for high values of damping, one can determine the base shear of a structure through modal synthesis. It should be noted that when using q_k in conjunction with absolute acceleration spectra, the resulting base shear includes both elastic restoring and damping forces. This base shear is different from the code-based one coming from pseudo-acceleration spectra and including only elastic restoring forces. It becomes practically equal to the code-based one only for low values of damping. Since equivalent modal damping ratios ξ_k can reach very high values, the modal damping reduction factors $B_{d,\ k}$ and hence the modal behavior factors q_k are expressed by Eqs. (9.1) and (9.2), respectively, in terms of absolute values of the modal acceleration spectra. However, current seismic codes, such as EC8 (2004), use pseudo response/design acceleration spectra for seismic structural design and engineers are more familiar with the concept of those acceleration spectra. That implies that modal behavior factors should be defined and determined in such a way so that they can be used with pseudo-acceleration response/design spectra. The modal damping reduction factors $\overline{B}_{d,k}$ for pseudo-acceleration response spectra are defined as

$$\bar{B}_{d,k} = PS_{a,k}(T_k, \xi_{eq,k})/PS_{a,k}(T_k, \xi_{5\%}) \tag{9.3}$$

where $PS_{a,\ k}$ denotes pseudo-acceleration spectral values for mode k, $\xi_{eq,\ k}$ denotes equivalent damping at mode k, $\xi_{5\%}$ stands for damping equal to 5% and T_k is the undamped natural period at mode k. The modal behavior factor $\overline{q}_k$ for mode k in view of Eq. (9.2) and in conjunction with the pseudo-acceleration spectrum concept can be expressed as

$$\bar{q}_k = \bar{V}_{el,k}/\bar{V}_{y,k} = M_k^* PS_{a,k}(T_k, \xi_{5\%,k})/M_k^* PS_{a,k}(T_k, \xi_{eq,k}) \tag{9.4}$$

where $\overline{V}_{el,k}$ and $\overline{V}_{y,k}$ denote modal base shear forces for the structure being elastic and at first yield, respectively. Equation (9.4), on account of Eq. (9.3), can also be written as

$$\bar{q}_k = PS_{a,k}(T_k, \xi_{5\%,k})/PS_{a,k}(T_k, \xi_{eq,k}) = 1/\bar{B}_{d,k} \tag{9.5}$$

Thus, the values of the modal behavior factors $\overline{q}_k$ can be obtained by inverting those of the modal damping reduction factors $\overline{B}_{d,k}$ associated to pseudo-acceleration spectra. Equation (9.5) provides the behavior factor for a given period and damping. Using the response databanks constructed by extensive parametric studies on various types of plane steel frames under the 100 seismic motions of Tables 8.2–8.5, one can determine mean values for $\overline{q}$ as functions of period, soil type and damping ratio ξ/ξ_{eq}. Figure 9.1 depicts the dependence of $\overline{q}$ on those parameters for the case of plane steel MRFs and soil types B and C. However, these figures cannot be directly used for design purposes since they do not show any deformation/damage dependence. Combination of the analytical expressions for the equivalent damping ratios ξ_k presented in the previous chapter with the aforementioned figures can produce expressions for modal behavior factors controlling directly the deformation of the structure and the damage of its structural members.

Conversion of the absolute modal behavior factors q_k into the pseudo ones $\overline{q}_k$ is accomplished as follows: The absolute behavior factor q_k defined by Eq. (9.2) can be written as

$$q_k = S_{a,k}(T_{d,k}, \xi_{5\%,k})/S_{a,k}(T_{d,k}, \xi_{eq,k}) \tag{9.6}$$

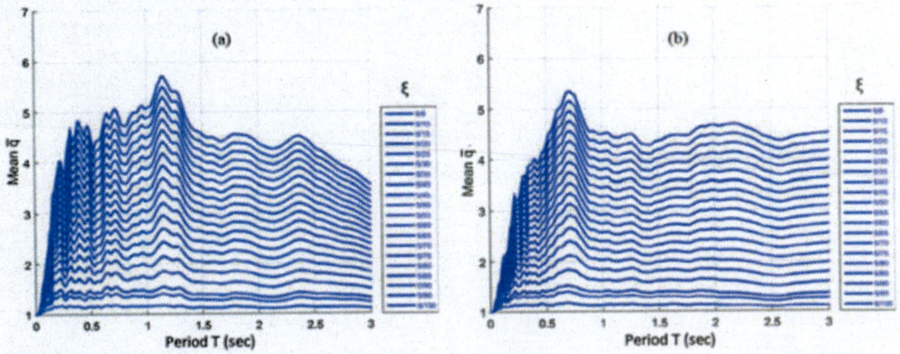

Fig. 9.1 Mean behavior factors $\bar{q}$ versus period T for various values of damping ratio 5 % /ξ_{eq} and (**a**) soil type B; (**b**) soil type C (after Loulelis 2015, reprinted with permission from UPCE)

where $T_{d,\ k}$ is the damped period of the kth mode computed as $T_{d,k} = T_k/\sqrt{1-\xi_k^2}$ with $\xi_k = \xi_{eq,\ k}$. To relate $T_{d,\ k}$ with T_k at 5% damping and any other damping ratio ξ_{eq}, one defines $S_{a,\ k}(T_{d,\ k}, \xi_{5\ \%\ ,\ k}) = k_1 S_{a,\ k}(T_k, \xi_{5\ \%\ ,\ k})$ and $S_{a,\ k}(T_{d,\ k}, \xi_{eq,\ k}) = k_2 S_{a,\ k}(T_k, \xi_{eq,\ k})$. Furthermore, according to Papagiannopoulos et al. (2013), absolute acceleration and pseudo-acceleration for any damping ratio are related as $S_{a,\ k}(T_{d,\ k}, \xi_{5\ \%\ ,\ k}) = \lambda_1 PS_{a,\ k}(T_k, \xi_{5\ \%\ ,\ k})$ and $S_{a,\ k}(T_{d,\ k}, \xi_{eq,\ k}) = \lambda_2 PS_{a,\ k}(T_k, \xi_{eq,\ k})$. Thus Eq. (9.6) in view of Eq. (9.5) can be transformed into

$$q_k = \frac{k_1\lambda_1 PS_{a,k}(T_k, \xi_{5\%,k})}{k_2\lambda_2 PS_{a,k}(T_k, \xi_{eq,k})} = \frac{k_1\lambda_1}{k_2\lambda_2}\bar{q}_k \tag{9.7}$$

and hence

$$\bar{q}_k = \frac{k_2\lambda_2}{k_1\lambda_1} q_k \tag{9.8}$$

9.2.2 *Modal Behavior Factors for Plane Steel MRFs*

Using Eqs. (9.2) and (9.3) as well as computed values for k_1, k_2, λ_1 and λ_2, one can obtain values and hence expressions for the behavior factors $\bar{q}_k$ to be used in pseudo-acceleration spectra. Figs. 9.2 and 9.3 depict the variation of the behavior factors $\bar{q}_k$ versus period T for the first two modes, soil type B and two performance levels (IDR = 1.5%, 2.5%) for the plane steel MRFs of Table 8.1 of the previous chapter. Similar figures for all the remaining combinations of modes, soil types and performance levels can be found elsewhere (Loulelis 2015).

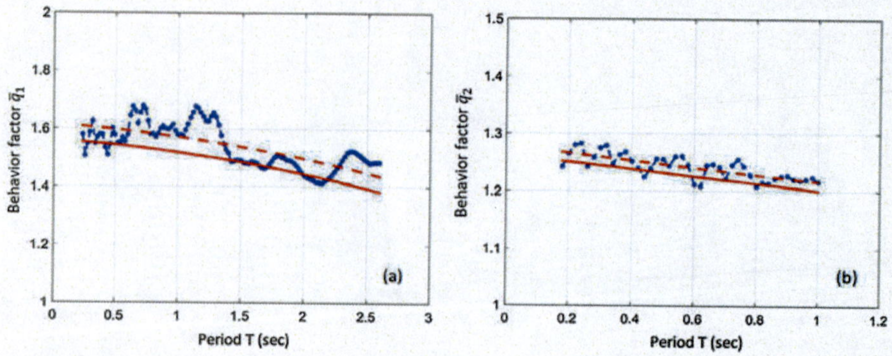

Fig. 9.2 Behavior factor for soil type B and performance level with $IDR = 1.5\%$: (**a**) first mode; (**b**) second mode (dashed lines: mean values, solid lines: mean minus one deviation values) (after Loulelis 2015, reprinted with permission from UPCE)

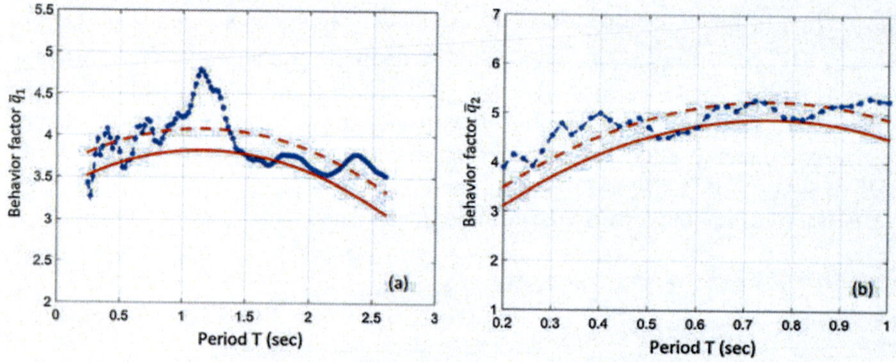

Fig. 9.3 Behavior factor for soil type B and performance level with $IDR = 2.5\%$: (**a**) first mode; (**b**) second mode (dashed lines: mean values, solid lines: mean minus one deviation values) (after Loulelis 2015, reprinted with permission from UPCE)

The dashed curves (mean values) in Figs. 9.2 and 9.3 have been constructed using polynomial regression with MATLAB (2009) in order to best fit the data points. Solid line curves in Figs.9.2 and 9.3 correspond to mean minus one deviation of values and represent lower and hence conservative bounds. Tables 9.12, 9.2, 9.3, and 9.4 have been constructed based on these conservative $\overline{q}_k$ solid line curves. The tables provide explicit expressions of $\overline{q}_k$ for steel plane MRFs in terms of the period T, the first four significant modes, four soil types (A, B, C, D) and the four performance levels of Table 8.6 of the previous chapter characterized by deformation (IDR) and damage (θ_p). These factors can be used in conjunction with pseudo-acceleration response/design elastic spectra to seismically design steel plane MRFs in a more rational way than current seismic codes, like EC8 (2004).

It should be made clear that the $\overline{q}_k$ factors can be applied not only to MRFs used for their construction (Table 8.1) but to any plane steel regular MRF with values of its three characteristic parameters T_1, ρ and α, where T_1 is the first natural period in

Table 9.1 Modal behavior factors $\overline{q}_k$ for MRFs with various values of *IDR* and θ_p and soil type A (after Loulelis 2015, reprinted with permission from UPCE)

Mode	IDR = 0.7% $\theta_p = 0$	IDR = 1.5% $\theta_p = \theta_y$	IDR = 2.5% $\theta_p = 3.5\theta_y$	IDR = 5.0% $\theta_p = 6.5\theta_y$
1	$\overline{q} = 1$ for $0.50 \leq T \leq 2.75$	$\overline{q} = 0.04T^2 - 0.24\,T + 1.50$ for $0.50 \leq T \leq 2.50$	$\overline{q} = 0.23T^2 - 1.23\,T + 3.56$ for $0.17 \leq T \leq 2.75$	$\overline{q} = 0.13T^2 - 1.31\,T + 4.41$ for $0.50 \leq T \leq 2.75$
2	$\overline{q} = 1$ for $0.25 \leq T \leq 1.0$	$\overline{q} = 0.03T^2 - 0.11\,T + 1.16$ for $0.25 \leq T \leq 1.0$	$\overline{q} = -1.74T^2 + 3.16\,T + 0.15$ for $0.25 \leq T \leq 1.0$	$\overline{q} = 0.20T^2 - 1.91\,T + 5.86$ for $0.25 \leq T \leq 1.0$
3	$\overline{q} = 1$ for $0.25 \leq T \leq 0.60$	$\overline{q} = 1.15 - 0.01(T - 0.25)$ for $0.25 \leq T \leq 0.60$	$\overline{q} = 2.33T^2 - 1.04(T - 0.35)$ for $0.25 \leq T \leq 0.60$	$\overline{q} = 0.20T^2 - 1.91\,T + 5.86$ for $0.25 \leq T \leq 0.60$
4	$\overline{q} = 1$ for $0.20 \leq T \leq 0.40$	$\overline{q} = 1.07 + 0.30(T - 0.25)$ for $0.20 \leq T \leq 0.34$ $\overline{q} = 1.10 - 0.26(T - 0.34)$ for $0.34 \leq T \leq 0.40$	$\overline{q} = 0.20T^2 - 1.91\,T + 5.86$ for $0.20 \leq T \leq 0.40$	$\overline{q} = 0.20T^2 - 1.91\,T + 5.86$ for $0.20 \leq T \leq 0.40$

Table 9.2 Modal behavior factors $\overline{q}_k$ for MRFs with various values of *IDR* and θ_p and soil type B (after Loulelis 2015, reprinted with permission from UPCE)

Mode	IDR = 0.7% $\theta_p = 0$	IDR = 1.5% $\theta_p = \theta_y$	IDR = 2.5% $\theta_p = 3.5\theta_y$	IDR = 5.0% $\theta_p = 6.5\theta_y$
1	$\overline{q} = 1$ for $0.50 \leq T \leq 2.75$	$\overline{q} = -0.02T^2 - 0.03\,T + 1.56$ for $0.50 \leq T \leq 2.75$	$\overline{q} = -0.10T^2 + 0.23\,T + 3.3$ for $0.50 \leq T \leq 2.75$	$\overline{q} = -0.36T^2 + 1.86\,T + 1.64$ for $0.50 \leq T \leq 2.75$
2	$\overline{q} = 1$ for $0.25 \leq T \leq 1.0$	$\overline{q} = -0.06\,T + 1.26$ for $0.20 \leq T \leq 1.0$	$\overline{q} = -0.38T^2 - 0.82\,T + 2.22$ for $0.20 \leq T \leq 1.0$	$\overline{q} = -6.0T^2 + 8.86\,T + 1.64$ for $0.20 \leq T \leq 1.0$
3	$\overline{q} = 1$ for $0.25 \leq T \leq 0.60$	$\overline{q} = 2.71 + 3.61(T - 0.13)$ for $0.15 \leq T \leq 0.39$	$\overline{q} = 1.52 + 0.04(T - 0.34)$ for $0.20 \leq T \leq 0.60$	$\overline{q} = -6.0T^2 + 8.86\,T + 1.64$ for $0.20 \leq T \leq 0.60$
4	$\overline{q} = 1$ for $0.20 \leq T \leq 0.40$	$\overline{q} = 1.01 + 0.10(T - 0.21)$ for $0.20 \leq T \leq 0.34$	$\overline{q} = -6.0T^2 + 8.86\,T + 1.64$ for $0.20 \leq T \leq 0.40$	$\overline{q} = -6.0T^2 + 8.86\,T + 1.64$ for $0.20 \leq T \leq 0.40$

Table 9.3 Modal behavior factors $\overline{q}_k$ for MRFs with various values of *IDR* and θ_p and soil type C (after Loulelis 2015, reprinted with permission from UPCE)

Mode	IDR = 0.7% $\theta_p = 0$	IDR = 1.5% $\theta_p = \theta_y$	IDR = 2.5% $\theta_p = 3.5\theta_y$	IDR = 5.0% $\theta_p = 6.5\theta_y$
1	$\overline{q} = 1$ for $0.50 \le T \le 2.75$	$\overline{q} = -0.08\ T + 1.48$ for $0.50 \le T \le 2.75$	$\overline{q} = 0.13T^2 + 0.47\ T + 3.10$ for $0.50 \le T \le 2.75$	$\overline{q} = -0.18T^2 + 0.50\ T + 3.19$ for $0.50 \le T \le 2.75$
2	$\overline{q} = 1$ for $0.25 \le T \le 1.0$	$\overline{q} = -0.19T^2 + 0.22\ T + 1.07$ for $0.25 \le T \le 1.0$	$\overline{q} = 2.69T^2 - 5.65\ T + 4.76$ for $0.25 \le T \le 1.0$	$\overline{q} = -0.32T^2 + T + 3.67$ for $0.25 \le T \le 1.0$
3	$\overline{q} = 1$ for $0.25 \le T \le 0.60$	$\overline{q} = 1.06 + 0.03(T - 0.18)$ for $0.25 \le T \le 0.60$	$\overline{q} = -0.32T^2 + T + 3.67$ for $0.25 \le T \le 0.60$	$\overline{q} = -0.32T^2 + T + 3.67$ for $0.25 \le T \le 0.60$
4	$\overline{q} = 1$ for $0.20 \le T \le 0.40$	$\overline{q} = 1.19 + 0.04(T - 0.20)$ for $0.20 \le T \le 0.40$	$\overline{q} = -0.32T^2 + T + 3.67$ for $0.20 \le T \le 0.40$	$\overline{q} = -0.32T^2 + T + 3.67$ for $0.20 \le T \le 0.40$

Table 9.4 Modal behavior factors $\overline{q}_k$ for MRFs with various values of *IDR* and θ_p and soil type D (after Loulelis 2015, reprinted with permission from UPCE)

Mode	IDR = 0.7% $\theta_p = 0$	IDR = 1.5% $\theta_p = \theta_y$	IDR = 2.5% $\theta_p = 3.5\theta_y$	IDR = 5.0% $\theta_p = 6.5\theta_y$
1	$\overline{q} = 1$ for $0.5 \leq T \leq 2.75$	$\overline{q} = -0.04\,T + 1.44$ for $0.50 \leq T \leq 2.75$	$\overline{q} = -0.32T^2 + 0.82\,T + 2.43$ for $0.50 \leq T \leq 2.75$	$\overline{q} = -0.80T^2 + 2.18\,T + 2.82$ for $0.50 \leq T \leq 2.75$
2	$\overline{q} = 1$ for $0.25 \leq T \leq 1.0$	$\overline{q} = -0.15T^2 + 0.14\,T + 1.11$ for $0.25 \leq T \leq 1.0$	$\overline{q} = 11.92T^2 - 19.25\,T + 9.53$ for $0.25 \leq T \leq 1.0$	$\overline{q} = -1.30T^2 + 3.70\,T + 3.0$ for $0.25 \leq T \leq 1.0$
3	$\overline{q} = 1$ for $0.25 \leq T \leq 0.60$	$\overline{q} = 1.18$ for $0.25 \leq T \leq 0.60$	$\overline{q} = 2.29$ for $0.25 \leq T \leq 0.60$	$\overline{q} = -1.30T^2 + 3.70\,T + 3.00$ for $0.25 \leq T \leq 0.60$
4	$\overline{q} = 1$ for $0.20 \leq T \leq 0.40$	$\overline{q} = 1.29 + 0.80(T - 0.25)$ for $0.20 \leq T \leq 0.40$	$\overline{q} = -1.3T^2 + 3.70\,T + 3.00$ for $0.20 \leq T \leq 0.40$	$\overline{q} = -1.30T^2 + 3.70\,T + 3.00$ for $0.20 \leq T \leq 0.40$

sec and ρ and α are the stiffness and strength ratios, respectively of the frame, defined by Eq. (5.7). The range of these parameters for the MRFs of Table 8.1 is: $0.37 \leq T \leq 2.42$ s, $0.06 \leq \rho \leq 0.20$ and $1.32 \leq \alpha \leq 7.45$.

Values of the behavior factor $\overline{q}_1$ for the first mode of plane steel MRFs and the LS performance level with maximum $IDR = 2.5\%$, vary from 3 to 4, irrespectively of soil type, while the EC8 (2004) proposed behavior factor q, the same for all modes, varies from 4 to 6.5. Factor $\overline{q}_1 < q$ because it is directly dependent on damage and deformation and corresponds to the first mode. In general, the difference between modal $\overline{q}_k$ factors and the single valued factor q proposed by EC8 (2004) is more evident for the performance level with IDR equal to 1.5%, where deformation requirements are stricter. This results in smaller values for the modal $\overline{q}_k$ factors.

9.2.3 *Modal Behavior Factors for Plane Steel EBFs and CBFs*

The procedure described in the previous Sect. 9.2.2 for the case of plane steel MRFs is also used here for determining $\overline{q}_k$ factors for plane steel EBFs of chevron type with short, long and intermediate seismic links (Fig. 8.7a) as well as for plane steel CBFs with BRBs of chevron type (Fig. 8.7b). Figure 9.4 depicts the variation of the $\overline{q}$ factor versus the natural period T for various amounts of equivalent damping ξ_{eq} and soil types B and C for EBFs of the chevron type and long links. Figure 9.5 shows the relation between $\overline{q}_k$ and natural period T for various performance levels (defined by IDR), soil type B and the two first modes ($k = 1, 2$) of CBFs of the chevron type with BRBs (Fig. 9.5a, b) and of EBFs of the chevron type with long links (Fig. 9.5c, d). This relation is given not only in the form of rough curves, but also in the form of dashed refined curves constructed through polynomial regression. The complete set of figures of the types of Fig. 9.5 covering all soil types and frames of Fig. 8.7a,

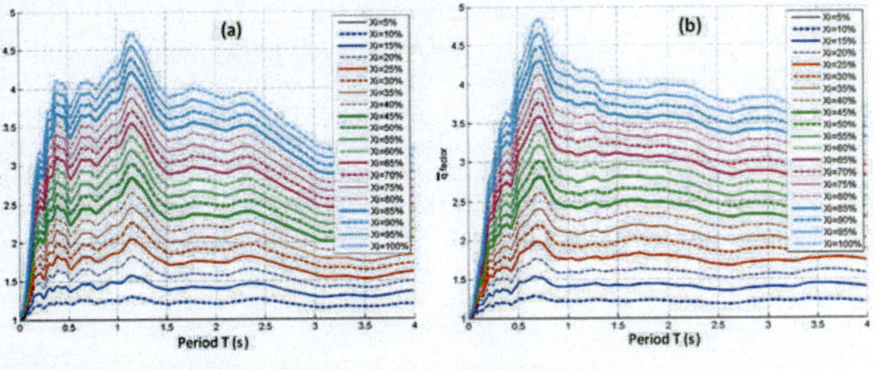

Fig. 9.4 Mean behavior factors $\overline{q}$ versus T for various values of $\xi_{eq}(X_i)$ and (**a**) soil type B; (**b**) soil type C (after Kalapodis 2017, reprinted with permission from UPCE)

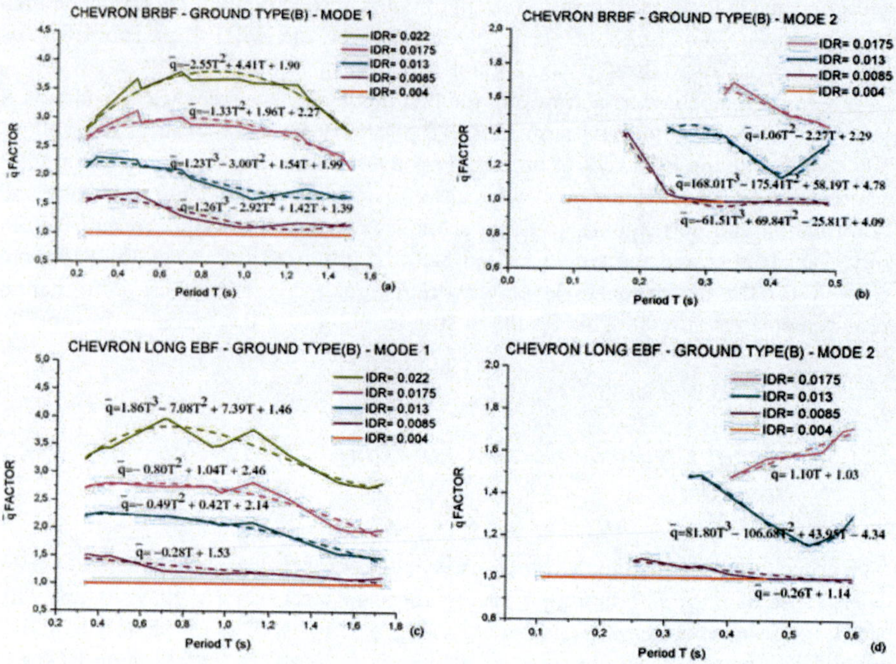

Fig. 9.5 Behavior factors $\overline{q}$ versus period T, for the first and second mode, chevron CBF with BRBs (**a**, **b**) and EBF (**c**, **d**) configurations, soil type B and various performance levels in terms of *IDR* (after Kalapodis 2017, reprinted with permission from UPCE)

(b) can be found in Kalapodis (2017). Plane eccentrically braced frames (EBFs) with diagonal braces and concentrically braced frames (CBFs) with diagonal BRBs have been also considered but they are not presented in this chapter. The interested reader may consult Kalapodis (2017) and Loulelis (2015) for these two types of frames.

Tables 9.5, 9.6, 9.7, 9.8, 9.9, 9.10, 9.11, 9.12, 9.13, 9.14, 9.15, 9.16, 9.17, 9.18, 9.19, and 9.20 provide analytical expressions for $\overline{q}_k$ in terms of T for the first four modes, four performance levels and soil types A, B, C and D for the types of braced frames of Fig. 8.7a, b. For the cases of all soil types and EBFs with diagonal BRBs one can look at Kalapodis (2017), whereas for all soil types and CBFs with diagonal BRBs one can look at Loulelis (2015). It should be noted that performance levels in Tables 9.5–9.20 are defined by SP1 to SP4 (following Table 8.15). In case that one wishes to design an EBF or a CBF with BRBs for the CP (SP4) level, the use of $\xi_k = 100\%$ is obligatory for all k modes (Kalapodis 2017) and the corresponding $\overline{q}_k$ are calculated by using figures like Figs. 9.4 for $\xi = 100\%$.

Table 9.5 Modal behavior factors for EBFs (chevron type) with long links and soil type A (after Kalapodis 2017, reprinted with permission from UPCE)

SP Level	Mode 1	Mode 2	Mode 3	Mode 4
SP1-IO	$\bar{q}=1$ $(0.30 \leq T_1 \leq 1.70)$	$\bar{q}=1$ $(0.10 \leq T_2 \leq 0.60)$	$\bar{q}=1$ $(0.14 \leq T_3 \leq 0.35)$	$\bar{q}=1$ $(0.09 \leq T_4 \leq 0.24)$
SP2-DL	$\bar{q}=1.0T^2 - 2.98\,T + 3.49$ $(0.30 \leq T_1 \leq 1.70)$	$\bar{q}= - 1.35\,T + 1.87$ $(0.35 \leq T_2 \leq 0.60)$	$\bar{q}= - 3.52\,T + 2.26$ $(0.24 \leq T_3 \leq 0.35)$	$\bar{q}= - 115.01\,T^2 + 48.60\,T - 1.15$ $(0.09 \leq T_4 \leq 0.24)$
SP3-LS	$\bar{q}=0.52T^2 - 2.05\,T + 4.33$ $(0.30 \leq T_1 \leq 1.70)$	$\bar{q}= - 24.86\,T^2 + 23.10\,T - 0.26$ $(0.10 \leq T_2 \leq 0.60)$	$\bar{q}=38.59T^2 - 14.73\,T + 5.30$ $(0.14 \leq T_3 \leq 0.35)$	$\bar{q}= - 115.01\,T^2 + 48.60\,T - 1.15$ $(0.09 \leq T_4 \leq 0.24)$
SP4-CP	$\bar{q}= - 1.41\,T + 5.42$ $(0.30 \leq T_1 \leq 1.70)$	$\bar{q}= - 24.86\,T^2 + 23.10\,T - 0.26$ $(0.10 \leq T_2 \leq 0.6)$	$\bar{q}=38.59T^2 - 14.73\,T + 5.30$ $(0.14 \leq T_3 \leq 0.35)$	$\bar{q}= - 115.01\,T^2 + 48.60\,T - 1.15$ $(0.09 \leq T_4 \leq 0.24)$

Table 9.6 Modal behavior factors for EBFs (chevron type) with long links and soil type B (after Kapapodis 2017, reprinted with permission from UPCE)

SP Level	Mode 1	Mode 2	Mode 3	Mode 4
SP1-IO	$\bar{q}=1$ $(0.30 \leq T_1 \leq 1.70)$	$\bar{q}=1$ $(0.10 \leq T_2 \leq 0.60)$	$\bar{q}=1$ $(0.14 \leq T_3 \leq 0.35)$	$\bar{q}=1$ $(0.09 \leq T_4 \leq 0.24)$
SP2-DL	$\bar{q}=-0.49\,T^2+0.42\,T+2.14$ $(0.30 \leq T_1 \leq 1.70)$	$\bar{q}=81.80T^3-106.60T^2+43.90\,T-4.34$ $(0.34 \leq T_2 \leq 0.60)$	$\bar{q}=-1.73\,T+1.99$ $(0.24 \leq T_3 \leq 0.35)$	$\bar{q}=4.20\,T+2.18$ $(0.09 \leq T_4 \leq 0.24)$
SP3-LS	$\bar{q}=1.86T^3-7.08T^2+7.39\,T+1.46$ $(0.30 \leq T_1 \leq 1.70)$	$\bar{q}=-14.85\,T^2+14.10\,T+0.69$ $(0.10 \leq T_2 \leq 0.60)$	$\bar{q}=6.52\,T+1.64$ $(0.14 \leq T_3 \leq 0.35)$	$\bar{q}=4.20\,T+2.18$ $(0.09 \leq T_4 \leq 0.24)$
SP4-CP	$\bar{q}=-1.47\,T^2+3.07\,T+3.38$ $(0.30 \leq T_1 \leq 1.70)$	$\bar{q}=-14.85\,T^2+14.10\,T+0.69$ $(0.10 \leq T_2 \leq 0.6)$	$\bar{q}=6.52\,T+1.64$ $(0.14 \leq T_3 \leq 0.35)$	$\bar{q}=4.20\,T+2.18$ $(0.09 \leq T_4 \leq 0.24)$

Table 9.7 Modal behavior factors for EBFs (chevron type) with long links and soil type C (after Kapapodis 2017, reprinted with permission from UPCE)

SP Level	Mode 1	Mode 2	Mode 3	Mode 4
SP1-IO	$\bar{q}=1$ $(0.30 \le T_1 \le 1.70)$	$\bar{q}=1$ $(0.10 \le T_2 \le 0.60)$	$\bar{q}=1$ $(0.14 \le T_3 \le 0.35)$	$\bar{q}=1$ $(0.09 \le T_4 \le 0.24)$
SP2-DL	$\bar{q}=1.30T^3-3.95T^2+2.90\,T+1.42$ $(0.30 \le T_1 \le 1.70)$	$\bar{q}=0.53T^2-1.11\,T+1.59$ $(0.34 \le T_2 \le 0.60)$	$\bar{q}=-1.83\,T+1.70$ $(0.24 \le T_3 \le 0.35)$	$\bar{q}=8.16\,T+0.92$ $(0.09 \le T_4 \le 0.24)$
SP3-LS	$\bar{q}=2.72T^3-9.18T^2+9.0\,T+0.63$ $(0.30 \le T_1 \le 1.70)$	$\bar{q}=4.21\,T+1.95$ $(0.10 \le T_2 \le 0.60)$	$\bar{q}=7.70\,T+0.96$ $(0.14 \le T_3 \le 0.35)$	$\bar{q}=8.16\,T+0.92$ $(0.09 \le T_4 \le 0.24)$
SP4-CP	$\bar{q}=-1.98\,T^2+4.17\,T+2.85$ $(0.30 \le T_1 \le 1.70)$	$\bar{q}=4.21\,T+1.95$ $(0.10 \le T_2 \le 0.6)$	$\bar{q}=7.70\,T+0.96$ $(0.14 \le T_3 \le 0.35)$	$\bar{q}=8.16\,T+0.92$ $(0.09 \le T_4 \le 0.24)$

Table 9.8 Modal behavior factors for EBFs (chevron type) with long links and soil type D (after Kalapodis 2017, reprinted with permission from UPCE)

SP Level	Mode 1	Mode 2	Mode 3	Mode 4
SP1-IO	$\bar{q}=1$ $(0.30 \leq T_1 \leq 1.70)$	$\bar{q}=1$ $(0.10 \leq T_2 \leq 0.60)$	$\bar{q}=1$ $(0.14 \leq T_3 \leq 0.35)$	$\bar{q}=1$ $(0.09 \leq T_4 \leq 0.24)$
SP2-DL	$\bar{q}=-0.37\,T^2+0.08\,T+2.30$ $(0.30 \leq T_1 \leq 1.70)$	$\bar{q}=87.09T^3-126.27T^2+58.59\,T-7.34$ $(0.35 \leq T_2 \leq 0.6)$	$\bar{q}=-1.19\,T+1.53$ $(0.24 \leq T_3 \leq 0.35)$	$\bar{q}=10.77\,T+0.72$ $(0.09 \leq T_4 \leq 0.24)$
SP3-LS	$\bar{q}=-0.99\,T^2+2.02\,T+2.47$ $(0.30 \leq T_1 \leq 1.70)$	$\bar{q}=-25.75\,T^2+27.12\,T-2.23$ $(0.10 \leq T_2 \leq 0.60)$	$\bar{q}=7.88\,T+1.18$ $(0.14 \leq T_3 \leq 0.35)$	$\bar{q}=10.77\,T+0.72$ $(0.09 \leq T_4 \leq 0.24)$
SP4-CP	$\bar{q}=-2.13\,T^2+5.00\,T+2.69$ $(0.30 \leq T_1 \leq 1.70)$	$\bar{q}=-25.75\,T^2+27.12\,T-2.23$ $(0.10 \leq T_2 \leq 0.60)$	$\bar{q}=7.88\,T+1.18$ $(0.14 \leq T_3 \leq 0.35)$	$\bar{q}=10.77\,T+0.72$ $(0.09 \leq T_4 \leq 0.24)$

Table 9.9 Modal behavior factors for EBFs (chevron type) with intermediate links and soil type A (after Kalapodis 2017, reprinted with permission from UPCE)

SP Level	Mode 1	Mode 2	Mode 3	Mode 4
SP1-IO	$\bar{q}= 1$ $(0.20 \leq T_1 \leq 1.90)$	$\bar{q}_k= 1$ $(0.08 \leq T_2 \leq 0.65)$	$\bar{q}= 1$ $(0.13 \leq T_3 \leq 0.35)$	$\bar{q}= 1$ $(0.09 \leq T_4 \leq 0.24)$
SP2-DL	$\bar{q}=$ $0.32T^3-1.03T^2 + 0.78\ T + 2.06$ $(0.20 \leq T_1 \leq 1.90)$	$\bar{q}= -1.86\ T^2 + 1.49\ T + 1.19$ $(0.35 \leq T_2 \leq 0.65)$	$\bar{q}= 11.34T^2 + 0.27\ T + 3.33$ $(0.19 \leq T_3 \leq 0.35)$	$\bar{q}= -\ 116.60\ T^2 + 70.75\ T -6.74$ $(0.09 \leq T_4 \leq 0.24)$
SP3-LS	$\bar{q}=$ $1.14T^3-4.00T^2 + 3.26\ T + 2.80$ $(0.20 \leq T_1 \leq 1.90)$	$\bar{q}= -20.78\ T^2 + 18.94\ T + 0.73$ $(0.08 \leq T_2 \leq 0.65)$	$\bar{q}= 11.34T^2 + 0.27\ T + 3.33$ $(0.13 \leq T_3 \leq 0.35)$	$\bar{q}= -116.60\ T^2 + 70.75\ T - 6.74$ $(0.09 \leq T_4 \leq 0.24)$
SP4-CP	$\bar{q}=$ $1.71T^3-5.85T^2 + 4.77\ T + 3.40$ $(0.20 \leq T_1 \leq 1.90)$	$\bar{q}= -20.78\ T^2 + 18.94\ T + 0.73$ $(0.08 \leq T_2 \leq 0.65)$	$\bar{q}= 11.34T^2 + 0.27\ T + 3.33$ $(0.13 \leq T_3 \leq 0.35)$	$\bar{q}= -116.60\ T^2 + 70.75\ T - 6.74$ $(0.09 \leq T_4 \leq 0.24)$

Table 9.10 Modal behavior factors for EBFs (chevron type) with intermediate links and soil type B (after Kapapodis 2017, reprinted with permission from UPCE)

SP Level	Mode 1	Mode 2	Mode 3	Mode 4
SP1-IO	$\bar{q}= 1$ $(0.20 \le T_1 \le 1.90)$	$\bar{q}= 1$ $(0.08 \le T_2 \le 0.65)$	$\bar{q}= 1$ $(0.13 \le T_3 \le 0.35)$	$\bar{q}= 1$ $(0.09 \le T_4 \le 0.24)$
SP2-DL	$\bar{q}=0.24T^2 + 0.52\ T + 1.99$ $(0.20 \le T_1 \le 1.90)$	$\bar{q}=0.07\ T + 1.22$ $(0.32 \le T_2 \le 0.65)$	$\bar{q}=6.52\ T + 1.64$ $(0.19 \le T_3 \le 0.35)$	$\bar{q}=4.20\ T + 2.18$ $(0.09 \le T_4 \le 0.24)$
SP3-LS	$\bar{q}= -\ 0.74\ T^2 + 1.65\ T+ 2.70$ $(0.20 \le T_1 \le 1.90)$	$\bar{q}=10.40T^2 + 10.56\ T + 1.36$ $(0.08 \le T_2 \le 0.65)$	$\bar{q}=6.52\ T + 1.64$ $(0.13 \le T_3 \le 0.35)$	$\bar{q}=4.20\ T + 2.18$ $(0.09 \le T_4 \le 0.24)$
SP4-CP	$\bar{q}= -\ 1.41\ T^2 + 3.30\ T+ 3.03$ $(0.20 \le T_1 \le 1.90)$	$\bar{q}=10.40T^2 + 10.56\ T + 1.36$ $(0.08 \le T_2 \le 0.65)$	$\bar{q}=6.52\ T + 1.64$ $(0.13 \le T_3 \le 0.35)$	$\bar{q}=4.20\ T + 2.18$ $(0.09 \le T_4 \le 0.24)$

Table 9.11 Modal behavior factors for EBFs (chevron type) with intermediate links and soil type C (after Kapapodis 2017, reprinted with permission from UPCE)

SP Level	Mode 1	Mode 2	Mode 3	Mode 4
SP1-IO	$\bar{q}=1$ $(0.20 \leq T_1 \leq 1.90)$	$\bar{q}=1$ $(0.08 \leq T_2 \leq 0.65)$	$\bar{q}=1$ $(0.13 \leq T_3 \leq 0.35)$	$\bar{q}=1$ $(0.09 \leq T_4 \leq 0.24)$
SP2-DL	$\bar{q}=0.98T^3 - 3.38T^2 + 3.44\,T + 1.28$ $(0.20 \leq T_1 \leq 1.90)$	$\bar{q}=3.05T^2 - 2.14\,T + 1.66$ $(0.32 \leq T_2 \leq 0.65)$	$\bar{q}=7.70\,T + 0.96$ $(0.19 \leq T_3 \leq 0.35)$	$\bar{q}=8.16\,T + 0.92$ $(0.09 \leq T_4 \leq 0.24)$
SP3-LS	$\bar{q}=2.03T^3 - 7.01T^2 + 7.37\,T + 1.16$ $(0.20 \leq T_1 \leq 1.90)$	$\bar{q}=4.26\,T + 1.92$ $(0.08 \leq T_2 \leq 0.65)$	$\bar{q}=7.70\,T + 0.96$ $(0.13 \leq T_3 \leq 0.35)$	$\bar{q}=8.16\,T + 0.92$ $(0.09 \leq T_4 \leq 0.24)$
SP4-CP	$\bar{q}=3.29T^3 - 12.04T^2 + 13.14\,T + 0.66$ $(0.20 \leq T_1 \leq 1.90)$	$\bar{q}=4.26\,T + 1.92$ $(0.08 \leq T_2 \leq 0.65)$	$\bar{q}=7.70\,T + 0.96$ $(0.13 \leq T_3 \leq 0.35)$	$\bar{q}=8.16\,T + 0.92$ $(0.09 \leq T_4 \leq 0.24)$

Table 9.12 Modal behavior factors for EBFs (chevron type) with intermediate links and soil type D (after Kalapodis 2017, reprinted with permission from UPCE)

SP Level	Mode 1	Mode 2	Mode 3	Mode 4
SP1-IO	$\bar{q}=1$ $(0.20 \leq T_1 \leq 1.90)$	$\bar{q}=1$ $(0.08 \leq T_2 \leq 0.65)$	$\bar{q}=1$ $(0.13 \leq T_3 \leq 0.35)$	$\bar{q}=1$ $(0.09 \leq T_4 \leq 0.24)$
SP2-DL	$\bar{q}=0.76T^3 - 2.79T^2 + 3.02\,T + 1.56$ $(0.20 \leq T_1 \leq 1.90)$	$\bar{q}=76.24T^3 - 112.36T^2 + 54.09\,T - 6.90$ $(0.32 \leq T_2 \leq 0.65)$	$\bar{q}= - 12.58\,T^2 + 14.19\,T + 0.45$ $(0.19 \leq T_3 \leq 0.35)$	$\bar{q}= - 79.77\,T^2 + 53.72\,T - 5.83$ $(0.09 \leq T_4 \leq 0.24)$
SP3-LS	$\bar{q}$ $=0.99T^3-4.21T^2 + 5.35\,T + 1.80$ $(0.20 \leq T_1 \leq 1.90)$	$\bar{q}= - 14.00\,T^2 + 16.66\,T + 0.02$ $(0.08 \leq T_2 \leq 0.65))$	$\bar{q}= - 12.58\,T^2 + 14.19\,T + 0.45$ $(0.13 \leq T_3 \leq 0.35)$	$\bar{q}= - 79.77\,T^2 + 53.72\,T - 5.83$ $(0.09 \leq T_4 \leq 0.24)$
SP4-CP	$\bar{q}=2.30T^2 + 5.78\,T + 2.17$ $(0.20 \leq T_1 \leq 1.90)$	$\bar{q}= - 14.00\,T^2 + 16.66\,T + 0.02$ $(0.08 \leq T_2 \leq 0.65)$	$\bar{q}= - 12.58\,T^2 + 14.19\,T + 0.45$ $(0.13 \leq T_3 \leq 0.35)$	$\bar{q}= - 79.77\,T^2 + 53.72\,T - 5.83$ $(0.09 \leq T_4 \leq 0.24)$

Table 9.13 Modal behavior factors for EBFs (chevron type) with short links and soil type A (after Kalapodis 2017, reprinted with permission from UPCE)

SP Level	Mode 1	Mode 2	Mode 3	Mode 4
SP1-IO	$\bar{q}=1$ $(0.22 \leq T_1 \leq 1.70)$	$\bar{q}=1$ $(0.09 \leq T_2 \leq 0.55)$	$\bar{q}=1$ $(0.11 \leq T_3 \leq 0.30)$	$\bar{q}=1$ $(0.14 \leq T_4 \leq 0.22)$
SP2-DL	$\bar{q}=-0.49\,T+2.21$ $(0.22 \leq T_1 \leq 1.70)$	$\bar{q}=-1.09\,T+1.70$ $(0.38 \leq T_2 \leq 0.55)$	$\bar{q}=-1.61\,T+1.58$ $(0.20 \leq T_3 \leq 0.30)$	$\bar{q}=-1.61\,T+1.58$ $(0.18 \leq T_4 \leq 0.22)$
SP3-LS	$\bar{q}=2.53T^3-9.11T^2+8.50\,T+1.31$ $(0.22 \leq T_1 \leq 1.70)$	$\bar{q}=-0.40\,T+1.72$ $(0.09 \leq T_2 \leq 0.55)$	$\bar{q}=3.88\,T+3.09$ $(0.11 \leq T_3 \leq 0.30)$	$\bar{q}=3.54\,T+2.89$ $(0.14 \leq T_4 \leq 0.22)$
SP4-CP	$\bar{q}=2.58T^3-8.7T^2+7.57\,T+2.77$ $(0.22 \leq T_1 \leq 1.70)$	$\bar{q}=-0.40\,T+1.72$ $(0.09 \leq T_2 \leq 0.55)$	$\bar{q}=3.88\,T+3.09$ $(0.11 \leq T_3 \leq 0.30)$	$\bar{q}=3.54\,T+2.89$ $(0.14 \leq T_4 \leq 0.22)$

Table 9.14 Modal behavior factors for EBFs (chevron type) with short links and soil type B (after Kalapodis 2017, reprinted with permission from UPCE)

SP Level	Mode 1	Mode 2	Mode 3	Mode 4
SP1-IO	$\overline{q}=1$ $(0.22 \leq T_1 \leq 1.70)$	$\overline{q}=1$ $(0.09 \leq T_2 \leq 0.55)$	$\overline{q}=1$ $(0.11 \leq T_3 \leq 0.30)$	$\overline{q}=1$ $(0.14 \leq T_4 \leq 0.22)$
SP2-DL	$\overline{q}=-0.49\,T+2.23$ $(0.22 \leq T_1 \leq 1.70)$	$\overline{q}=-1.62\,T+1.98$ $(0.38 \leq T_2 \leq 0.55)$	$\overline{q}=15.3T^2-9.17\,T+2.53$ $(0.20 \leq T_3 \leq 0.30)$	$\overline{q}=-2.01\,T+1.56$ $(0.14 \leq T_4 \leq 0.22)$
SP3-LS	$\overline{q}=-1.43\,T^2+2.56\,T+2.26$ $(0.22 \leq T_1 \leq 1.70)$	$\overline{q}=-22.38\,T^2+19.43\,T+0.27$ $(0.09 \leq T_2 \leq 0.55)$	$\overline{q}=7.46\,T+1.77$ $(0.11 \leq T_3 \leq 0.30)$	$\overline{q}=8.62\,T+0.98$ $(0.14 \leq T_4 \leq 0.22)$
SP4-CP	$\overline{q}=-1.82\,T^2+3.88\,T+2.93$ $(0.22 \leq T_1 \leq 1.70)$	$\overline{q}=-22.38\,T^2+19.43\,T+0.27$ $(0.09 \leq T_2 \leq 0.55)$	$\overline{q}=7.46\,T+1.77$ $(0.11 \leq T_3 \leq 0.30)$	$\overline{q}=8.62\,T+0.98$ $(0.14 \leq T_4 \leq 0.22)$

Table 9.15 Modal behavior factors for EBFs (chevron type) with short links and soil type C (after Kalapodis 2017, reprinted with permission from UPCE)

SP Level	Mode 1	Mode 2	Mode 3	Mode 4
SP1-IO	$\bar{q}=1$ $(0.22 \leq T_1 \leq 1.70)$	$\bar{q}=1$ $(0.09 \leq T_2 \leq 0.55)$	$\bar{q}=1$ $(0.11 \leq T_3 \leq 0.30)$	$\bar{q}=1$ $(0.14 \leq T_4 \leq 0.22)$
SP2-DL	$\bar{q}=1.78T^3 - 5.29T^2 + 4.0\,T + 1.22$ $(0.22 \leq T_1 \leq 1.70)$	$\bar{q}= -0.87\,T + 1.58$ $(0.38 \leq T_2 \leq 0.55)$	$\bar{q}= -1.70\,T + 1.65$ $(0.20 \leq T_3 \leq 0.30)$	$\bar{q}= -2.47\,T + 1.65$ $(0.14 \leq T_4 \leq 0.22)$
SP3-LS	$\bar{q}=2.40T^3 - 8.16T^2 + 7.61\,T + 1.17$ $(0.22 \leq T_1 \leq 1.70)$	$\bar{q}=6.57\,T + 1.36$ $(0.09 \leq T_2 \leq 0.55)$	$\bar{q}=7.54\,T + 1.39$ $(0.11 \leq T_3 \leq 0.30)$	$\bar{q}=11.50\,T - 0.39$ $(0.14 \leq T_4 \leq 0.22)$
SP4-CP	$\bar{q}=4.33T^3 - 15.30T^2 + 16.08\,T + 0.08$ $(0.22 \leq T_1 \leq 1.70)$	$\bar{q}=6.57\,T + 1.36$ $(0.09 \leq T_2 \leq 0.55)$	$\bar{q}=7.54\,T + 1.39$ $(0.11 \leq T_3 \leq 0.30)$	$\bar{q}=11.50\,T - 0.39$ $(0.14 \leq T_4 \leq 0.22)$

Table 9.16 Modal behavior factors for EBFs (chevron type) with short links and soil type D (after Kalapodis 2017, reprinted with permission from UPCE)

SP Level	Mode 1	Mode 2	Mode 3	Mode 4
SP1-IO	$\bar{q}=1$ $(0.22 \leq T_1 \leq 1.70)$	$\bar{q}=1$ $(0.09 \leq T_2 \leq 0.55)$	$\bar{q}=1$ $(0.11 \leq T_3 \leq 0.30)$	$\bar{q}_k=1$ $(0.14 \leq T_4 \leq 0.22)$
SP2-DL	$\bar{q}=-0.30\,T+2.05$ $(0.22 \leq T_1 \leq 1.70)$	$\bar{q}=-3.31\,T^2+1.90\,T+1.08$ $(0.20 \leq T_2 \leq 0.55)$	$\bar{q}=1.13\,T+1.06$ $(0.26 \leq T_3 \leq 0.30)$	$\bar{q}=-39.10\,T^2+13.03\,T+0.10$ $(0.14 \leq T_4 \leq 0.22)$
SP3-LS	$\bar{q}=-1.02\,T^2+2.00\,T+2.26$ $(0.22 \leq T_1 \leq 1.70)$	$\bar{q}=-2.79\,T+3.28$ $(0.38 \leq T_2 \leq 0.55)$	$\bar{q}=105.41T^2-51.23\,T+9.38$ $(0.11 \leq T_2 \leq 0.30)$	$\bar{q}=9.35\,T+0.20$ $(0.14 \leq T_4 \leq 0.22)$
SP4-CP	$\bar{q}=-0.83\,T^3+0.56\,T^2+1.94\,T+3.34$ $(0.22 \leq T_1 \leq 1.70)$	$\bar{q}=-15.23\,T^2+17.26\,T$ $(0.09 \leq T_2 \leq 0.55)$	$\bar{q}=105.41T^2-51.23\,T+9.38$ $(0.11 \leq T_2 \leq 0.30)$	$\bar{q}=9.35\,T+0.20$ $(0.14 \leq T_4 \leq 0.22)$

Table 9.17 Modal behavior factors for CBFs with BRBs (chevron type) and soil type A (after Kalapodis 2017, reprinted with permission from UPCE)

SP Level	Mode 1	Mode 2	Mode 3	Mode 4
SP1-IO	$\bar{q}=1$ $(0.22 \le T_1 \le 1.50)$	$\bar{q}=1$ $(0.11 \le T_2 \le 0.48)$	$\bar{q}=1$ $(0.10 \le T_3 \le 0.27)$	$\bar{q}=1$ $(0.11 \le T_4 \le 0.19)$
SP2-DL	$\bar{q}=2.78T^3 - 6.71T^2 + 3.31\,T + 2.14$ $(0.22 \le T_1 \le 1.50)$	$\bar{q}=2.04T^2 + 0.82\,T + 1.24$ $(0.25 \le T_2 \le 0.48)$	$\bar{q}=5.74T^2 - 4.34\,T + 1.83$ $(0.14 \le T_3 \le 0.27)$	$\bar{q}= -321.24\,T^2 + 108.67\,T - 5.46$ $(0.11 \le T_4 \le 0.19)$
SP3-LS	$\bar{q}=4.17T^3 - 12.49T^2 + 9.70\,T + 1.74$ $(0.22 \le T_1 \le 1.50)$	$\bar{q}=5.13\,T + 2.80$ $(0.11 \le T_2 \le 0.48)$	$\bar{q}= -14.27\,T^2 + 9.92\,T + 2.44$ $(0.10 \le T_3 \le 0.27)$	$\bar{q}= -321.24\,T^2 + 108.67\,T - 5.46$ $(0.11 \le T_4 \le 0.19)$
SP4-CP	$\bar{q}=6.06T^3 - 17.44T^2 + 13.77\,T + 1.63$ $(0.22 \le T_1 \le 1.50)$	$\bar{q}=5.13\,T + 2.80$ $(0.11 \le T_2 \le 0.48)$	$\bar{q}= -14.27\,T^2 + 9.92\,T + 2.44$ $(0.10 \le T_3 \le 0.27)$	$\bar{q}= -321.24\,T^2 + 108.67\,T - 5.46$ $(0.11 \le T_4 \le 0.19)$

Table 9.18 Modal behavior factors for CBFs with BRBs (chevron type) and soil type B (after Kalapodis 2017, reprinted with permission from UPCE)

SP Level	Mode 1	Mode 2	Mode 3	Mode 4
SP1-IO	$\overline{q}=1$ $(0.22 \leq T_1 \leq 1.50)$	$\overline{q}_k=1$ $(0.11 \leq T_2 \leq 0.48)$	$\overline{q}=1$ $(0.10 \leq T_3 \leq 0.27)$	$\overline{q}=1$ $(0.11 \leq T_4 \leq 0.19)$
SP2-DL	$\overline{q}=1.23T^3 - 3.00T^2 + 1.54\ T + 1.99$ $(0.22 \leq T_1 \leq 1.50)$	$\overline{q}=168.01T^3 -$ $175.41T^2 + 58.18\ T - 4.77$ $(0.25 \leq T_2 \leq 0.48)$	$\overline{q}= -14.36T^2 + 5.63\ T + 0.72$ $(0.14 \leq T_3 \leq 0.27)$	$\overline{q}=16.93\ T + 0.08$ $(0.11 \leq T_4 \leq 0.19)$
SP3-LS	$\overline{q}= -2.55\ T^2 + 4.41\ T + 1.90$ $(0.22 \leq T_1 \leq 1.50)$	$\overline{q}=3.72\ T + 2.42$ $(0.11 \leq T_2 \leq 0.48)$	$\overline{q}=6.91\ T + 1.57$ $(0.10 \leq T_3 \leq 0.27)$	$\overline{q}=16.93\ T + 0.08$ $(0.11 \leq T_4 \leq 0.19)$
SP4-CP	$\overline{q}= -2.19\ T^2 + 4.51\ T + 2.67$ $(0.22 \leq T_1 \leq 1.50)$	$\overline{q}=3.72\ T + 2.42$ $(0.11 \leq T_2 \leq 0.48)$	$\overline{q}=6.91\ T + 1.57$ $(0.10 \leq T_3 \leq 0.27)$	$\overline{q}=16.93\ T + 0.08$ $(0.11 \leq T_4 \leq 0.19)$

Table 9.19 Modal behavior factors for CBFs with BRBs (chevron type) and soil type C (after Kalapodis 2017, reprinted with permission from UPCE)

SP Level	Mode 1	Mode 2	Mode 3	Mode 4
SP1-IO	$\bar{q}=1$ $(0.22 \leq T_1 \leq 1.50)$	$\bar{q}=1$ $(0.11 \leq T_2 \leq 0.48)$	$\bar{q}=1$ $(0.10 \leq T_3 \leq 0.27)$	$\bar{q}=1$ $(0.11 \leq T_4 \leq 0.19)$
SP2-DL	$\bar{q}=3.53T^3-9.39T^2+6.40\,T+1.25$ $(0.22 \leq T_1 \leq 1.50)$	$\bar{q}=98.60T^3-103.41T^2+33.97\,T-2.16$ $(0.25 \leq T_2 \leq 0.48)$	$\bar{q}=1082.60T^3-683.65T^2+141.26\,T-8.31$ $(0.17 \leq T_3 \leq 0.27)$	$\bar{q}=11.13\,T+0.38$ $(0.11 \leq T_4 \leq 0.19)$
SP3-LS	$\bar{q}=4.93T^3-14.58T^2+12.37\,T+0.52$ $(0.22 \leq T_1 \leq 1.50)$	$\bar{q}=4.40\,T+1.85$ $(0.11 \leq T_2 \leq 0.48)$	$\bar{q}=8.72\,T+0.77$ $(0.10 \leq T_3 \leq 0.27)$	$\bar{q}=11.13\,T+0.38$ $(0.11 \leq T_4 \leq 0.19)$
SP4-CP	$\bar{q}=4.94T^3-16.38T^2+16.44\,T+0.01$ $(0.22 \leq T_1 \leq 1.50)$	$\bar{q}=4.40\,T+1.85$ $(0.11 \leq T_2 \leq 0.48)$	$\bar{q}=8.72\,T+0.77$ $(0.10 \leq T_3 \leq 0.27)$	$\bar{q}=11.13\,T+0.38$ $(0.11 \leq T_4 \leq 0.19)$

Table 9.20 Modal behavior factors for CBFs with BRBs (chevron type) and soil type D (after Kalapodis 2017, reprinted with permission from UPCE)

SP Level	Mode 1	Mode 2	Mode 3	Mode 4
SP1-IO	$\bar{q}=1$ $(0.22 \le T_1 \le 1.50)$	$\bar{q}=1$ $(0.11 \le T_2 \le 0.48)$	$\bar{q}_k=1$ $(0.10 \le T_3 \le 0.27)$	$\bar{q}=1$ $(0.11 \le T_4 \le 0.19)$
SP2-DL	$\bar{q}= -11.47\,T^4 + 43.48\,T^3 - 57.03T^2 + 28.81\,T - 2.04$ $(0.22 \le T_1 \le 1.50)$	$\bar{q}=10.59T^2 - 8.69\,T + 3.11$ $(0.25 \le T_2 \le 0.48)$	$\bar{q}=35.80T^2 - 16.72\,T + 3.10$ $(0.18 \le T_3 \le 0.27)$	$\bar{q}=10.90\,T + 0.62$ $(0.11 \le T_4 \le 0.19)$
SP3-LS	$\bar{q}= -15.56\,T^4 + 56.34\,T^3 - 73.03T^2 + 39.25\,T - 3.34$ $(0.22 \le T_1 \le 1.50)$	$\bar{q}=8.54\,T+ 1.05$ $(0.11 \le T_2 \le 0.48)$	$\bar{q}= -60.13\,T^2 + 33.08\,T - 1.30$ $(0.10 \le T_3 \le 0.27)$	$\bar{q}=10.90\,T + 0.62$ $(0.11 \le T_4 \le 0.19)$
SP4-CP	$\bar{q}= -3.41\,T^2 + 7.47\,T + 1.64$ $(0.22 \le T_1 \le 1.50)$	$\bar{q}=8.54\,T+ 1.05$ $(0.11 \le T_2 \le 0.48)$	$\bar{q}= -60.13\,T^2 + 33.08\,T - 1.30$ $(0.10 \le T_3 \le 0.27)$	$\bar{q}=10.90\,T + 0.62$ $(0.11 \le T_4 \le 0.19)$

9.3 Numerical Examples

In this section, the proposed design method using modal behavior factors is used for the seismic design of a number of plane steel moment resisting and braced frames for illustration purposes. The advantages and limitations of the method with respect to code-based methods are also discussed with the aid of NLTH analyses. The effects of panel zone and deterioration on seismic design for the case of MRFs are also assessed in this section.

9.3.1 Ten-Storey Three-Bay Plane Steel MRF

Consider a regular orthogonal plane steel MRF with ten stories of 3.0 m height and three bays of 5.0 m span, as shown in Fig. 8.11. HEB and IPE profiles are used for the columns and beams, respectively. The gravity load combination on beams consisting of dead plus 0.30 of live loads is equal to 27.5 kN/m^2. The seismic action is that corresponding to the elastic response spectrum of EC8 (2004) for $PGA = 0.36$ g and soil type B. The steel material is assumed to be of grade S275 for both columns and beams. The frame is seismically designed first by the proposed method using modal behavior factors in conjunction with the 5%-damped elastic S_{a} spectrum of Fig. 8.3 (modified for 0.36 g) and then by the method of EC8 (2004). Both designs are finally assessed by using NLTH analyses.

The design by the proposed method is done first for the LS performance level corresponding to the DBE. The initial sections of the frame are assumed to be 340/340/340/340–330 (1–6) and 320/320/320/320–300 (7–10) with the meaning of these numbers as explained in Sect. 8.3.1. For this section selection, the first four natural periods are found to be $T_1 = 2.069$ s, $T_2 = 0.693$ s, $T_3 = 0.376$ s and $T_4 = 0.243$ s. Use of these period values in Table 9.2 provides the corresponding modal behavior factors $\overline{q}_1 = 3.35$, $\overline{q}_2 = 1.47$, $\overline{q}_3 = 1.52$ and $\overline{q}_4 = 3.44$. From the mean pseudo-acceleration spectra of Fig. 8.3a for the above values of $\overline{q}$ and T, one obtains $S_{\mathrm{a1}} = 0.074$ g, $S_{\mathrm{a2}} = 0.534$ g, $S_{\mathrm{a3}} = 0.721$ g and $S_{\mathrm{a4}} = 0.309$ g. These spectral acceleration values lead to a design base shear of 362.34 kN with the aid of the SAP 2000 (2016) computer program. It should be noted that since SAP 2000 (2016) works with only one value of behavior factor, a modified design spectrum is created in the program with ordinates the four previously computed S_{a} values corresponding to the four natural periods, as shown in Fig. 9.6.

At this point one should note that use of the pseudo-acceleration elastic response spectrum with 5% damping of Fig. 8.3a was used instead of the analogous one of EC8 (2004) in order to be able to make more consistent comparisons between the present method of $\overline{q}_{\,\mathrm{k}}$ factors and the method of ξ_k ratios discussed in the previous Chap. 8.

Distributing this base shear value of 362.34 kN along the height of the frame and performing a strength checking that takes into account the capacity design rules, one

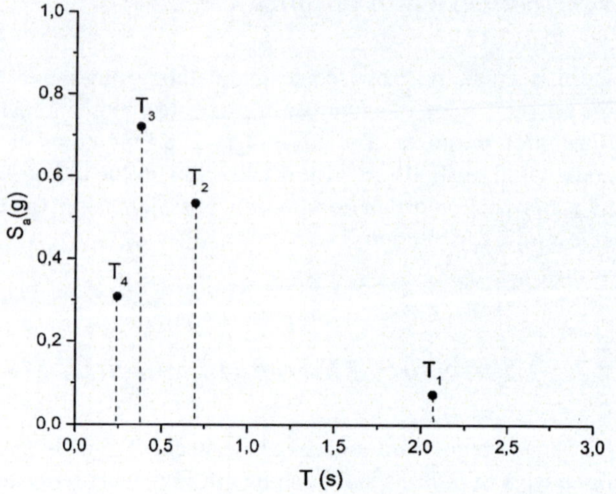

Fig. 9.6 Modified response spectrum

can arrive at the section selection 360/400/400/360–360 (1–3), 360/400/400/360–330 (4), 340/400/400/340–330 (5–7) and 340/360/360/340–330 (8–10). For these sections, the first four natural periods and corresponding modal behavior factors are $T_1 = 1.858$ s and $\overline{q}_1 = 3.38$, $T_2 = 0.60$ s and $\overline{q}_2 = 1.59$, $T_3 = 0.333$ s and $\overline{q}_3 = 1.52$, $T_4 = 0.214$ s and $\overline{q}_4 = 3.26$, respectively. For these values of T and $\overline{q}$, one obtains with the aid of SAP 2000 (2016) a base shear equal to 416.67kN. For this new base shear value, the strength checking that takes into account the capacity design rules, leads to the section selection 400/450/450/400–400 (1–2), 400/450/450/400–360 (3), 400/400/400/400–360 (4–5), 400/400/400/400–330 (6–7) and 360/360/360/360–330 (8–10). For these sections, the first four natural periods and corresponding modal behavior factors are $T_1 = 1.698$ s and $\overline{q}_1 = 3.40$, $T_2 = 0.566$ s and $\overline{q}_2 = 1.64$, $T_3 = 0.307$ s and $\overline{q}_3 = 1.52$, $T_4 = 0.200$ s and $\overline{q}_4 = 3.18$, respectively. For these values of T and $\overline{q}$, one obtains with the aid of SAP 2000 (2016) a base shear equal to 439.01 kN. For this new base shear value, the strength checking is repeated and is found to be satisfied for the sections obtained in the previous step indicating that these sections are the final ones for the LS performance level. No displacement checking is required for the LS level because displacement requirements are automatically satisfied by using the displacement dependent modal behavior factors $\overline{q}_k$.

Next the section selection for the LS level is checked according to the proposed method for strength satisfaction for the IO and CP performance levels corresponding to the FOE and MCE, respectively. The S_a spectra for the IO and CP performance levels are derived by multiplying the 5%-damped and modified for 0.36 g ordinates of Fig. 8.3a that correspond to DBE, by 0.5 and 1.5, respectively. On the basis of the aforementioned first four natural periods, the modal behavior factors and corresponding S_a values for the IO performance level are $\overline{q}_1 = 1.0$, $\overline{q}_2 = 1.0$, $\overline{q}_3 = 1.0$ and $\overline{q}_4 = 1.0$, whereas those for the CP performance level are $\overline{q}_1 = 3.76$, $\overline{q}_2 = 4.73$, $\overline{q}_3 = 3.80$ and $\overline{q}_4 = 3.17$. Application of the proposed method provides base shears 558.42 and 492.76 kN for the IO and CP levels, respectively. Taking into account that strength checking of all selected sections for the LS level is also satisfied for the IO and CP levels, indicates that among the three levels, the IO level, associated with the highest base shear, controls the design.

Nonlinear dynamic analyses with the aid of SAP 2000 (2016) of the three designs (for IO, LS and CP levels) by the proposed method are performed by employing 10 seismic motions compatible with the elastic design spectra for the IO, LS and CP levels, obtained on the basis of the modified for 0.36 g S_a elastic spectrum with 5% damping of Fig. 8.3a that corresponds to the LS level. The values of this spectrum are multiplied by 0.5 and 1.5 in order to account for the IO and CP levels, respectively. The mean values of maximum base shear (defined at first yielding for the LS and CP levels and as the maximum one for the IO level) and maximum *IDR* are 561.07 kN and 0.68%, 497.82 kN and 2.10%, 543.44 kN and 3.81%, respectively for the IO, LS and CP levels. It is observed that in all cases, the obtained from non-linear dynamic analyses *IDR* values do not exceed the limits of 0.7%, 2.5% and 5.0% for the IO, LS and CP levels of Table 8.6, as expected because of the displacement dependence of the employed modal behavior factors. However, the values of *IDR* for the LS and CP are significantly lower than the limit ones indicating a conservative design. This can be attributed to the fact that the used values of modal behavior factors $\overline{q}_k$ have been constructed as lower bounds and thus lead to higher design seismic forces and hence to stiffer frames and smaller drifts. Nevertheless, the proposed method even though results in conservative design, satisfies all three seismic performance levels constituting thereby a truly performance-based seismic design method.

A comparison of the frame designed by the proposed method is finally done against that designed by the method of EC8 (2004) for the case of a design spectrum defined for $PGA = 0.36$ g and soil type B. Because the frame and its loading are the same as those in Sect. 8.5.1, the EC8 (2004) frame design will be also the same with sections 360/360/360/360–360 (1–4) and 340/340/340/340–330 (5–10), for which one can determine that there corresponds a value of $q = 4.0$. The above sections being heavier than the initial ones coming from using $q = 6.5$ and the fact that they correspond to $q < 6.5$ indicates that the damage limitation state controls the design.

A nonlinear dynamic analysis of the EC8 (2004) design for the previously considered 10 seismic motions now compatible to the elastic design spectrum with 5% damping of EC8 (2004) for $PGA = 0.36$ g and soil type B, provides a base shear (at first yielding) of 401.24 kN and *IDR* of 2.07%. Thus, the *IDR* limit of 2.50% is not exceeded for the LS level as in the case of the design by the proposed method. However, the design by the EC8 method is lighter than the one by the proposed method as this is evident by looking at the respective section selections and the *IDR* values.

Finally, comparing the proposed design method with the one presented in the previous chapter that makes use of modal damping ratios ξ_k, one can observe that the design method with ξ_k leads to lighter sections. This is attributed to the fact that the used values of modal behavior factors $\overline{q}_k$ have been constructed not only as lower bounds but in addition are derived from the lower bounds assigned to the modal damping ratios ξ_k.

9.3.2 Seven-Storey Three-Bay Plane Steel EBF

Consider a seven-storey three-bay plane steel chevron EBF with long flexural beam links of the type shown in Fig. 8.7a. The length of each bay is 5.0 m and the height of each storey is 3.0 m. HEB, IPE and CHS sections made of S275 grade steel are used for columns, beams and braces, respectively. The gravity load combination is G + 0.3Q = 27.5 kN/m with G and Q being the dead and live loads, respectively. The seismic action is that corresponding to the elastic response spectrum of EC8 (2004) with $PGA = 0.24$ g and soil type B. The EBF is seismically designed first (Kalapodis et al. 2018) by the proposed method using $\overline{q}_k$ in conjunction with the 5%-damped S_a elastic spectra of Fig. 8.3 and then by the method of EC8 (2004) with $q = 4.0$ (Table 3.6).

The design by the proposed method is done here only for the LS performance level associated with a target $IDR = 2.2\%$ (Table 8.15), even though such a design may not be the controlling one in a performance-based design framework with three or four performance levels. The initial sections of the frame are assumed to be 260–300–193.7 × 4.5 (1–2) and 240–270–168.3 × 4 (3–5) with the meaning of these numbers as explained in Sect. 8.4.1. For this section selection, the first four natural periods are found to be $T_1 = 0.655$ s, $T_2 = 0.230$ s, $T_3 = 0.136$ s and $T_4 = 0.097$ s. Use of these period values in Table 9.16 for the LS level provide the corresponding modal behavior factors $\overline{q}_1 = 3.79$, $\overline{q}_2 = 3.15$, $\overline{q}_3 = 2.53$ and $\overline{q}_4 = 2.59$. From the mean pseudo-acceleration spectra of Fig. 8.3a for the above values of $\overline{q}$ and T, one obtains $S_{a1} = 0.144$ g, $S_{a2} = 0.232$ g, $S_{a3} = 0.261$ g and $S_{a4} = 0.20$ g. These spectral acceleration values lead to a design base shear of 357.59 kN with the aid of the SAP 2000 (2016) computer program. It is recalled that since SAP 2000 (2016) works with only one value of behavior factor, a modified design spectrum is created in SAP 2000 (2016) with ordinates the four previously computed S_a values corresponding to the four natural periods.

Distributing this base shear value of 357.59 kN along the height of the frame and performing a strength checking that takes into account the capacity design rules, one can arrive at the section selection of 300–330–168.3 × 4.0 (1–2), 280–300–152.4 × 4.0 (2), 260–270–152.4 × 4.0 (3), 260–270–152.4 × 4.0 (4–5), 260–270–139.7 × 4.0 (6) and 240–270–114.3 × 3.6 (7). For these sections, the first four natural periods and corresponding modal behavior factors are $T_1 = 0.910$ s and $\overline{q}_1 = 4.14$, $T_2 = 0.320$ s and $\overline{q}_2 = 3.75$, $T_3 = 0.200$ s and $\overline{q}_3 = 3.20$, $T_4 = 0.150$ s and $\overline{q}_4 = 2.85$, respectively. For these values of T and $\overline{q}$, one obtains with the aid of SAP 2000 (2016) a base shear equal to 375.63 kN. For this new base shear value, the strength checking is repeated and found to be satisfied for the sections of the previous step indicating that these sections are the final ones for the LS performance level. No displacement checking is required for the LS level because displacement requirements are automatically satisfied by using the displacement dependent modal behavior factors $\overline{q}_k$.

Nonlinear dynamic analyses with the aid of SAP 2000 (2016) of the designed EBF by the proposed method are performed by employing 10 seismic motions compatible with the elastic design spectrum obtained on the basis of the modified for 0.24 g S_a elastic spectrum with 5% damping of Fig. 8.3a that corresponds to the LS level. The mean value of maximum base shear (defined at first yielding), maximum *IDR* and θ_{link} (link rotation) are found to be 392.08 kN, 0.67% and 0.0191 rad, respectively. It is observed that the obtained from non-linear dynamic analyses maximum *IDR* value does not exceed the limit of 2.2% for the LS level of Table 8.6, as expected, because of the displacement dependence of the employed modal behavior factors $\overline{q}_k$. However, this value of *IDR* is significantly lower than the limit one indicating a conservative design. This can be attributed to the fact that the used values of modal behavior factors $\overline{q}_k$ have been constructed as lower bounds and thus lead to higher design seismic forces and hence to stiffer frames and smaller drifts.

A comparison of the frame designed by the proposed method is finally done against that designed by the method of EC8 (2004) for the case of a design spectrum defined for $PGA = 0.24$ g and soil type B. Selecting an initial value of $q = 4.0$ and performing the necessary strength and damage limit state (limit *IDR* of 0.7%) checkings according to Chap. 3, one needs two iterations to finally end up with sections 280–300–152.4 × 4.0 (1), 280–270–139.7 × 4.0 (2), 260–270–139.7 × 4.0 (3–5), 240–270–127.0 × 4.0 (6), 240–270–114.3 × 3.6 (7) and 240–270–114.3 × 3.6 (7).

Nonlinear dynamic analyses of the EC8 (2004) design for the previously considered 10 seismic motions now compatible with the elastic design spectrum with 5% damping of EC8 (2004) for $PGA = 0.24$ g and soil type B, provide the mean values 372. 75 kN, 0.74% and 0.0192 rad for the maximum base shear (defined at first yielding), maximum *IDR* and θ_{link} (link rotation), respectively. Thus, the *IDR* limit of 2.2% is not exceeded for the LS level as in the case of the design by the proposed method. The designs by the proposed and the EC8 (2004) methods are different as this is evident by looking at the respective section selections with the one of EC8 (2004) being lighter. However, the EC8 (2004) design also satisfies the damage limit state (IO) for which the design by the proposed method has not been checked, even though such a checking can be easily done by using behavior factor values for the DL level from Table 9.16. This checking has been done and found to be satisfied. On the other hand, the proposed method also satisfies the deformation limits for the LS level for which the EC8 (2004) design cannot be checked. It should be also noted that for both methods the mean θ_{link} value is marginally lower than the value of 0.02 rad permitted for the targeted *IDR* level according to Table 8.15.

9.3.3 Seven-Storey Three-Bay Plane Steel CBF

The proposed method using modal behavior factors $\overline{q}_k$ in conjunction with the S_a spectra of Fig. 8.3 is now employed to seismically design a seven-storey and three-bay CBF with chevron type BRBs of the type shown in Fig. 8.7b. The length of each bay is 5.0 m and the height of each storey is 3.0 m. The gravity load combination is

G + 0.3Q = 27.5 kN/m with G and Q being the dead and live loads, respectively. The seismic action is that corresponding to the elastic response spectrum of EC8 (2004) with $PGA = 0.36$ g and soil type B. The length of each bay is 5.0 m and the height of each storey is 3.0 m. HEB and IPE are used for columns and beams, whereas the section of the BRB is the one shown in Fig. 8.8b. All steel sections are of S275 grade. Dimensioning of the BRBs follows the work of Bosco et al. (2015).

The design by the proposed method (Kalapodis et al. 2018) is done here only for the LS performance level associated with a target $IDR = 2.2\%$ (Table 8.15), even though such a design may not be the controlling one in a performance-based design framework with three or four performance levels. The initial sections of the frame are assumed to be 320–330–20 (1–2), 300–330–18 (3–4) and 280–270–17 (5–7) with the meaning of these numbers as explained in Sect. 8.4.1. For this section selection, one computes by the proposed method a design base shear of 622.17 kN. Distributing this base shear along the height of the frame and performing a strength checking that takes into account the capacity design rules, one can arrive at the section selection of 360–330–20.6 (1), 320–330–19.5 (2), 300–330–17 (3), 280–330–16.2 (4), 260–330–15.5 (5), 260–330–12.5 (6) and 240–330–12.5 (7). For these sections, the first four natural periods and corresponding modal behavior factors are $T_1 = 0.740$ s and $\overline{q}_1 = 3.75$, $T_2 = 0.260$ s and $\overline{q}_2 = 3.35$, $T_3 = 0.140$ s and $\overline{q}_3 = 2.80$, $T_4 = 0.110$ s and $\overline{q}_4 = 2.10$, respectively. For these values of T and $\overline{q}$, one obtains with the aid of SAP 2000 (2016) a base shear equal to 641.50 kN. For this new base shear value, the strength checking is repeated and found to be satisfied for the sections obtained in the previous step indicating that these sections are the final ones for the LS performance level. No displacement checking is required for the LS level because displacement requirements are automatically satisfied by using the displacement dependent modal behavior factors $\overline{q}_k$.

Nonlinear dynamic analyses with the aid of SAP 2000 (2016) of the designed frame by the proposed method are performed by employing 10 seismic motions compatible with the elastic design spectrum obtained on the basis of the modified for 0.36 g S_a elastic spectrum with 5% damping of Fig. 8.3a that corresponds to the LS level. The mean values of maximum base shear (defined at first yielding), maximum IDR and μ_δ (brace ductility) were found to be 709.64 kN, 0.91% and 3.94, respectively. It is observed that the obtained from non-linear dynamic analyses maximum IDR value does not exceed the limit of 2.2% for the LS level of Table 8.15, as expected, because of the displacement dependence of the employed modal behavior factors $\overline{q}_k$. However, this value of IDR is significantly lower than the limit one, indicating a conservative design. This can be attributed to the fact that the used values of $\overline{q}_k$ have been constructed as lower bounds and thus lead to higher design seismic forces and hence to a stiffer frame and smaller drifts.

A comparison of the frame designed by the proposed method is finally done against that designed by the method of EC8 (2004) for the case of a design spectrum defined for $PGA = 0.36$ g and soil type B. Selecting an initial value of $q = 4.0$ and performing the necessary strength and damage limit state (limit IDR of 0.7%) checkings according to Chap. 3, one needs two iterations to finally end up with sections 340–330–19.5 (1), 320–330–17 (2), 300–330–16.2 (3), 280–330–15.5 (4), 260–330–15.5 (5), 260–330–12.5 (6) and 240–330–12.5 (7).

Nonlinear dynamic analyses of the EC8 (2004) design for the previously considered 10 seismic motions now compatible with the elastic design spectrum with 5% damping of EC8 (2004) for $PGA = 0.36$ g and soil type B, provide as mean values for maximum base shear (defined at first yielding), maximum *IDR* and μ_δ (brace ductility), the values 665.20 kN, 0.98% and 4.23,respectively. Thus, the *IDR* limit of 2.2% is not exceeded for the LS level as in the case of the design by the proposed method. The designs by the proposed and the EC8 (2004) methods are different as this is evident by looking at the respective section selections with the one of EC8 (2004) being lighter. However, one should have in mind the comments about the two methods made at the end of Sect. 9.3.2.

9.4 Conclusions

On the basis of the preceding developments, one can draw the following conclusions:

(1) A new seismic design method for plane steel MRFs, EBFs and CBFs with BRBs has been presented. According to this method, the base shear of the framed structure can be determined through response spectrum analysis and modal synthesis using modal behavior (strength reduction) factors. These factors are obtained through extensive parametric studies involving many frames under many far-fault seismic motions.

(2) Explicit empirical expressions which provide modal behavior factors as functions of period, deformation/damage and soil type are derived. These expressions are used in conjunction with design pseudo-acceleration elastic spectra to calculate the design base shear of the structure from which its member dimensions can be determined. The fact that these expressions are defined for three or four performance levels, renders the proposed method a truly performance-based seismic design method.

(3) The proposed method is applied to the seismic design of various types of plane steel frames and is compared with the seismic design method of EC8. Nonlinear time history analyses with seismic motions compatible with the corresponding elastic design spectra, are conducted for checking the performance of the method with respect to the seismic response (base shear, *IDR* and plastic rotation or axial ductility) of the designed structure. It is found that the proposed method is more rational than the EC8 method because it uses different behavior factors for each mode instead of a single value of this factor for all modes and controls deformation automatically. However, the proposed method appears to be more conservative than the EC8 method.

(4) Design results between the present method of modal behavior factors and the method of modal damping ratios are close but not the same with the former method to be more conservative than the latter one. This is attributed to the fact that modal behavior factors have been constructed not only as lower bounds but in addition are derived from the lower bounds assigned to the modal damping ratios ξ_k. The present method is a better choice than that of modal damping ratios because it uses the familiar to engineers behavior factors instead of damping ratios and elastic spectra with 5% damping instead of spectra with high amounts of damping.

References

Bosco M, Marino EM, Rossi PP (2015) Design of steel frames equipped with BRBs in the framework of Eurocode 8. J Constr Steel Res 113:43–57

Cuesta I, Aschheim M, Fajfar P (2003) Simplified R – factor relationships for strong ground motions. Earthquake Spectra 19:25–45

EC8 (2004) Eurocode 8, Design of structures for earthquake resistance, Part 1: general rules, seismic actions and rules for buildings, EN 1998-1-1. European Committee for Standardization (CEN), Brussels

IBC (2018) International building code. International Code Council, Washington, DC

Kalapodis NA (2017) Seismic design of plane steel braced frames with the use of three new methods. PhD Thesis, Department of Civil Engineering, University of Patras, Patras, Greece (in Greek)

Kalapodis NA, Papagiannopoulos GA (2020) Seismic design of plane steel braced frames using equivalent modal damping ratios. Soil Dyn Earthq Eng 129:105947

Kalapodis NA, Papagiannopoulos GA, Beskos DE (2018) Modal strength reduction factors for seismic design of plane steel braced frames. J Constr Steel Res 147:549–563

Kalapodis NA, Papagiannopoulos GA, Beskos DE (2020) A comparison of three performance-based seismic design methods for plane steel braced frames. Earthq Struct 18:27–44

Kalapodis NA, Muho EV, Beskos DE (2021) Seismic design of plane steel MRFs, EBFs and CBFs by improved direct displacement-based design method. Soil Dyn Earthq Eng (submitted)

Karavasilis TL, Bazeos N, Beskos DE (2007) Behavior factor for performance-based seismic design of plane steel moment resisting frames. J Earthq Eng 11:531–559

Loulelis DG (2015) Seismic design of planar steel frames with modal strength reduction factors for three performance levels. PhD Thesis, Department of Civil Engineering, University of Patras, Patras, Greece (in Greek)

Loulelis DG, Papagiannopoulos GA, Beskos DE (2018) Modal strength reduction factors for seismic design of steel moment resisting frames. Eng Struct 154:23–37

MATLAB (2009) The language of technical computing, Version 2009a. The Mathworks Inc, Natick, MA

Mazzolani FM, Piluso V (1996) Theory and design of seismic resistant steel frames. E & FN Spon, Chapman and Hall, London

Miranda E, Bertero VV (1994) Evaluation of strength reduction factors for earthquake resistant design. Earthquake Spectra 10:357–379

Papagiannopoulos GA (2018) Jacobsen's equivalent damping concept revisited. Soil Dyn Earthq Eng 115:82–89

Papagiannopoulos GA, Beskos DE (2010) Towards a seismic design method for plane steel frames using equivalent modal damping ratios. Soil Dyn Earthq Eng 30:1106–1118

Papagiannopoulos GA, Beskos DE (2011) Modal strength reduction factors for seismic design of plane steel frames. Earthq Struct 2:65–88

Papagiannopoulos GA, Hatzigeorgiou GD, Beskos DE (2013) Recovery of spectral absolute acceleration and spectral relative velocity from their pseudo-spectral counterparts. Earthq Struct 4:489–508

Priestley MJN (2003) Myths and fallacies in earthquake engineering revisited, the 9th Mallet-Milne lecture. IUSS Press, Pavia

SAP 2000 (2016) Structural analysis program 2000, static and dynamic finite element analysis of structures, Version 20. Computers and Structures Inc, Berkeley, CA

Chapter 10
Design Using Advanced Analysis

Abstract A rational and efficient seismic design method for plane and space steel moment resisting frames using advanced methods of analysis is presented. This method employs an advanced dynamic finite element analysis working in time domain that takes into account geometrical and material nonlinearities and member and frame imperfections. Seismic actions are in the form of accelerograms compatible with the elastic response spectra of EC8 for three performance levels. The design starts with assumed member sections for the frame, proceeds with the checking of drifts, member plastic rotation, damage and plastic hinge pattern for the three performance levels considered here and ends with the adjustment of member sizes iteratively so as the above response parameters to satisfy their limit values for every level. Thus, the method can sufficiently capture the limit states of displacements, strength, stability and damage of the structure and its members so that separate member capacity checks through the interaction equations of EC3 or the use of the approximate behavior factor of EC8 are not required. Numerical examples dealing with the seismic design of plane and space steel moment resisting frames are presented to illustrate the method and demonstrate its advantages.

Keywords Advanced analysis in seismic design · Material and geometric nonlinearities · Performance-based seismic design · Plane and space frames · Steel moment resisting frames

10.1 Introduction

Current codes for the design of steel building frames under static loads, such as EC3 (2009) or the AISC (2005) present two significant disadvantages. The first is that they do not take into account the interaction of strength and stability between the members and the whole structure directly but indirectly through the effective length concept. The second and most significant disadvantage is that the determination of internal member design forces is done by first order elastic analysis and inelasticity is

G. A. Papagiannopoulos et al., *Seismic Design Methods for Steel Building Structures*, Geotechnical, Geological and Earthquake Engineering 51,
https://doi.org/10.1007/978-3-030-80687-3_10

accounted for indirectly through the interaction equations for separate strength checking of every member of the structure.

Use of advanced methods of analysis accounting for geometric and material nonlinearities can eliminate the above disadvantages and lead to a design method that can sufficiently capture strength and stability of a structure and its members at the limit state without separate member strength checking. One can mention here the works of Ziemian et al. (1992a, b), Kim and Chen (1996a, b), Chen (1998) and Kim et al. (2001) as well as the books of Chen and Kim (1997) and Surovek (2012) on the static design of plane and space steel frames using advanced methods of analysis.

Vasilopoulos and Beskos (2006, 2009) have been able to extend this idea of using advanced methods of analysis in the design of plane and space steel frames from the static to the dynamic (seismic) case. Modern seismic codes such as EC8 (2004), in addition to the standard elastic spectral analysis in conjunction with the behavior or strength reduction factor, permit the use of nonlinear (static or dynamic) analysis for determining design forces and deformations. However, these analyses, even though take, in general, into account material and geometrical nonlinearities (including P-Δ effects) they are not refined enough to take into account in a practical and simple way some additional aspects like member and structure imperfections, residual stresses, P-δ effects, out of plane buckling or lateral torsional buckling. This implies that some or most of the member strength interaction equations of EC3 (2009) should be checked for possible satisfaction and this defeats the goal of avoiding member checking separately. Of course, one may find or develop special computer programs than can analyze a framed structure by taking into account all the aforementioned aspects but such programs are impractical for design purposes and usually not easily available.

Phenomena such as P-δ and P-Δ effects due to large deflections at the member and structure level, respectively and the associated stability problem, member and structure imperfections, residual stresses in member sections, out of plane buckling or lateral torsional buckling have been studied in the framework of creating practical advanced static analysis for design purposes. This information has been extended in the works of Vasilopoulos and Beskos (2006, 2009) from the static to the dynamic (seismic) case and implemented into the two- and three- dimensional versions of the DRAIN-2DX & 3DX (Prakash et al. 1993, 1994) software.

Use of the modified DRAIN-2D & 3D (Prakash et al. 1993, 1994) programs for nonlinear time history (NLTH) analyses of steel plane and space moment resisting frames (MRFs) enables the designer to determine for three performance levels the structural response to a number of seismic accelerograms compatible to the elastic response spectra of EC8 (2004) for every performance level. This response consists of interstorey drift ratios (*IDR*), member plastic rotations, member damage and plastic hinge patterns. The maximum values of the first three response quantities are checked against their limit values for every performance level. The plastic hinge pattern is used to check if a global collapse mechanism has been formed in order to satisfy capacity design requirements.

In the following sections, after a brief evaluation of the EC3 (2009) and EC8 (2004) provisions in order to clearly see their shortcomings and two sections discussing the theoretical aspects of the method for plane and space steel MRFs, respectively, the basic steps of the method are listed together with a commentary and the chapter closes with detailed numerical examples for both plane and space frames and a list of conclusions.

10.2 Brief Evaluation of EC3 and EC8 Provisions

In this section the most important aspects of the EC3 (2009) and EC8 (2004) provisions are briefly discussed with the goal of detecting their most important shortcomings for subsequent elimination or at least significant reduction by the employment of the proposed method of this chapter.

10.2.1 EC3 Design Procedure

The static design of steel MRFs according to EC3 (2009) involves, for each load combination, the execution of the basic steps of:

1. Consideration of imperfection effects at the local and global level.
2. Global elastic analysis of the first or second order.
3. Determination of internal forces and moments from the above global analysis.
4. Strength checking of the members separately with the aid of the interaction equations.

Thus, use of EC3 (2009) is characterized by some important shortcomings. Firstly, this code does not consider directly the interaction of strength between the member and the structure and employs the effective length concept in order to approximately consider the influence of the structure on the individual members with respect to stability. Secondly, the code uses elastic analysis to determine member forces and moments and accounts for inelasticity approximately through the design interaction equations (for combined deformations) used separately for every member and without verifying the compatibility between the isolated member and the whole structure.

10.2.2 EC8 Design Procedure

EC8 (2004) seismic design code requires the satisfaction of two performance requirements, the no-collapse requirement and the damage limitation requirement, which are associated with the ULS and the DLS, as described in Chap. 3. These two limit states correspond to the LS and IO levels of performance-based design.

The main type of analysis used by EC8 (2004) is dynamic elastic spectral analysis in conjunction with the design inelastic spectrum corresponding to the ULS or LS level. This spectrum is obtained from the corresponding elastic by dividing its ordinates by the behavior factor q, in order to approximately account for inelastic effects in an indirect manner. Use of the above dynamic spectral analysis, leads to the determination of the seismic design base shear and the subsequent dimension of the structure through strength checking based in EC3 (2009). Then displacements at the DLS or IO level are checked against their limit values and sectional dimensions are modified iteratively. The collapse prevention requirement is supposed to be always satisfied through capacity design guidelines and careful detailing procedures.

It is apparent that the above only two performance levels, are not clearly defined with respect to specific damage indices, displacements or plastic rotations and a picture of the plastic hinge pattern. Moreover, inelasticity is considered indirectly and approximately through the behavior factor q provided for various types of steel structures without considering their specific dynamic and inelastic characteristics. On the other hand, axial forces in exterior columns can also exceed the values predicted by modal synthesis and inelastic spectra and thus lead to a reduction of plastic moment capacity and an increase of required ductility.

10.3 Advanced Analysis Fundamentals

A seismic design method for steel building structures based on advanced methods of dynamic analysis involving material and geometric nonlinearities is developed in this chapter. The method is used in conjunction with the EC3 (1992) and EC8 (2004) design codes. The basic characteristics of the method are the following:

10.3.1 Selection and Application of Earthquake Loading

Because the method employs NLTH analysis, earthquake loading is in the form of accelerograms. Selection of earthquake ground motions is done by following the EC8 (2004) guidelines. Thus, three recorded earthquake ground motions are selected and scaled so as their elastic spectrum to be compatible to the EC8 (2004) elastic spectrum, which corresponds to moderate intensity earthquakes with return period T = 475 years (or probability of exceedance 10% in 50 years) and is appropriate for the LS performance level. Because only three ground motions are used, the maximum of the three response values is considered for design purposes (EC8 2004).

A performance-based seismic design method with three performance levels (IO, LS and CP) is employed here. The three selected earthquake motions are also made compatible to the IO level (5% in 50 years or T = 949 years) and to the CP level (5% in 50 years or T = 949 years). Design spectra for the IO and CP performance levels are obtained by multiplying the ordinates of the LS design spectrum by 0.3 and 1.5, respectively.

For the case of the space MRFS, the earthquake ground motions have two horizontal components, which are made spectrum compatible. Then they are applied to the frames, either on the basis of the $x + 0.3z$ and $z + 0.3x$ design combinations with x and z being the two horizontal and y the vertical directions of the frame according to DRAIN-3DX (Prakash et al. 1994), or directly as a pair in an alternate way. Since the frames considered here are regular, only accidental eccentricities of $\pm 0.05L_i$ are taken into account in each horizontal direction, where L_i is the floor dimension perpendicular to the direction of the seismic motion.

10.3.2 Inelastic Modeling of Members

The refined plastic hinge model of Chen and Kim (1997) used for simulating the inelastic behavior of a beam-column element is an enhanced plastic-hinge model capable of simulating gradual yielding effects due to a number of mechanical phenomena, such as strain hardening, residual stresses and interactions between bending moment and shear or axial force including flexural and/or lateral-torsional buckling.

Thus, for the two-dimensional case, the force-displacement relation for a beam-column finite element with refined elastic-plastic hinges at its two ends A and B, connects the load vector rate $\{\dot{F}\}$ of nodal bending moments and axial force with the deformation vector rate $\{\dot{d}\}$ of nodal rotations and displacements (Kim and Lee 2002)

$$\{\dot{F}\} = ([K_f] + [K_g])\{\dot{d}\} \tag{10.1}$$

where $[K_f]$ is the flexural stiffness matrix and $[K_g]$ is the geometric stiffness matrix. Matrix $[K_g]$ is a function of the axial load N and element length L, while $[K_f]$ is a function of the cross-sectional moment of inertia I, length L, tangent modulus E_t and the two scalar parameters η_A and η_B at the two ends of the element simulating the gradual inelastic bending stiffness reduction and having the form (Chen and Kim 1997)

$$\begin{aligned} \eta &= 1 \text{ for } \alpha \leq 0.5 \\ \eta &= 4\alpha(1-\alpha) \text{ for } \alpha > 0.5 \end{aligned} \tag{10.2}$$

with α being the force–state parameter. This parameter is defined on the basis of the limit state (plastic strength) curve for the combined axial force N and bending moment M at every element end (potential hinge position) and has the form (EC3 1992)

$$a = \frac{M_{sd}}{M_{pl.Rd}} + \left(\frac{N_{sd}}{N_{pl.Rd}}\right)^2 \leq 1 \tag{10.3}$$

where M_{sd} and N_{sd} denote the design moment and axial force, respectively, while $M_{pl.\ Rd}$ and $N_{pl.\ Rd}$ the plastic moment and axial force resistances, respectively. These resistances have the form (EC3 1992)

$$M_{pl.Rd} = f_y W_{pl}/\gamma_{M0}, N_{pl.Rd} = f_y A/\gamma_{M0} \tag{10.4}$$

where f_y and W_{pl} denote the yielding stress and plastic modulus, respectively, while $\gamma_{M0} = 1.1$ is the factor of safety. For $a = 1$ one has the plastic strength curve of Eq. (10.3), while for $a = 0.5$ the initial yield curve of the same shape as that of the plastic strength curve. Thus, for $0 \leq a \leq 0.5$ the element remains elastic ($\eta = 1$ from Eq. $(10.2)_1$), while for $0.5 < a \leq 1.0$ the stiffness is continuously reduced accordingly to Eq. $(10.2)_2$ ($0 \leq \eta \leq 1.0$). At this point, it should be stated that here and in the following use is made of the EC3 (1992) provisions and not of the current EC3 (2009) ones, simply because theory and examples come from the works of Vasilopoulos and Beskos (2006, 2009), which were performed before the release of EC3 (2009).

The tangent modulus E_t is used to account for gradual yielding effects due to residual stresses along the length of members under axial loads between two plastic hinges and has the form (Chen and Kim 1997)

$$\begin{aligned} E_t &= E \text{ for } N_{sd} \leq 0.5N_y \\ E_t &= 4E\frac{N_{sd}}{N_y}\left(1 - \frac{N}{N_y}\right) \text{ for } N_{sd} > 0.5N_y \end{aligned} \tag{10.5}$$

where $N_y = Af_y$ is the axial load at yield and E is the elastic modulus.

The influence of the in-plane and out of plane flexural buckling as well as of lateral torsional buckling can be accurately taken into account with the employment of a space finite beam element with 14 degrees of freedom in order that all possible nodal deformations to be represented (three translations, three rotations and one warping at each end) (Trahair 1993; Wongkaew and Chen 2002). This is in agreement with the advanced analysis philosophy but very complicated theoretically and computationally. Alternatively, instability effects caused by flexural and lateral torsional buckling can be approximately accounted for by code-based interaction equations (Kim and Lee 2002).

The flexural buckling effect can be easily taken into account by modifying the definition of the force-state parameter α given by Eq. (10.2). The new parameter a_{fb}, which includes the combined effect of M and N as well as the flexural buckling effect and replaces α, is found with the aid of EC3 (1992) to have the form

$$a_{fb} = \left(\frac{N_{sd}}{\chi_{min} N_{pl.Rd}}\right)^2 + \frac{M_{ysd}}{M_{pl.y.Rd}} \leq 1.0 \qquad (10.6)$$

In the above, y and z denote the strong and weak cross-sectional axes, respectively, $\chi_{min} = \; min\;(\chi_y, \chi_z)$ with χ_y and χ_z being the buckling reduction factors for in-plane and out-of-plane buckling effects, respectively and M_{ysd} and N_{sd} denote the design moment and axial force, respectively. The values of χ_y and χ_z are explicitly given in EC3 (1992) in terms of the corresponding normalized slenderness $\overline{\lambda}_y$ and $\overline{\lambda}_z$ of the member with the imperfection factor taken equal to 1.0 since imperfections are accounted for through the E_t modulus reduction (Sect. 10.3.3) and with effective length factors $k_y = k_z = 1.0$, provided the two ends of the member are laterally supported (EC3 1992). When the minimum normalized slenderness $\overline{\lambda}_{min} \leq 0.2$, no flexural buckling checking is necessary (EC3 1992) and the force-state parameter a is computed from Eq. (10.3), otherwise a is computed from Eq. (10.6), which includes flexural buckling effects (Chen and Kim 1997; Kim and Lee 2002).

The lateral torsional buckling effect can be accounted for by computing the new force-state parameter a_{LT} not by Eq. (10.3) but, in accordance with Kim and Lee (2002), by

$$a_{LT} = \left(\frac{N_{sd}}{N_{pl.Rd}}\right)^2 + \frac{M_{ysd}}{\chi_{LT} M_{pl.y.Rd}} \leq 1.0 \qquad (10.7)$$

In the above, χ_{LT} is the lateral-torsional reduction factor given in EC3 (1992) in terms of the normalized member slenderness $\overline{\lambda}_{LT}$, various member cross-sectional properties, the member end bending moments and the effective length factor k_w for warping (due to lateral torsional deformation). It should be noted that $k_w = 1$, unless special provision for warping fixity is made and that the imperfection factor here is assumed to be 1.0 because all imperfections are accounted for through the E_t reduction (Sect. 10.3.3). The original refined plastic hinge model implies that lateral torsional buckling is prevented by adequate lateral bracing. In case of insufficient lateral bracing, lateral torsional buckling has to be taken into account as described above.

All the above modeling developments dealing with the refined plastic-hinge model as applied to plane steel MRFs have been implemented into the computer program DRAIN-2DX (Prakash et al. 1993) in connection with the beam-column element E02 associated with the simple plastic-hinge model. According to EC3 (1992), when use is made of plastic analysis, the structural members should possess sufficient ductility. This implies that the cross sections of those members should be able to develop their full plastic moment capacity and experience large hinge rotations before the appearance of local buckling. These requirements are satisfied only by class 1 sections. Thus, only member sections of class 1 are used in connection with the proposed method.

For the case of steel space MRFs, the DRAIN-3DX (Prakash et al. 1994) computer program is employed and all the structural members are modeled by the fiber beam-column E15 element. This element combines the concept of concentrated plasticity with that of the fiber model at every plastic hinge. Residual stresses are taken into account by using the expressions of Eq. (10.5) for E_t as in the case of plane frames and this is implemented in the DRAIN-3DX (Prakash et al. 1994) computer program.

The force-displacement relation for a space beam-column finite element is again of the form of Eq. (10.1) but in a three-dimensional context (Liew et al. 2000; Kim et al. 2001), with the matrix $[K_f]$ being a function of the moments of inertia I_y and I_z, member length L, tangent modulus E_t, shear modulus G, torsional modulus J and the scalar parameters η_A and η_B at the two ends of the element, depending on the force-state parameter α according to Eq. (10.2). Parameter a allows for the gradient inelastic stiffness reduction due to the bending moment-axial force interaction equation, or the flexural buckling and lateral-torsional buckling interaction equations, which are the three-dimensional counterparts of Eqs. (10.3), (10.6) and (10.7) and can be found in EC3 (1992). One could take into account lateral-torsional buckling effects by adding one more degree of freedom (the warping one) per node in the space beam-column finite element considered here by following Kwak et al. (2001) and Wongkaew and Chen (2002). Even though this would be consistent with the advanced analysis philosophy, its implementation in DRAIN-3DX (Prakash et al. 1994) was not done because of its complexity. One should note that when $\overline{\lambda}_{LT} \leq 0.4$, the lateral-torsional buckling effect can also be neglected when the member is laterally adequately supported. All the above features of the method have been implemented in the DRAIN-3DX (Prakash et al. 1994) software.

10.3.3 Geometric Nonlinearity Effects

Geometric nonlinearities in frame analysis come from P-δ and P-Δ effects as well as imperfection phenomena. The P-δ effect is defined at a member level and represents the influence of the axial force on bending. This effect can be accounted for exactly through the stability functions or approximately through the geometric stiffness coefficients, both in a finite element framework. DRAIN-2DX & 3DX (Prakash et al. 1993, 1994) employs geometric stiffness coefficients to account for the P-δ effect. As it has been shown by Beskos (1977), two finite elements per member can lead to results of satisfactory accuracy. The replacement of the distributed mass in a member by concentrated masses at its two ends in dynamic analysis, is also an approximation. This approximation is acceptable if every member is discretized in at least two finite elements (Beskos 1976). Thus, a discretization of every member of the frame into two finite elements can provide buckling and vibration results of satisfactory accuracy and this is what is done here in the context of advanced analysis.

The P-Δ effect is defined at the structural level and comes from secondary moments due to the lateral displacement of gravity loads at every storey of the frame. This effect can be accounted for in the framework of nonlinear finite element analysis including large displacements (Bathe 1996). The P-Δ effect can also be taken into account in an approximate manner (El Hafez and Powell 1973) by the addition of a geometric stiffness based on the column axial force due to gravity load to the elastic stiffness of the columns (El Hafez and Powell 1973). This is the approach followed in DRAIN-2DX & 3DX (Prakash et al. 1993, 1994).

Geometric imperfections in the structure and its members reduce their strength. EC3 (2009), accounts for the effects of imperfections approximately by the imperfection factor built-in the reduction factors χ of the buckling strength formulae and by equivalent lateral loads at every storey of the building frame. In this work, imperfections are accounted for by simply multiplying the tangent modulus E_t by the reduction factor of 0.85 according to the suggestion of Chen and Kim (1997).

10.3.4 Seismic Damage Index

Damage is defined as the degradation of a member or a structure that reduces its load bearing capacity (Powell and Allahabadi 1988; Lemaitre 1996). Seismic damage is quantified by the damage index (of value 0 at no-damage and 1 at failure), which offers a comprehensive design measure for the various performance levels.

The damage index is associated with the plastic rotation or the plastic moments at the plastic hinges and can be defined at member, storey, or structural level. Thus, one has:

1. Member damage index I_{dm}, defined as function of the required plastic rotation (θ_r) or moment (M_{pr}) and the available plastic rotation (θ_a) or moment (M_{pa}) as (Lemaitre 1996; Gioncu and Mazzolani 2002) and having the form

$$I_{dm} = \langle M_{pr} - M_{pa} \rangle / (M_{ult} - M_{pa}) \tag{10.8}$$

where M_{ult} is the ultimate moment and $\langle\rangle$ denote the McCauley brackets.

2. Storey damage index I_{ds}, defined in terms of member damage indices (Powell and Allahabadi 1988) and having the form

$$I_{ds} = \sum_{1}^{n_b} I_{dm}^2 / \sum_{1}^{n_b} I_{dm} \tag{10.9}$$

where n_b is the number of frame bays and I_{dm} is defined by Eq. (10.8)

3. Global damage index I_{dg}, defined in terms of the m member damage indices of the whole structure (Powell and Allahabadi 1988) and having the form

Table 10.1 Global damage index values for various damage levels

No damage	$I_{dg} < 0.05$	Collapse prevention	$I_{dg} < 0.80$
Minor damage	$I_{dg} < 0.15$	Near collapse	$I_{dg} > 0.80$
Repairable damage	$I_{dg} < 0.50$	Structural collapse	$I_{dg} = 1.00$

$$I_{dg} = \sum_{1}^{m} I_{dm}^{2} / \sum_{1}^{m} I_{dm} \tag{10.10}$$

The above expressions have been implemented in the program DRAIN-2DX (Prakash et al. 1993) for damage determination at the member, storey and global structural levels. The correlation between the global damage index and different damage levels due to Gioncu and Mazzolani (2002) is shown in Table 10.1.

10.4 Basic Steps of the Design Method

10.4.1 Brief Description of the Basic Steps

This section briefly describes the basic steps of the seismic design method based on advanced analysis for plane or space steel MRFs, which are as follows:

Step 1: Types of loads and design load combinations.

These include dead (G) and live (Q) vertical loads, lateral seismic loads (E) applied as accelerograms and design load combinations G+0.3Q+E as described in EC8 (2004).

Step 2: Design performance levels and seismic load selection.

Three performance levels are considered here: The IO (Immediate Occupancy) under the FOE (Frequent Occurred Earthquake), the LS (Life Safety) under the DBE (Design Basis Earthquake) and the CP (Collapse Prevention) under the MCE (Maximum Considered Earthquake). Three or six recorded earthquake motions of moderate to strong intensity are selected and made compatible to the elastic design spectra of EC8 (2004) for the three aforementioned performance levels. The design is done on the basis of three ground motions and further verified by using three more ground motions.

Step 3: Preliminary member sizing.

This depends on the experience of the designer or some simplified analysis. For example, beam sections can be selected on the assumption that beams are simply supported and under gravity loads only and column sections on the basis of the overall drift requirements. The capacity design rule of "strong columns-weak beams" should be also observed during this design step.

Step 4: Load application and seismic analysis.

Nonlinear time history (NLTH) analyses are conducted using the selected spectrum compatible seismic records of step 2. For space frames, every seismic

Table 10.2 Performance levels and corresponding limit response values

Performance levels	IDR	θ_p	μ_θ	Damage	Plastic hinge formation
IO	1.5% (SEAOC 1999) 0.6% (Grecea et al. 2002) 0.5% (Vasilopoulos and Beskos 2006, 2009) 0.7% transient & negligible permanent (FEMA-273 1997)	$\leq\theta_y$ (FEMA-273 1997)	2.0 (FEMA-273 1997)	≤5% (Vasilopoulos and Beskos 2006, 2009) 0.1–10% (ATC13 1985)	Nowhere
LS	3.2% (SEAOC 1999) 1.5% (Vasilopoulos and Beskos 2006, 2009) 2.5% transient & 1% permanent (FEMA-273 1997)	$\leq 6.0\theta_y$ (FEMA-273 1997)	7.0 (FEMA-273 1997)	≤20% (Vasilopoulos and Beskos 2006, 2009) 10–30% (ATC13 1985)	Only in beams
CP	3.8% (SEAOC 1999) 3.0% (Vasilopoulos and Beskos 2006, 2009) 5.0% transient & 5.0% permanent (FEMA-273 1997)	$\leq 8.0\theta_y$ (FEMA-273 1997)	9.0 (FEMA-273 1997)	≤50–60% (Vasilopoulos and Beskos 2006, 2009) 30–60% (ATC13 1985)	In beams and columns without collapse

$\theta_y = W_{pl}f_yL_b/6EI_b$, $\theta_y = W_{pl}f_yL_c(1 - N/N_y)/6EI_c$ for beams (b) and columns (c), respectively

pair should be applied twice. One with its strong component parallel to the strong building axis and a second time with this component perpendicular to that axis. Columns and beams are modeled by two finite elements.

Step 5: Checking for satisfaction of the LS level limit values.

The designed frame in Step 3 is analyzed for three ground motions spectrum compatible to the LS level and its response values are checked against their limit values for that level in Table 10.2. If the response values are much lower than the limit values of Table 10.2, indicating an overdesigned structure, the few possible members with plastic hinges remain the same, while those without plastic hinges are replaced by lighter ones. When some of the response values exceed the limit values of Table 10.2, indicating an under-designed structure, the few possible members without plastic hinges remain the same, while those with plastic hinges are replaced by stronger members. Many such analyses may have to be done iteratively in order to reach, by proper member adjustments, response values which are close to the limits of Table 10.2 without exceeding them. Using three more seismic motions spectrum compatible with the LS level, NLTH analyses are performed in order to further check if the response values of the designed frame in step 5 satisfy again the limit ones in Table 10.2.

Step 6: Checking for satisfaction of the IO level limit values.

The six seismic records of step 2, spectrum compatible with the IO level, are used for NLTH analyses to check if the limit response values of Table 10.2 for this level are satisfied or not by the LS design.

Step 7: Checking for satisfaction of the CP level limit values.

The six seismic records of step 2, spectrum compatible with the CP level, are used for NLTH analyses to check if the limit response values of Table 10.2 for this level are satisfied or not by the LS and IO designs.

10.4.2 Some Comments on the Procedure of the Method

In this section, some comments concerning the fundamental concepts and design equations of Sect. 10.3 for plane and space frames as well as the stepwise procedure of the method presented in the previous section are outlined. These comments are as follows:

1. The proposed seismic design method is associated with the three performance levels of IO, LS and CP. The allowable limits of three performance quantities (*IDR*, θ_p, damage index) as well as the plastic hinge formation for the abovementioned performance levels as obtained from various sources are presented in Table 10.2. In the examples presented in Sect. 10.5 the limit values of Vasilopoulos and Beskos (2006, 2009) are used, which somehow appear to be between maximum and minimum values proposed by other investigators.
2. In the numerical examples of Sect. 10.5 the design starts with an effort to satisfy the requirements of performance level LS, which is associated with the design basis earthquake (DBE). Three accelerograms compatible with the elastic design spectrum of EC8 (2004) are used. This is the minimum number of accelerograms required by EC8 (2004). Furthermore, the maximum values of the deformation/damage response quantities to those three accelerograms are checked for satisfaction of the limit values of Table 10.2 and not their mean values as it is usually done in cases when more than or equal to seven accelerograms are employed. Using only three accelerograms, one reduces the computational load during the iterative procedure of member size adjustment. Finally, analyses for three additional accelerograms are performed for verification, which usually involves no iterations.
3. The proposed design method, which employs nonlinear dynamic analysis, can also be used in conjunction with nonlinear static (pushover) analysis. In this case the whole design procedure is greatly simplified at the expense of losing a degree of accuracy. Vasilopoulos and Kamaris (2020) have successfully applied this approach to the seismic design of steel space MRFs utilizing an improved fixed multimodal lateral loading distribution due to Pavlidis et al. (2003).

10.5 Application Examples

In this section two numerical examples are presented in detail for explaining the application of the method and showing its merits. These examples deal with one plane steel MRF with four stories and one bay and one space steel MRF with three stories and one by one bay taken from Vasilopoulos and Beskos (2006, 2009), respectively. More examples involving one plane steel MRF with seven stories and two bays and one space steel MRF with seven stories and two by three bays can be found in Vasilopoulos and Beskos (2006, 2009), respectively. Two more examples involving space steel MRFs with irregularities (natural eccentricities and setbacks) can be found in Vasilopoulos et al. (2008).

10.5.1 Seismic Design of a Steel Plane MRF

This section deals with the seismic design of a plane steel frame of four stories and one bay, as shown in Fig. 10.1 together with its finite element discretization. The storey height and bay length of the frame are 3.0 and 4.0 m, respectively. The

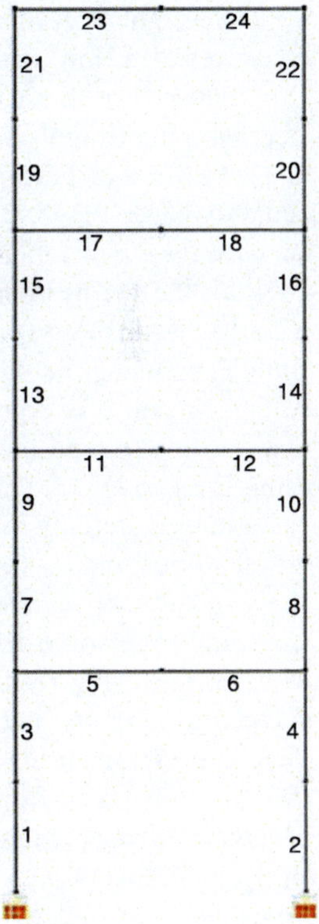

Fig. 10.1 Structural geometry and finite element numbering of the four stories and one bay steel plane MRF considered in Sect. 10.5.1 (after Vasilopoulos 2005, reprinted with permission from UPCE)

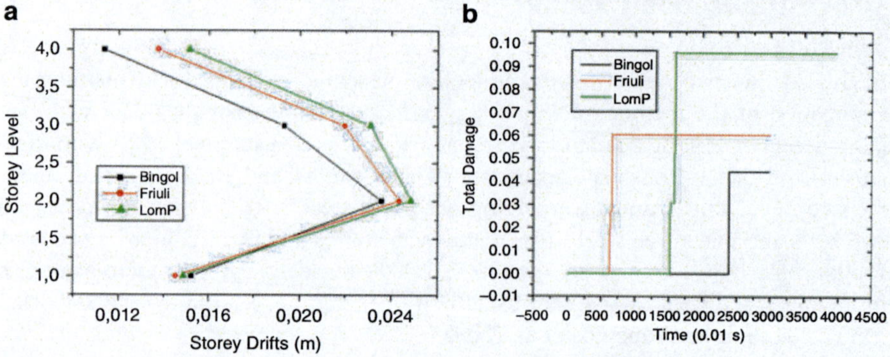

Fig. 10.2 Maximum storey drifts (**a**) and maximum total damage indices (**b**) for frame of Fig. 10.1 at LS level and first section selection (after Vasilopoulos 2005, reprinted with permission from UPCE)

connections and the supports in the frames are considered rigid and fixed, respectively. The grade of steel is assumed to be S275, while the modulus of elasticity *E* and shear modulus *G* of steel are taken as 200 GPa and 81 GPa, respectively. The steel strain hardening is taken as 3%.

The vertical load combination G+0.3Q consists of the deal load G coming from the self-weight of beams and columns equal to 78.50 kN/m^3, the self-weight of slabs equal to 5.50 kN/m^2 and the cladding weight equal to 2.00 kN/m^2 and the live load Q = 2.00 kN/m^2. Six recorded accelerograms taken from PEER (2009) are considered here: the Bingol of 2003, Friuli of 1976, Loma Prieta of 1989, Imperial Valley of 1979, Kobe of 1995 and Parkfield of 2004 with peak ground acceleration (*PGA*) equal to 0.515, 0.375, 0.644, 0.315, 0.611 and 0.476 g, respectively. The first three are used for the member adjustment phase of the design procedure, while the second three for the verification phase. These seismic records have been made compatible with the elastic design spectra of EC8 (2004) for the three performance levels IO, LS and CP by using the software of Karabalis et al. (1993). For the LS level *PGA* = *0.30* g and soil is of type B. Damping in all dynamic analyses is assumed 5% for the first two modes and thus the Rayleigh damping matrix can be easily obtained by using Eq. (2.18).

Assuming initially member sections HEB280 for columns and IPE360 for beams for all stories, the maximum seismic response values of the frame to the first three above mentioned earthquakes (Bingol, Friuli, Loma Prieta) in their EC8 (2004) spectrum compatible for the LS level forms are determined by NLTH analyses with the following results: The maximum drift value occurs at the second storey, is equal to 2.55 cm and comes from the Loma Prieta seismic motion (Fig. 10.2a). This value is much lower than the limit value of 1.5%x300 = 4.50 cm, with the *IDR* = 1.5% taken from Table 10.2. The maximum total damage indices due to Bingol, Friuli and Loma Prieta earthquakes have been plotted in Fig. 10.2b and read $D_{Bingol} = 4.43\%$, $D_{Friuli} = 6.03\%$ and $D_{LomP} = 9.58\%$, respectively. The maximum storey damage indices are due to the Loma Prieta earthquake and are equal to

Table 10.3 Member damage indices for frame of Fig. 10.1 (LS level) and first section selection: HEB280 (columns)-IPE360 (beams)

Member	Seismic motion	*i*-end damage (%)	*j*-end damage (%)
5	Bingol	0.48	–
6		–	4.82
11		1.22	–
12		–	4.89
5	Friuli	2.67	–
6		–	5.88
11		4.34	–
12		–	8.04
5	Loma Prieta	5.69	–
6		–	3.06
11		9.58	–
12		–	5.82

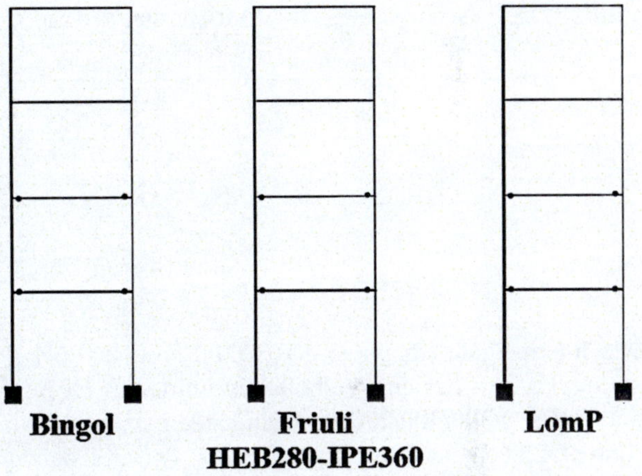

Fig. 10.3 Plastic hinge formation in frame of Fig. 10.1 at LS level and first section selection (after Vasilopoulos 2005, reprinted with permission from UPCE)

$d_{s1} = 4.93\%$, $d_{s2} = 8.16\%$, $d_{s3} = 0.0\%$ and $d_{s4} = 0.0\%$ for the first, second, third and fourth storeys, respectively. Table 10.3 provides member damage indices for the three earthquakes with the maximum one being 9.58% and coming from the Loma Prieta earthquake. One should note that here and throughout this example member numbers in reality denote finite element members, as shown in Fig. 10.1. It is observed that all the above damage values are well below the 20% damage limit of the LS level (Table 10.2) indicating a very low damage. Plastic hinges appear only in the beams of the two lower storeys, as shown in Fig. 10.3. The maximum base shear for the most unfavorable earthquake case (Loma Prieta) is found to be equal to 250.66 kN.

The above results indicate that the originally selected sections can be reduced in size in an iterative manner. The final member selection consists of HEB240 columns and IPE330 beams for the first storey, HEB220 columns and IPE300 beams for the

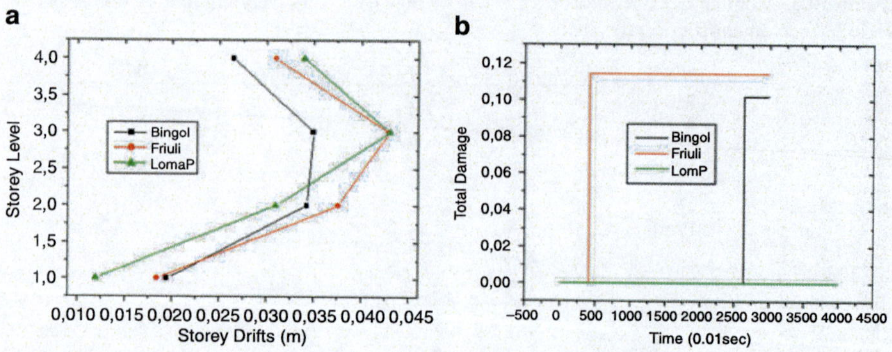

Fig. 10.4 Maximum storey drifts (**a**) and maximum total damage indices (**b**) for frame of Fig. 10.1 at LS level and final section selection (after Vasilopoulos 2005, reprinted with permission from UPCE)

Table 10.4 Member damage indices for frame of Fig. 10.1 (LS level) and final section selection: HEB 240/220/220/200 (columns)-IPE 330/300/300/270 (beams)

Member	Seismic motion	*i*-end damage (%)	*j*-end damage (%)
5	Bingol	3.81	–
6		–	11.97
12		–	0.44
5	Friuli	11.44	–
6		–	3.16
12		–	1.75

two intermediate storeys and HEB200 columns and IPE270 beams for the fourth storey. For this design, symbolically written as HEB 240/220/220/200 for columns and IPE 330/300/300/270 for beams, the maximum storey drift values approach the limit of 4.50 cm, as shown in Fig. 10.4a. The total damage is shown in Fig. 10.4b and for the most unfavorable seismic cases of Bingol and Friuli reaches the values of $D = 10.21\%$ and $D = 11.44\%$, respectively.

The member damage indices are shown in Table 10.4. Thus, the damage indices in structural and member level do not exceed the value of 20% (Table 10.2), meaning that the damage is small and repairable. For the Loma Prieta earthquake, no damage is induced to the members of the frame. Plastic hinges are developed only in beams, as shown in Fig. 10.5. Finally, Table 10.5 provides maximum values of normalized member plastic rotations (θ_p/θ_y), which are all lower than the limit value of 6.0 for the LS level (Table 10.2). Finally, the maximum value of the base shear is equal to 176.07 kN.

A verification of the finally selected member sections for frame of Fig. 10.1 is done by using three additional earthquakes (Imperial Valley, Kobe and Parkfield) scaled according to the EC8 elastic design spectrum for the LS level. The storey relative drift values shown in Fig. 10.6a approach the limit value of 4.50 cm. The maximum total damage for the frame is equal to 8.05% coming from the Kobe

Fig. 10.5 Plastic hinge formation in frame of Fig. 10.1 at LS level and final section selection (after Vasilopoulos 2005, reprinted with permission from UPCE)

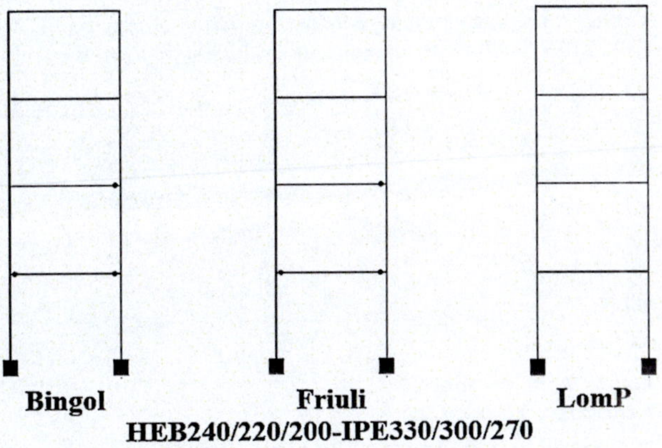

Table 10.5 Maximum member θ_p/θ_y ratios (LS level) for frame of Fig. 10.1 and final section selection: HEB 240/220/220/200 (columns)-IPE 330/300/300/270 (beams)

Member	Seismic motion	*i*-end θ_p/θ_y	*j*-end θ_p/θ_y
5	Bingol	0.95	–
6		–	2.95
12		–	0.20
5	Friuli	2.85	–
6		–	0.80
12		–	0.81

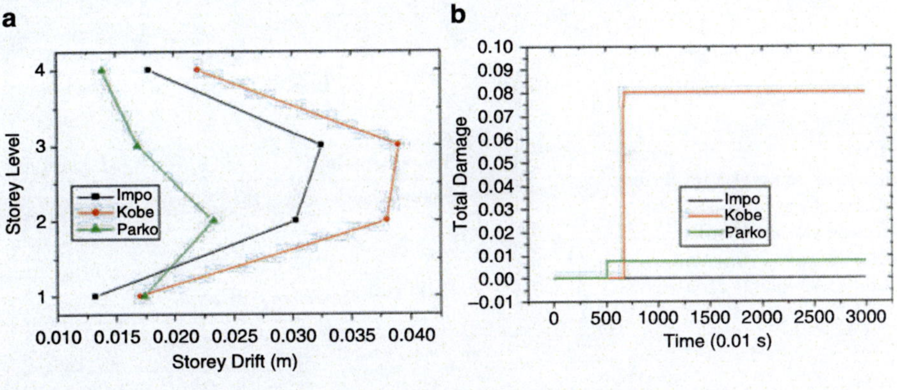

Fig. 10.6 Storey drifts (**a**) and total damage index (**b**) at LS level for final design of frame of Fig. 10.1 (after Vasilopoulos 2005, reprinted with permission from UPCE)

earthquake (Fig. 10.6b). The storey damage indices are $d_{s1} = 8.05\%$, $d_{s2} = 0.0\%$, $d_{s3} = 0.0\%$ and $d_{s4} = 0.0\%$. The member damage indices are provided in Table 10.6 and their values in all seismic cases do not exceed the allowable limit of 20% (small and repairable damage). No damage takes place in the members of the frame for the Imperial Valley earthquake.

Table 10.6 Member damage indices for frame of Fig. 10.1 (LS level) and final section selection: HEB 240/220/220/200 (columns)-IPE 330/300/300/270 (beams)

Member	Seismic motion	*i*-end damage (%)	*j*-end damage (%)
5	Kobe	0.39	–
6		–	8.41
11		0.75	–
5	Parkfield	1.12	–
6		–	0.76
11		–	–

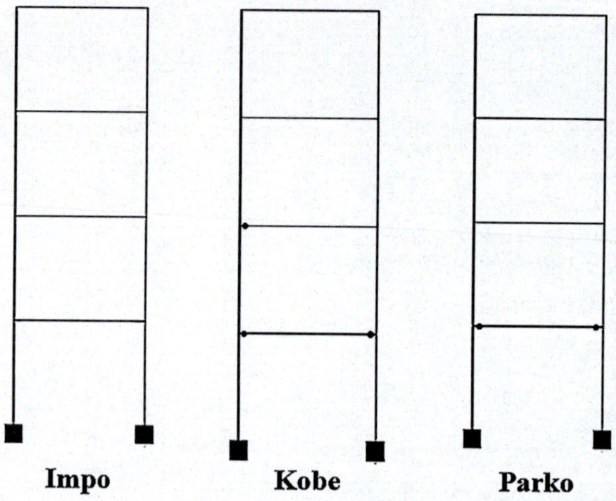

Fig. 10.7 Plastic hinge formation at LS level and final design of frame of Fig. 10.1 (after Vasilopoulos 2005, reprinted with permission from UPCE)

Table 10.7 Maximum member θ_p/θ_y ratios (LS level) for frame of Fig. 10.1 and final section selection: HEB 240/220/220/200 (columns)-IPE 330/300/300/270 (beams)

Member	Seismic motion	*i*-end θ_p/θ_y	*j*-end θ_p/θ_y
5	Kobe	0.20	–
6		–	2.82
11		0.51	–
5	Parkfield	1.25	–
6		–	0.60
11		–	–

From the plastic hinge formation in Fig. 10.7, one can observe that the capacity design rule is satisfied. Finally, Table 10.7 provides the maximum values of the ratio θ_p/θ_y, which are all lower than 6.0 for the LS level (Table 10.2). Regarding the IO level, it is observed that for all six earthquakes, the relative storey drifts do not exceed the limit value (Table 10.2) of $0.5 \cdot h = 0.5 \cdot 300 = 1.50$ cm.

A final checking is made by performing analyses using earthquakes compatible with the CP level in order to determine maximum response values and compare them against the limits of Table 10.2. Thus, for the frame with section members HEB 240/220/220/200 for columns and IPE 330/300/300/270 for beams, the storey relative

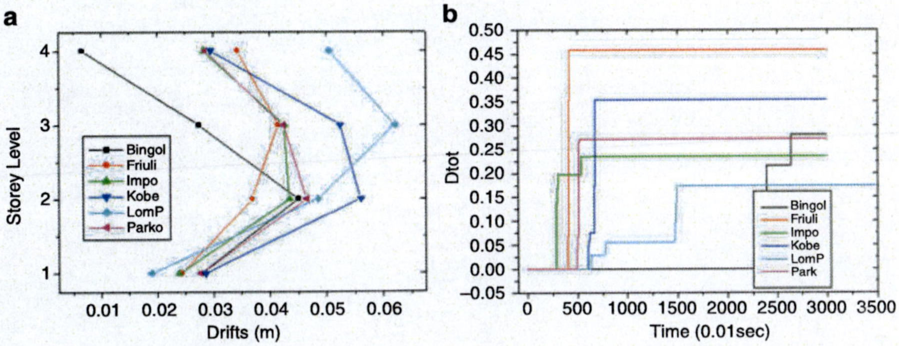

Fig. 10.8 Storey drifts (**a**) and total damage index (**b**) at CP level for final section selection (after Vasilopoulos 2005, reprinted with permission from UPCE)

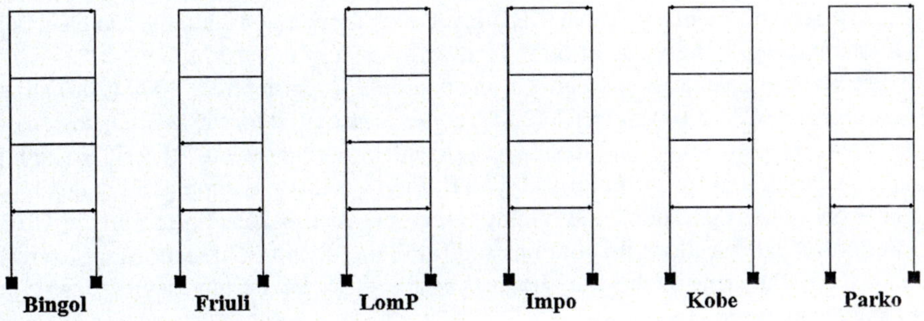

Fig. 10.9 Plastic hinge formation coming from seismic excitations at CP level for final section selection (after Vasilopoulos 2005, reprinted with permission from UPCE)

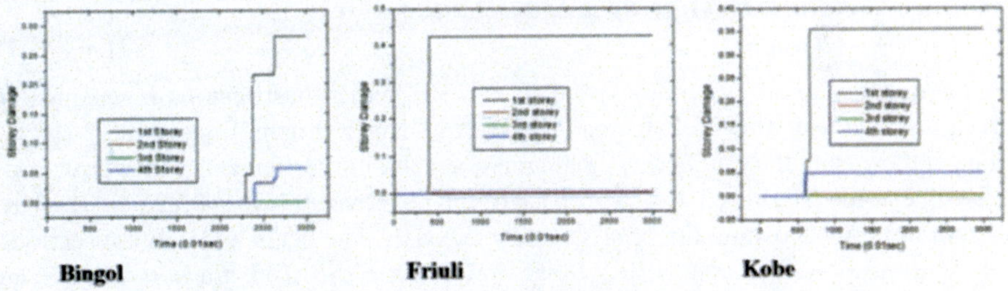

Fig. 10.10 Storey damage indices at CP level for final section selection (only the three worst cases are presented) (after Vasilopoulos 2005, reprinted with permission from UPCE)

drifts (Fig. 10.8a) are increased in relation to those of the LS level but they are still below the new drift limit (Table 10.2) of 3.0 % $\cdot h = 3.0$ % $\cdot 300 = 9.0$ cm for the present CP level. Plastic hinges appear only in beams in agreement with the capacity design rule (Fig. 10.9). The damage index values at total (Fig. 10.8b), storey

Table 10.8 Maximum member damage indices at CP level for final section selection (only the three worst cases are presented)

Member	Seismic motion	*i*-end damage (%)	*j*-end damage (%)
5	Bingol	23.41	–
6		–	31.54
24		5.75	–
5	Friuli	45.91	–
6		–	38.21
11		0.21	–
5	Kobe	31.16	–
6		–	38.90
12		–	0.27
23		4.90	–

(Fig. 10.10) and member (Table 10.8) level also show a considerable increase but without exceeding the value of 60% for the CP level (Table 10.2).

Finally, for reasons of comparison, the computer program SAP 2000 (2007) that incorporates EC8 (2004) and EC3 (2009) codes is also used for the seismic design of the frame of Fig. 10.1. The design is done in accordance with the EC8 (2004) acceleration design spectrum for soil type B, $PGA = 0.30$ g, damping 5% and $q = 4$. Under the above conditions, the optimum member section selection results in HEB 240/220/220/200 sections for columns and IPE 330/300/300/270 sections for beams, i.e., exactly the same as the one obtained by the proposed design method based on advanced analysis.

10.5.2 Seismic Design of a Steel Space MRF

A three-storey one by one-bay space steel MRF, which has been both seismically designed and tested by Kakaliagos (1994), is considered here. Figure 10.11 shows the geometry and the numbering of the members and finite elements of the structure. The first storey height is 3.4 m, while that of the next two storeys is 3.225 m. The bay length in both horizontal directions x and z is 5.0 m. The beam and column sections of the frame are IPE and HEB, respectively, while the steel grade is S275. The modulus of elasticity E and the shear modulus G are equal to 205 GPa and 85.4 GPa, respectively, while the steel strain hardening is equal to 3.0%. The soil type and the damping ratio are assumed to be B and $\xi = 0.05$, respectively. Accidental eccentricities of 5% are considered as described in Sect. 10.3.1. The weight density γ_s of steel beams and columns is 78.50 kN/m^3, while the concrete slab and secondary beams self-weigh G_s is 6.25 and 5.57 kN/m^2 for the first storey and the two upper storeys, respectively. The live loads Q for the first storey and the next two storeys are 2.00 and 1.50 kN/m^2, respectively. The effective seismic mass is associated with the gravity load combination G+0.3Q and is considered to be concentrated at the center

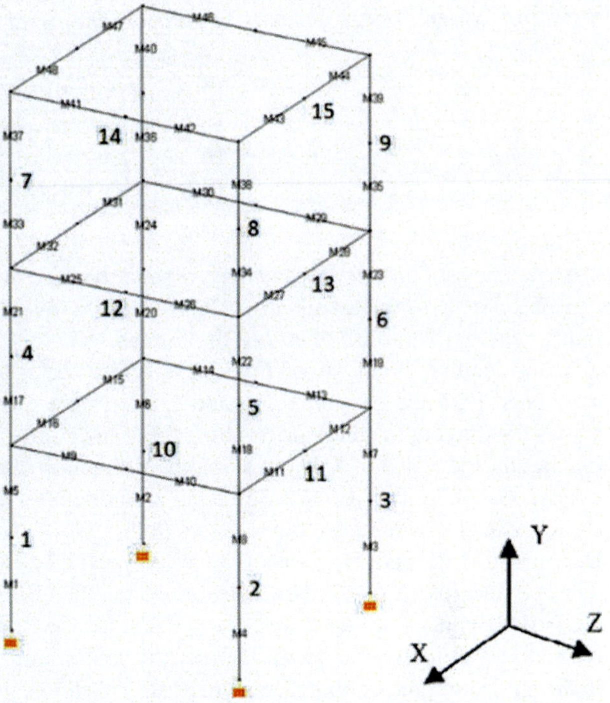

Fig. 10.11 Geometry, member numbering and finite element numbering of the three storey and one by one bay steel space MRF (after Vasilopoulos 2005, reprinted with permission from UPCE)

of each floor. Floor diaphragm action is simulated by allowing two horizontal translational and one torsional components of motion with respect to the *x*, *z* and *y* axes, respectively. The Kakaliagos (1994) frame, designed according to EC8 (1994), has HEB400 columns, IPE400 beams along the *z* direction for all stories, IPE500 beams along the *x* direction for the first storey and IPE450 beams along the *x* direction for the next two stories.

The ground motions of Bingol, Friuli, Loma Prieta, Imperial Valley, Kobe and Parkfield used in the previous section are also used here but with both of their horizontal components due to the three-dimensionality of the structure. These seismic motions are made first compatible with the elastic design spectrum of EC8 (2004) for soil type B and $PGA = 0.30$ g, which is associated with the LS performance level. The same six motions are also made compatible to the elastic design spectra corresponding to the IO and CP performance levels. The damping in all dynamic analyses is equal to 5% for the first two modes and thus the Rayleigh damping matrix can be easily determined. In accordance with EC8 (2004), two seismic loading combinations, $x + 0.3z$ and $z + 0.3x$, are applied. The *x* and *z* horizontal axes of the frame are assumed to denote the strong and weak axes of the building, respectively. Table 10.9 includes four alternative member section selections of the space frame, A, B, C and D, which come from the employed iterative design approach. In that table, three numbers separated by slashes denote section sizes for the three stories (from first to third) and two numbers separated by a

Table 10.9 Member section selections for the three-storey space steel frame

Space frame	Member sections
A	HEB 280/240/220 (columns)-IPE 300/270/240 (beams)
B	HEB 260/240/220 (columns)-IPE 270/240 (beams)
C	HEB 240/220/200 (columns)-IPE 270/240 (beams)
D	HEB 220/200 (columns)-IPE 270/240 (beams)

slash denote section sizes for the first two and the third stories. From Table 10.9, one can observe that the Kakaliagos (1994) frame is overdesigned as its sections are much heavier than those of any of the frames A–D.

Consider first Frame A of Fig. 10.12 under the $x + 0.3z$ loading combination associated with the Bingol, Friuli and Loma Prieta ground motions. Figure 10.13 shows the maximum storey drifts along the x and z directions. It is observed that the maximum drift value of 2.90 cm along the x direction due to the Bingol motion does not exceed the limit of 1.5 % · 322.5 cm = 4.90 cm for the LS level. The maximum values of the total damage indices shown in Fig. 10.14 are equal to $D_{Bingol} = 11.84\%$, $D_{Friuli} = 1.31\%$ and $D_{LomP} = 3.11\%$, respectively, while the maximum storey damage indices (not shown here) due to the Bingol motion are equal to $d_{s1} = 0.00\%$, $d_{s2} = 11.84\%$ and $d_{s3} = 0.0\%$ for the first, second and third storeys, respectively. Finally, Table 10.10 presents maximum values of member damage indices for the three earthquakes, which are much lower than the 20% limit for the LS level. Plastic hinges appear only in one beam of the second storey, as shown in Fig. 10.15. The maximum base shears for the worst earthquake case (Loma Prieta) are found to be equal to $V_x = 222.81$ kN and $V_z = 57.24$ kN. Analogous results are found for the $z + 0.3x$ loading combination.

On the basis of the above results, one realizes that frame A has been designed very conservatively and almost all its members can be replaced by lighter ones. Thus, one can proceed with the section selections corresponding in turn to frame designs B, C and D of Table 10.9. Here only the final section selection corresponding to frame design D of Table 10.9 is presented and discussed.

Thus, frame D under the $x + 0.3z$ loading combination associated with the Bingol, Friuli and Loma Prieta earthquakes is considered. For this frame storey drifts (Fig. 10.16), total damage indices (Fig. 10.17) and member damage indices (Table 10.11) do not exceed the LS limits of Table 10.2, while plastic hinges appear only in beams (Fig. 10.18). Frame D also satisfies the LS limits of Table 10.2 under the $z + 0.3x$ load combination. Indeed, storey drifts (Fig. 10.19), total damage indices (Fig. 10.20) and member damage indices (Table 10.12) do not exceed the LS limits, while plastic hinges develop only in beams in agreement with the capacity design rule (Fig. 10.21). It is also found that the maximum ratio θ_p/θ_y in the members of frame D is 3.79, 3.86 and 3.86 for the Bingol, Friuli and Loma Prieta earthquakes, respectively, i.e., smaller than the limit value of 6.0 for the LS level in Table 10.2.

A verification of the frame design D is also done by using three additional scaled ground motions, namely those of Imperial Valley, Kobe and Parkfield. Figures 10.22 and 10.23 and Table 10.13 exhibit storey drifts, total damage indices and member

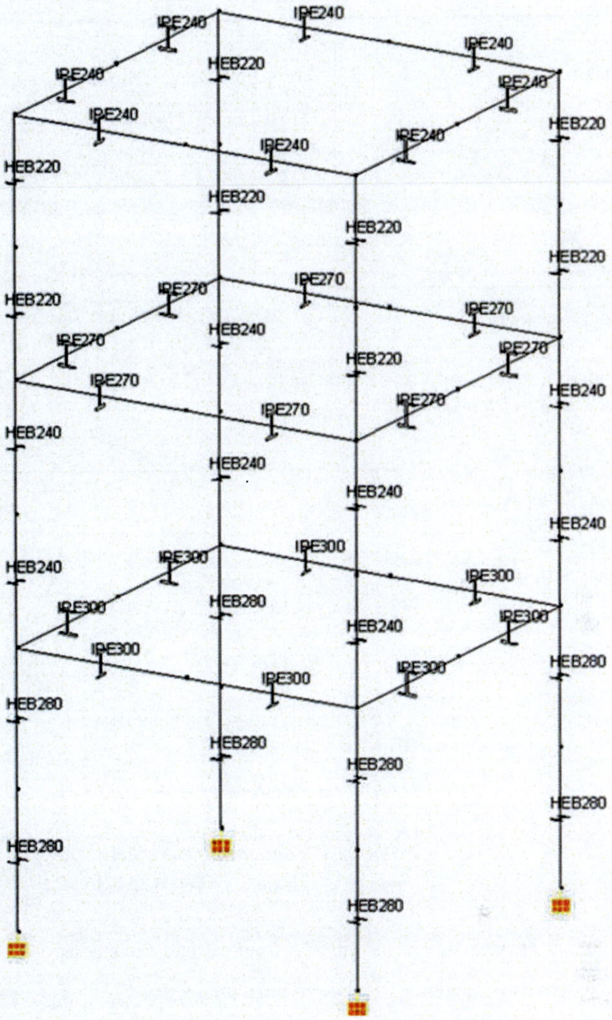

Fig. 10.12 Section selection representation for frame A (after Vasilopoulos 2005, reprinted with permission from UPCE)

damage indices, respectively, under both load combinations $x + 0.3z$ and $z + 0.3x$. It is observed that all the response values are within the limits of the LS level.

Concerning base shear forces V_x and V_z along the x and z directions, respectively, it is also found that for all six seismic motions considered here, these shears have maximum values very close for both load combinations. Indeed, 210 kN $\leq V_x \leq$ 245 kN and 65 kN $\leq V_z \leq$ 73 kN for the $x + 0.3z$ load combination and 100 kN $\leq V_x \leq$ 113 kN and 150 kN $\leq V_z \leq$ 175 kN for the $z + 0.3x$ load combination. It should be stated that frame D satisfies the drift limit of Table 10.2 for the IO performance level (0.5 % $\cdot$ 322.5 = 1.61 cm).

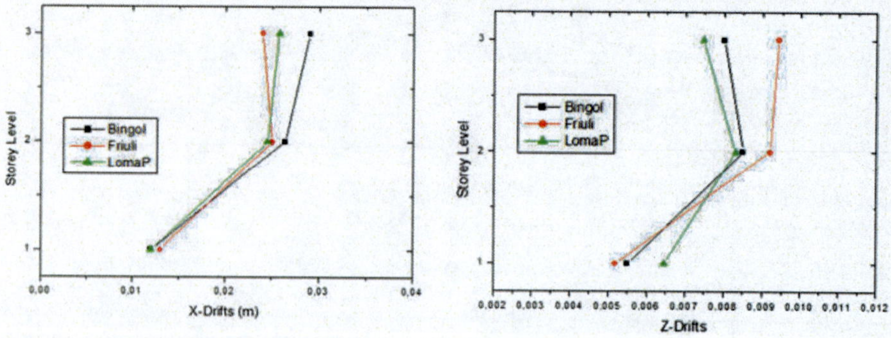

Fig. 10.13 Maximum storey drifts for Frame A (x + $0.3z$/LS level) (after Vasilopoulos 2005, reprinted with permission from UPCE)

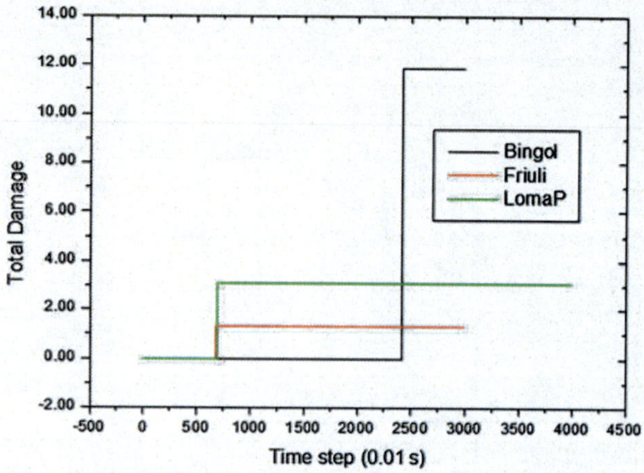

Fig. 10.14 Seismic total damage index for Frame A (x + $0.3z$/LS level) (after Vasilopoulos 2005, reprinted with permission from UPCE)

Table 10.10 Member damage indices for frame A from seismic excitations (x + $0.3z$/LS level)

Member	Seismic motion	*i*-end damage (%)	*j*-end damage (%)
31	Bingol	11.80	–
32			10.60
31	Friuli	1.30	–
32		–	–
31	Loma Prieta	3.10	–
32		1.49	–

A final checking is made by performing NLTH analyses of Frame D using all six seismic motions in a form compatible with the design spectrum of the CP performance level. Thus, the storey drifts in Fig. 10.24 do not exceed the allowable CP limits of $3.0\ \% \cdot h = 3.0\ \% \cdot 322.5 = 9.675cm$ for both seismic combinations $x + 0.3z$ and $z + 0.3x$. The total damage indices for both seismic combinations (Fig. 10.25) approach from below the value of 20%, while the member damage

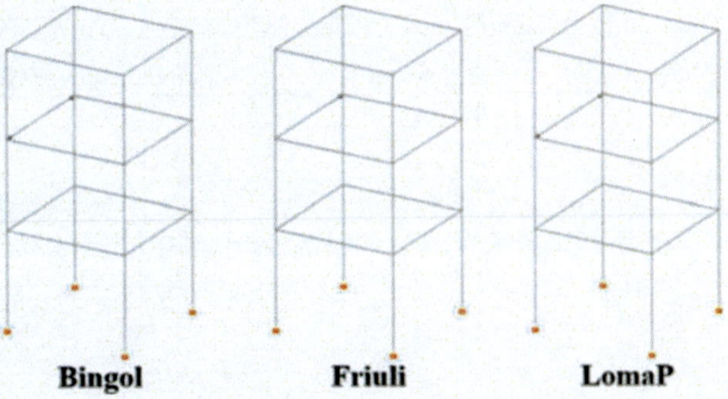

Fig. 10.15 Plastic hinge formation for frame A from seismic excitations (*x* + *0.3z*/LS level) (after Vasilopoulos 2005, reprinted with permission from UPCE)

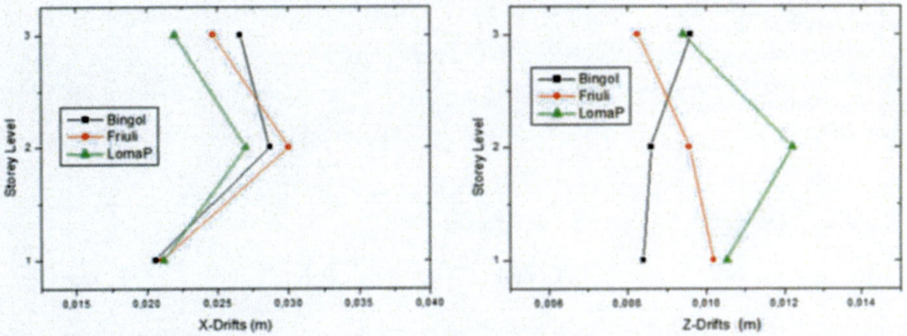

Fig. 10.16 Seismic storey drifts for frame D (*x* + *0.3z*/LS level) (after Vasilopoulos 2005, reprinted with permission from UPCE)

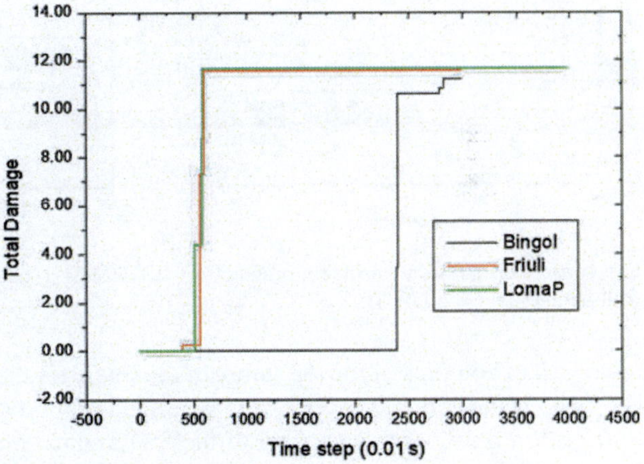

Fig. 10.17 Seismic total damage index for frame D (*x* + *0.3z*/LS level) (after Vasilopoulos 2005, reprinted with permission from UPCE)

indices are presented in Table 10.14 (only for the worst cases of Bingol, Kobe and Loma Prieta earthquakes) with values well below the limit of 50%.

Figures 10.26 and 10.27 depict the plastic hinge pattern at the CP level for the two load combinations, respectively. One can observe that plastic hinges develop at the

Table 10.11 Member damage indices for frame D from seismic excitations ($x + 0.3z$/LS level)

Member	Seismic motion	i-end damage (%)	j-end damage (%)
12	Bingol		11.88
15		11.82	
12	Friuli		11.92
15		11.83	
15	Loma Prieta	11.83	
16		7.67	

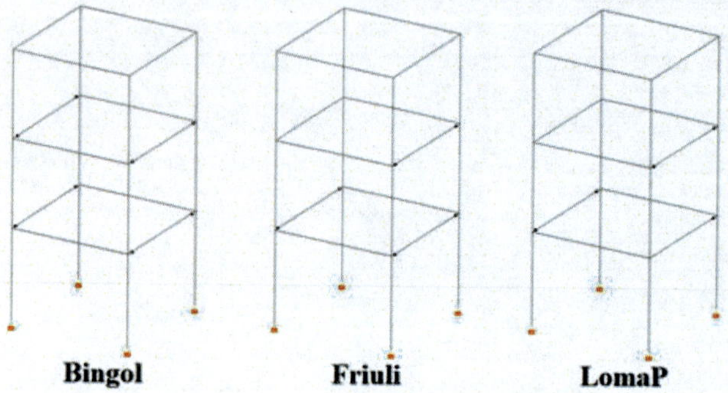

Fig. 10.18 Plastic hinge formation for frame D from seismic excitations ($x + 0.3z$/LS level) (after Vasilopoulos 2005, reprinted with permission from UPCE)

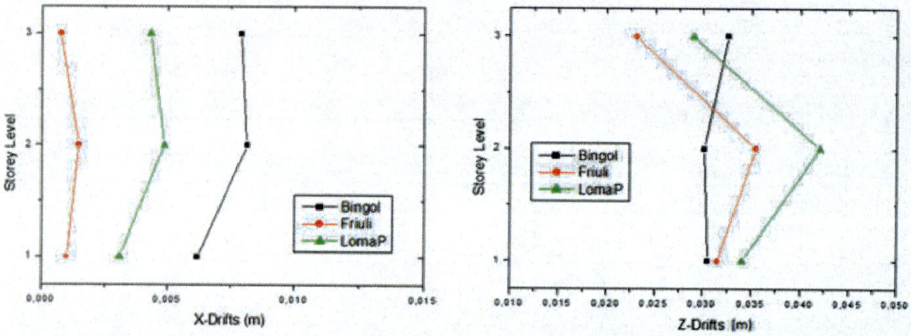

Fig. 10.19 Seismic storey drifts for frame D ($z + 0.3x$/LS level) (after Vasilopoulos 2005, reprinted with permission from UPCE)

lower end of columns of the top storey of the frame for the z + 0.3x load combination (Fig. 10.27) indicating that the capacity design is not satisfied there. However, this is restricted to only one top column of the frame, thus, precluding any partial storey collapse. Of course, one could replace all top HEB200 columns by HEB220 ones, but this may be too conservative.

For reasons of comparison, the computer program SAP 2000 (2007) that incorporates EC8 (2004) and EC3 (1992) codes is also used for the seismic design of the frame considered here. Thus, elastic spectrum analyses using the EC8 (2004)

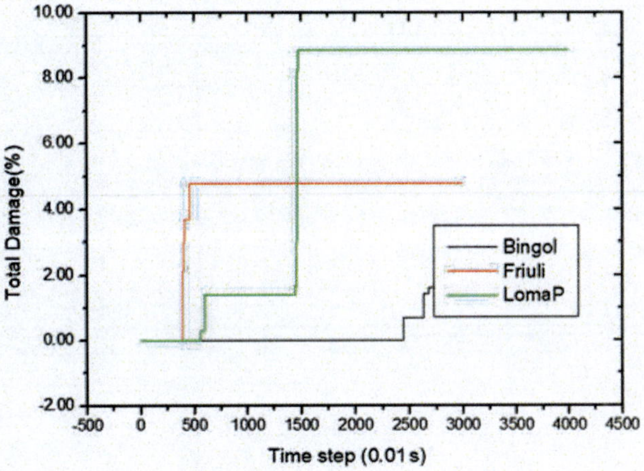

Fig. 10.20 Seismic total damage index for frame D ($z + 0.3x$/LS level) (after Vasilopoulos 2005, reprinted with permission from UPCE)

Table 10.12 Member damage indices for frame D from seismic excitations ($z + 0.3x$/LS level)

Member	Seismic motion	*i*-end damage (%)	*j*-end damage (%)
1	Bingol	3.00	–
4		3.00	–
4	Friuli	4.54	–
10		–	6.46
13	Loma Prieta	11.90	–
29		7.03	–

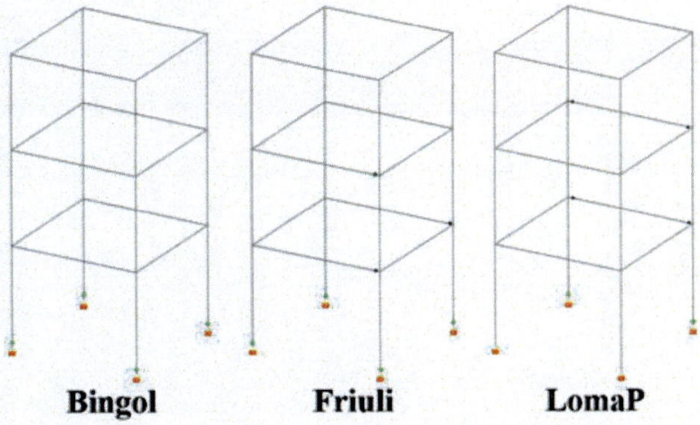

Fig. 10.21 Plastic hinge formation for frame D from seismic excitations ($z + 0.3x$/LS level) (after Vasilopoulos 2005, reprinted with permission from UPCE)

acceleration design spectrum for soil type B, $PGA = 0.30$ g, $q = 4$ and damping 5% are performed for all the frames of Table 10.9 and the best solution is again found to be frame D. The section selection of frame D satisfies all the code-based requirements as the capacity ratios for all members approach 1.0 without exceeding it.

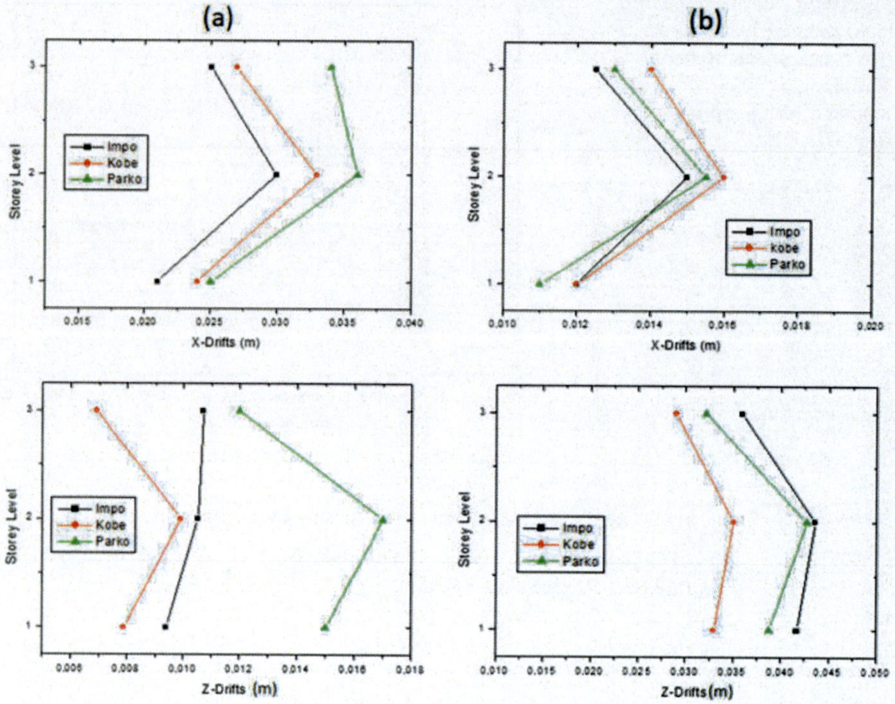

Fig. 10.22 Seismic storey drifts for frame D: $x + 0.3z$/LS level (**a**) and $z + 0.3x$/LS level (**b**) (after Vasilopoulos 2005, reprinted with permission from UPCE)

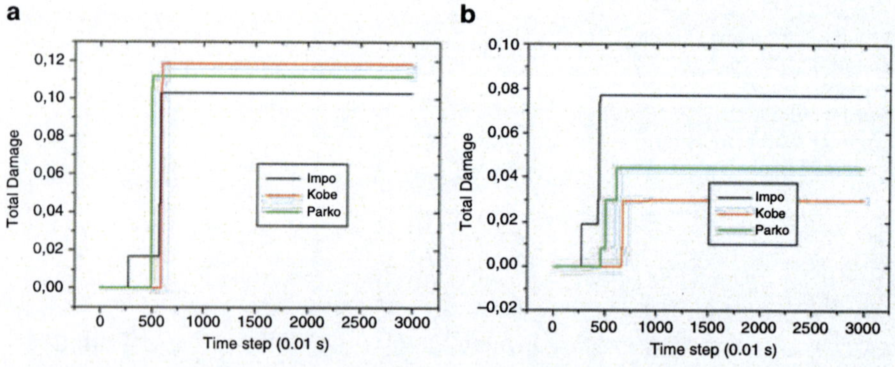

Fig. 10.23 Seismic total damage index for frame D: $x + 0.3z$/LS level (**a**) and $z + 0.3x$/LS level (**b**) (after Vasilopoulos 2005, reprinted with permission from UPCE)

Table 10.13 Member damage indices for frame D from seismic excitations: (a) $x + 0.3z$/LS level and (b) $z + 0.3x$/LS level

Member	Seismic motion (a)	i-end damage (%)	j-end damage (%)
(a)			
12	Imperial Valley	–	11.80
28		–	10.10
15	Kobe	11.80	–
31		11.90	–
32		–	11.90
11	Parkfield	11.80	–
12		–	11.90
27		11.80	–
28		–	11.90
31		11.80	–
32		–	11.80
(b)			
1	Imperial Valley	5.13	–
10		–	10.00
1	Kobe	2.97	–
4		2.98	–
10	Parkfield	–	5.40
13		8.44	–

10.6 Conclusions

On the basis of the preceding sections the following conclusions can be stated:

1. A seismic design method for regular plane and space steel moment resisting frames has been presented. The method uses an advanced finite element dynamic analysis taking into account material and geometric nonlinearities and works in conjunction with Eurocodes EC3 and EC8.
2. The method is a performance-based seismic design one with three performance levels, namely immediate occupancy, life safety and collapse prevention. It checks at every performance level the satisfaction of performance objectives in terms of interstorey drift ratios, member plastic rotations, structural, storey and member damage indices and plastic hinge pattern.
3. Because of the employment of advanced nonlinear analysis, the limit state of strength and stability and their member-structure interactions are captured in a direct manner. Thus, the use of the behavior factor of EC8 and the member interaction equations of EC3 for taking into account in an indirect and approximate way material and geometric nonlinearities is avoided.
4. Designs performed by the proposed method result in member sections sizes close to those coming from the EC3/EC8 method because the advanced analysis is calibrated against the interaction equations of EC3 and the design spectrum of

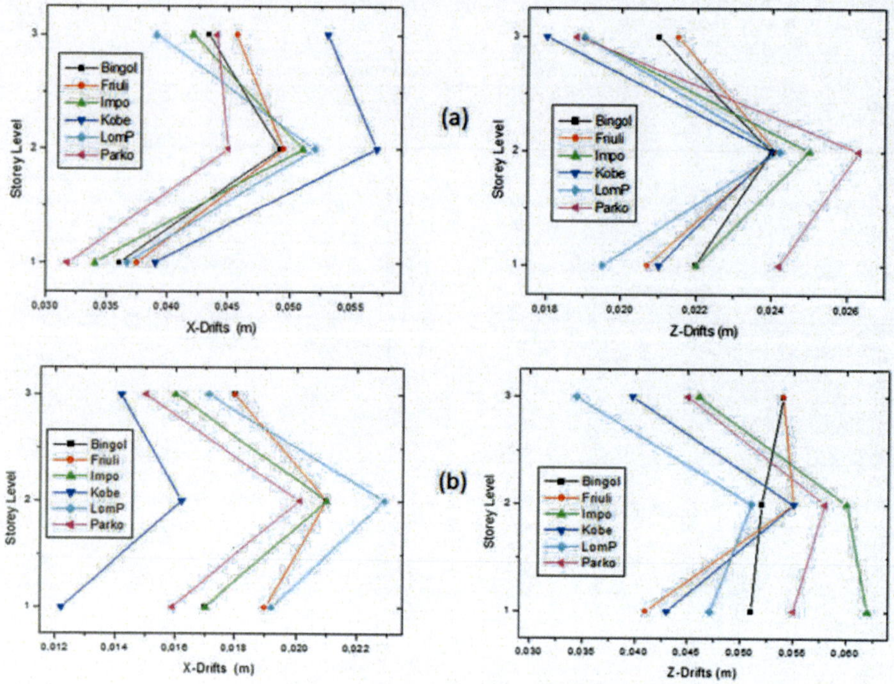

Fig. 10.24 Seismic storey drifts for frame D: $x + 0.3z$/CP level (a) and $z + 0.3x$/CP level (b) (after Vasilopoulos 2005, reprinted with permission from UPCE)

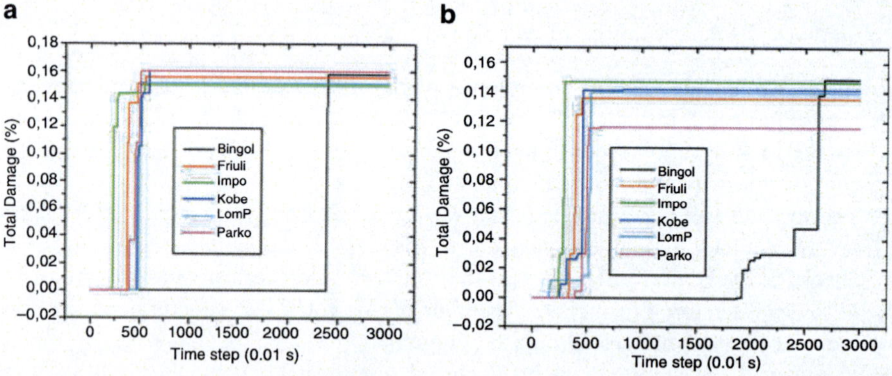

Fig. 10.25 Seismic total damage index for frame D: $x + 0.3z$/CP level (a) and $z + 0.3x$/CP level (b) (after Vasilopoulos 2005, reprinted with permission from UPCE)

EC8. The execution of nonlinear time history analyses instead of the simpler linear spectrum analyses is a disadvantage of the method. In conclusion, the proposed method provides at least a more rational and accurate alternative to the EC3/EC8 design method and points towards the future design direction.

Table 10.14 Member damage indices for frame D from seismic excitations: (a) $x + 0.3z$/CP level and (b) $z + 0.3x$/CP level.

Member	Seismic motion (a)	*i*-end damage (%)	*j*-end damage (%)
(a)			
11	Bingol	16.11	–
12		–	16.42
27		16.09	–
28		–	16.21
15	Kobe	21.53	–
16		–	20.24
31		21.73	–
32		–	22.61
11	Loma Prieta	20.54	–
15		21.25	–
16		–	21.98
(b)			
10	Bingol	–	15.65
13		15.19	–
26		–	15.48
29		15.36	–
10	Kobe	–	21.24
13		22.86	–
9	Loma Prieta	21.42	–
13		23.20	–
14		–	21.44

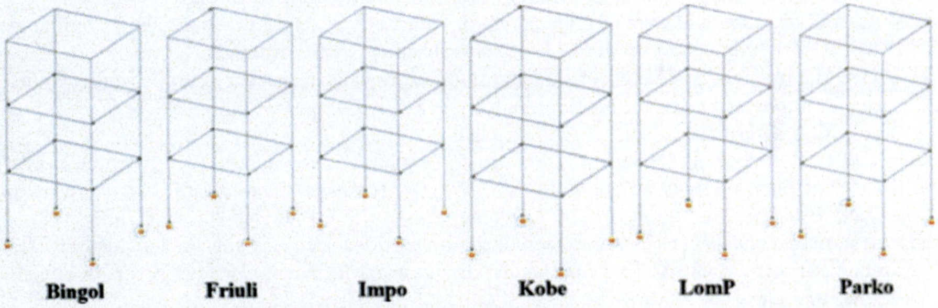

Fig. 10.26 Seismic plastic hinge formation for frame D under ($x + 0.3z$/CP level) (after Vasilopoulos 2005, reprinted with permission from UPCE)

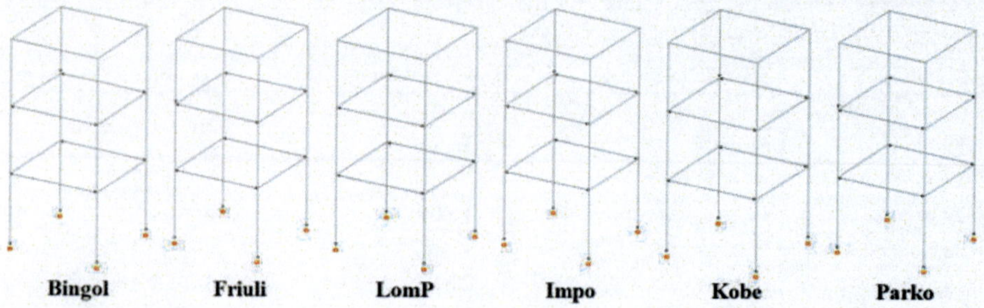

Fig. 10.27 Seismic plastic hinge formation for frame D under (z + $0.3x$/CP level) (after Vasilopoulos 2005, reprinted with permission from UPCE)

References

AISC (2005) Specification for structural steel buildings. American Institute of Steel Construction, Chicago, IL

ATC 13 (1985) Earthquake damage evaluation for California. Applied Technology Council, Redwood City, CA

Bathe KJ (1996) Finite element procedures. Prentice Hall, Englewood Cliffs, NJ

Beskos DE (1976) The lumping mass effect on frequencies of beam-columns. J Sound Vib 47:139–142

Beskos DE (1977) Framework stability by finite element method. J Struct Div ASCE 103:2273–2276

Chen WF (1998) Implementing advanced analysis for steel frame design. Prog Struct Eng Mater 1:323–328

Chen WF, Kim SE (1997) LRFD steel design using advanced analysis. CRC Press, Boca Raton, FL

EC3 (1992) Eurocode 3, Design of steel structures – Part 1-1: general rules and rules for buildings, ENV 1993-1-1. European Committee for Standardization (CEN), Brussels

EC3 (2009) Eurocode 3, Design of steel structures – Part 1-1: general rules and rules for buildings, EN 1993-1-1. European Committee for Standardization (CEN), Brussels

EC8 (1994) Eurocode 8, Design of structures for earthquake resistance, Part 1: general rules, seismic actions and rules for buildings, ENV 1998-1-1. European Committee for Standardization (CEN), Brussels

EC8 (2004) Eurocode 8, Design of structures for earthquake resistance, Part 1: general rules, seismic actions and rules for buildings, EN 1998-1-1. European Committee for Standardization (CEN), Brussels

El Hafez MB, Powell GH (1973) Computer aided ultimate load design of unbraced multistorey steel frames, Report no. EERC 73-3. Earthquake Engineering Research Center, University of California, Berkeley, CA

FEMA 273 (1997) NEHRP guidelines for the seismic rehabilitation of buildings. Federal Emergency Management Agency, Washington, DC

Gioncu V, Mazzolani F (2002) Ductility of seismic resistant steel structures. Spon Press, London

Grecea D, Dinu F, Dubina D (2002) Performance criteria for MR steel frames in seismic zones. In: Lamas A, Da Silva LS (eds) Proceedings of EUROSTEEL 2002 Conference, Coimbra, Portugal. Multicomp, Lisbon, pp 1269–1278

Kakaliagos A (1994) Pseudo-dynamic testing of a full scale three storey one bay steel moment resisting frame: experimental and analytical results. Report EUR 15605 EN. ELSA Laboratory, Safety Technology Institute, Joint Research Centre, Commission of the European Communities, Ispra

Karabalis DL, Cokkinides GJ, Rizos DC, Mulliken JS (1993) An interactive computer code for generation of artificial earthquake records. In: Khozeimeh K (ed) Computing in civil engineering. American Society of Civil Engineers, New York, pp 1122–1155
Kim SE, Chen WF (1996a) Practical advanced analysis for unbraced steel frame design. J Struct Eng ASCE 122:1259–1265
Kim SE, Chen WF (1996b) Practical advanced analysis for braced steel frame design. J Struct Eng ASCE 122:1266–1274
Kim SE, Lee J (2002) Improved refined plastic-hinge analysis accounting for lateral torsional buckling. J Constr Steel Res 58:1431–1453
Kim SE, Park MH, Choi SH (2001) Direct design of three-dimensional frames using practical advanced analysis. Eng Struct 23:1491–1502
Kwak HG, Kim DY, Lee HW (2001) Effect of warping in geometric nonlinear analysis of spatial beams. J Constr Steel Res 57:729–751
Lemaitre J (1996) A course on damage mechanics. Springer, Berlin
Liew JYR, Chen WF, Chen H (2000) Advanced inelastic analysis of frame structures. J Constr Steel Res 55:245–265
Pavlidis G, Bazeos N, Beskos DE (2003) Effects of higher modes and seismic frequency content on the accuracy of pushover analysis of steel frames. In: Mazzolani FM (ed) Behaviour of Steel Structures in Seismic Areas (STESSA 2003). Swets & Zeitlinger, Lisse, pp 547–550
PEER (2009) Pacific Earthquake Engineering Research Center, Strong Ground Motion Database, Berkeley, CA. http://peer.berkeley.edu/
Powell GH, Allahabadi R (1988) Seismic damage prediction by deterministic methods: concepts and procedures. Earthq Eng Struct Dyn 16:719–734
Prakash V, Powell GH, Campbell S (1993) DRAIN-2DX, Base program description and user guide, Version 1.10, Report No UCB/SEMM-93/17. University of California, Berkeley, CA
Prakash V, Powell GH, Campbell S (1994) DRAIN-3DX, base program description and user guide, Version 1.10, Report No UCB/SEMM-94/08. University of California, Berkeley, CA
SAP 2000 (2007) Structural analysis program 2000, static and dynamic finite element analysis of structures, Version 11. Computers and Structures Inc, Berkeley, CA
SEAOC (1999) Recommended lateral force requirements and commentary, 7th edn. Structural Engineers Association of California, Sacramento, CA
Surovek AE (2012) Advanced analysis in steel frame design. ASCE, Reston, VA
Trahair NS (1993) Flexural-torsional buckling of structures. CRC Press, Boca Raton, FL
Vasilopoulos AA (2005) Seismic design of steel structures using advanced methods of analysis. Ph. D. Thesis, Department of Civil Engineering, University of Patras, Patras, Greece (in Greek)
Vasilopoulos AA, Beskos DE (2006) Seismic design of plane steel frames using advanced methods of analysis. Soil Dyn Earthq Eng 26:1077–1100
Vasilopoulos AA, Beskos DE (2009) Seismic design of space steel frames using advanced methods of analysis. Soil Dyn Earthq Eng 29:194–218
Vasilopoulos AA, Kamaris GS (2020) Seismic design of space steel frames using advanced static inelastic (pushover) analysis. Soil Dyn Earthq Eng 137:106309
Vasilopoulos AA, Bazeos N, Beskos DE (2008) Seismic design of irregular space steel frames using advanced methods of analysis. Steel Compos Struct 8:53–83
Wongkaew K, Chen WF (2002) Consideration of out of plane buckling in advanced analysis for planar steel frame design. J Constr Steel Res 58:943–965
Ziemian RD, McGuire W, Deierlein GG (1992a) Inelastic limit states design. Part I: planar frame studies. J Struct Eng ASCE 118:2532–2549
Ziemian RD, McGuire W, Deierlein GG (1992b) Inelastic limit states design. Part II: three dimensional frame study. J Struct Eng ASCE 118:2550–2568

Chapter 11
Direct Damage-Controlled Design

Abstract The direct damage controlled design method for seismic design of plane steel moment resisting framed structures is described. This method in its dynamic or static (pushover) version is capable of controlling damage at all levels of performance. The proposed method can be used not only for designing a structure for a given seismic load and desired level of damage, but also for determining damage in a designed structure locally or globally due to any seismic load or evaluating the maximum seismic load a structure can undertake for a desired level of damage. The above are accomplished by introducing a new seismic damage index and constructing appropriate damage performance levels. The seismic damage index takes into account axial force-bending moment interaction, strength and stiffness deterioration and low cycle fatigue. The damage performance levels are established with the aid of extensive parametric studies on a large number of frames under a large number of seismic motions in conjunction with the above damage index for damage evaluation. Numerical examples are provided to illustrate the proposed method in its dynamic and static versions and demonstrate its merits against other methods of seismic design.

Keywords Damage index · Damage controlled design · Plane steel frames · Damage performance levels · Nonlinear dynamic analysis · Pushover analysis

11.1 Introduction

In seismic design of building structures, various design methods have been proposed and employed in practice. One can mention here, the force-based design (FBD) described in Chap. 3, the displacement-based design (DBD) described in Chap. 4 and the hybrid force/displacement based design (HFD) described in Chap. 5.

The FBD method, which is adopted by current seismic codes, uses seismic forces as the main design parameters. This method performs design in two steps involving strength checking at the life safety (LS) level and displacement checking at the

G. A. Papagiannopoulos et al., *Seismic Design Methods for Steel Building Structures*, Geotechnical, Geological and Earthquake Engineering 51,
https://doi.org/10.1007/978-3-030-80687-3_11

immediate occupancy (IO) level. A response spectrum elastic analysis with the aid of the behavior (strength) reduction factor to take into account inelasticity effects is used for the base seismic shear determination of the first step.

The DBD is an alternative seismic design method that employs displacements as the main design parameters. The method performs design in essentially one step, which starts with target displacements and proceeds with the determination of the stiffness and base shear of the structure required to limit the displacements to the target ones. This is accomplished by the construction of an equivalent linear single-degree-of-freedom (SDOF) system to the original nonlinear structure and the use of a displacement spectrum with high amounts of damping.

The HFD is a new seismic design method, which combines the advantages of both of the aforementioned FBD and DBD methods. The method starts with both target displacements and ductilities and proceeds with the response spectrum elastic analysis and a deformation and period dependent behavior factor to determine the base shear force required to limit the deformation to the target one.

The performance based design (PBD), as discussed in Chap. 1, provides a general framework in seismic design by introducing performance levels and objectives to corresponding seismic hazard levels. Thus, three to five performance levels are defined with objectives mainly referring to damage in a direct or an indirect (through deformation) way.

Seismic damage is quantified through indices, such as the interstorey drift ratio (*IDR*), or the member plastic rotations (θ_{p}), or many other indices (Powell and Allahabadi 1988; Cosenza et al. 1993; Ghobarah et al. 1999). These indices are expressed in terms of deformation, dissipated energy or a combination of deformation and dissipated energy.

Damage-based seismic design methods have been developed by, e.g., Park et al. (1987) using explicit damage levels and Panyakapo (2008) using a damage-based capacity-demand method, Kawashima and Aizawa (1986), Ballio and Castiglioni (1994), Tiwari and Gupta (2000), Kunnath and Chai (2004) and Lu and Wei (2008) using inelastic spectra obtained with the aid of a damage dependent behavior factor and Bozorgnia and Bertero (2004), Panyakapo (2004) and Ghobarah and Safar (2010) using damage or cyclic demand spectra.

In this chapter, the Direct Damage Controlled Design (DDCD) method for the seismic design of steel plane moment resisting frames (MRFs) is presented on the basis of the works of Kamaris et al. (2009, 2015). The lateral seismic loading can be applied dynamically in the framework of a nonlinear time history (NLTH) analysis, or statically in the framework of a pushover analysis. The basic advantage of the method is the seismic design of structures with direct damage control at both local and global levels. Thus, the desired level of damage in a structural member or the whole structure can be selected at the start and the design can proceed with the aid of NLTH analysis in order to arrive at a structure that can experience seismic damage not exceeding the preselected one. In addition, the method can determine (i) the damage in any member or the whole structure under any seismic load and (ii) the maximum seismic load a structure can sustain without exceeding a desired level of damage. The method employs a new seismic damage index developed by Kamaris et al. (2013) and limit values for three damage performance levels determined by extensive parametric studies involving a large number of frames and seismic motions. The DDCD method controls damage directly and not indirectly through deformation as the DBD or the HFD methods and can be used in three different ways, i.e., for damage determination, member dimensioning or maximum seismic load determination. The DDCD method in its dynamic version requires NLTH analysis, which is more involved than elastic/inelastic spectrum analysis. However, it takes into account material and geometrical nonlinearities directly and leads to more accurate results. In that sense, the present method is a method using advanced analysis as the one presented in Chap. 10. If a simpler approach is desired, the static (pushover) version of the DDCD method can be employed (Kamaris et al. 2009).

11.2 Damage Indices

Damage theory models the progressive mechanical degradation or deterioration of materials under different stages of loading. This material degradation process is governed by a damage variable d, the local damage index, which is defined pointwise as (Lemaitre 1996)

$$d = \lim_{S_n \to 0} \frac{S_n - \overline{S_n}}{S_n} \tag{11.1}$$

where S_n denotes the whole section in a damaged material volume, $\overline{S_n}$ the effective or undamaged area, while $S_n - \overline{S_n}$ denotes the inactive area of defects, cracks and voids, as shown in Fig. 11.1. This index is equal to zero when the material is in the undamaged state and equal to one at complete failure.

Damage mechanics mainly determines the initiation and evolution of the damage index d during the deformation. On the assumption that damage evolution for steel

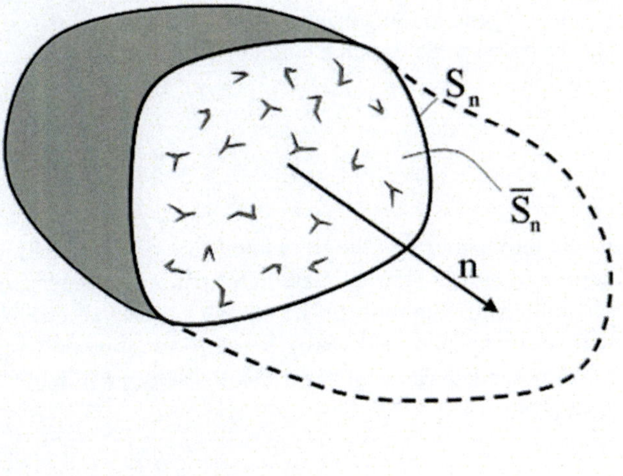

Fig. 11.1 Cross section of a damaged material

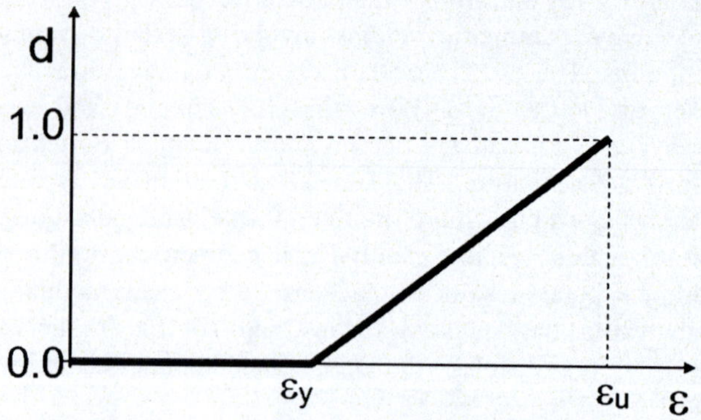

Fig. 11.2 Damage-strain curve for steel

takes place only during plastic loading, Lemaitre (1996) proposed a simple damage evolution law, as shown in Fig. 11.2. This damage evolution law is expressed as

$$d = 0 \; for \; \varepsilon \leq \varepsilon_y \;, \quad d = \frac{\varepsilon - \varepsilon_y}{\varepsilon_u - \varepsilon_y} \; for \; \varepsilon_y < \varepsilon \leq \varepsilon_u \tag{11.2}$$

where ε denotes strain and subscripts y and u stand for yielding and ultimate, respectively.

Local damage is usually referred to a point or even a section of a structure, while global damage to a section, member, substructure, or the whole structure. Both are expressed in terms of damage indices receiving values between zero and one for the undamaged and failure states, respectively, and constitute the most suitable indicators of the loading capacity at a point, section, member, or structure level.

Damage in buildings can be determined in terms of deformation at the local level by using, e.g., Eq. (11.2), or the member rotation/curvature ductility and at the global level by using, e.g., the interstorey drift ratio (*IDR*). However, during seismic deformation, damage due to the cyclic nature of the loading has also to be taken into account. Thus, the damage index of Park and Ang (1985), which is a linear combination of the damage caused by excessive deformation and that contributed by repeated cyclic loading effects, is probably the most widely accepted damage index for structures under seismic loading. This index is defined at a member section by the equation

$$D_s = \frac{\delta_m}{\delta_u} + \frac{\beta}{V_y \delta_u} \int dE \tag{11.3}$$

where the first term at the right hand side of Eq. (11.3) is the ratio of the maximum deformation δ_m to the ultimate deformation δ_u under monotonic loading, while the second term is the ratio of the energy of dissipation $\int dE$ to the expression $V_y \delta_u / \beta$ with V_y being the yield strength and β a non-negative parameter determined from tests. A typical value of β for steel structures is 0.025 (Castiglioni and Pucinotti 2009).

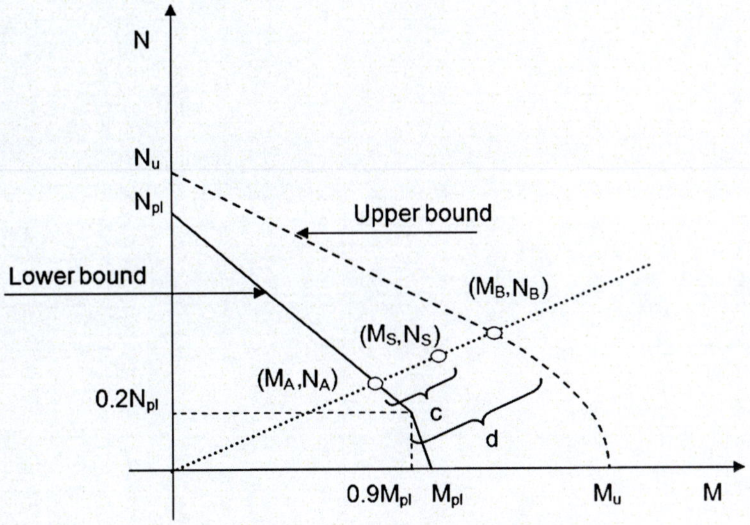

Fig. 11.3 Bending moment-axial force interaction diagram and definition of parameters of proposed damage index (after Kamaris 2011, reprinted with permission from UPCE)

In this work, use is made of the damage index proposed by Kamaris et al. (2009, 2013) because of its merits both for static and seismic loading. This damage index is defined at a section S of a structural member (beam or column) by the equation

$$D_s = \frac{c}{d} = \frac{\sqrt{(M_S - M_A)^2 + (N_S - N_A)^2}}{\sqrt{(M_B - M_A)^2 + (N_B - N_A)^2}} \tag{11.4}$$

which takes into account the interaction between the bending moment M_S and axial force N_S acting at that section, as shown in Fig. 11.3. This figure displays a lower bound damage curve, which is the limit between elastic and inelastic material behavior and an upper bound damage curve, which is the limit between inelastic behavior and complete failure. Thus, damage at the former curve is zero, while at the latter curve is one. Equation (11.4) reflects the assumption that damage evolution varies linearly between the above two damage bounds. Points (M_A, N_A) and (M_B, N_B), can be found as the intersections of these two bound curves with a straight line connecting (M_S , N_S) to the origin of the M and N axes.

The lower bound curve of Fig. 11.3 is associated with concentrated plasticity and the formation of a plastic hinge at a member. In Ruaumoko (Carr 2005) software, used herein, the lower bound curve is described as

$$
\begin{aligned}
&\frac{0.88M}{fM_{pl}} + \frac{N}{fN_{pl}} = 1 \; for \; M \le 0.9M_{pl} \; and \; N \le 0.2N_{pl} \\
&\frac{M - M_{pl}}{fM_{pl}} + \frac{N}{fN_{pl}} = 1 \; for \; M > 0.9M_{pl} \; and \; N > 0.2N_{pl}
\end{aligned}
\tag{11.5}
$$

where N_{pl} and M_{pl} are given by the expressions

$$
M_{pl} = f_y W_{pl}, \; N_{pl} = f_y A \tag{11.6}
$$

with f_y being the yield stress, W_{pl} the section plastic modulus and A the sectional area.

The upper bound curve of Fig. 11.3 is assumed to have the form

$$
\frac{M}{fM_u} + \left(\frac{N}{fN_u}\right)^2 = 1 \tag{11.7}
$$

where N_u and M_u are the ultimate axial force and bending moment, respectively, which cause failure of the section and are equal to

$$
M_u = f_u W_{pl}, \; N_u = f_u A \tag{11.8}
$$

with f_u being the ultimate stress of steel. It is observed that Eq. (11.7) can be thought of as coming from the M-N interaction equation of EC3 (1992) by replacing f_y by f_u , i.e., by assuming that there exists strain hardening. The factor f in Eqs. (11.5) and (11.7) is a scale factor taking into account strength and stiffness degradation and has the form (Carr 2005)

$$
f = \frac{S_r - 1}{n_2 - n_1}(n - n_1) + 1 \tag{11.9}
$$

where n is the number of cycles, n_1 and n_2 denote the cycle at which degradation begins and stops, respectively, and S_r is the residual strength factor that multiplies the initial yield strength to produce the residual strength. For typical values of n_1 , n_2 and S_r equal to 3, 55 and 0.55, respectively, Eq. (11.9) for $n = 8$ yields $f = 0.96$. It should be noted that complete elimination of strength is not possible in Ruaumoko (Carr 2005) where a 1% residual strength value for S_r is practically considered as close enough to zero.

The increase of damage due to strength reduction coming from low-cycle fatigue is taken into account by following the idea of Sucuoğlu and Erberik (2004) and using results of extensive parametric studies from Kamaris et al. (2013). This increase of damage ΔD_s can be obtained by using the empirical expression

$$\Delta D_s = 0.56n^{0.292} D_s^{0.914} \tag{11.10}$$

where D_s is the damage index of Eq. (11.4) at loading cycle n.

In conclusion, for a combination of moment M_s and axial force N_s at a member section, the damage index there can be easily evaluated by using Eqs. (11.4) and (11.10) at every time step of a nonlinear dynamic analysis. The calculation of M-N pairs is accomplished with the aid of the Ruaumoko 2D (Carr 2005) finite element program, which takes into account material and geometric nonlinearities. Thus, a bilinear moment-rotation model with strength and stiffness degradation in the framework of concentrated plasticity is used for beam-column members in conjunction with large displacement effects.

The above damage index of Kamaris et al. (2013) after calibration by tests, has been found to correlate well with five of the most well-known damage indices. For the case of static load, the damage index D_s is computed from Eq. (11.4) without considering the ΔD_s of Eq. (11.10) and by using Eqs. (11.5) and (11.7) with $f = 1$. This static damage index of Eq. (11.4) at a section can be thought of as an extension of Eq. (11.2) from strains (or stresses) to stress resultants, i.e., forces and moments.

The member damage index D_M is taken as the largest section damage index along the member, i.e.,

$$D_M = max\,(D_S) \tag{11.11}$$

The overall damage index, which represents the damage of the whole structure, can be determined by rationally combining the member damage indices in order to take into account both the severity of the member damage and the geometric distribution of damage within the whole structure. Thus, for a structure consisting of m-members, the overall damage index, D_O, can be expressed in the form

$$D_O = \left(\frac{\sum_{i=1}^{m} D_{M,i}^2 W_i}{\sum_{i=1}^{m} W_i} \right)^{1/2} \tag{11.12}$$

where $D_{M,\,i}$ and W_i denote the damage and weighting factor of the ith member, respectively. The weighting factor W_i can be selected as equal to the volume Ω_i of the ith member or simply to 1. In the former case, both the severity of the member damage and the geometric distribution of damage in the structure are taken into account (Cervera et al. 1995), while in the latter case simplicity is the main factor (Powell and Allahabadi 1988).

11.3 Direct Damage Controlled Steel Design: Dynamic

The application of the dynamic version of the DDCD method to the seismic design of plane steel moment-resisting frames (MRFs) is done with the aid of the Ruaumoko 2D (Carr 2005) software working in the time domain and taking into account material and geometric nonlinearities (concentrated plasticity with stress-strain bilinear modeling including cyclic strength and stiffness degradation and large deflection effects). In connection with damage controlled steel design the following three options ae available:

11.3.1 Damage Determination in a Structure Under Given Seismic Load

Damage determination can be done with the use of Eq. (11.3) for the case of Park and Ang (1985) damage index and of Eqs. (11.4) and (11.10) for the case of the Kamaris et al. (2013) damage index in conjunction with Ruaumoko 2D (Carr 2005) program. The former index is in the damage index library of the program, while the latter one has been added to the program together with Eqs. (11.10) and (11.12) with $\Omega_i = 1$ to accommodate damage determination at the member or structure level. For the case of maximum damage determination on the basis of the Park and Ang (1985) damage index, one can also use empirical explicit expressions, which are derived in the next section and read as follows (Kamaris et al. 2012):

$$D_{S,c} = 0.169 n_s^{0.181} \rho^{0.019} \alpha^{-0.051} (S_a/g)^{0.191} \tag{11.13}$$

for columns (their bases) and

$$D_{S,b} = 0.262 n_s^{0.326} \rho^{-0.013} \alpha^{-0.059} (S_a/g)^{0.351} \tag{11.14}$$

for beams (their ends), where n_s is the number of stories in the frame ρ and a are the stiffness and strength ratios of the frame as defined by Eq. (5.7) and S_a/g is the dimensionless spectral acceleration at the fundamental period of the frame.

11.3.2 Structural Dimensioning for Given Seismic Load and Desired Level of Damage

This can be done iteratively with the aid of the Ruaumoko 2D (Carr 2005) program equipped with the Kamaris et al. (2013) damage index and the information in Table 11.1 providing limit values of damage for columns and beams for the three

Table 11.1 Proposed damage scale for the performance levels IO, LS and CP

Performance Levels	Maximum column damage	Maximum beam damage
IO	≤1%	≤4%
LS	≤39%	≤74%
CP	≤99%	100%

performance levels of FEMA 273 (1997) and a library of sectional properties (HEB, HEA, IPE) for standard European profiles of columns and beams. Table 11.1 has been constructed on the basis of the Kamaris et al. (2013) damage index as described in the next section. The proposed design is capable of easily applying capacity design (strong columns-weak beams). Thus, by assuming at the beginning, in accordance with Table 11.1, that the desired maximum level of damage for columns and beams at the LS performance level is e.g., 30% and 70%, respectively, one can determine iteratively the appropriate sections for columns and beams satisfying the above criterion.

11.3.3 *Maximum Seismic Load a Structure Can Sustain for a Desired Level of Damage*

For a steel plane MRF properly designed and a given earthquake characterized by its peak ground acceleration (*PGA*), one can perform a series of NLTH analyses with Ruaumoko 2D (Carr 2005) for a sequence of values of *PGA* of that earthquake until he determines iteratively that value of *PGA*, which creates the desired level of damage in columns and beams of the structure.

11.4 Damage Expressions and Performance Levels: Dynamic

Damage is used in the proposed design method of this chapter as a performance objective. Thus, the establishment of the proposed method requires, not only a method for damage determination, but also damage limit values for various performance levels, or in general a damage scale. Available damage scales in the literature can be used for an easy selection of desired damage levels related to the strength degradation and carrying load capacity of a structure. Table 11.2 has been assembled based on existing data in the literature dealing mainly with steel frames and presents three performance levels (IO, LS and CP) with their corresponding limit response values pertaining to $IDR =$ interstorey drift ratio, $\theta_p =$ plastic end rotation, $\mu_\theta =$ local ductility and $d =$ damage.

In this section, the explicit expressions in Eqs. (11.13) and (11.14) for the Park and Ang (1985) damage and a new damage scale based on the Kamaris et al. (2013)

Table 11.2 Performance levels and corresponding limit response values including damage *d*

Performance levels	IDR	θ_p	μ_θ	*d*
IO	1.5% (SEAOC 1999) 0.2% (Ghobarah 2004) 0.7% transient & negligible permanent (FEMA-2731997)	$\leq\theta_y$ (FEMA-273 1997)	2 (FEMA-273 1997)	10–20% (Ghobarah 2004) 0.1–10% (ATC13 1985)
LS	0.4–1.0% (Ghobarah 2004) 3.2% (SEAOC 1999) 2.5% transient & 1% permanent (FEMA-2731997)	$\leq 6\theta_y$ (FEMA-273 1997)	7 (FEMA-273 1997)	20–40% (Ghobarah 2004) 10–30% (ATC13 1985)
CP	3.8% (SEAOC 1999) 1.8% (Ghobarah 2004) 5.0% transient & 5.0% permanent (FEMA-2731997)	$\leq 8\theta_y$ (FEMA-273 1997)	9 (FEMA-273 1997)	40–80% (Ghobarah 2004) 30–60% (ATC13 1985)

$\theta_y = W_{p\ell} f_y L_b / 6EI_b$, $\theta_y = W_{p\ell} f_y L_c(1 - N/N_y)/6EI_c$ for beams (b) and columns (c), respectively

damage index for plane steel MRFs for the IO, LS and CP performance levels are constructed based on extensive parametric studies conducted on a large number of frames and seismic motions. In the following, the frames and seismic motions considered are presented and the methods employed to derive the damage expressions and construct the damage scale are described in detail.

11.4.1 Steel Frames Considered

A set of 36 plane regular and orthogonal steel MRFs with storey heights and bay widths equal to 3.0 m and 5.0 m, respectively, was used for the parametric studies of this work. Figure 11.4 displays the general geometry of a typical frame from this set, while Table 11.3 provides information on all the important characteristics of the frames of the set. In Table 11.3 n_s and n_b denote the number of stories and bays, respectively, a is the strength ratio as defined by Eq. (5.7) and T_1 and T_2 stand for the first two natural periods of the frames. Furthermore, expressions of the form, e.g., 260–360(1–4) + 240–330(5–6) in that table mean that the first four stories have columns with HEB260 sections and beams with IPE360 sections, whereas the fifth and sixth stories have columns with HEB240 sections and beams with IPE330 sections. It should be noted that the number of bays in the frames was limited to only 3 and 6 because, as it has been found in Karavasilis et al. (2007, 2008), this number has a very small significance on the seismic response. This was also verified here with respect to damage (Kamaris et al. 2015). The frames made of steel grade S235 have been designed according to Eurocodes EC3 (2009) and EC8 (2004) on the assumptions of gravity load (dead plus 0.3 of live) equal to 27.5 kN/m on beams

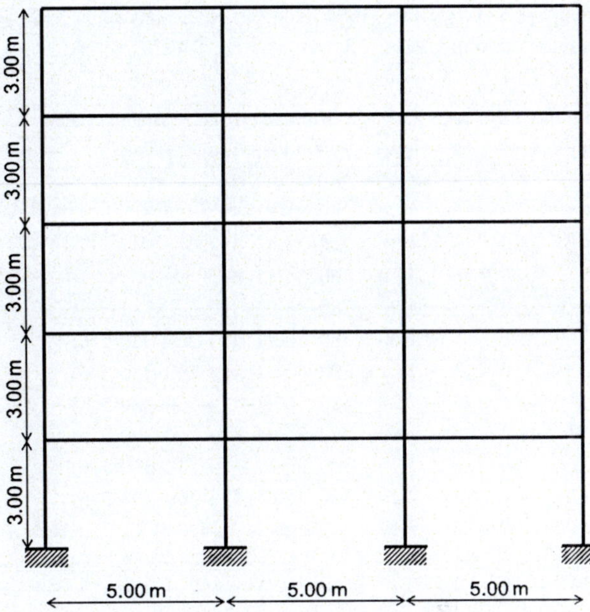

Fig. 11.4 Geometry of a typical plane steel MRF considered in parametric studies

and lateral seismic load defined by the acceleration elastic design spectrum of EC8 (2004) with $PGA = 0.35$ g and soil type B.

11.4.2 Ground Motions Considered

In this work, 40 physical ground motions, selected from the PEER (2009) ground motion database, have been used for the NLTH analyses. Only far-fault ground motions, i.e., motions recorded at a distance more than 15 km from the causative fault have been considered. Furthermore, the motions were selected so as their mean spectrum to be as close as possible to the response spectrum of EC8 (2004). Table 11.4 provides information on the date, the record name, the excitation component and the *PGA* of the motions considered here, while their elastic response spectra are depicted in Fig. 11.5, where their median spectrum is shown by a thick line. In order to study the whole seismic deformation range of the frames from elastic behavior up to collapse, all the ground motions of Table 11.4 were scaled appropriately, using as intensity measure the spectral acceleration at the first natural period $S_a(T_1)$ and conducting an incremental dynamic analysis (Vamvatsikos and Cornell 2002).

Table 11.3 Plane steel MRFs considered in parametric studies (after Kamaris 2011, reprinted with permission from UPCE)

General data				Sections	Periods	
Frame	n_s	n_b	α	Columns: (HEB) & Beams: (IPE)	T_1(s)	T_2(s)
1	3	3	1.30	240–330(1–3)	0.73	0.26
2	3	3	1.60	260–330(1–3)	0.69	0.21
3	3	3	1.90	280–330(1–3)	0.65	0.19
4	3	6	1.30	240–330(1–3)	0.75	0.23
5	3	6	1.60	260–330(1–3)	0.70	0.21
6	3	6	1.90	280–330(1–3)	0.66	0.20
7	6	3	1.60	280–360(1–4) + 260–330(5–6)	1.22	0.41
8	6	3	1.97	300–360(1–4) + 280–330(5–6)	1.17	0.38
9	6	3	2.27	320–360(1–4) + 300–330(5–6)	1.13	0.37
10	6	6	1.60	280–360(1–4) + 260–330(5–6)	1.25	0.42
11	6	6	1.97	300–360(1–4) + 280–330(5–6)	1.19	0.40
12	6	6	2.27	320–360(1–4) + 300–330(5–6)	1.15	0.38
13	9	3	2.19	340–360(1) + 340–400(2–5) + 320–360 (6–7) + 300–330(8–9)	1.55	0.54
14	9	3	2.43	360–360(1) + 360–400(2–5) + 340–360 (6–7) + 320–330(8–9)	1.52	0.53
15	9	3	2.93	400–360(1) + 400–400(2–5) + 360–360 (6–7) + 340–330(8–9)	1.46	0.51
16	9	6	2.19	340–360(1) + 340–400(2–5) + 320–360 (6–7) + 300–330(8–9)	1.57	0.55
17	9	6	2.43	360–360(1) + 360–400(2–5) + 340–360 (6–7) + 320–330(8–9)	1.53	0.53
18	9	6	2.93	400–360(1) + 400–400(2–5) + 360–360 (6–7) + 340–330(8–9)	1.47	0.51
19	12	3	2.60	400–360(1) + 400–400(2–3) + 400–450 (4–5) + 360–400(6–7) + 340–400(8–9) + 340–360 (10) + 340–330(11–12)	1.90	0.66
20	12	3	3.00	450–360(1) + 450–400(2–3) + 450–450 (4–5) + 400–450(6–7) + 360–400(8–9) + 360–360 (10) + 360–330(11–12)	1.78	0.62
21	12	3	3.63	500–360(1) + 500–400(2–3) + 500–450 (4–5) + 450–450(6–7) + 400–400(8–9) + 400–360 (10–11) + 400–330(12)	1.72	0.60
22	12	6	2.60	400–360(1) + 400–400(2–3) + 400–450 (4–5) + 360–400(6–7) + 340–400(8–9) + 340–360 (10) + 340–330(11–12)	1.90	0.67
23	12	6	3.00	450–360(1) + 450–400(2–3) + 450–450 (4–5) + 400–450(6–7) + 360–400(8–9) + 360–360 (10) + 360–330(11–12)	1.78	0.63
24	12	6	3.63	500–360(1) + 500–400(2–3) + 500–450 (4–5) + 450–450(6–7) + 400–400(8–9) + 400–360 (10–11) + 400–330(12)	1.72	0.61

(continued)

Table 11.3 (continued)

General data				Sections	Periods	
Frame	n_s	n_b	α	Columns: (HEB) & Beams: (IPE)	T_1(s)	T_2(s)
25	15	3	3.87	500–300(1) + 500–400(2–3) + 500–450 (4–5) + 450–400(6–7) + 400–400(8–12) + 400–360 (13–14) + 400–330(15)	2.29	0.78
26	15	3	4.49	550–300(1) + 550–400(2–3) + 550–450 (4–5) + 500–400(6–7) + 450–400(8–12) + 450–360 (13–14) + 450–330(15)	2.22	0.75
27	15	3	4.76	600–300(1) + 600–400(2–3) + 600–450 (4–5) + 550–450(6–7) + 500–450(8–9) + 500–400 (10–12) + 500–360(13–14) + 500–330(15)	2.10	0.72
28	15	6	3.87	500–300(1) + 500–400(2–3) + 500–450 (4–5) + 450–400(6–7) + 400–400(8–12) + 400–360 (13–14) + 400–330(15)	2.30	0.78
29	15	6	4.49	550–300(1) + 550–400(2–3) + 550–450 (4–5) + 500–400(6–7) + 450–400(8–12) + 450–360 (13–14) + 450–330(15)	2.21	0.75
30	15	6	4.76	600–300(1) + 600–400(2–3) + 600–450 (4–5) + 550–450(6–7) + 500–450(8–9) + 500–400 (10–12) + 500–360(13–14) + 500–330(15)	2.10	0.72
31	20	3	4.54	600–300(1) + 600–400(2–3) + 600–450 (4–5) + 550–450(6–10) + 500–450(11–13) + 500–400 (14–16) + 450–400(17) + 450–360(18–19) + 450–330 (20)	2.82	0.97
32	20	3	5.19	650–300(1) + 650–400(2–3) + 650–450 (4–5) + 600–450(6–10) + 550–450(11–13) + 550–400 (14–16) + 500–400(17) + 500–360(18–19) + 500–330 (20)	2.76	0.94
33	20	3	5.90	700–300(1) + 700–360(2) + 700–400(3) + 700–450 (4–5) + 650–450(6–10) + 600–450(11–13) + 600–400 (14–16) + 550–400(17) + 550–360(18–19) + 550–330 (20)	2.73	0.93
34	20	6	4.54	600–300(1) + 600–400(2–3) + 600–450 (4–5) + 550–450(6–10) + 500–450(11–13) + 500–400 (14–16) + 450–400(17) + 450–360(18–19) + 450–330 (20)	2.75	0.96
35	20	6	5.16	650–300(1) + 650–400(2–3) + 650–450 (4–5) + 600–450(6–10) + 550–450(11–13) + 550–400 (14–16) + 500–400(17) + 500–360(18–19) + 500–330 (20)	2.70	0.93
36	20	6	5.90	700–300(1) + 700–360(2) + 700–400(3) + 700–450 (4–5) + 650–450(6–10) + 600–450(11–13) + 600–400 (14–16) + 550–400(17) + 550–360(18–19) + 550–330 (20)	2.67	0.92

Table 11.4 Characteristics of ground motions used in parametric studies (after Kamaris 2011, reprinted with permission from UPCE)

No.	Date	Record Name	Comp.	Station Name	*PGA* (g)
1	1992/04/25	Cape Mendocino	NS	89509 Eureka	0.154
2	1992/04/25	Cape Mendocino	EW	89509 Eureka	0.178
3	1980/06/09	Victoria, Mexico	N045	6604 Cerro Prieto	0.621
4	1980/06/09	Victoria, Mexico	N135	6604 Cerro Prieto	0.587
5	1992/04/25	Cape Mendocino	EW	89324 Rio Dell Overpass	0.385
6	1992/04/25	Cape Mendocino	NS	89324 Rio Dell Overpass	0.549
7	1978/08/13	Santa Barbara	N048	283 Santa Barbara Courthouse	0.203
8	1978/08/13	Santa Barbara	N138	283 Santa Barbara Courthouse	0.102
9	1992/06/28	Landers	NS	12149 Desert Hot Springs	0.171
10	1992/06/28	Landers	NS	12149 Desert Hot Springs	0.154
11	1979/08/06	Coyote Lake	N213	1377 San Juan Bautista	0.108
12	1979/08/06	Coyote Lake	N303	1377 San Juan Bautista	0.107
13	1994/01/17	Northridge	NS	90021 LA—N Westmoreland	0.361
14	1994/01/17	Northridge	EW	90021 LA—N Westmoreland	0.401
15	1986/07/08	Northern Palm Springs	NS	12204 San Jacinto—Soboba	0.239
16	1986/07/08	Northern Palm Springs	EW	12204 San Jacinto—Soboba	0.250
17	1970/09/12	Lytle Creek	N115	290 Wrightwood	0.162
18	1970/09/12	Lytle Creek	N205	290 Wrightwood	0.200
19	1989/10/18	Loma Prieta	NS	58065 Saratoga—Aloha Ave	0.324
20	1989/10/18	Loma Prieta	EW	58065 Saratoga—Aloha Ave	0.512
21	1992/06/28	Landers	NS	22170 Joshua Tree	0.284
22	1992/06/28	Landers	EW	22170 Joshua Tree	0.274

(continued)

Table 11.4 (continued)

No.	Date	Record Name	Comp.	Station Name	*PGA* (g)
23	1976/09/15	Friuli, Italy	NS	8014 Forgaria Cornino	0.212
24	1976/09/15	Friuli, Italy	EW	8014 Forgaria Cornino	0.260
25	1992/06/28	Landers	N045	24577 Fort Irwin	0.114
26	1989/10/22	Loma Prieta	EW	678 Golden Gate Bridge	0.233
27	1994/01/17	Northridge	EW	24389 LA—Century City CC North	0.256
28	1994/01/17	Northridge	EW	24538 Santa Monica City Hall	0.883
29	1994/01/17	Northridge	N279	90013 Beverly Hills- 14,145 Mulhol	0.516
30	1994/01/17	Northridge	EW	24278 Castaic—Old Ridge Route	0.568
31	1994/01/17	Northridge	NS	90018 Hollywood—Willoughby Ave	0.245
32	1994/01/17	Northridge	EW	24303 LA—Hollywood Stor FF	0.231
33	1994/01/17	Northridge	N070	17-90015 LA—Chalon Rd	0.225
34	1994/01/17	Northridge	EW	24400 LA—Obregon Park	0.355
35	1994/01/17	Northridge	NS	24157 LA—Baldwin Hills	0.168
36	1994/01/17	Northridge	EW	127 Lake Hughes #9	0.165
37	1994/01/17	Northridge	N177	90063 Glendale—Las Palmas	0.357
38	1994/01/17	Northridge	N035	90014 Beverly Hills—12,520 Mulhol	0.617
39	1994/01/17	Northridge	NS	90047 Playa Del Rey—Saran	0.136
40	1994/01/17	Northridge	EW	24401 San Marino, SW Academy	0.116

11.4.3 Method for Damage Scale Determination

In this work, extensive parametric studies were performed pertaining to damage evaluation for the 36 plane steel MRFs of Table 11.3, under the 40 ground motions of Table 11.4. The frames were analyzed with the program Ruaumoko 2D (Carr 2005) using NLTH analysis. Thus, 23,040 analyses (= 36 frames x 40 ground motions x 16 analyses on average for every frame) were performed in this

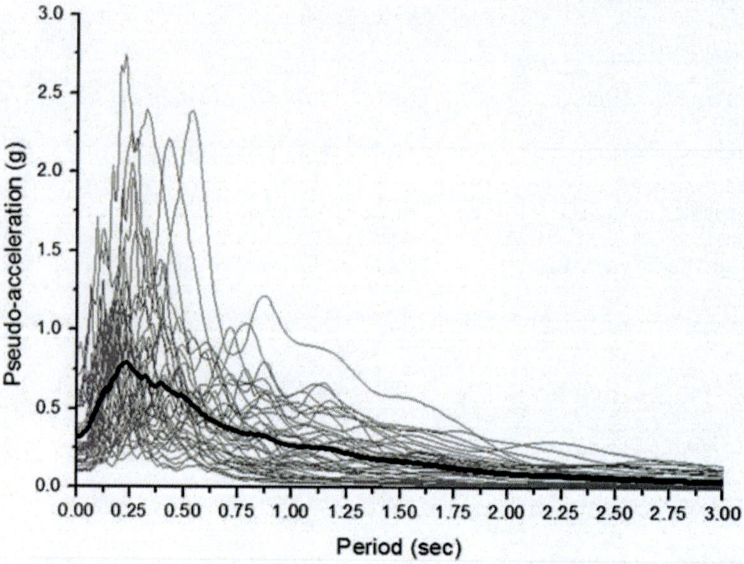

Fig. 11.5 Response spectra of ground motions considered in parametric studies (after Kamaris 2011, reprinted with permission from UPCE)

investigation. The frames were modeled on the basis of centerline representations with inelastic material behavior simulated by bilinear (hysteretic) point plastic hinges with 3% hardening and cyclic strength and stiffness degradation. No panel zone effects were considered in connections, which were assumed to be rigid. Finally, diaphragm action was considered at every floor due to the presence of the slab.

Maximum seismic response values, such as *IDR*s, plastic rotations, number of cycles and damage indices (computed by Eq. (11.3) for the Park and Ang (1985) index and Eqs. (11.4) and (11.7) for the Kamaris et al. (2013) index) for all the members of the frames and the whole range of seismic intensity for every motion were recorded to create a large response databank.

The results of the parametric studies were subjected to non-linear regression analysis with the Levenberg-Marquardt algorithm (MATLAB 2009), producing two expressions, one for the columns and one for the beams, that provide the maximum damage observed at a member as follows:

(1) Equations (11.13) and (11.14) for the Park and Ang (1985) damage index determination in terms of structural and seismic characteristics of the frame and
(2) The following equations for the Kamaris et al. (2013) damage index to be used for the establishment of damage performance levels (damage scale) (Kamaris et al. 2015):

Table 11.5 Statistical parameters of the proposed expressions

Empirical expression	Correlation R^2	Standard deviation σ
Eq. (11.13)	0.75	0.300
Eq. (11.14)	0.91	0.260
Eq. (11.15)	0.74	0.304
Eq. (11.16)	0.91	0.226
Eq. (11.17)	0.72	0.339
Eq. (11.18)	0.87	0.272

$$D_{S,c} = \frac{IDR^3}{132.903 + 4.695 IDR^3}\left(1 - 0.162n + 1.887\sqrt{n}\right) \leq 1.0 \qquad (11.15)$$

for columns (their bases) and

$$D_{S,b} = \frac{IDR^3}{27.692 + 3.947 IDR^3}\left(1 - 0.104n + 1.281\sqrt{n}\right) \leq 1.0 \qquad (11.16)$$

for beams (their ends), where *IDR* is the interstorey drift ratio and n is the number of cycles. Ignoring the effect of the number of cycles in the expressions for damage, Eqs. (11.15) and (11.16) can be replaced by the much simpler equations

$$D_{S,c} = \frac{IDR^3}{27.463 + 0.788 IDR^3} \leq 1.0 \qquad (11.17)$$

$$D_{S,b} = \frac{IDR^3}{8.126 + 0.838 IDR^3} \leq 1.0 \qquad (11.18)$$

On the basis of the above equations and definition of performance levels in terms of *IDR* (FEMA 273 1997), Table 11.1 has been constructed. Table 11.5 shows the basic statistical parameters for the proposed expressions (11.13)–(11.18).

Figure 11.6 shows the maximum damage of columns using the empirical expressions of Eqs. (11.15) and (11.17) together with the 'exact' values coming from dynamic inelastic analyses. The 16% and 84% confidence levels (Vamvatsikos and Cornell 2002), corresponding to the median plus/minus one standard deviation, are also shown in those figures by heavy dashed lines.

Likewise, Fig. 11.7 displays the maximum damage of beams using the empirical expressions of Eqs. (11.16) and (11.18) together with the 'exact' values coming from dynamic inelastic analyses. Finally, Fig. 11.8 shows the response databank 'exact' values and Eqs. (11.15)–(11.18) for column and beam damage as functions of *IDR*. Increase of damage for increasing values of *IDR* is apparent and as a matter of fact, the variation of damage versus *IDR* has the distinct sigmoid form of fragility curves used to estimate the damage probabilities of structures under seismic load, as shown in Fig. 1.12.

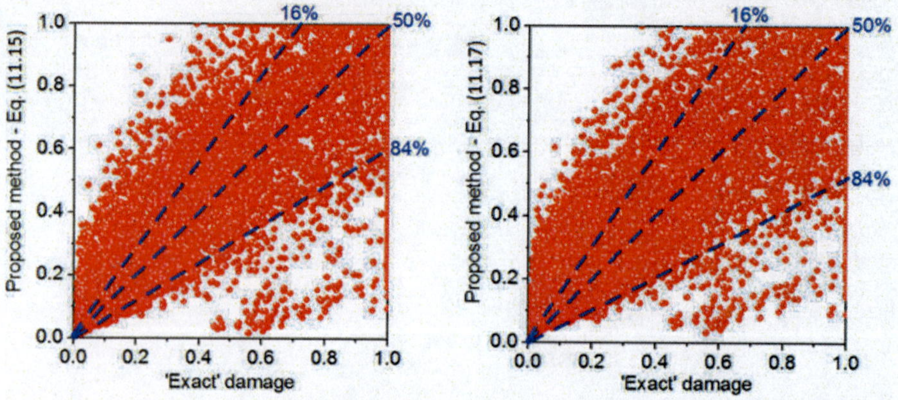

Fig. 11.6 Maximum damage of columns: proposed method versus 'exact' values (after Kamaris 2011, reprinted with permission from UPCE)

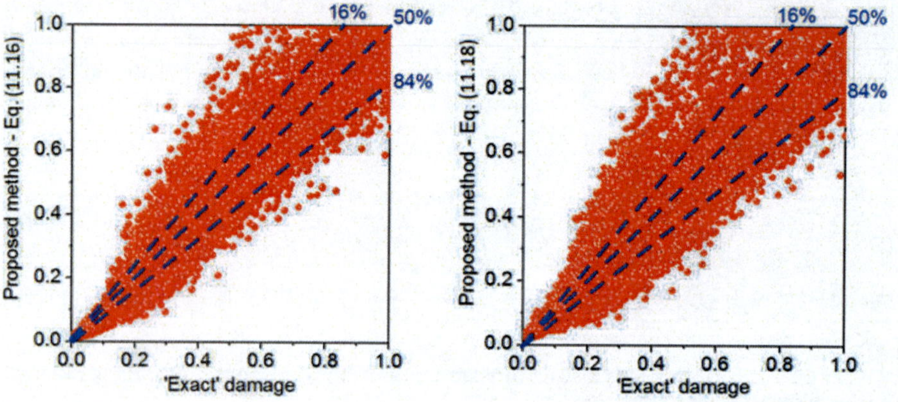

Fig. 11.7 Maximum damage of beams: proposed method versus 'exact' values (after Kamaris 2011, reprinted with permission from UPCE)

Inspection of Figs. 11.6 and 11.7, reveals that the dispersion is lower for the case of beam damage than that for the case of column damage. This scatter is probably due to the fact that the sample of data for columns is much smaller than that for beams. In addition, plastic hinges and hence damage in columns is concentrated in column bases, while damage in beams is more uniformly distributed giving a better sample with less scatter. However, in both cases the accuracy is found to be satisfactory. In any case, in seismic structural analysis, the scatter behavior of results is unavoidable and this has to do with ground motion uncertainties (Lam et al. 1998; Hatzigeorgiou 2010).

Figure 11.9 shows the variation of damage in columns and beams with the number of cycles for two values of *IDR*, as obtained by using Eqs. (11.15) and (11.16). It is observed that increasing values of the number of cycles results in

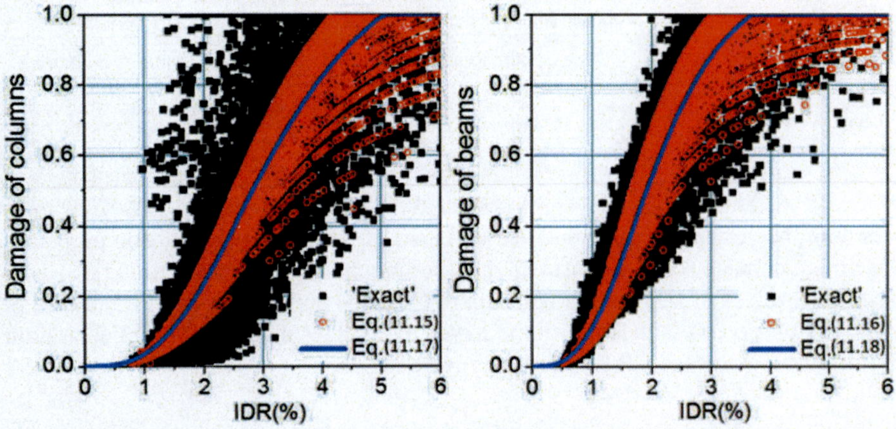

Fig. 11.8 Damage of columns and beams versus *IDR* (after Kamaris 2011, reprinted with permission from UPCE)

Fig. 11.9 Damage of columns and beams versus number of cycles (after Kamaris 2011, reprinted with permission from UPCE)

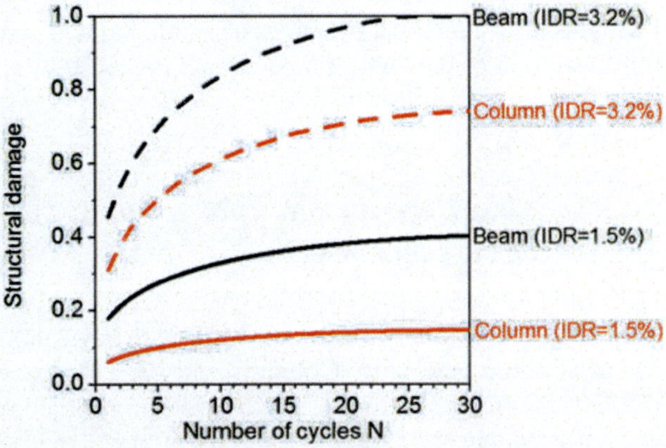

increasing values of structural damage (more in beams than in columns) and that this increase is more pronounced for increasing values of *IDR*. For example, the damage of steel MRFs in the Northridge and Kobe earthquakes was attributed to low cycle fatigue (Raghunandan and Liel 2013). Moreover, the number of cycles increases damage even more in cases of multiple earthquakes (Hatzigeorgiou and Beskos 2009; Hatzigeorgiou 2010; Loulelis et al. 2012).

If, for simplicity reasons, the effect of the number of cycles on damage is not taken into account, Eqs. (11.17) and (11.18) can be used instead of Eqs. (11.15) and (11.16), respectively. Using Eqs.(11.17) and (11.18) in conjunction with the limit values of the maximum *IDR* available in FEMA-273 (1997), a damage scale for beams and columns for the three performance levels of these guidelines, can be constructed. This damage scale is shown in Table 11.1. One can observe that damage is very high for the CP performance level, especially for columns. This comes as a

Table 11.6 Damage scale proposed here for the performance levels of SEAOC (1999)

Performance Levels	Maximum column damage	Maximum beam damage
IO	≤1%	≤2%
LS	≤11%	≤31%
CP	≤39%	≤74%

result of the very high *IDR* value of 5% for the CP level. For comparison purposes, one more damage scale is constructed by combining Eqs. (11.17) and (11.18) with the mean values of *IDR* available in SEAOC (1999) and is provided in Table 11.6. One can observe now that due to the lower *IDR* values used for this scale, column damage values for the CP level are also lower.

11.5 Examples of Application: Dynamic

In this section, numerical examples pertaining to two plane steel MRFs are presented for illustration of the DDCD method and demonstration of its merits.

11.5.1 *First Design Option for a Three-Storey Three-Bay Plane Steel MRF*

A three-storey three-bay plane steel MRF with bay width of 5.0 m and story height of 3.0 m is considered here. Columns and beams consist of HEB240 sections and IPE330 sections, respectively, while the steel material is of S235 grade. The frame is under a gravity load 27.5 kN/m (dead plus 0.3 of live load) and a lateral seismic load. The frame has been designed by SAP 2000 (2007) software according to the EC3 (2009) and EC8 (2004) codes for $PGA = 0.35$ g and soil type B. Its first natural period is equal to 0.73 s, while its maximum *IDR* value was found to be 1.6 % $< 2.5\%$ for the LS performance level. Even though this clearly indicates that the frame has been overdesigned, this creates no problem for the purpose of this example.

The above frame, under the ground motion No. 3 of Table 11.4, is analyzed by the proposed DDCD method and found to exhibit a damage distribution as the one depicted in Fig. 11.10a, characterized by low damage values everywhere. However, the damage pattern consisting of concentrated damages only at the beam ends and column bases, indicates that, for a higher enough seismic intensity, the resulting plastic collapse mechanism will be of the global type and the capacity design rule of "weak beams–strong columns" will be satisfied. Indeed, this can be accomplished by scaling up the above ground motion by a scale factor of 2.25, as Fig. 11.10b clearly shows.

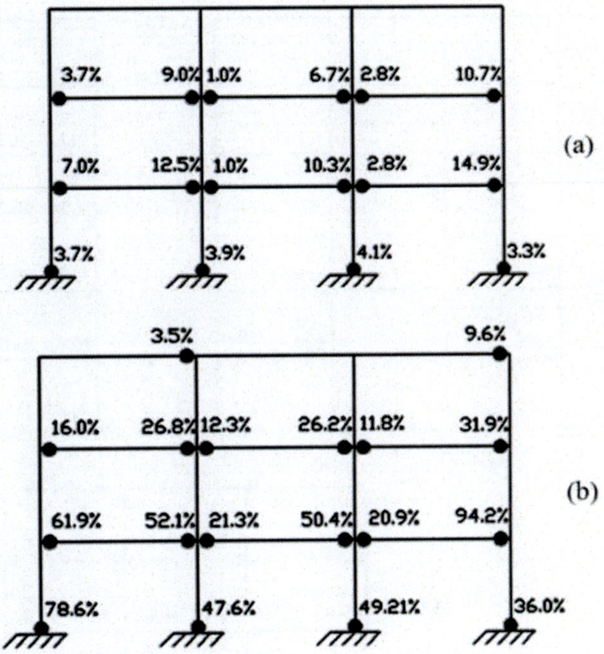

Fig. 11.10 Damage indices in members of the frame of Sect. 11.5.1: (**a**) for ground motion No. 3 of Table 11.4; (**b**) for scaled up ground motion No. 3 by a factor of 2.25 (after Kamaris 2011, reprinted with permission from UPCE)

11.5.2 Second Design Option for a Six-Storey Three-Bay Plane Steel MRF

Consider the six-storey three-bay steel MRF of Fig. 11.4 with columns and beams having HEB and IPE sections, respectively, of steel grade S235. Gravity load is assumed to be 27.5 kN/m, while the seismic load is characterized by a $PGA = 0.35$ g and soil type B. This frame will be dimensioned by the proposed DDCD method so as to satisfy the damage limit values of the three performance levels: IO under the FOE, LS under the DBE and CP under the MCE. The FOE, DBE and MCE are expressed through design spectra of EC8 (2004) which have *PGA* values equal to $0.3 \times PGA_{\mathrm{DBE}}$, $1.0 \times PGA_{\mathrm{DBE}}$ and $1.5 \times PGA_{\mathrm{DBE}}$, respectively, where PGA_{DBE} is equal to 0.35 g. NLTH analyses of the frame are performed by using 8 semi-artificial accelerograms compatible with the spectrum corresponding to each performance level. Thus, a level of damage that is in agreement with the limits of Table 11.1 is selected for the beams and the columns for every performance level. The same limits can be computed by Eqs. (11.17) and (11.18) on the basis of the values of the maximum *IDR* provided in FEMA 273 (1997).

For the IO performance level, the sections were selected to be (HEB260-IPE330) for the first storey and (HEB240-IPE330) for the rest of stories, giving maximum damage values of 3.0% and 0.0% for the beams and the columns, respectively, for the case of the eighth accelerogram, as shown in Fig. 11.11a. These values do not exceed the limit values of 4% and 1%, respectively of Table 11.1.

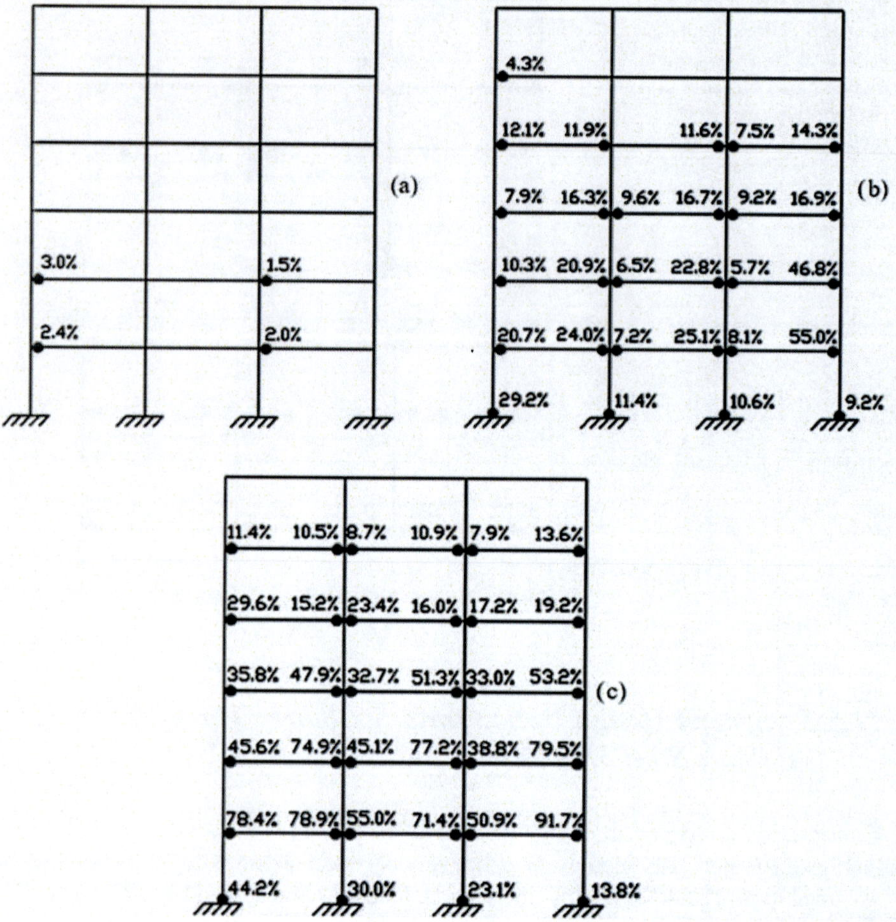

Fig. 11.11 Distribution of damage in the frame of Sect. 11.5.2: (**a**) IO performance level; (**b**) LS performance level; (**c**) CP performance level (after Kamaris 2011, reprinted with permission from UPCE)

The sections IPE330 for beams and HEB240 for columns in all stories were selected for the LS performance level giving maximum damage values of 55.0% and 29.2% for the beams and the columns, respectively, for the case of the second accelerogram, as shown in Fig. 11.11b. These values do not exceed the limit values of 74% and 39% for beams and columns, respectively, of Table 11.1.

The design for the CP performance level resulted in (HEB260-IPE360) sections for the first storey and (HEB260-IPE330) sections for the rest of the frame, leading to maximum damage values of 91.7% and 44.2% for the beams and the columns, respectively, for the case of the third accelerogram, as shown in Fig. 11.11c. These values do not exceed the limit values of 100% and 99%, for beams and columns, respectively, of Table 11.1.

Thus, the CP performance level, resulting in the heaviest sections with a total weight of 109.9 kN, controls the design. This design is lighter by 7.5% than the one coming from the use of EC3 (2009) and EC8 (2004) codes and resulting in a frame with (HEB280-IPE360) sections for the first four stories and (HEB260-IPE330) sections for the next two higher ones and total weight of 118.7 kN. It also satisfies the capacity design as plastic hinges are formed only at the ends of beams and column bases of the first floor, giving weak-beam-strong-column ratios in the range of 1.25–3.0.

11.5.3 Third Design Option for a Three-Storey Three-Bay Plane Steel MRF

The frame of the example of Sect. 11.5.1 consisting of HEB240 sections for columns and IPE330 sections for beams is subjected now to the Cape Mendocino-1992 ground motion of Table 11.4 with $PGA = 0.154$ g. For this frame, the *PGA* value of the above motion required for the development of 30% and 55% maximum damage in its columns and beams, respectively, is determined by the proposed DDCD method. To this end, the frame is analyzed with the aid of the Ruaumoko 2D (Carr 2005) software by performing a NLTH analysis with increasing *PGA* values until values of damage less or equal to the desired ones are developed. It is found that for a value of $PGA = 0.80$ g, the maximum values of damage at columns and beams become equal to 29.8% and 54.8%, respectively, indicating that this is the required seismic intensity to develop the desired level of damage in the frame.

For this $PGA = 0.80$ g, the maximum number of inelastic cycles is found to be equal to 14 for the columns and 17 for the beams, while the maximum *IDR* is 2.0%. Employment of Eqs. (11.17) and (11.18), results in maximum values of damage in columns and beams 23.7% and 53.9%, respectively. These values are very close to those computed by NLTH analysis. Application of the more accurate Eqs. (11.15) and (11.16), which take into account the number of cycles, gives the values 27.2% and 60.9%, which are also very close to the computed ones. In addition, the estimation of the maximum damage in columns calculated by Eq. (11.15) is much more accurate than the one given by Eq. (11.17). Finally, the distribution of damage in this frame is shown in Fig. 11.12.

11.6 Direct Damage-Controlled Design: Static (Pushover)

11.6.1 Damage Expressions and Performance Levels

The DDCD method in its static version can also be used for the seismic design of plane steel MRFs by assuming the seismic forces to be static and act laterally on the

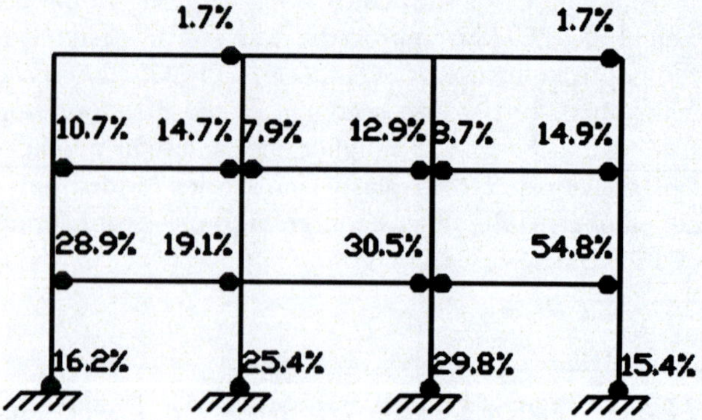

Fig. 11.12 Distribution of damage in frame of Sect. 11.5.3 (after Kamaris 2011, reprinted with permission from UPCE)

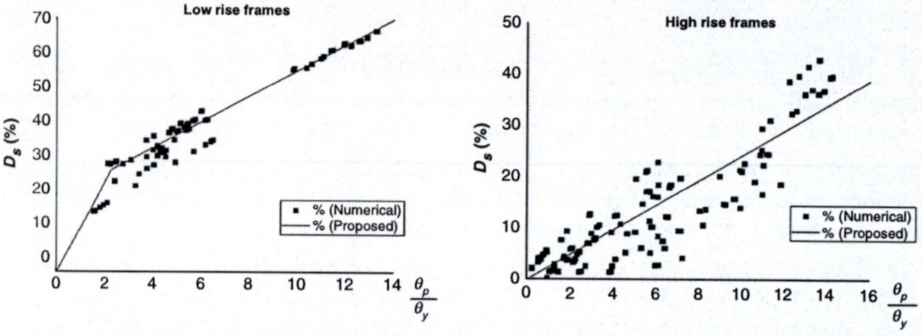

Fig. 11.13 D_s versus θ_p/θ_y curves for low- and high-rise frames (after Kamaris et al. 2009, reprinted with permission from MSP)

frame in a pushover fashion. These lateral forces are progressively increased until collapse of the frame, which is also subjected to the vertical gravity load combination $G + 0.3Q$, where G and Q stand for dead and live loading, respectively. The method can be used in conjunction with any nonlinear analysis program, like Ruaumoko 2D (Carr 2005) or DRAIN-2DX (Prakash et al. 1993), which can take into account material and geometric nonlinearities. As in the case of the dynamic version of the method, the user has three design options at his disposal in connection with damage-controlled steel design: (i) determine damage in any member of a given structure under given lateral load; (ii) dimension the structure for given loading and given target damage; (iii) determine the maximum lateral loading a given structure can sustain for a given target damage.

For all these options one needs expressions for damage at the local and global levels as well as limit values for various damage performance levels.

Damage determination can be done on the basis of Kamaris et al. (2013) damage index described by Eq. (11.4) in conjunction with Eqs. (11.5) and (11.7) with $f = 1$. Using the 36 frames of Table 11.3 under vertical and lateral load, one can construct a response databank including deformation and damage (computed as described above). Figure 11.13 shows the variation of the section damage index D_s versus

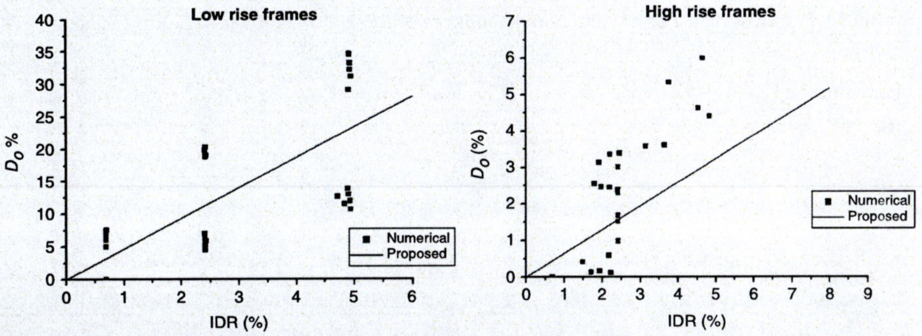

Fig. 11.14 D_o versus *IDR* curves for low- and high-rise frames (after Kamaris et al. 2009, reprinted with permission from MSP)

the ratio θ_p/θ_y for low-rise (3 and 6 stories) and high-rise (9, 12, 15 and 20 stories) frames, where θ_p and θ_y are the plastic and yield rotation at member end, respectively. The rotation at yield θ_y is given by (FEMA 273 1997)

$$\theta_y = M_{pl}L/6EI \tag{11.19}$$

where M_{pl} is the plastic moment, L and I are the length and second moment of inertia of the section of the member and E is the modulus of elasticity. When there is an axial compressive force P in the member, the right hand side of Eq. (11.19) is multiplied by the factor $1 - P/P_y$, where P_y is the axial force at yield of the member.

Fig. 11.14 shows the variation of the overall damage index D_o versus *IDR* for low- and high-rise frames. Using the method of least squares the mean values of these variations were determined and plotted as straight line segments in Figs. 11.13 and 11.14. The analytical expressions of these lines are of the form (Kamaris et al. 2009):

For the low-rise frames:

$$\begin{aligned} D_s &= 12.526(\theta_p/\theta_y) \; for \quad \theta_p/\theta_y \leq 2.2 \\ D_s &= 3.54(\theta_p/\theta_y) + 20.14 \; for \quad \theta_p/\theta_y > 2.2 \end{aligned} \tag{11.20}$$

$$D_O = 4.67IDR \tag{11.21}$$

For the high rise frames:

$$D_s = 2.42(\theta_p/\theta_y) \tag{11.22}$$

$$D_O = 0.94IDR \tag{11.23}$$

Table 11.7 Performance levels and corresponding section and overall damage limit values

	D_s		D_o	
Performance Levels	Low rise frames	High rise frames	Low rise frames	High rise frames
IO	≤13%	≤3%	≤3%	≤1%
LS	≤40%	≤15%	≤12%	≤2%
CP	≤50%	≤20%	≤24%	≤5%

The coefficient of determination R^2 of Eqs. (11.20) and (11.22) is 0.96 and 0.79, respectively, indicating that there is a good correlation between the section damage and the plastic hinge rotation. On the contrary, the correlation between structure damage and the *IDR* is not so good as the coefficient of determination is 0.53 and 0.72 for Eqs. (11.21) and (11.23), respectively.

Use of the θ_p and *IDR* values available in FEMA 273 (1997) for the three performance levels of Table 11.2 into Eqs. (11.20)–(11.23), limit values for section and overall damage performance levels are determined for low and high rise frames and presented in Table 11.7. It is clear from Eq. (11.12) that even if there are large values of section damage in a few sections, the overall damage will have a small value because of the small or zero values in other sections. For this reason, the overall damage index is not considered as a representative one, and the section damage index is mostly used in the applications.

11.6.2 Example of Seismic Design of a Plane Steel MRF by Pushover Analysis

Consider a S235 plane steel MRF of three bays and three storeys. The bay width is assumed equal to 5.0 m and the storey height equal to 3.0 m. The gravity load combination $G + 0.3Q$ is equal to 27.5 kN/m. Column and beams are assumed to have HEB and IPE profiles, respectively. The frame was designed according to EC3 (2009) and EC8 (2004) codes for a $PGA = 0.40$ g, a soil type D and a behavior factor $q = 4$ with the aid of SAP 2000 (2007). Thus, for a design base shear of 355 kN, the following column and beam sections were obtained for the three stories: (HEB280-IPE360) + (HEB260-IPE330) + (HEB240-IPE300) going from the first to the third storey. For a maximum elastic roof displacement found equal to 0.0465 m, use of the equal displacement rule, can provide the corresponding inelastic one as $0.0465q = 0.186$ m.

The above frame is now analyzed with the aid of static inelastic (pushover) analysis using an inverted triangle type of heightwise distribution of the lateral forces. These lateral forces are progressively increased until the maximum inelastic roof displacement of the frame becomes equal to the previously computed value of 0.186 m.

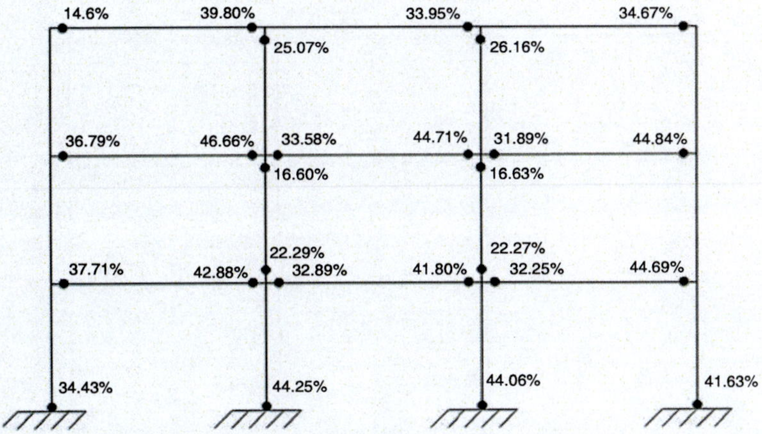

Fig. 11.15 Damage distribution in the frame of Sect. 11.6.2 designed according to EC3 (2009) and EC8 (2004) codes (after Kamaris et al. 2009, reprinted with permission from MSP)

Figure 11.15 provides the damage distribution in the frame for this maximum value of the roof displacement. It is observed that plastic hinges are formed both in beams and columns, clearly indicating that the capacity design rule is not satisfied. Damage values are up to about 47% in the beams and up to 26% in columns (44% at their bases). The DDCD method can avoid this undesirable formation of plastic hinges in the columns, because it can directly control damage and plastic hinge formation in the frame. Thus, this frame is designed here for the CP performance level of Table 11.7 by assuming target damage of 45% in the beams and 0% in all columns except those of the first floor where the target damage at their bases is 40%. For this target damage distribution and design base shear computed with the aid of the EC8 (2004) spectrum, the sections of the frame are obtained. For the resulting frame the pushover curve is constructed and used to determine the elastic displacement for the above base shear. This displacement is multiplied by $q = 4$ in order to find the maximum inelastic one and hence the corresponding base shear from the pushover curve. For this base shear the distribution of damage is obtained. If this distribution is in accordance with the target one, the selected sections are acceptable. Otherwise, the sections are changed and the previous procedure is repeated. Thus, for the damage distribution of Fig. 11.16 with damage values up to about 44% in the beams and up to 37% in column bases, the column and beam sections for the three stories of the frame were found to be (HEB300-IPE330) + (HEB300-IPE330) + (HEB280-IPE300). This selection results in a global collapse mechanism satisfying completely the capacity design rule. The total weight of this design by using the pushover method is determined to be 55.9 kN, i.e., heavier by only 1.8% than the corresponding one of 54.7 kN for the design according to the EC8 (2004) method.

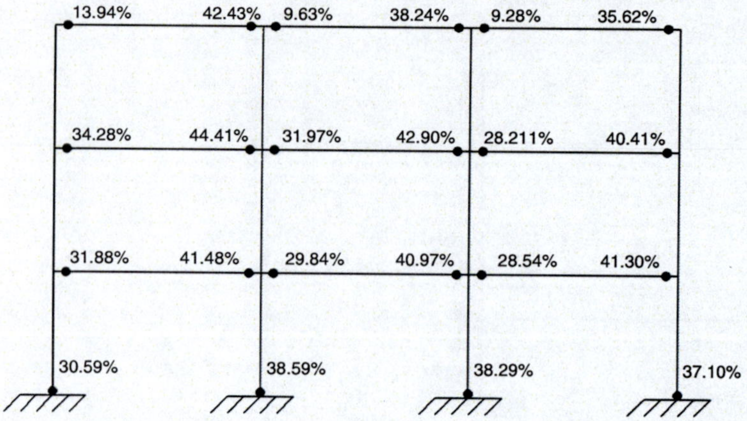

Fig. 11.16 Damage distribution in the frame of Sect. 11.6.2 designed according to DDCD (after Kamaris et al. 2009, reprinted with permission from MSP)

11.7 Conclusions

On the basis of the preceding discussion, the following conclusions can be drawn:

(1) In this chapter, the direct damage-controlled design (DDCD) method for the seismic design of plane steel moment resisting frames subjected to far-fault ground motions has been presented. This method in its dynamic or static (pushover) version can directly control damage in a structure at the local or global level. The method is employed with the aid of the finite element method taking into account material and geometric nonlinearities and works in its dynamic (time domain) or static (pushover) version.

(2) It employs a damage index that takes into account axial force-bending moment interaction at a member section as well as cyclic strength and stiffness degradation in conjunction with low-cycle fatigue. The same damage index can also be used for the static (pushover) version of the method without degradation and fatigue. The damage index of Park and Ang can also be employed for the dynamic version of the method.

(3) It considers three damage performance levels with damage limit values determined based on extensive parametric studies involving many frames under many ground motions for the creation of a large response databank. This databank also serves to construct empirical expressions for damage at a member section. Damage expressions and damage limit values are different for columns and beams. Thus, selection of very small damage values for columns and larger ones for beams is possible for an easy satisfaction of capacity design rules.

(4) The designer is allowed to follow either one of the following three design options: determine the damage level for a given structure under any given seismic load, or dimension a structure for given seismic load and desired level

of damage, or determine the maximum seismic load a designed structure can sustain in order to exhibit a desired level of damage.

(5) The method controls damage in a structure in a more direct and accurate way than other methods and, at least for the examples considered here, results in designs very close to the ones coming from the use of the EC8 code. However, the collapse mechanism of the method is always of the global type, which is not always the case with the EC8 design. A disadvantage of the method is that it requires nonlinear material and geometrical analyses under static or dynamic (seismic) lateral loading.

References

ATC 13 (1985) Earthquake damage evaluation for California. Applied Technology Council, Redwood, CA

Ballio G, Castiglioni CA (1994) An approach to the seismic design of steel structures based on cumulative damage criteria. Earthq Eng Struct Dyn 23:969–986

Bozorgnia Y, Bertero VV (2004) Damage spectrum and its applications to performance-based earthquake engineering. In: Proceedings of 13th World Conference on Earthquake Engineering, Vancouver B.C., Canada, August 1–6, Paper No 1497

Carr AJ (2005) Ruaumoko 2D and 3D: programs for inelastic dynamic analysis. Theory and user guide to associated programs. Department of Civil Engineering, University of Canterbury, Christchurch

Castiglioni CA, Pucinotti R (2009) Failure criteria and cumulative damage models for steel components under cyclic loading. J Constr Steel Res 65:751–765

Cervera M, Oliver J, Faria R (1995) Seismic evaluation of concrete dams via continuum damage models. Earthq Eng Struct Dyn 24:1225–1245

Cosenza E, Manfredi G, Ramasco R (1993) The use of damage functionals in earthquake engineering: a comparison between different methods. Earthq Eng Struct Dyn 22:855–868

EC3 (1992) Eurocode 3, Design of steel structures – Part 1-1: general rules and rules for buildings, ENV 1993-1-1. European Committee for Standardization (CEN), Brussels

EC3 (2009) Eurocode 3, Design of steel structures – Part 1-1: general rules and rules for buildings, EN 1993-1-1. European Committee for Standardization (CEN), Brussels

EC8 (2004) Eurocode 8, Design of structures for earthquake resistance, Part 1: general rules, seismic actions and rules for buildings, EN 1998-1-1. European Committee for Standardization (CEN), Brussels

FEMA 273 (1997) NEHRP guidelines for the seismic rehabilitation of buildings. Federal Emergency Management Agency, Washington, DC

Ghobarah A (2004) On drift limits associated with different damage levels. In: Fajfar P, Krawinkler H (eds) Performance-based seismic design concepts and implementation, PEER Report 2004/05. University of California, Berkeley, CA, pp 321–332

Ghobarah A, Safar M (2010) A damage spectrum for performance-based design. In: Fardis MN (ed) Advances in performance-based earthquake engineering. Springer Science, Berlin, pp 193–201

Ghobarah A, Abou-Elfath H, Biddah A (1999) Response-based damage assessment of structures. Earthq Eng Struct Dyn 28:79–104

Hatzigeorgiou GD (2010) Behaviour factors for nonlinear structures subjected to multiple near-fault earthquakes. Comput Struct 88:309–321

Hatzigeorgiou GD, Beskos DE (2009) Inelastic displacement ratios for SDOF structures subjected to repeated earthquakes. Eng Struct 31:2744–2755

Kamaris GS (2011) New method of seismic design of steel structures with direct damage control. Ph.D. Thesis, Department of Civil Engineering, University of Patras, Patras, Greece (in Greek)
Kamaris GS, Hatzigeorgiou GD, Beskos DE (2009) Direct damage controlled design of plane steel-moment resisting frames using static inelastic analysis. J Mech Mater Struct 4:1375–1393
Kamaris GS, Vallianatou YM, Beskos DE (2012) Seismic damage estimation of in-plane regular steel moment resisting and X-braced frames. Bull Earthq Eng 10:1745–1766
Kamaris GS, Hatzigeorgiou GD, Beskos DE (2013) A new damage index for plane steel frames exhibiting strength and stiffness degradation under seismic motion. Eng Struct 46:727–736
Kamaris GS, Hatzigeorgiou GD, Beskos DE (2015) Direct damage controlled seismic design of plane steel degrading frames. Bull Earthq Eng 13:587–612
Karavasilis TL, Bazeos N, Beskos DE (2007) Behavior factor for performance-based seismic design of plane steel moment resisting frames. J Earthq Eng 11:531–559
Karavasilis TL, Bazeos N, Beskos DE (2008) Drift and ductility estimates in regular steel MRF subjected to ordinary ground motions: a design oriented approach. Earthquake Spectra 24:431–451
Kawashima K, Aizawa K (1986) Earthquake response spectra taking account of number of response cycles. Earthq Eng Struct Dyn 14:185–197
Kunnath SK, Chai YH (2004) Cumulative damage-based inelastic cyclic demand spectrum. Earthq Eng Struct Dyn 33:499–520
Lam N, Wilson J, Hutchinson G (1998) The ductility reduction factor in the seismic design of buildings. Earthq Eng Struct Dyn 27:749–769
Lemaitre J (1996) A course on damage mechanics. Springer, Berlin
Loulelis DG, Hatzigeorgiou GD, Beskos DE (2012) Moment resisting steel frames under repeated earthquakes. Earthq Struct 3:231–248
Lu Y, Wei J (2008) Damage-based inelastic response spectra for seismic design incorporating performance considerations. Soil Dyn Earthq Eng 28:536–549
MATLAB (2009) The language of technical computing, Version 2009a. The Mathworks Inc., Natick, MA
Panyakapo P (2004) Evaluation of site-dependent constant-damage design spectra for reinforced concrete structures. Earthq Eng Struct Dyn 33:1211–1231
Panyakapo P (2008) Seismic capacity diagram for damage based design. In: Proceedings of 14th World Conference on Earthquake Engineering, Beijing, China
Park YJ, Ang AHS (1985) Mechanistic seismic damage model for reinforced concrete. J Struct Eng ASCE 111:722–739
Park YJ, Ang AHS, Wen YK (1987) Damage-limiting aseismic design of buildings. Earthquake Spectra 3:1–26
PEER (2009) Pacific Earthquake Engineering Research Center, Strong Ground Motion Database, Berkeley, CA. http://peer.berkeley.edu/
Powell GH, Allahabadi R (1988) Seismic damage prediction by deterministic methods: concepts and procedures. Earthq Eng Struct Dyn 16:719–734
Prakash V, Powell GH, Campbell S (1993) DRAIN-2DX, Base program description and user guide, Version 1.10, Report No UCB/SEMM-93/17. University of California, Berkeley, CA
Raghunandan M, Liel AB (2013) Effect of ground motion duration on earthquake-induced structural collapse. Struct Saf 41:119–133
SAP 2000 (2007) Structural analysis program 2000, static and dynamic finite element analysis of structures, Version 11. Computers and Structures Inc., Berkeley, CA
SEAOC (1999) Recommended lateral force requirements and commentary, 7th edn. Structural Engineers Association of California, Sacramento, CA
Sucuoğlu H, Erberik A (2004) Energy-based hysteresis and damage models for deteriorating systems. Earthq Eng Struct Dyn 33:69–88
Tiwari AK, Gupta VK (2000) Scaling of ductility and damage-based strength reduction factors for horizontal motions. Earthq Eng Struct Dyn 29:969–987
Vamvatsikos D, Cornell CA (2002) Incremental dynamic analysis. Earthq Eng Struct Dyn 31:491–514

Chapter 12
Design Using Seismic Isolation

Abstract This chapter presents various methods for the seismic design of steel building structures equipped at their base by seismic isolation devices. The most well-known isolation devices are the lead rubber bearings and the friction pendulum bearings. These isolation devices or isolators succeed to uncouple the seismic response of the structure from the ground motion and thus to reduce the structural seismic forces. There are basically two kinds of design methods for base isolated steel building frames: the force-based and the displacement-based ones. The design methods according to ASCE provisions are either the equivalent lateral force method or methods based in dynamic analysis including the response spectrum analysis and the nonlinear time-history analysis. The design methods according to Eurocode 8 are analogous to the aforementioned ones of the ASCE provisions. A method using an improved linear analysis is also presented. In addition to these force-based methods of design employing acceleration spectra, a displacement-based design method employing displacement spectra for high damping values is also presented. Three numerical examples involving steel building frames are presented to illustrate the design methods of ASCE and Eurocode 8 as well as the method using the improved simplified linear analysis and demonstrate the effectiveness of base isolation in seismic design.

Keywords Base isolation · Isolators · Lead rubber bearings · Friction pendulum bearings · Equivalent lateral force procedure · Dynamic response spectrum analysis procedure · Simplified linear analysis procedure · Displacement-based design approach

12.1 Introduction

Seismic isolation or base isolation is a design method employed to uncouple the response of a structure from the earthquake motion in order to reduce the seismic forces. Seismic isolation has been evolved over the last 40 years to a mature technology and specialized guidelines for the design and construction of seismically

G. A. Papagiannopoulos et al., *Seismic Design Methods for Steel Building Structures*, Geotechnical, Geological and Earthquake Engineering 51,
https://doi.org/10.1007/978-3-030-80687-3_12

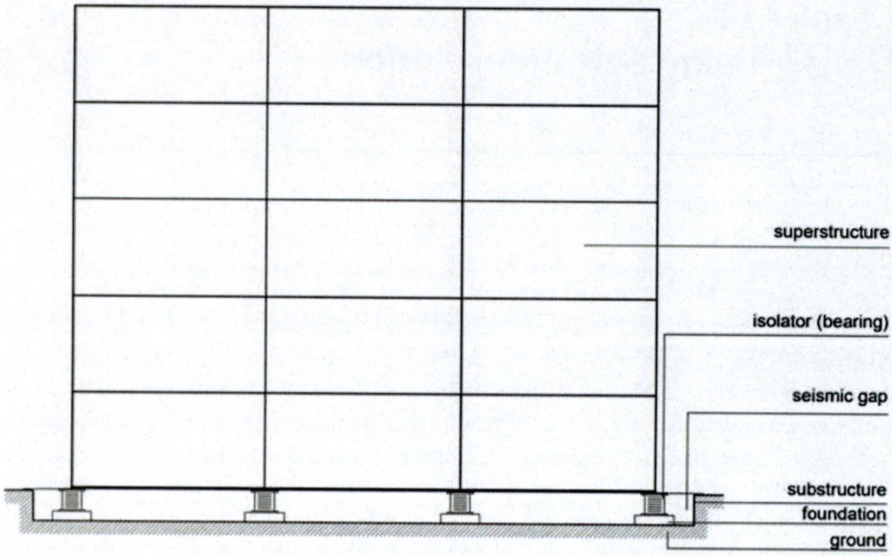

Fig. 12.1 Base-isolated steel structure

isolated structures have been codified along with standardized testing procedures of isolation devices. The origins and early developments of base isolation can be found in Makris (2019), whereas recent state-of-the art reviews on base isolation have been presented by Warn and Ryan (2012) and De Luca and Guidi (2019). For a more in depth study on base isolation one can look at the books of Kelly (1997), Naeim and Kelly (1999), Komodromos (2000), Cheng et al. (2008), Kelly and Konstantinidis (2011) and Elghazouli (2017) as well as the provisions of ASCE/SEI 7-16 (2017) and EC8 (2004).

Figure 12.1 displays a steel building structure equipped with a base isolation system. The building structure above the isolation level is termed as superstructure. The isolation system consists of a number of isolators (bearings or isolation devices) supported by foundations. Rigid diaphragms provided above and under the isolation system, consisting of a reinforced concrete slab and a grid of tie-beams, respectively, constitute the substructure. The isolators are fixed at both ends to the rigid diaphragms either directly or by means of vertical elements. The elements above and below the isolation system have to be sufficiently rigid in both horizontal and vertical directions in order to minimize the effect of differential seismic ground displacements. A seismic gap is left at the isolation level that permits the unrestricted lateral deformation of the isolators and accommodates the drift of the superstructure. During seismic shaking the superstructure experiences essentially a rigid-body motion, whereas the isolators are significantly strained.

In general, the design of seismically isolated structures is expected to satisfy the following two requirements: (i) resist minor or moderate earthquake motion without damage to structural elements and non-structural components and (ii) resist major or severe earthquake motion without failure of the isolation system and without significant or extensive damage to structural elements and non-structural components. The first requirement is satisfied by limiting the interstorey drift to both superstructure and substructure. The second requirement is satisfied by ensuring that the ultimate capacity of the bearings in terms of strength and deformation, is not exceeded, while the superstructure and substructure should remain in the elastic range of seismic response. A minimum separation (seismic gap) between the isolated structure and other structures is also a mandatory design requirement with respect to the overall expected behavior of a base-isolated structure during the major or severe earthquake motion. The large displacement demand of isolated structures is certainly a disadvantage of the method. Use of supplemental dampers can reduce these demands (Makris and Chang 2000; Chang et al. 2002; Providakis 2008, 2009). However, the added damping in the seismic isolation system should be viewed with caution as it can increase interstorey drifts and floor accelerations of the superstructure (Kelly 1999).

An additional design requirement needed to be satisfied by the isolation system is that of a sufficient restoring force at the design displacement. This requirement ensures that the isolation system limits the residual seismic displacements and mitigates structural damage or collapse that may occur by the earthquake aftershocks. To satisfy this requirement, the isolating system should possess a significant post-yielding stiffness. However, the restoring force may reduce the beneficial uncoupling effect offered by the isolation system to the structure. Nevertheless, this uncoupling, even though is only partially achieved in praxis, may result in a structural response that is considerably reduced in comparison to that of a conventional structure which is fully attached to the ground.

The effectiveness of an isolation system essentially depends on the size, number and type of the bearings chosen. Letting aside the issues of size and number, the choice behind the type of bearings employed depends on the anticipated seismic performance and on several requirements with respect to their maintenance, robustness and economy of use (Cheng et al. 2008). Several types of bearings are nowadays available and they can be roughly separated into elastomeric bearings and sliding (friction-based) systems. Common elastomeric bearings, shown in Fig. 12.2a, are comprised of alternating horizontal layers of elastomeric material (rubber) and steel plates, which are vulcanized together under high pressure and temperature (laminated rubber bearing). To increase their damping capacity, use of high-damping rubber (HDR) is made, whereas by replacing the central portion of the rubber with a lead plug, the initial stiffness of the bearing is increased. Thus, the lead-rubber bearings (LRB), shown in Fig. 12.2b, possess enough flexibility, damping capacity and the initial rigidity needed to sustain minor to moderate wind and earthquake actions.

On the other hand, the concept of the sliding bearings is to permit the structure to slide on a controlled surface by dissipating energy through friction. A single friction pendulum bearing (FPB), shown in Fig. 12.3a, makes use of an articulated slider on a spherical sliding surface and by using gravity as a restoring force, reduces the movement of the structure without sacrificing the effectiveness of the isolation. To

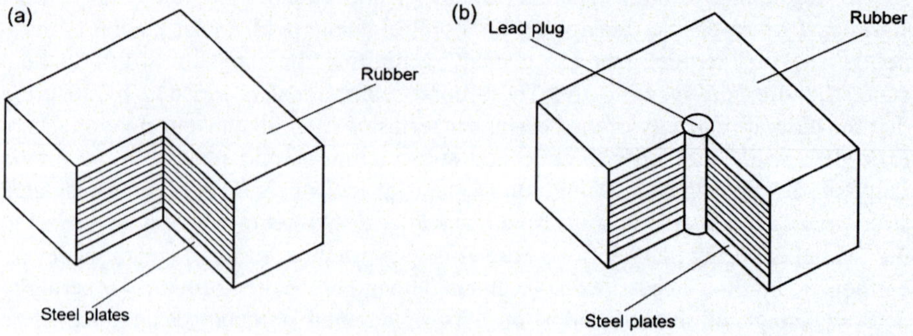

Fig. 12.2 Elastomeric (**a**) and lead-rubber (**b**) bearings

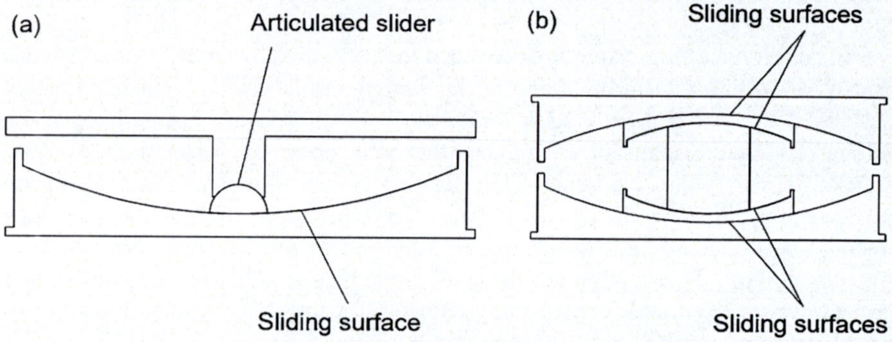

Fig. 12.3 Single (**a**) and triple (**b**) friction pendulum bearings

address the issue of reducing the dimensions of the isolator as well as that of accommodating larger displacement demands, multi-spherical sliding bearings have been proposed (Fenz and Constantinou 2008). Among them, the triple friction pendulum bearing, shown in Fig. 12.3b, is a vastly improved version of FPBs and it is able to control small, moderate or large displacements independently. For more details on the mechanics, engineering properties and models of elastomeric and sliding bearings one can consult Kelly and Konstantinidis (2011), Skinner et al. (2011) and Sarlis and Constantinou (2016). Figure 12.4 shows a typical force F versus displacement D relation for both LRB and FPB isolators, which is based on a bilinear hysteretic modeling.

In the following, certain procedures to design seismically isolated steel structures are presented in detail. For a detailed overview of the major international codes for the design of seismically isolated buildings the reader may consult, e.g., Pietra et al. (2015) and Yenidogan and Erdik (2016).

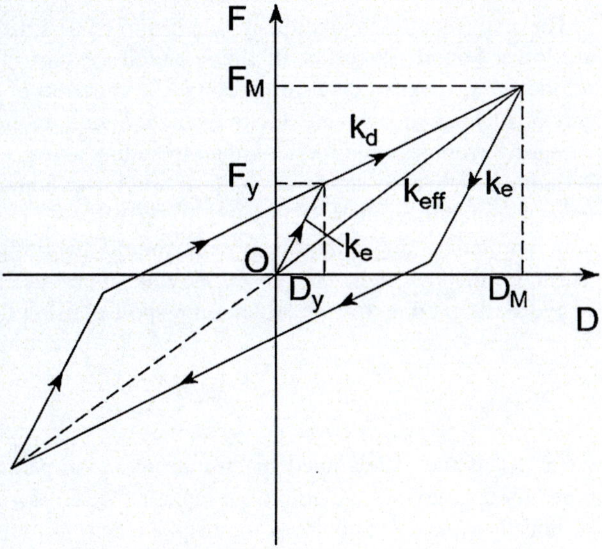

Fig. 12.4 Bilinear hysteretic force-displacement relation for LRB and FPB isolators

12.2 The Design Procedures of ASCE/SEI 7-16

12.2.1 Equivalent Lateral Force Procedure

This procedure is permitted to be used for the design of seismically isolated steel structures upon satisfaction of the following requirements: (i) the structure is founded on a site class A, B, C or D site (this site classification is different than that of EC8 2004); (ii) the effective period of the isolated structure at the maximum displacement D_M is less than or equal to 5.0 s; (iii) the structure above the isolation system is less than or equal to four stories of 19.8 m in height measured from the base level; (iv) the effective damping of the isolation system is less than or equal to 30%; (v) the effective period of the isolated structure is greater than three times the elastic fixed-base period of the structure above the isolation system; (vi) the structure does not have any irregularity and (vii) the isolation system meets certain criteria regarding its effective stiffness at the maximum displacement and its total maximum displacement. These criteria have as follows: (a) the effective stiffness of the isolation system at the maximum displacement is greater than one-third of the effective stiffness at 20% of the maximum displacement; (b) the isolation system should produce a restoring force such that the lateral force at the corresponding maximum displacement is at least 0.025 W greater than the lateral force at 50% of the corresponding maximum displacement, where W is the effective seismic weight of the structure above the isolation interface; (c) the isolation system does not limit the maximum earthquake displacement to less than the total maximum displacement D_{TM} which is defined in the following.

The analysis of the isolation system has to be performed separately for the upper and lower bound properties of the isolation devices. These bound properties are introduced in order to take into account for variation of the nominal design parameters of each isolator device due to cyclic dynamic motion, loading rate, variability of bearing properties in production, temperature, aging, environmental exposure etc. Default multipliers to define these upper and lower bound properties of the isolation devices are tabulated in ASCE/SEI 7-16 (2017).

In particular, the isolation system should be designed to accommodate the maximum displacement D_M, by employing upper and lower bound properties, in the most critical direction of structural response, using the expression

$$D_M = \frac{gS_{M1}T_M}{4\pi^2 B_M} \tag{12.1}$$

where S_{M1} is the 5%-damped spectral acceleration parameter at 1.0 s period determined for the maximum credible earthquake (MCE), T_M is the effective period of the seismically isolated structure at the displacement D_M and B_M is a numerical coefficient that depends on the effective damping ratio β_M of the isolation system at the displacement D_M. The range of the values of the B_M coefficient is 0.8–2.0 for a corresponding range of values 2–50% of the effective damping ratio (ASCE/SEI 7-16 2017). The effective period T_M at the displacement D_M is given by

$$T_M = 2\pi\sqrt{\frac{W}{k_M g}} \tag{12.2}$$

where W is the effective seismic weight of the structure above the isolation interface and k_M is the effective stiffness of the isolation system at the maximum displacement D_M. The latter is determined by the equation

$$k_M = \frac{\sum|F_M^+| + \sum|F_M^-|}{2D_M} \tag{12.3}$$

where $\sum|F_M^+|$ and $\sum|F_M^-|$ is the sum of all isolator devices at the positive and the negative displacement D_M, respectively. The total maximum displacement of the elements of the isolation system D_{TM} is given as

$$D_{TM} = D_M\left[1 + \left(\frac{y}{P_T^2}\right)\frac{12e}{b^2+d^2}\right] \geq 1.15D_M \tag{12.4}$$

where y is the distance between the centers of rigidity of the isolation system and the element of interest measured perpendicular to the direction of seismic loading under consideration, e is the actual eccentricity measured in plan between the center of mass of the structure above the isolation interface and the center of rigidity of the isolation system plus an accidental eccentricity of 5%, b and d are the shortest and longest plan dimensions of the structure, respectively, and P_T is given by

$$P_T = \frac{1}{r_I}\sqrt{\frac{\sum_{i=1}^{N}(x_i^2 + y_i^2)}{N}} \tag{12.5}$$

Thus, Eq. (12.4) includes the additional displacement caused by actual and accidental torsion taking into account the most critical location of the eccentric mass. In Eq. (12.5), x_i and y_i are horizontal distances from the center of mass to the ith isolator unit, N is the number of the isolation units and $r_I = [(b^2 + d^2)/12]^{0.5}$ is the radius of gyration of the isolation system.

The isolation system, the foundation as well as all structural elements below the base level should be designed for a minimum lateral seismic force V_b defined as

$$V_b = k_M D_M \tag{12.6}$$

The structure above the base level should be designed as a non-isolated structure for a minimum shear force

$$V_s = \frac{V_{st}}{R_I} \tag{12.7}$$

where V_{st} is the total unreduced seismic design shear force above the base level and R_I is a coefficient related to the type of seismic force resisting system above the isolation system. R_I equals $3R/8$ with R being the well-known response modification factor, and its minimum and maximum values are 1.0 and 2.0, respectively. The value of R_I is permitted to be taken greater than 2.0, provided the strength of the structure above the base level in the direction of interest (as determined by nonlinear static analysis at a prescribed displacement) is not less than $1.1V_b$.

The total unreduced seismic design force V_{st} in Eq. (12.7) is given by

$$V_{st} = V_b\left(\frac{W_s}{W}\right)^{(1-2.5\beta_M)} \tag{12.8}$$

where β_M is the effective damping of the isolation system and W_s is the effective seismic weight of the structure above the isolation interface excluding the effective seismic weight of the base level. The exponent term in Eq. (12.8) holds only for isolation systems with a hysteretic behavior not characterized by an abrupt transition before and after yielding or slip, otherwise it has to be replaced by $1 - 3.5\beta_M$.

Returning to Eq. (12.7), there are certain limits on the value of V_s needed to be satisfied. In particular, V_s should not be less than: (a) the lateral seismic force for a fixed-base structure of the same effective seismic weight W_s and a period of the isolation system using the upper bound properties for T_M; (b) the base shear corresponding to the factored design wind load and (c) the lateral seismic force V_{st} calculated by Eq. (12.7) and with V_b equal to the force required to fully activate the isolation system using the greater and upper bound properties or 1.5 times the nominal properties for the yield level of a softening system.

The shear force V_s is distributed over the height of the structure above the base level, considering the upper and lower bound properties of the isolation system, on the basis of the equations

$$F_1 = \frac{(V_b - V_{st})}{R_I} \tag{12.9}$$

$$F_x = C_{vx} V_s \tag{12.10}$$

$$C_{vx} = \frac{w_x h_x^k}{\sum_{i=2}^{n} w_i h_i^k} \tag{12.11}$$

where F_1 is the lateral seismic force induced at the base level (level 1), F_x is the lateral seismic force at level x ($x > 1$) and C_{vx} is the vertical distribution factor. To calculate this factor, one makes use of w_i and w_x, i.e., the portions of W_s assigned to levels i and x, and of h_i and h_x, i.e., the heights of levels i and x above the isolation level, respectively. The exponent k in Eq. (12.11) equals $14\beta_M T_{fb}$, where T_{fb} is the fundamental period of the structure above the isolation level determined by modal analysis on the assumption of fixed-base conditions.

The maximum allowable storey drift of the structure above the isolation level is $0.015h_{sx}$, where h_{sx} is the storey height below level x. This storey drift is calculated using the displacement modification factor C_d that corresponds to the value of R_I.

12.2.2 Dynamic Analysis Procedures

Response spectrum analysis procedure can be used for the design of seismically isolated steel structures if the criteria (i)–iv) and (vi) of Sect. 12.2.1 are met. If these criteria are not met, one should make use of a non-linear time-history (NLTH) analysis procedure for the design of seismically isolated steel structures. Upper and lower bound properties of the isolation system should be considered separately in the aforementioned analysis procedures and the dominant case for each response parameter has to be used in design. Moreover, the MCE is used to calculate the lateral forces and displacements of the isolated structure as well as the maximum forces and displacements at the isolation system. The analysis should be performed in three dimensions utilizing both horizontal and vertical seismic motion in terms of their corresponding spectra or acceleration records. Dynamic analysis procedures are mandatory for near-fault sites and site-specific ground motion conditions.

In the response spectrum analysis, one makes use of a modal damping ratio value corresponding to the fundamental mode in the direction of interest. This value cannot be greater than the effective damping of the isolation system or 30% of critical, whichever is less. Modal damping values for higher modes should be consistent with those usually employed in response spectrum analysis of a fixed-base structure. On the other hand, NLTH analysis should be performed for a set of seismic motion pairs appropriately scaled. Each pair is applied simultaneously to the numerical model of the isolated structure considering also the effects of eccentric masses.

The isolation system, foundation and all structural elements below the base level should be designed for the forces obtained from dynamic analysis without any reduction. However, the design lateral force should not be taken as less than $0.9V_b$. The total maximum displacement of the isolation system should not be taken as less than $0.8D_{TM}$, but D_M can be replaced by D'_M given as

$$D'_M = \frac{D_M}{\sqrt{1 + (T/T_M)^2}} \tag{12.12}$$

where T is the fixed-base period of the structure above the isolation system. Structural elements above the base level should be designed as non-isolated ones with forces obtained by dynamic analysis but reduced by a factor of R_I, as described after Eq. (12.7). For response spectrum analysis, the design shear at any storey should not be less than the storey shear resulting from Eq. (12.10) and a value of V_b equal to the base shear obtained from the response spectrum analysis in the direction of interest. For NLTH analysis, if the structure is regular, V_b shall not be less than 80% or 100% of that determined by Eq. (12.6), for the case of regular or irregular structure, respectively.

Finally, when response spectrum analysis is used, the maximum allowable storey drift of the structure above the isolation level is $0.015h_{sx}$. This allowable storey drift is $0.02h_{sx}$ when NLTH analysis is employed.

On the basis of comparison studies between the seismic responses of conventional and seismically isolated steel moment resisting framed (MRF) and concentrically braced framed (CBF) structures (Sayani et al. 2011; Erduran et al. 2011), it has been concluded that: (i) seismically isolated steel structures designed by ASCE/SEI 7-16 (2017) meet the typical seismic performance objectives of the design basis and frequent earthquake events; (ii) storey drifts and peak floor accelerations of seismically isolated steel structures are significantly reduced compared to those of conventional steel structures; (iii) a favorable response reduction of storey drifts and floor accelerations for seismically isolated steel structures when compared to conventional steel structures is attested in the MCE event. However, controlling or limiting the deformation demands of the isolation system in the MCE remains still a problem. This can be resolved by e.g., adding supplemental dampers to the isolated structure even though this is at the expense of increased deformation of the superstructure, as it was mentioned in the introduction.

12.3 The Design Procedures of Eurocode 8

In the context of EC8 (2004), the following methods can be used to design seismically isolated steel structures: (i) simplified linear analysis; (ii) modal simplified linear analysis and (iii) NLTH analysis. The first two methods of analysis are used when the isolation system can be converted into an equivalent linear SDOF system. Only full isolation is considered in EC8 (2004), i.e., the structure above the isolation level remains in the elastic range.

At the damage limitation state (DLS), all lifelines crossing the joints around the isolated steel structure should remain within the elastic range. At the ultimate limit state (ULS), the capacity of the isolation devices (in terms of strength and deformation), upon consideration of a magnification factor $\gamma_x = 1.2$, defined in the relevant section of EC8 (2004), should not be exceeded.

Capacity design and global or local ductility conditions are not needed and thus the low ductility class is adopted for the steel structure. In view of this, the resistance condition of the members of the steel structure should be checked against seismic action by using a behavior factor $q \leq 1.5$. The isolation system may attain its ultimate capacity but the steel structure above and below the isolation system must remain in the elastic range.

The two horizontal and the vertical components of seismic action are assumed to act simultaneously. In building structures of importance class IV, a site-specific spectrum including near-source effects has to be taken into account, if the structure is located at a distance less than 15 km from the nearest potentially active fault. If response history analysis is required, a set of at least three seismic motions should be used. To control differential ground motions, a rigid diaphragm should be provided above and below the isolation system.

The upper and lower bound properties of the isolation system are determined by EN 15129 (2018). Values of these properties to be used in the analysis must be the most unfavorable ones to be expected in the lifetime of the structure. Acceleration and inertia forces induced by the earthquake should be evaluated taking into account the maximum value of the stiffness and the minimum value of damping and friction coefficients, whereas earthquake-induced displacements should be evaluated by considering the minimum value of stiffness, damping and friction coefficients. For building structures of importance class I or II, mean values of properties of the isolation system can be used, provided that the extreme (maximum or minimum) values do not differ by more than 15% from the mean values.

An equivalent linear model of the original nonlinear isolator is used in conjunction with simplified linear analysis or modal simplified linear analysis. This model makes use of: (i) the effective stiffness of each isolator unit, i.e., its secant stiffness at the total design displacement d_{db} (Fig. 12.4) and (ii) an equivalent (effective) viscous damping ξ_{eff} of the isolation system. The effective stiffness K_{eff} of the isolation system is the sum of the effective stiffness of the isolator units. When the effective stiffness or damping of certain isolator units depend on the total design displacement d_{db}, an iterative procedure is applied, until the difference between the assumed and computed values of d_{db} does not exceed 5%. A damping correction factor η is used to define the seismic action through an elastic spectrum. It should be noted that the behavior of the isolation system may be considered as being linear if specific conditions are met.

The simplified linear analysis method considers two horizontal dynamic translations and superimposes static torsional effects. Thus, the effective period for a structure is defined as

$$T_{eff} = 2\pi\sqrt{\frac{M}{K_{eff}}} \tag{12.13}$$

where M is the mass of the superstructure. Torsional movement about the vertical axis can be neglected in the evaluation of K_{eff}, if in each of the two horizontal directions the total eccentricity (including the accidental one) between the stiffness center of the isolation system and the vertical projection of the center of mass of the structure does not exceed 7.5% of the length of the structure transverse to the horizontal direction considered.

The simplified linear analysis method that makes use of equivalent linear damped behavior for the isolation system, should also conform to the following conditions: (i) the distance from the site to the nearest potentially active fault is greater than 15 km; (ii) the largest dimension of the structure in plan is not greater than 50 m; (iii) effects of differential ground motion are negligible; (iv) all isolation devices are located above elements of the substructure which support vertical loads; (v) $3T_f \leq T_{eff} \leq 3.0s$, where T_f is the fundamental period of the fixed-base structure (usually obtained by Eq. (2.4)); (vi) the lateral-load resisting system of the structure is regularly and symmetrically arranged along the two main axes of the structure in plan; (vii) the ratio of the vertical stiffness of the isolation system over the horizontal one is at least equal to 150; (viii) rocking rotation of the structure is negligible and (ix) the fundamental period in the vertical direction is less than 0.1 s.

The displacement of the stiffness center due to seismic action (calculated in each horizontal direction) can be found by the spectral displacement at the effective period T_{eff} as

$$d_{dc} = \frac{MS_e(T_{eff}, \xi_{eff})}{K_{eff,min}} \tag{12.14}$$

where $S_e(T_{eff}, \xi_{eff})$ is the spectral acceleration considering the effective damping ratio ξ_{eff}. The horizontal forces applied at each level of the structure are calculated in each horizontal direction as

$$f_j = m_j S_e(T_{eff}, \xi_{eff}) \tag{12.15}$$

where m_j is the mass at level j. Eq. (12.15) implies that due to the rigid body response of the structure, inertia forces are uniformly distributed over the height of the structure. If the aforementioned condition for neglecting torsional movement is not satisfied, i.e., the total eccentricity (including the accidental one) exceeds 7.5%, then torsional effects in the isolator units may be taken into account by amplifying the action effects in the x direction by a factor δ_{xi} that reads

$$\delta_{xi} = 1 + y_i \frac{e_{tot,y}}{r_y^2} \tag{12.16}$$

In the above, y_i is the horizontal direction transverse to the direction x under consideration, $e_{tot,\ y}$ is the total eccentricity in the y direction and r_y is the torsional radius of the isolation system in the y direction given by the expression

$$r_y^2 = \sum \left(x_i^2 K_{yi} + y_i^2 K_{xi}\right) / \sum K_{xi} \tag{12.17}$$

in which x_i and y_i are the coordinates and K_{xi} and K_{yi} are the effective stiffness, respectively, of an isolator i in the directions x and y, respectively. Torsional effects in the superstructure are estimated in accordance with Eq. (3.16).

If one of the previously mentioned conditions of the simplified linear analysis is not met, a modal response spectrum analysis has to be performed. On the other hand, if all of the previously mentioned conditions of simplified linear analysis are met, except for the maximum horizontal eccentricity of 7.5%, then a simplified modal analysis can be used but with consideration of total (including accidental) eccentricity of the mass of the structure in the evaluation of eigenmodes. Displacements at every point of the structure should be calculated combining translational and rotational displacements, especially for the evaluation of the effective stiffness of each isolator unit.

If the isolation system cannot be represented by an equivalent linear model, then the seismic response of the isolated structure can only be computed with the aid of NLTH analysis utilizing a constitutive law for the devices. This constitutive law should appropriately reproduce the behavior of the system in the range of deformations anticipated in the seismic design situation.

As a final note to the presentation of the EC8 (2004) design method for seismically isolated structures, attention should be given to the evaluation of the resistance of the isolator units at the ultimate limit states. This is accomplished by using either the maximum possible vertical and horizontal seismic forces (including overturning effects) or the total horizontal seismic displacements (including effects of shrinkage, creep, temperature etc.). Closing this section, one should mention that design charts have been developed by Losanno et al. (2019) for the EC8-based design of elastomeric seismic isolation systems.

12.4 Design by the Improved Simplified Linear Analysis Method

A seismic design method for base isolated steel buildings using an improved simplified linear analysis (ISLA) method has been proposed by Weitzmann et al. (2006). The ISLA method reminds one of the simplified linear analysis of EC8 (2004) described in the previous section. Here, improved (more accurate) expressions for the equivalent stiffness (period) and equivalent damping of the equivalent linear SDOF system representing the nonlinear isolator obeying a bi-linear force-displacement relation as in Fig. 12.4 as well as for the damping reduction factor for constructing correctly acceleration spectra for high amounts of damping are employed. In addition, modal response spectrum analysis is used for taking into account higher mode effects, even though such effects are not significant for base isolated structures. The design method of Weitzmann et al. (2006) is carried out by means of the following steps:

1. Selection of a set of ground motions for which the standard 5%-damped elastic pseudo-acceleration spectra are derived. A mean 5%-damped pseudo-acceleration spectrum is then constructed.
2. The linear structure is modeled and an initial value for the maximum base displacement is chosen.
3. Equivalent viscous damping and period (stiffness) values ξ_{eq} and T_{eq}, respectively, for the equivalent linear SDOF isolator are calculated using the equations

$$\xi_{\mathrm{eq}} = \frac{2(\mu - 1)(1 - \gamma)}{\pi\mu(1 + \gamma\mu - \gamma)} \mathrm{v}_1 \mathrm{v}_2 \tag{12.18}$$

$$T_{eq} = T_0 \left(\frac{\mu}{1 + \gamma\mu - \gamma} \right)^{0.5} \tag{12.19}$$

where T_0 denotes the initial period of the isolator, γ denotes the ratio of the post-elastic stiffness to the initial stiffness, μ is the ductility ratio (ultimate over yield displacement) and v_1, v_2 are modification factors defined as

$$\mathrm{v}_1 = \mathrm{T}_{\mathrm{ex}}/\mathrm{T}_{\mathrm{eq}} \tag{12.20}$$

$$\mathrm{v}_2 = 1 - \kappa[(\mathrm{T}_{\mathrm{d}} - \mathrm{T})/\mathrm{T}_{\mathrm{d}}]^2 \tag{12.21}$$

where T is the structural period variable, T_{ex} is the excitation period of the isolator, the coefficient κ is close to 1.0 for the period range from 2.0 to 8.0 s, while T_d is the plateau period of the displacement spectrum. An approximate expression for T_{ex} is $T_{ex} = 0.6T - 0.4\mathrm{T}_{\mathrm{av}}$ where $\mathrm{T}_{\mathrm{av}} = (\mathrm{T}_{\mathrm{a}} + \mathrm{T}_{\mathrm{v}})/2$ with T_{a} and T_{v} being the plateau periods of pseudo-acceleration and pseudo-velocity spectra.

4. The equivalent stiffness K_{eff}, obtained from $T_{eq} = 2\pi\sqrt{m/K_{eff}}$, for known T_{eq} from step 3, is assigned to the base isolation elements and an eigenvalue analysis is executed to find the modal periods and modal shapes of the base isolated structure with m being the overall mass.
5. The pseudo-acceleration (PSA) spectrum due to the equivalent plus inherent viscous damping ξ is reduced by multiplying its ordinates by a new reduction factor

$$\eta_n = 1 + \left[\left(\frac{10}{100\xi + 5}\right)^{-0.5} - 1\right] \frac{\mathrm{PS}_{5\%}(\mathrm{T})}{\mathrm{PS}_{5\%}(\mathrm{T}_{\mathrm{av}})} \frac{\mathrm{T}}{\mathrm{T}_{\mathrm{av}}} \tag{12.22}$$

where T_{av} is the dominant period of the ground motion defined in step 3.

6. Spectral deformation is derived and the corresponding value of the deformation at the base is calculated utilizing the modal periods and modal shapes obtained in step 4).
7. Steps 3–6 are repeated with the updated value of the base level deformation until the solution converges within acceptable error limits for the deformation.
8. Forces are calculated. Spectral pseudo-acceleration $PS_{\mathrm{a},\,\xi}(\mathrm{T})$ at period $T_{ac} \approx T(T_0/T_{0,\,s})$ with $T_{0,s} = 2\pi\sqrt{m/\left(V_y/d_y^*\right)}$ is determined and converted into total acceleration $S_{\mathrm{a},\,\xi}(\mathrm{T})$ using the ratio

$$\mathrm{r} = \frac{\mathrm{S}_{\mathrm{a},\xi}(\mathrm{T})}{\mathrm{PS}_{\mathrm{a},\xi}(\mathrm{T})} = \mathrm{n}_2\mathrm{T}(2\xi)^{5/3} + 1 \tag{12.23}$$

where $n_2 = 0.33\ \mathrm{s}^{-1}$, m is the overall mass, V_y is the elastic limit base shear and d_y^* is the elastic limit deformation at the top of the structure.

It should be noted that according to Weitzmann et al. (2006), the necessity of incorporating several modes in the aforementioned calculations is not needed since the influence of higher modes on deformations and shear forces is negligible for base-isolated structures. On the other hand, to estimate accelerations, higher modes are worth considering only for periods lower than about 3.0 s. This can be roughly done by a multiplication of the acceleration derived for the first mode by a factor of about 1.1-1.3, depending on the period of the isolated structure.

12.5 Displacement-Based Design of Base Isolated Buildings

The direct displacement-based design (DDBD) method and its advantages over the force-based design (FBD) approach have been described in Chap. 4. In this section, the DDBD method as applied to base isolated buildings is very briefly described on the basis of the work of Cardone et al. (2008, 2010). This work, even though considers only reinforced concrete building application, it is quite general and its applications to steel structures is straightforward. The interested reader can also consult the Model Code DBD12 (Sullivan et al. 2012) for additional information on seismic isolation in the framework of DDBD.

The proposed by Cardone et al. (2008, 2010) DDBD method for base isolated n-storey shear type of buildings with uniform mass and stiffness distribution heightwise using mainly lead-rubber bearings (LRB) or friction pendulum bearings (FPB) consists of the following steps:

1. Definition of input data. This includes geometry and basic building characteristics, such as storey height from base h_i, total building height H, total storey mass m_i and fixed base period T_{fb} obtained by Eq. (2.45). The target displacement Δ_i at storey i of the base isolated building is expressed as

$$\Delta_i = D_d + (IDR)_d c_1 \Phi_i = D_d + D_i \tag{12.24}$$

where Φ_i is the assumed deformed shape of the superstructure of the form

$$\Phi_i = cos\left[(\pi/2I_r)(1 - h_i/H)\right] - cos\,(\pi/2I_r) \tag{12.25}$$

In the above equations, D_d is the target displacement of the isolation system, the interstorey drift ratio of the superstructure $(IDR)_d = D_1/h_1$, $c_1 = h_1/\Phi_1$ and $I_r = T_{IS}/T_{fb}$ with T_{IS} being the effective period of the isolation system.

2. Selection of isolation type and target displacements. Here the isolation type (LRB or FPB) and the target displacement D_d and target $(IDR)_d$ are selected. Appropriate graphical tools in Cardone et al. (2008, 2010) help the designer for this target displacement selection, since these target values cannot be selected arbitrarily. These graphical tools require the use of the design acceleration S_{ad} for which explicit expressions are provided in Cardone et al. (2008, 2010) for both LRB and FPB isolators. Since I_r is unknown, a trial value of it is initially used to determine Φ_i and hence Δ_i.
3. Conversion of the base isolated building to an equivalent single-degree-of-freedom (SDOF) system. This system is defined by a design displacement Δ_d and an effective mass m_e of the form

$$\Delta_d = \sum\nolimits_{i=0}^{n} \left(m_i \Delta_i^2\right) / \sum\nolimits_{i=0}^{n} (m_i \Delta_i) \tag{12.26}$$

$$m_e = \sum\nolimits_{i=0}^{n} (m_i \Delta_i) / \Delta_d \tag{12.27}$$

4. Computation of the equivalent damping ratio ξ_{eq} of the SDOF system. This is done by using the equation

$$\xi_{eq} = [\xi_{IS}D_d + \xi_s(\Delta_d - D_d)]/\Delta_d \tag{12.28}$$

where ξ_{IS} and ξ_s are the equivalent viscous damping of the isolation system and the superstructure, respectively. For LRB isolators (Fig. 12.4) one has (Sullivan et al. 2012)

$$\xi_{IS} = \beta_1 - \beta_2 ln\mu \tag{12.29}$$

where μ is the ductility demand on the isolator at the design displacement D_d and β_1 and β_2 are two parameters depending on the post-yielding stiffness r_Δ of the isolator (e.g., for $r_\Delta = 5\%$, $\beta_1 = 0.4830$ and $\beta_2 = 0.0735$). For FPB isolators (Fig. 12.4) one has (Sullivan et al. 2012)

$$\xi_{IS} = 0.4635/(\Lambda + 0.65)^{0.8} \leq 0.35 \tag{12.30}$$

where the parameter Λ is given by

$$\Lambda = D_d/R\mu_{FR} \tag{12.31}$$

with R and μ_{FR} being the radius of curvature and the friction coefficient, respectively, of the FPB isolator. On the other hand, since the superstructure responds elastically, ξ_s can be taken as 5% (2–3% for steel structures).

5. Determination of the design base shear. With known equivalent damping ξ_{eq} and design displacement Δ_d, already determined in step 3, one can use the displacement response spectrum with high amounts of damping to obtain the equivalent period T_{eq} of the equivalent SDOF. Then the equivalent stiffness

$$K_{eq} = m_e(2\pi/T_{eq})^2 \tag{12.32}$$

and hence the design base shear of the base isolated building is computed as

$$V_b = K_{eq}\Delta_d \tag{12.33}$$

Finally, the effective stiffness K_{IS} of the base isolated building is obtained as

$$K_{IS} = V_b/D_d \tag{12.34}$$

6. Evaluation of the mechanical characteristics of the isolation system. These are finally specified based on their equivalent linear characteristics K_{IS} and ξ_{IS} and those assumed at the start of the design process (post-yield hardening, friction coefficient, viscous damping ratio, etc.).
7. Dimensioning of the members of the building. The design base shear V_b is distributed heightwise for determining the lateral seismic forces F_i per storey i as

$$F_i = V_b(m_i\Delta_i)/\sum\nolimits_{i=0}^{n}(m_i\Delta_i) \tag{12.35}$$

A linear static analysis of the building under these lateral forces and gravity load is conducted for member force determination and dimensioning using strength checking. No displacement checking is required as this is automatically satisfied.

12.6 Effect of Isolator Parameters on Response and their Optimum Design

It is apparent that the seismic performance of base isolated steel structures depends on the type of isolators used for their design. The variation of the isolation parameters is also a key factor on the effectiveness of the isolation system.

There are a few studies available regarding the assessment of the seismic performance of steel structures designed with seismic isolation where the type and properties of isolators vary. In particular, Deringöl and Bilgin (2018) and Deringöl and Güneyisi (2019) examined the seismic performance of steel MRFs isolated by LRBs or FPBs by means of non-linear time-history analyses, considering various values for the isolation period and of the ratio of strength to weight (yield strength ratio) supported by the isolators. They concluded that the seismic response of seismically isolated by LRBs or FPBs steel MRFs, is influenced by the characteristics of the seismic motion and that the degree of reduction in displacements, floor accelerations and base shear depends on the values chosen for the design of the LRBs or FPBs. As a general trend, the greater the values of the isolation parameters, i.e., isolation period, effective damping ratio and yield strength ratio, the more favorable the seismic performance of the base isolated steel MRFs in terms of seismic displacement and seismic drift demands.

In a different study on base isolated steel MRFs and CBFs by FPBs, Bao and Becker (2018) demonstrated that the stiffness of steel MRFs or CBFs has a large influence on the overall seismic performance. More specifically, by taking into account the modeling of failure types of the FPB, i.e., the impact and uplift behavior, in conjunction with the degrading behavior of steel MRFs or CBFs, they concluded the following: (i) for MRFs, designed to remain elastic under the MCE or to yield only after impact, the superstructure-base isolation collapse mode involves both bearing uplift and yielding of the MRF. If larger isolator design is adopted, failure is dominated by yielding of the MRF; (ii) for CBFs, no matter which of the aforementioned design options is selected, the superstructure-base isolation collapse mode is governed by excessive yielding of the structure. On the other hand, increasing both the strength of the steel MRF and the displacement capacity of the FPB has beneficial effects for the base isolated MRF. However, only an increase of the displacement capacity of the FPB has some favorable effect for the base isolated CBF.

A performance-based optimum design method for mid-rise base isolated steel MRFs has been presented by Zhang and Shu (2018). This method originates from the earlier work of Sayani and Ryan (2009) and it is based on the development of a seismic performance index that estimates the losses due to structural and non-structural damage and seeks for their minimization. This index incorporates the type of isolation devices along with their various mechanical properties in a probabilistic framework. To demonstrate their methodology, Zhang and Shu (2018) employ a 5-storey base isolated steel MRF with FPBs. The pre-yield, post-yield and characteristic strength values of the FPBs vary in order to find out the optimal design range of the isolation system. For each design of the isolation system, fragility curves are generated and the seismic performance index is computed. The optimal isolation parameters are those corresponding to the minimum value of the seismic performance index. The methodology proposed by Zhang and Shu (2018) permits the optimal selection of the isolation devices taking into account the variability of structural properties and the uncertainties of the expected ground motion into a fragility function framework.

12.7 Numerical Examples

12.7.1 *Design of a Base-Isolated Steel Building Using ASCE/SEI 7-16*

The six-storey steel building with MRFs and CBFs of Sect. 3.6.3 equipped with elastomeric isolators at its base is considered in this example. On the basis of the column layout shown in Fig. 3.17, 20 elastomeric isolators are placed under the base of the 20 columns of the building. The effective seismic weight W of the building above the isolation level is 24605 kN (24.61MN). The building is assumed to be of occupancy category II and is constructed at a site of class C according to the ASCE/SEI 7-16 (2017) specifications. The seismic action is assumed to correspond to a MCE with $S_1 = 0.5$ and $S_S = 1.0$. It is considered that the isolation system does not limit the MCE displacement to less than the total maximum displacement. Each isolator has a diameter equal to 650 mm and is characterized by a minimum (maximum) stiffness value of 1.05 (1.30) kN/mm at a maximum displacement (target value) of 0.350 m. The effective damping ratio provided by the isolators is assumed to be $\beta_M = 15\%$. The isolated building is first designed on the basis of the equivalent lateral force procedure assuming the MCE to act along the x direction (Fig. 3.17). Following the ASCE/SEI 7-16 (2017) provisions of Sect. 12.2.1, one first determines the effective period of the isolation system T_M from Eq. (12.2) for its minimum stiffness $k_M = 20 \cdot 1.05 = 21.0$ kN/mm $= 21.0$ MN/m as $T_M = 2\pi\sqrt{W/(g \cdot k_M)} = 2 \cdot 3.14 \cdot \sqrt{24.61/(9.81 \cdot 21.0)} = 2.17 \text{ s} < 3.0 \text{ s}$. For site class C and $S_1 = 0.5$, $S_{M1} = F_V \cdot S_1 = 1.5 \cdot 0.5 = 0.75$. On the other hand, for a damping ratio of the isolation system equal to $\beta_M = 15\%$, the damping factor $B_M = 1.35$.

Thus, the maximum displacement D_M is computed from Eq. (12.1) and reads $D_M = (gS_{M1}T_M)/(4\pi^2 B_M) = (9.81 \cdot 0.75 \cdot 2.17)/(4 \cdot 3.14^2 \cdot 1.35) = 0.30$ m. The total maximum displacement D_{TM} given by Eq. (12.4) requires the computation of the distance y, the total eccentricity e, the sum $b^2 + d^2$ and the factor P_T. Since the seismic motion is along the x direction, $y = 20/2 = 10.0$ m and the accidental eccentricity is $0.05 \cdot 20 = 1.0$ m. The actual eccentricity between the mass center above the isolation system and the rigidity center of that system is zero due to lack of horizontal and vertical irregularities of the building. Hence the total eccentricity $e = 1.0$ m. With b and d being the two maximum in plan building dimensions, one has $r_I = [(b^2 + d^2)/12]^{0.5} = [(18^2 + 20^2)/12]^{0.5} = 7.77$ m. Thus, for $N = 20$ and $\sum_{i=1}^{N}\left(x_i^2 + y_i^2\right) = 1900$, one obtains from Eq. (12.5) $P_T = (1/7.77)\sqrt{1900/20} = 1.25$. The total maximum displacement D_{TM} is thus obtained from Eq. (12.4) as $D_{TM} = 0.300[1 + (10/1.25^2)(12 \cdot 1.0/(18^2 + 20^2))] = 0.332$ m. Since this value is less than $1.15 \cdot 0.300 = 0.345$ m, $D_{TM} = 0.345$ m.

No further iterations are performed herein with respect to the design of the isolators since the values of 0.300 m and 0.345 m for D_M and D_{TM} are considered to be close to the target ones of 0.350 m. If this is not the case, the design is repeated, most presumably by changing the stiffness of the isolators until a convergence between target and obtained values of D_M and D_{TM} is achieved. On the other hand, it should be noted that only the minimum stiffness of the isolators has been used because this essentially leads to the maximum expected displacements. The complete design process should also involve the maximum stiffness of the isolators as well as a lower damping value than 15% assigned to the isolator, e.g., 10%.

To design the isolation system, the foundation as well as all structural elements below the base level, one makes use of Eq. (12.6) in which the maximum value of 1.30 kN/mm for the stiffness of the isolation system has to be employed. Thus, the minimum lateral seismic force V_b for $D_M = 0.30$ m is $V_b = 20 \cdot 1.30 \cdot 1000 \cdot 0.30 = 7800$ kN. Overtuning loads on the isolation system, the foundation as well as all structural elements below the base level caused by lateral seismic force V_s are calculated by Eq. (12.7) with V_{st} from Eq. (12.8).

The structure above the base level is designed using Eq. (12.7) with $R_I = 1.0$. The value of V_{st} in Eq. (12.8) is calculated by considering $W = 24605$ kN, $W_s = 20504$ kN and $\beta_M = 0.15$. Thus, one has $V_{st} = 7800 \cdot (20504/24605)^{(1 - 2.5 \cdot 0.15)} = 6960$ kN and $V_s = 6960/1.0 = 6960$ kN. This value of V_s is checked against the one derived by the following two cases: a) the fixed-base structure with an effective period T_M and the yielding strength of the isolation system multiplied by a factor of 1.5. It is assumed that wind load does not govern the lateral force, and, thus, this case is omitted. Following ASCE/SEI 7-16 (2017), to define the base shear for a fixed-base structure with an effective period T_M, the seismic response coefficients $C_S = S_{DS}/(R/I)$ and $C_S = S_{D1}/(T_M(R/I))$ are calculated, where $S_{DS} = (2/3) \cdot S_{MS} = (2/3) \cdot S_S \cdot F_a = (2/3) \cdot 1.0 \cdot 1.2 = 0.8$ and $S_{D1} = (2/3) \cdot S_{M1} = (2/3) \cdot 0.75 = 0.5$. Therefore, considering $R = 3.25$ and $I = 1.0$, one has $C_S = 0.80/(3.25/1.0) = 0.25$ and $C_S = 0.50/(2.17(3.25/1.0)) = 0.071 > 0.01$.

Then, $V_s = C_s W = 0.071 \cdot 24605 = 1744.41$ kN < 6960 kN. At the yield level of the isolation system, one finds the lateral force as $V_s = 1.5 \cdot k_{e.\ max} \cdot D_{Dy.\ max} = 1.5 \cdot 20 \cdot 3 \cdot 1000 \cdot 0.01 = 900$ kN < 6960 kN, where $k_{e.\ max}$ is the maximum (elastic) value of the stiffness of each isolator, taken equal to 3.0 kN/mm and $D_{Dy.\ max}$ is the maximum value of the isolator yielding displacement, taken equal to 0.01 m. Therefore, $V_s = 6960$ kN for the design of the structural elements above the isolation level.

Returning to check the satisfaction of the requirements (i)–(vii), mentioned at the beginning of Sect. 12.2.1, regarding the application of the equivalent lateral force procedure, one realizes that only requirements (v) and (vii) (criteria (a) and (b)) remain to be checked. In particular, from Table 2.4 and Eq. (2.45), one has $T = C \cdot H^x = 0.0488 \cdot 18^{0.75} = 0.426$ s and $3 \cdot 0.426 = 1.28$ s < 2.17 s and thus requirement (v) is satisfied. For requirement (vi), one has that that the 20% of the maximum displacement is $0.2 \cdot 0.30 = 0.060$ m and that the effective stiffness at the maximum displacement, i.e., 1.30 kN/mm is greater than one-third of the effective stiffness at 20% of the maximum displacement calculated as $((0.060 - 0.01) \cdot 1.30 + 0.01 \cdot 3.0)/(3 \cdot 0.060) = 0.53$ kN/mm. A similar conclusion can be obtained for the minimum stiffness by checking the aforementioned inequality. To check if the isolation system can produce the restoring force needed, one calculates the 50% of the total maximum displacement as $0.5 \cdot 0.345 = 0.173$ m and the lateral force of the isolator at 50% of the total maximum displacement as $((0.173 - 0.01) \cdot 1.30 + 3 \cdot 0.01) \cdot 1000 = 242$ kN. Thus, the lateral force of the isolation system at 50% of the total maximum displacement is $20 \cdot 242/24605 = 0.197$W. Considering the lateral force at the total maximum displacement as $20 \cdot (3 \cdot 0.01 + 1.30 \cdot (0.345 - 0.01)) \cdot 1000/24605 = 0.378$W, one finally has that 0.378W − 0.197W = 0.181W > 0.025W, and, therefore, the isolation system can produce the restoring force needed.

Having designed the isolation system, one makes use of Eqs. (12.9)–(12.11) to distribute the lateral seismic force at stories and by application of this lateral force distribution checks the design (sectional and member checkings) of the steel structure. It is recalled that the maximum allowable storey drift of the structure above the isolation level is $0.015h_{sx}$, where h_{sx} is the storey height below level x. The storey drift is calculated using a displacement modification factor equal to 1.0 which is equal to the value of R_I used.

Application of response spectrum analysis procedure is now presented. According to this procedure, the MCE is used to calculate the lateral forces and displacements of the isolated structure as well as the maximum forces and displacements at the isolation system. The analysis is performed with the aid of SAP 2000 (2020) in three dimensions utilizing both horizontal and vertical seismic motion in terms of their corresponding spectra employing the 100%/30%/30% directional combination. Assuming the sections for the superstructure to be the same with those mentioned in Sect. 3.6.3, i.e., IPE330 and HEM320, for beams and columns respectively, and secondary IPE270 beams pinned to the main IPE330 beams per 2.0 m along the x direction of the structure following Fig. 3.17, one performs an eigenvalue analysis of the fixed-base superstructure and finds the modal periods of Table 3.16. These sections are expected to be larger if the fixed-base structure were to be designed for the MCE. By appropriately isolating the fixed-base structure designed for the DBE, one succeeds in making it capable of resisting MCE forces.

For isolation purposes, only the period value of the fundamental (isolation) mode is employed, i.e., $T_1 = 1.21$ s. The value of damping ratio assigned to the fundamental (isolation) mode is equal to 15%, whereas the damping ratio for higher modes is the usually employed 5% in spectrum analysis of a fixed-base structure. The minimum stiffness $k = 1.05$ kN/mm of the isolators is used in order to obtain the maximum displacements in the superstructure, whereas the maximum stiffness $k = 1.30$ kN/mm of the isolators is used to calculate the maximum base shear of the superstructure and forces in the isolators.

The maximum displacement of the isolation system is calculated by Eq. (12.12) considering $D_M = 0.30$ m, $T = 1.21$ s and $T_M = 2.17$ s. Thus, $D'_M = 0.30/\sqrt{1 + (1.21/2.17)^2} = 0.262$ m. The total maximum displacement using D'_M instead of D_M in Eq. (12.4) is $D_{TM} = 0.262 \cdot [1 + (10/1.25^2) \cdot (12 \cdot 1.0/(20^2 + 18^2)] = 0.29$ m with a lower bound limit equal to $0.8D_{TM} = 0.8 \cdot 0.29 = 0.232$ m. The minimum lateral force at and below the isolation system is $0.9V_b = 0.9 \cdot 7800 = 7020$ kN. To design the structural elements above the base level, one makes use of Eq. (12.10) in which, similarly to the equivalent lateral force procedure presented previously, $V_s = 6960/1.0 = 6960$ kN. Results of response spectrum analysis should not be less than the limit values $0.8D_{TM}$, $0.9V_b$ and V_s mentioned above. For comparison purposes, the base shear of the elastic fixed-base structure designed for the MCE is 17112 kN, i.e., about 2.45 times greater than that of the $V_s = 6960$ kN of the isolated structure.

After performing a response spectrum analysis for the MCE, the following values are found: (i) the maximum design base shear at the isolation level is $7103kN > 0.9V_b = 7020kN$; (ii) the maximum displacement of the isolator is $0.269m > 0.8D_{TM} = 0.8 \cdot 0.29 = 0.232$ m; (iii) the maximum uplift forces to the isolators with reference to points (x_1, y_1), (x_5, y_1), (x_1, y_4), (x_4, y_4), (x_4, y_1), (x_2, y_1), (x_2, y_4), (x_5, y_4) in Fig. 3.17, are 1999 kN, 1961.25 kN, 1960.28 kN, 1663.73 kN, 1458.36 kN, 1427.29 kN, 1386.01 kN and 1075.96 kN. These values are greater than the permissible value recommended in FEMA P-1051 (2016), i.e., $3GA = 3 \cdot 1000 \cdot 0.332 = 996$ kN, where $G = 1.0$ MPa is the effective dynamic shear modulus and $A = 3.14 \cdot 0.65^2/4 = 0.332$ m is the area of the isolator; iv) the maximum storey drift calculated using a displacement modification factor equal to 1.0 (because $R_I = 1.0$) is 0.029 m, less than the maximum allowable storey drift of the structure above the isolation level reading $0.015h_{sx} = 0.015 \cdot 3 = 0.045m$. Therefore, the only deficiency in the overall design of the base-isolated structure is the exceedance of the critical value $3GA$ by the maximum uplift forces in 8 isolators, as stated above. Taking into account that uplift forces, under the critical seismic combination, acts into 8 out of the 20 isolators, one has two options to confront this problem under the assumption that the geometry and dead load distribution of the steel structure does not change: (i) either to design an uplift restraining system or

(ii) to perform a detailed non-linear time-history analysis and check if this uplift creates deflections that do not cause overstress or instability to the members of the isolated structure. It should be noted that for the steel building studied, an uplift force to isolators occurs only for those seismic combinations in which the seismic motion acts 100% in its y direction.

The design of the steel structure is then performed by the corresponding sectional and member checkings according to AISC 360 (2016) and ANSI/AISC 341-16 (2016). One last thing to note is that for the steel building studied, due to its increased collapse potential (ASCE/SEI 41-17 2017), the horizontal clearance (seismic gap) at the isolation level has to be taken as $1.2D_{TM}$, where D_{TM} is the value found previously for each one of the type of analysis employed (0.345 m from the equivalent lateral force analysis and 0.290 m from the modal response spectrum analysis).

12.7.2 Design of a Base-Isolated Steel Building Using Eurocode 8

The six-storey steel building with MRFs and CBFs of Sect. 3.6.3 is equipped with elastomeric isolators and is designed in the framework of EC8 (2004) by the simplified linear analysis method following Elghazouli (2017). On the basis of the column layout shown in Fig. 3.17, 20 elastomeric isolators are employed. The building is assumed to be founded in a site with $a_{gR} = 0.24$ g, soil type C and design spectrum of type 1.

Considering an effective period $T_{eff} > 2.0$ s and the equivalent damping ratio provided by the isolators 15%, implying $\eta = \sqrt{10/(5+15)} = 0.707$, one makes use of Eqs. (3.4) and (3.5) and Table 3.3 to calculate the spectral acceleration as $S_e = 0.24 \cdot \text{g} \cdot 1.15 \cdot 0.707 \cdot 2.5 \cdot 0.6 \cdot 2.0/T^2 = 0.585\ \text{g/T}^2$. The spectral displacement S_{De} is then found by using the equation $S_{De} = S_e \cdot (T/2\pi)^2$ of EC8 (2004). Thus, $S_{De} = (0.585\ \text{g/T}^2) \cdot (T/2 \cdot 3.14)^2 = 0.146$ m. This spectral displacement is essentially the design displacement d_{dc} of the stiffness center of the isolation system in each horizontal direction due to seismic action. Assuming that the actual eccentricity between the stiffness center of the isolation system and the mass center of the superstructure is zero due to lack of irregularities in the base isolated building and that the accidental eccentricity $e = 0.05 \cdot 20 = 1.0$ m, the total eccentricity is 1.0 m and this is less than 7.5 % $\cdot\ 20 = 1.5$ m, implying negligible torsional effects. However, in this example, for reasons of conservatism and illustration purposes, torsional effects are taken into account. Thus, assuming that $K_{xi} = K_{yi}$ implying from Eq. (12.17) that $r_y^2 = \sum(x_i^2 + y_i^2)/20 = 1900/20 = 95$ and hence $r_y = 9.75$ m, the maximum value of δ from Eq. (12.16) is obtained as $\delta = 1 + 20 \cdot 1/9.75^2 = 1.21$. Thus, the total design displacement d_{db} for each isolator, as obtained by combining Eqs. (12.13) and (12.14) and taking into account the factors $\delta = 1.21$ and $\gamma_x = 1.20$ (EC8 2004), is $d_{db} = S_{De} \cdot \delta \cdot \gamma_x = 0.146 \cdot 1.21 \cdot 1.20 = 0.212$ m. Targeting an effective period $T_{eff} = 2.1$ s, the effective stiffness of the isolation system can be found from Eq. (12.13) and reads $K_{eff} = M \cdot (2\pi/T_{eff})^2 = (24.61/9.81) \cdot (2 \cdot 3.14/2.1)^2 = 22.43$ MN/m. Equally distributing K_{eff} to the 20 isolators, implies that each isolator should have an effective stiffness 22.43/20 = 1.12 MN/m or 1.12 kN/mm at the total design displacement. Selecting a diameter equal to 650 mm and an effective stiffness equal to 1.05 kN/mm for each isolator, one finally has $T_{eff} = 2.17$ s and because of the shape of the displacement spectrum for periods greater than the corner period T_C (herein equal to 2.0 s), $d_{dc} = 0.146$ m. For the selected type of isolator, the maximum displacement that can accommodate is 0.35 m > 0.212 m. The aforementioned design of the isolation system has to also satisfy two requirements with respect to the vertical stiffness of the isolators. In particular, the ratio of the vertical to horizontal stiffness, both provided from the manufacturer of the isolator used herein, is 1220/1.05 = 1162 > 150 and the fundamental period in the vertical direction is $T_V = 2\pi\sqrt{M/K_V} = 2 \cdot 3.14 \cdot \sqrt{24.61/(9.81 \cdot 1220 \cdot 20)} = 0.064 \text{ s} < 0.1 \text{ s}$. Therefore, the two requirements are indeed satisfied.

The design of the superstructure is performed using Eq. (12.15) for lateral force determination at storey level j with $S_e(T_{eff}, \xi_{eff})$ computed from Eq. (3.4) for $T_{eff} = 2.17$ s and $\xi_{eff} = 15\%$ and reading $0.24 \cdot g \cdot 1.15 \cdot 0.707 \cdot 2.5 \cdot 0.6 \cdot 2.0/2.17^2 = 0.124$ g. The spectral value derived from Eq. (3.4) has to be divided by a behavior factor not greater than 1.5 and to be increased by the factor δ given by Eq. (3.16). Sectional and member checkings are then performed according to EC3 (2009) and lighter sections are now expected in comparison to those found in Chap. 3 for the same structure.

A last thing to note regarding the application of the simplified linear analysis procedure is that the design of the isolation system may change due to the stiffness required under wind loading or due to small drift limits associated with the frequent earthquake event.

Performing now modal analysis of the steel structure under study, even though none of the requirements for the application of simplified linear analysis method is violated, torsional effects are explicitly considered by making use of an accidental eccentricity for both horizontal directions as defined by Eq. (3.12). The analysis is performed in three dimensions utilizing both horizontal and vertical seismic motion in terms of their corresponding spectra employing the 100%/30%/30% directional combination. Values assigned to damping ratio of the fundamental (isolation) mode in the direction of interest are equal to 15%, whereas the damping ratio for higher modes is the usually employed 5% in spectrum analysis of a fixed-base structure. The minimum stiffness of the isolators is used in order to obtain the maximum displacements of the superstructure, whereas the maximum stiffness of the isolators is used to calculate the maximum base shear of the superstructure and forces to isolators. The maximum and minimum stiffness of the isolators are 1.30 kN/m and 1.05 kN/m, respectively. In contrast to ASCE/SEI 41-17 (2017) provisions, EC8 (2004) does not provide lower bound values for the values obtained from modal analysis regarding the maximum displacement of the isolation system and the design forces above and below the isolation system. Full isolation, i.e., elastic behavior of the superstructure is targeted, and thus, the behavior factor is taken as $q = 1.0$.

Table 12.1 First nine modal periods and participating mass ratio

Mode number	Period (s)	Participating mass %		
		Translational x	Translational y	Rotational z
1	2.149	98.2	0	0
2	1.933	0	99.7	0
3	0.558	1.64	0	0
4	0.475	0	0	81.7
5	0.353	0	0.25	0
6	0.260	0.01	0	0
7	0.169	0	0	0.46
8	0.163	0	0	5.71
9	0.156	0	0	2.17
Sum of participating mass %		99.9	100.0	90.1

Making use of the upper bound stiffness for the isolators and SAP 2000 (2020) software, the first nine modal periods and their associated participating mass ratios are found and shown in Table 12.1, while the deformed building shapes associated with the first two (isolated) modes are found and shown in Fig. 12.5. Similarly, one can assign the lower bound stiffness and obtain the corresponding modal periods and deformed building shapes.

After performing a response spectrum analysis in SAP 2000 (2020), the following values are found: (i) the maximum design base shear at the isolation level is 3960 kN; (ii) the maximum displacement of the isolator is 0.143 m < 0.35 m; (iii) the maximum uplift forces to the isolators with reference to points (x_1, y_1), (x_1, y_4), (x_5, y_1), (x_4, y_4), (x_5, y_4), (x_4, y_1), (x_2, y_1), (x_2, y_4) in Fig. 3.17, are 776.21 kN, 744.15 kN, 736.81 kN, 503.90 kN, 410.98 N, 373.89 kN, 340.51 kN and 319.06 kN. These values are lower than the permissible value recommended in FEMA P-1051 (2016), i.e., $3GA = 3 \cdot 1000 \cdot 0.332 = 996$ kN, where $G = 1.0$ MPa is the effective dynamic shear modulus and $A = 3.14 \cdot 0.65^2/4 = 0.332\ \text{m}^2$ is the area of the isolator; (iv) the maximum storey drift calculated using a displacement modification factor equal to 1.0 (because of $q = 1.0$) is 0.0154 m. Hence from Eq. (3.25) and assuming $v = 0.5$, one has $0.0154 \cdot 0.5 = 0.0077 < 0.01 \cdot 3 = 0.03$. Thus, the damage limitation criterion is satisfied under the assumption that the non-structural components are fixed to the structure and do not interfere with its deformation, which corresponds to an interstorey drift limit of 0.01.

The maximum uplift forces on the isolators found above is below the limit value of $3GA$ recommended in FEMA P-1051 (2016). Nevertheless, if a stricter permissible value for the uplift force is required, and in view of the fact that for the steel building studied under the critical seismic combination, uplift forces act into 8 out of the 20 isolators, one should either design an uplift restraining system or perform a detailed non-linear time-history analysis and check if this uplift creates deflections that do not cause overstress or instability to members of the isolated structure.

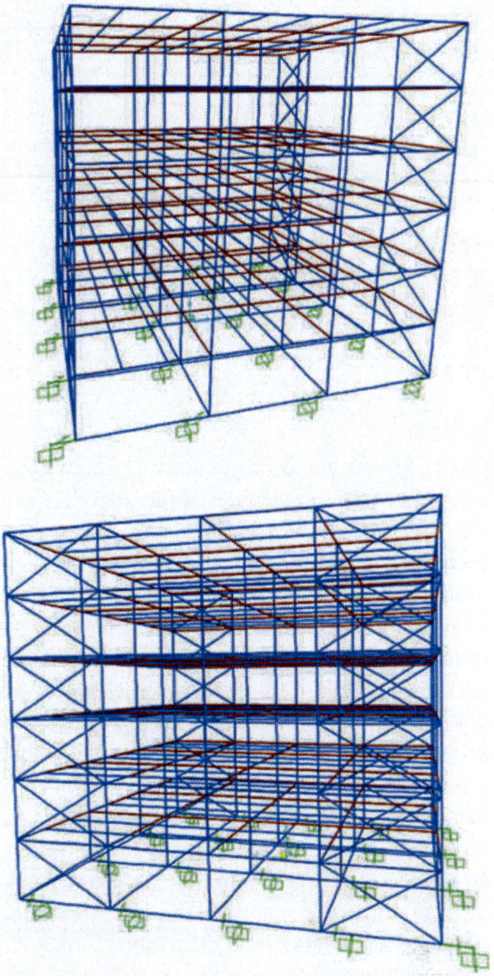

Fig. 12.5 Deformed building shapes for the (isolated) modes 1 (up) and 2 (bottom)

Sectional and member checkings of the steel structure are then performed according to EC3 (2009) and are found to be satisfied. Capacity design checks and satisfaction of dimensionless slenderness requirements as presented in Sect. 3.6.3 are redundant as the design of the superstructure is elastic. In order to judge the efficiency of the isolation system with respect to the steel members of the superstructure, one may compare the maximum values for axial force (N), shear force (V) and bending moment (M) of the critical beam, brace and column elements with those found for the case of the fixed-base steel structure and presented in Tables 3.19 and 3.20. These maximum values occur from the seismic combinations in which the two horizontal and the vertical earthquake actions have been considered in the base-isolated structure whereas only the two horizontal combinations have been taken into account in the fixed-base structure. It should be stressed that the earthquake design

Table 12.2 Maximum M, V, N values of the critical beam and column elements of the base-isolated structure

Column	Bottom end	Top end
M_2 (kNm)	21.40	9.73
M_3 (kNm)	379.29	91.63
V_2 (kN)	151.36	151.36
V_3 (kN)	10.92	10.92
N (kN)	1969.91	1962.71
Beam (rigidly connected)	**Left end**	**Right end**
M (kNm)	94.03	107.46
V (kN)	18.83	15.46
N (kN)	19.21	17.69

forces for the fixed-base structure in comparison to the base-isolated structure are different not only because of the different values of modal periods (Tables 3.16 and 12.1) of these structures but also because of the different values of the behavior factor ($q = 4$ for the fixed-base structure and $q = 1$ for the base isolated one) assigned to the design spectrum.

The critical brace and column elements are found to be in the first storey of the base-isolated steel structure and not to the second storey of the fixed-base structure as shown in Fig. 3.21. On the other hand, similarly to the fixed-base structure, the critical beam of the base-isolated structure is found to be in the second storey but in a different frame and bay. Therefore, it seems that a direct comparison between the critical brace, beam and column elements is meaningless and only the stress ratios can be compared. Therefore, Table 12.2 provides the maximum values for forces and moments of the critical beam and column elements of the base-isolated steel structure. The stress ratios of the critical beam and column elements are found to be 0.53 and 0.70, respectively. The corresponding values for stress ratios found in Sect. 3.6.3 for the critical beam and column elements are 0.43 and 0.58, respectively, in which the amplification factors θ and Ω have been included. Thus, from the point of view of stress ratio, an increase is observed to occur to the critical beam and column of the base-isolated structure in comparison to those of the fixed-base structure.

Table 12.3 provides the values of maximum compressive force N_{Ed} in the critical brace of every storey for the fixed-base and the base-isolated structure. From the results of Table 12.3, it is apparent that base-isolation reduces the compressive force to the critical brace of every storey with only exception the first storey, i.e., the storey just above the isolation system, where an increase in the compressive force of the critical brace takes place. On the basis of the $N_{b.\,Rd}$ value provided in Table 3.20, the stress ratio of the critical first storey brace is 0.98 and 0.81 for the base-isolated and the fixed-base structure, respectively. It should be recalled that since $q = 1.0$ for the base-isolated structure, braces in compression (in a diagonal configuration) can be taken into account, something which is not allowed for higher values of q, i.e., for $q = 4$ assigned to the design of the fixed-base structure.

Table 12.3 Maximum compressive forces for the critical brace of every storey for the fixed-base and the base-isolated structure

Storey	Section	N_{Ed} (kN) – base-isolated	N_{Ed} (kN) – fixed base
1	SHS 120 × 120 × 12.5	526.64	434.77
2	SHS 120 × 120 × 12.5	518.22	527.05
3	SHS 120 × 120 × 12.5	428.18	471.17
4	SHS 120 × 120 × 8	297.41	351.28
5	SHS 100 × 100 × 7.1	217.73	263.98
6	SHS 100 × 100 × 4	79.77	119.56

In general, the stress ratio, as a result of the sectional and stability checks of EC3 (2009), for the majority of the members of the base-isolated structure is found to be smaller than that of the fixed-base structure, thus demonstrating the effectiveness of the isolation system in reducing forces and moments in steel members. The reduction in stress ratios in more evident for the braces and columns of stories three to six of the base-isolated structure. These lower stress ratios may lead to lighter sections for some steel members under the assumption that they have to be checked against all other load combinations, i.e., wind etc. On the other hand, for the base-isolated structure, an increase of the stress ratio is observed to some steel members at the 1st storey, something that may render the use of increased sections there necessary.

One last thing to note is that regarding the axial capacity of the isolators for the vertical non-seismic (1.35G + 1.5Q) and the vertical seismic (G + ψ_2Q + earthquake) loading combinations. The vertical non-seismic combination is concurrent with zero horizontal displacement of the isolator, whereas the vertical seismic one with the maximum displacement that can be undertaken by the isolator. Therefore, the maximum axial force found for the vertical non-seismic combination is 3150.47 kN, whereas that for the vertical seismic is 2329.55 kN. These values are smaller than the maximum ones provided by the manufacturer of the isolators employed herein and reading as 8940 and 2230 kN for the vertical non-seismic and the vertical seismic combinations. Thus, the axial capacity of the isolators is not surpassed.

12.7.3 Design of a Base-Isolated Steel Building Using ISLA

A ten-storey one-bay plane steel MRF with base-level mass equal to $2 \cdot 10^6$ kg and storey masses equal to $1 \cdot 10^6$ kg in the remaining stories (first to ten) designed by Weitzmann et al. (2006) is briefly presented here. Its fixed-base period is found to be equal to 1.30 s. Base isolation is employed to this steel MRF by adjusting the stiffness and yield force in order to obtain an equivalent period T_{eq} defined by Eq. (12.19) ranging from 2.0 to 8.0 s. The base-isolated steel MRF is subjected to the Kobe near-fault data set (PEER 2009) and its non-linear response is computed. The maximum values for the displacement and acceleration at the isolation level, i.e.,

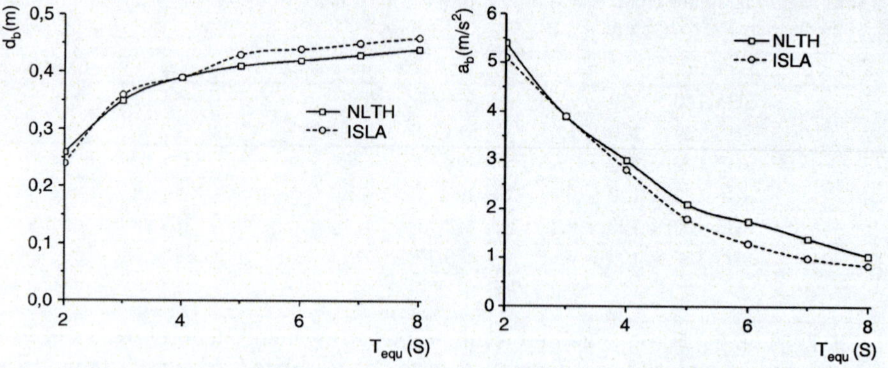

Fig. 12.6 Maximum displacements (left) and accelerations (right) at the base level

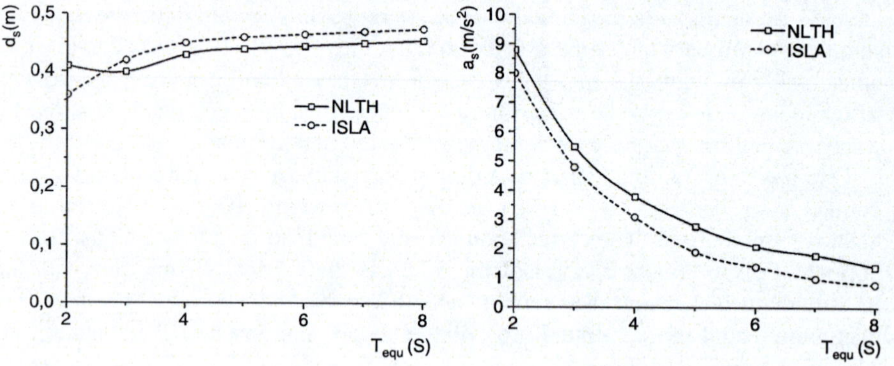

Fig. 12.7 Maximum displacements (left) and accelerations (right) at the superstructure level

d_b and a_b, respectively, as well as the displacement and acceleration at the superstructure, i.e., d_s and a_s, respectively, are obtained. These maximum values are compared to the corresponding ones obtained by Weitzmann et al. (2006) using their improved simplified linear analysis (ISLA) described in Sect. 12.4. The comparisons are shown in Figs. 12.6 and 12.7. Taking into account that the displacements computed by the ISLA method are from the safe side compared to the exact ones coming from non-linear analyses, one concludes that the ISLA method can be successfully used as a preliminary design tool of the isolation system. Regarding accelerations, the ISLA method gives estimates lower than those of non-linear analysis. However, for the purpose of a preliminary design of the isolation system, they can be considered acceptable.

12.8 Conclusions

The discussion presented in the previous sections of this chapter leads to the following conclusions:

1. Use of base seismic isolation in steel buildings frames is an effective way to considerably reduce seismic forces in the structure by creating a horizontally flexible and seismic energy dissipating base through isolating devices. These devices can be designed as a part of a new structure or as a means to retrofit already existing structures.
2. The most widely used types of isolating devices are the lead rubber bearings (LRB) and the friction pendulum bearings (FPB). Other types of isolators include high-damping rubber bearings (HDRB), elastomeric bearings with steel plates, double and triple friction pendulum bearings and combinations of flat sliding bearings with different re-centering and/or dissipative devices.
3. The usual design methods for base isolated steel structures are according to ASCE those employing equivalent lateral force analysis, response spectrum analysis and nonlinear time history analysis. Similar methods are proposed by EC8. In addition to those force-based design methods, a displacement-based design method can also be used. The most widely used method at least for preliminary design is the one employing the equivalent lateral force analysis. The emphasis is mainly on the design of the isolators since the superstructure should respond elastically during the seismic event.
4. Seismic design of new base isolated structures and retrofitting of old structures by base isolators has been proven to be successful even by using simplified methods. However, nonlinear time-history analysis employed for the seismic design of base isolated structures is the one recommended for final design of complicated structures. More research is needed towards improving existing simplified methods for the design of base isolated structures and structures equipped with more than one device for seismic force reduction, like those combining base isolation with supplemental dampers.

References

AISC 360 (2016) Specification for structural steel buildings. American Institute of Steel Construction, Chicago, IL

ANSI/AISC 341-16 (2016) Seismic provisions for structural steel buildings. American Institute of Steel Construction, Chicago, IL

ASCE/SEI 41-17 (2017) Seismic evaluation and retrofit of existing buildings. American Society of Civil Engineers, Reston, VI

ASCE/SEI 7-16 (2017) Minimum design loads and associated criteria for buildings and other structures. American Society of Civil Engineers, Reston, VI

Bao Y, Becker TC (2018) Effect of design methodology on collapse of friction pendulum isolated moment-resisting and concentrically braced frames. J Struct Eng of ASCE 144(04018203):1–14

Cardone D, Dolce M, Palermo G (2008) Force-based vs displacement-based design of buildings with seismic isolation. In: Proceedings of the 14th World Conference on Earthquake Engineering, Beijing, China
Cardone D, Palermo G, Dolce M (2010) Direct displacement-based design of buildings with different seismic isolation systems. J Earthq Eng 14:163–191
Chang S-P, Makris N, Whittaker AS, Thompson AST (2002) Experimental and analytical studies on the performance of hybrid isolation systems. Earthq Eng Struct Dyn 31:421–443
Cheng FY, Jiang H, Lou K (2008) Smart structures: innovative systems for seismic response control. CRC Press, Boca Raton, FL
De Luca A, Guidi LG (2019) State of art in the worldwide evolution of base isolation design. Soil Dyn Earthq Eng 125:105722
Deringöl AH, Bilgin H (2018) Effects of the isolation parameters on the seismic response of steel frames. Earthq Struct 15:319–334
Deringöl AH, Güneyisi EM (2019) Effect of friction pendulum bearing properties on behavior of buildings subjected to seismic loads. Soil Dyn Earthq Eng 125:105746
EC3 (2009) Eurocode 3, Design of steel structures – Part 1–1: general rules and rules for buildings, EN 1993-1-1. European Committee for Standardization (CEN), Brussels
EC8 (2004) Eurocode 8, Design of structures for earthquake resistance, Part 1: general rules, seismic actions and rules for buildings, EN 1998-1-1. European Committee for Standardization (CEN), Brussels
Elghazouli AY (2017) Seismic design of buildings to Eurocode 8. CRC Press, Boca Raton, FL
EN 15129 (2018) Anti-seismic devices. European Committee for Standardization (CEN), Brussels
Erduran E, Dao ND, Ryan KL (2011) Comparative response assessment of minimally compliant low-rise conventional and base-isolated steel frames. Earthq Eng Struct Dyn 40:1123–1141
FEMA P-1051 (2016) 2015 NEHRP recommended seismic provisions: design examples. Federal Emergency Management Agency, Washington, DC
Fenz DM, Constantinou MC (2008) Spherical sliding isolation bearings with adaptive behavior: theory. Earthq Eng Struct Dyn 37:163–183
Kelly JM (1997) Earthquake-resistant design with rubber, 2nd edn. Springer, London
Kelly JM (1999) The role of damping in seismic isolation. Earthq Eng Struct Dyn 28:3–20
Kelly JM, Konstantinidis DA (2011) Mechanics of rubber bearings for seismic and vibration isolation. Wiley, New York
Komodromos P (2000) Seismic isolation for earthquake resistant structures. WIT Press, Southampton
Losanno D, Hadad HA, Serino G (2019) Design charts for eurocode-based design of elastomeric seismic isolation systems. Soil Dyn Earthq Eng 119:488–498
Makris N (2019) Seismic isolation: early history. Earthq Eng Struct Dyn 48:269–283
Makris N, Chang S-P (2000) Effect of viscous, viscoplastic and friction damping on the response of seismic isolated structures. Earthq Eng Struct Dyn 29:85–107
Naeim F, Kelly JM (1999) Design of seismically Isolated structures: from theory to practice. Wiley, New York
PEER (2009) Pacific Earthquake Engineering Research Center, Strong Ground Motion Database, Berkeley, CA. http://peer.berkeley.edu/
Pietra D, Pampanin S, Meyes RL, Wetzel NG, Feng D (2015) Design of base-isolated buildings: an overview of international codes. Bull N Z Soc Earthq Eng 48:118–135
Providakis CP (2008) Effect of LRB isolators and supplemental viscous dampers on seismic isolated buildings under near-fault excitations. Eng Struct 30:1187–1198
Providakis CP (2009) Effect of supplemental damping on LRB and FPS seismic isolators under near-fault ground motions. Soil Dyn Earthq Eng 29:80–90
SAP 2000 (2020) Structural analysis program 2000, integrated software for structural analysis and design, Version 22. Computers and Structures Inc., Walnut Creek, CA
Sarlis AA, Constantinou MC (2016) A model of triple friction pendulum bearing for general geometric and frictional parameters. Earthq Eng Struct Dyn 45:1837–1853

Sayani PJ, Ryan KL (2009) Comparative evaluation of base-isolated and fixed-base buildings under a comprehensive response index. J Struct Eng ASCE 135:698–707

Sayani PJ, Erduran E, Ryan KL (2011) Comparative response assessment of minimally compliant low-rise base-isolated and conventional steel moment-resisting frame buildings. J Struct Eng ASCE 137:1118–1131

Skinner RI, Kelly TE, Robinson BWH (2011) Seismic isolation for designers and structural engineers, Robinson Seismic Ltd, Holmes Consulting Group, New Zealand

Sullivan TJ, Priestley MJN, Calvi GM (2012) A model code for the displacement-based seismic design of structures, DBD12. IUSS Press, Pavia

Warn GP, Ryan KL (2012) A review of seismic isolation for buildings: historical development and research needs. Buildings 2:300–325

Weitzmann R, Ohsaki M, Nakashima M (2006) Simplified methods for design of base-isolated structures in the long-period high-damping range. Earthq Eng Struct Dyn 35:497–515

Yenidogan C, Erdik M (2016) A comparative evaluation of design provisions for seismically isolated buildings. Soil Dyn Earthq Eng 90:265–286

Zhang J, Shu Z (2018) Optimal design of isolation devices for mid-rise steel moment frames using performance based methodology. Bull Earthq Eng 16:4315–4338

Chapter 13
Design Using Supplemental Dampers

Abstract This chapter describes various methods for the seismic design of new steel building structures equipped with supplemental dampers and the seismic retrofitting of existing structures by supplemental dampers. The most widely used dampers are the fluid viscous ones (linear or nonlinear), which enhance the seismic energy of dissipation of the structures and thus reduce the seismic forces on them. The result is lighter new structures or effective retrofitting of existing structures against stronger earthquakes. The most widely used design methods are the force-based simplified linear ones, including the equivalent lateral force method and the response spectrum one. The ASCE code provisions are based on such force-based simplified methods. Displacement-based simplified design methods are also presented. The arrangement of dampers for an optimum structural performance is also discussed briefly here by mentioning appropriate optimization procedures for that purpose. The limitations of the simplified methods are pointed out and the method using nonlinear time-history analysis is suggested as the appropriate method for final design of a structure with dampers. Four numerical examples are presented in some detail in order to illustrate methods based on force or displacement as well as methods characterized by new concepts.

Keywords Supplemental dampers · Viscous dampers · Viscoelastic dampers · Linear dampers · Nonlinear dampers · Equivalent damping · Optimum damper placement · Seismic design with dampers · Retrofitting with dampers

13.1 Introduction

The use of supplemental dampers (damping devices) in building structures has gained popularity in the last few decades because of its good performance in reducing seismic response through dissipation of energy. These damping devices supplement the already existing inherent structural damping, which, especially in steel structures, is anyhow small (2% in steel and 5% in reinforced concrete structures). More details about damping devices for seismic protection of buildings

G. A. Papagiannopoulos et al., *Seismic Design Methods for Steel Building Structures*, Geotechnical, Geological and Earthquake Engineering 51,
https://doi.org/10.1007/978-3-030-80687-3_13

can be found in the books of Soong and Dargush (1997), Christopoulos and Filiatrault (2006), Cheng et al. (2008) and Takewaki (2009), while summaries of the developments and state-of-practice in the application of damping devices for the seismic protection of structures can be found in Symans et al. (2008), Kitayama and Constantinou (2018) and De Domenico et al. (2019).

A variety of dampers are nowadays available (Christopoulos and Filiatrault 2006): fluid viscous dampers (they operate by forcing a viscous fluid or some form of oil through an orifice), viscoelastic fluid or solid dampers (they operate by shearing of highly viscous fluids or viscoelastic solids and are frequency and temperature dependent), metallic yielding dampers (they dissipate energy through yielding of their steel elements), friction dampers (they operate using preloaded sliding surfaces) and other types of dampers (with shape-memory alloys, friction-spring assemblies etc.) All these dampers when employed to a structure essentially increase its damping but may also alter its stiffness.

Design using supplemental dampers has the advantage of confining energy dissipation to the damping devices rather than in structural elements during strong ground motions. There are basically two kinds of dampers: velocity dependent including fluid viscous dampers and fluid or solid viscoelastic dampers and displacement dependent including friction dampers and metallic yielding dampers (Cheng et al. 2008). A comprehensive review on various types of dampers can be found in Lu et al. (2018). The most widely used dampers are the fluid viscous dampers with damping forces depending on velocity in a linear or nonlinear manner. The configuration or implementation (in a diagonal or chevron bracing or else) of damping devices in a structure depends on architectural considerations as well as on the type of the device employed. Figure 13.1 shows plane steel frames with various arrangements of dampers. The efficiency of supplemental dampers can be improved by providing a geometrical configuration of the bracing in order to amplify the displacement of the damper for a specified interstorey drift ratio (*IDR*). The material savings in a structure with supplemental dampers must be balanced with the cost of the damping devices as well as their testing, installation, maintenance and inspection. In general, supplemental damping can be used to retrofit structures with limited energy dissipation capacity (low ductility class according to EC8 2004) or to complement the energy dissipation of well detailed structures (medium or high ductility class according to EC8 2004).

A large number of studies has been carried out by various researchers regarding the seismic performance of structures with supplemental dampers. There are basically two strategies for the performance-based seismic design of steel structures with supplemental dampers: the force-based design procedure (e.g., Kasai et al. 1998; Fu and Kasai 1998; Ramirez et al. 2000; Lee et al. 2005; Tchamo and Zhou 2018) and the displacement-based design procedure (e.g., Lin et al. 2003, 2008; Kim and Choi 2006; Mazza and Vulcano 2014). To these methods, one can mention optimization methods for seismic design of structures with dampers in order to achieve the best performance through design, arrangement and number of dampers (e.g., Adachi et al. 2013; Cha et al. 2014; Wang et al. 2018). Most of the existing design methods are force-based linear simplified methods, like the equivalent lateral force method

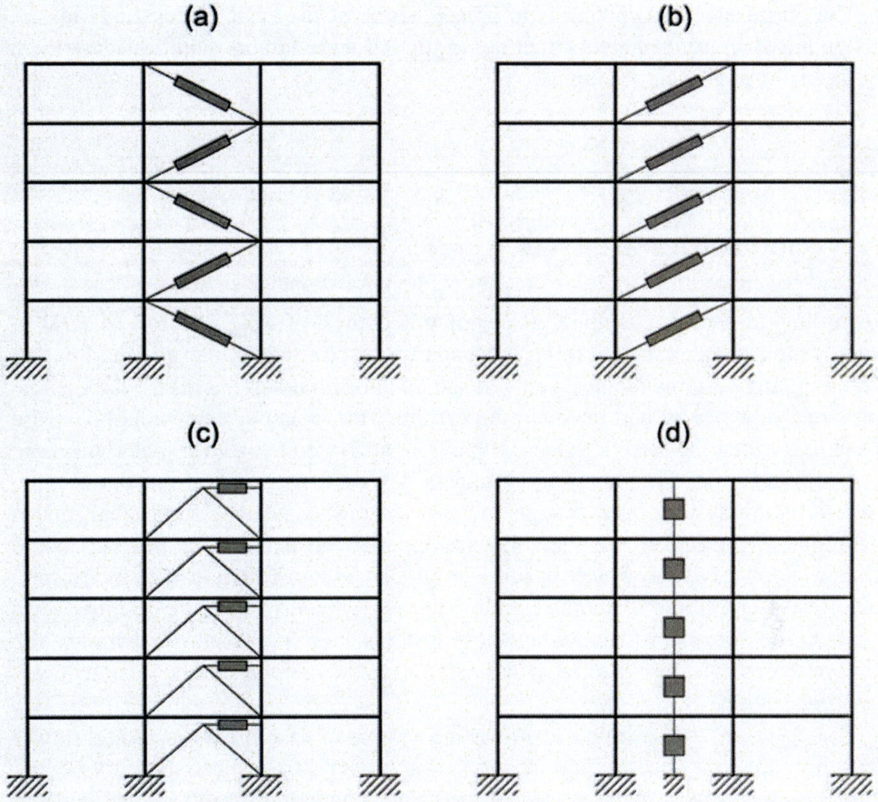

Fig. 13.1 Plane steel frames with supplemental dampers (shown by dark solid areas) in various arrangements: (**a**), (**b**) dampers in diagonal braces; (**c**) dampers in chevron braces; (**d**) wall dampers with vertical connections

and the response spectrum method. The ASCE/SEI 7-16 (2017) provisions, which are based on the work by Ramirez et al. (2000), discuss in detail these two simplified methods. However, application of these methods is only permitted, if a number of limitations is met. At this point, one should note that the current version of EC8 (2004) does not provide design guidelines for seismic design of structures with dampers. Nonlinear time-history (NLTH) analysis procedures are in general recommended for the design of all types of structures with dampers, especially nowadays when more designers are familiar with this type of analysis and there are easily available powerful computers. It is suggested that for practical applications, one should use simplified methods for a preliminary and approximate design and NLTH analysis for the final design (McVitty and Taylor 2016). Closing this discussion, one can point out that supplemental dampers can also be used together with base isolation devices to create an effective hybrid design scheme (Makris and Chang 2000; Chang et al. 2002; Providakis 2008, 2009).

The purpose of this chapter is to present some of the available methods for the design and/or retrofit of steel structures equipped with dampers and illustrate them by means of numerical examples.

13.2 Force-Based Design Procedures of ASCE/SEI 7-16 and ASCE/SEI 41-17

According to the two simplified design procedures of ASCE/SEI 7-16 (2017), namely the equivalent lateral force and response spectrum ones, design of a structure with damping systems (devices) should satisfy specific requirements for the seismic force-resisting system and the damping system. The former system should have the strength to resist the seismic forces, while the two systems together should meet the drift requirements. The damping system is designed separately from the seismic force-resisting system even though the two systems may have some elements in common. In particular, the damping system may be external or internal to the structure and, thus, may have no, some or all elements in common with the seismic force-resisting system. Elements common to these systems should be designed for a combination of loads, whereas when these systems have no elements in common, the damper forces must be collected and transferred to the members of the seismic force-resisting system.

The design of the seismic force-resisting system in each direction should satisfy the minimum base shear requirements, i.e., its seismic base shear shall not be less than V_{min}, where V_{min} is the greater of the values computed from the equations

$$V_{min} = \frac{V}{B_{V+1}} \tag{13.1}$$

$$V_{min} = 0.75V \tag{13.2}$$

where V is the seismic design base shear in the direction of interest without dampers and computed on the basis of the equivalent lateral force method and B_{V+1} is the damping coefficient that accounts for effective damping equal to the sum of viscous damping β_{V1} in the fundamental mode of vibration of the structure in the direction of interest plus inherent damping β_I. Having a seismic force-resisting system designed for a minimum of 75% of the strength required for structures without damping devices, provides safety in the event of malfunction of the damping system. It is stressed that the seismic base shear used for the design of the seismic force-resisting system should not be taken as less than $1.0V$ if: (i) in the direction of interest, the damping system has fewer than two damping devices on each floor level, configured to resist torsion and (ii) the seismic force-resisting system has horizontal or vertical irregularity. The analysis and design of the seismic force-resisting system under the base shear V_{min} from Eqs. (13.1) and (13.2) or under the unreduced base shear V

Fig. 13.2 Force F versus displacement u relation for velocity dependent damper (linear viscous for $\alpha = 1$ and nonlinear viscous for $\alpha < 1$)

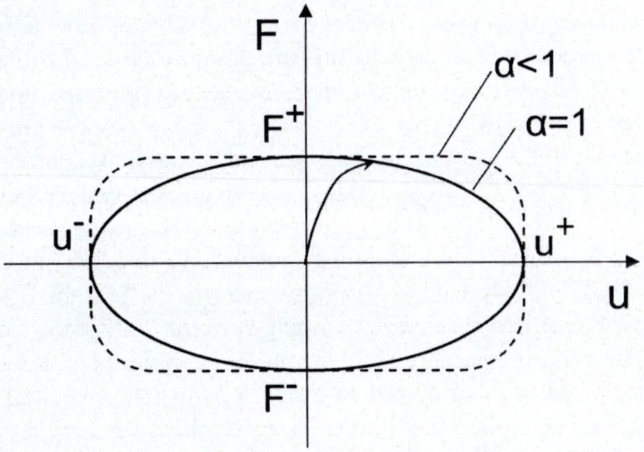

should be based on a model of the seismic force-resisting system that does not include the damping devices.

The damping devices must remain elastic for the maximum credible earthquake (MCE), whereas other elements of the damping system may be permitted to exhibit inelastic response under the condition that do not affect the function of the damping system. However, the inelastic response should be limited.

For each damping device, maximum and minimum property modification factors λ are established in analysis and design in order to account for the variation from nominal properties caused by aging (device wear) and environmental effects (ambient and design temperature), creep, fatigue as well as testing effects. The maximum and minimum values for λ are 1.2 and 0.85, respectively. On the basis of these factors, a maximum and minimum analysis and design property has to be established for each modeling parameter necessary for the selected method of analysis. The nominal properties of the damping devices should be confirmed by approved tests conducted prior to production of devices for construction. Similar tests are also mandatory by EN 15129 (2018).

The models used to describe the force-velocity-displacement characteristics of damping devices are given in detail in ASCE/SEI 41-17 (2017) and here only the case of fluid viscous dampers is presented. More specifically, in the absence of stiffness in the frequency range $0.5f_1$ to $2.0f_1$ (where f_1 is the fundamental frequency of the structure), the force in the viscous fluid device is computed as

$$F = C|\dot{u}|^a \cdot sgn\,(\dot{u}) \tag{13.3}$$

where C is the damping coefficient of the device, $\dot{u}$ is the relative velocity between each end of the device, sgn is the signum function that defines the sign of the relative velocity and a is the velocity exponent of the device. Figure 13.2 depicts the hysteretic force F versus displacement u relation for a velocity dependent fluid viscous damper for which the velocity exponent α is equal to 1 for linear and less than 1 for nonlinear viscous damper. The force-velocity-displacement characteristics and damping properties of the damping devices have to be confirmed by tests before the production of devices that are going to be used for retrofitting purposes.

If fewer than four damping devices are provided in any storey of a structure in either principal direction, or fewer than two devices are located on each side of the center of stiffness of any storey in each principal direction, all damping devices must be designed to accommodate displacements and velocities equal to 130% of the maximum calculated displacement and velocity in the device under the MCE.

Structures with a damping system have to be analyzed by nonlinear time-history (NLTH) analysis. The response spectrum is permitted to be used for analysis and design of structures with damping systems if all of the following conditions hold: (i) in each principal direction, the damping system has at least two damping devices in each storey, configured to resist torsion; (ii) the total effective damping of the fundamental mode β_{mD} ($m = 1$) of the structure in the direction of interest is $\leq 35\%$ and (iii) the S_1 value for the site is less than 0.6. The equivalent lateral force procedure is also permitted to be used for analysis and design of structures with damping systems if all of the following conditions hold: (i) in each principal direction, the damping system has at least two damping devices in each storey, configured to resist torsion; (ii) the total effective damping of the fundamental mode β_{mD} ($m = 1$) of the structure in the direction of interest is $\leq 35\%$; (iii) the seismic force-resisting system has no horizontal or vertical irregularity; (iv) floor diaphragms are rigid; (v) the height of the structure above the base does not exceed 30 m and (vi) the S_1 value for the site is less than 0.6.

When a NLTH analysis is used, the force-velocity-displacement characteristics of the damping devices should be explicitly modeled to account for device dependence on frequency, amplitude and duration of seismic loading. If the properties of the damping device change with time and/or temperature, such behavior should be also explicitly modeled. The flexible elements of the damping devices that connect them to the structure should be included in the model. NLTH analysis shall be performed for both the design basis earthquake (DBE) and the MCE. The analysis under the DBE does not need to include effects of accidental eccentricity. Results from the analysis under the DBE are used to design the seismic force-resisting system, whereas those from the analysis under the MCE are used to design the damping system. Inherent damping in the structure should not be taken greater than 3% unless test data consistent with the anticipated level of deformation support higher damping values. Computed drift ratio at the MCE should be less than or equal to the smallest of: 3%; 1.9 times the drift at the DBE; $1.5R/C_d$, where R and C_d are given in ASCE/SEI 7-16 (2017) in a table form depending on the building framing under consideration.

Considering the option of selecting a response spectrum analysis for the structure with damping devices, the seismic base shear V_D of the structure in a given direction is determined employing either the CQC or the SRSS modal combination rules (both presented in Chap. 2) and should satisfy $V_D \geq V_{min}$. Modal base shear of the mth mode of vibration V_m is calculated by

$$V_m = C_{Sm}\overline{W}_m \tag{13.4}$$

$$\overline{W}_m = \frac{\left(\sum_{i=1}^{n} w_i\varphi_{im}\right)^2}{\sum_{i=1}^{n} w_i\varphi_{im}^2} \tag{13.5}$$

where C_{Sm} is the seismic response coefficient of the mth mode of vibration of the structure, $\overline{W}_m$ is the effective seismic weight of the mth mode of vibration of the structure, w_i is the vertical weight at floor i, and φ_{im} is the displacement amplitude at the ith level of the structure in the mth mode of vibration, normalized to unity at the roof level.

The seismic response coefficient of the fundamental mode, C_{S1}, and higher modes, C_{Sm}, are then calculated. The expressions for the C_{S1} associated with the first (m = 1) mode, depending on the relationship between the first mode effective period of the structure T_{1D} with respect to the period $T_S = S_{D1}/S_{DS}$, as well as for the C_{Sm} associated with higher modes (m > 1) are given by

$$\begin{aligned} C_{S1} &= \left(\frac{R}{C_d}\right)\frac{S_{DS}}{\Omega_0 B_{1D}} \quad \text{for } T_{1D} < T_S \\ C_{S1} &= \left(\frac{R}{C_d}\right)\frac{S_{D1}}{T_{1D}\Omega_0 B_{1D}} \quad \text{for } T_{1D} \geq T_S \end{aligned} \tag{13.6}$$

$$\begin{aligned} C_{Sm} &= \left(\frac{R}{C_d}\right)\frac{S_{DS}}{\Omega_0 B_{mD}} \quad \text{for } T_m < T_S \\ C_{Sm} &= \left(\frac{R}{C_d}\right)\frac{S_{D1}}{T_m\Omega_0 B_{mD}} \quad \text{for } T_m \geq T_S \end{aligned} \tag{13.7}$$

where R is the response modification coefficient, C_d is the deflection amplification factor, Ω_0 is the overstrength factor, B_{1D} is a numerical coefficient for effective damping equal to $\beta_{mD}(m = 1)$ and effective period of structure equal to T_{1D}, B_{mD} is a numerical coefficient for effective damping equal to β_{mM} and period of structure equal to T_m (period of the mth mode of vibration of the structure) and S_{DS}, S_{D1} are the design spectrum acceleration parameters at short period and at 1.0 s, respectively. It is recalled that $S_{D1} = (2/3)S_{M1}$ and $S_{DS} = (2/3)S_{MS}$ where the subscript M at the spectral values S denotes the MCE response spectrum. The effective first mode period at the DBE and MCE are

$$T_{1D} = T_1\sqrt{\mu_D} \tag{13.8}$$

$$T_{1M} = T_1\sqrt{\mu_M} \tag{13.9}$$

where μ_D and μ_M are the effective ductility demands on the seismic force resisting system caused by the DBE and MCE and T_1 is the fundamental period of the structure in the direction under consideration.

The design lateral force F_{im} at level i for the mth mode of vibration of the structure is determined in accordance with the equation

$$F_{im} = w_i \varphi_{im} \frac{\Gamma_m}{\bar{W}_m} V_m \tag{13.10}$$

where the modal participation factor Γ_m for the mth mode of vibration is determined as

$$\Gamma_m = \frac{\bar{W}_m}{\sum_{i=1}^{n} w_i \varphi_{im}} \tag{13.11}$$

Design forces in damping devices and other elements of the damping system shall be determined on the basis of the floor displacement, storey drift and storey velocity response parameters. More specifically, floor deflections, δ_{iD} and δ_{iM}, storey drifts, Δ_{iD} and Δ_{iM}, and storey velocities, ∇_{iD} and ∇_{iM} should be calculated for both the DBE and MCE and this explains the presence of subscripts D and M, respectively, in the previous quantities. The displacement of the structure due to the DBE at level i in the mth mode of vibration δ_{mD} is determined as

$$\delta_{mD} = D_{mD} \varphi_{im} \tag{13.12}$$

The total design displacement at each floor of the structure can be calculated by the SRSS or CQC modal combination methods.

The expressions for the fundamental (m = 1) and higher mode (m > 1) roof displacements, D_{1D} and D_{mD}, respectively, are

$$\begin{aligned} D_{1D} &= \left(\frac{g}{4\pi^2}\right)\Gamma_1 \frac{S_{DS} T_{1D}^2}{B_{1D}} \geq \left(\frac{g}{4\pi^2}\right)\Gamma_1 \frac{S_{DS} T_1^2}{B_{1E}} \quad \text{for} \quad T_{1D} < T_S \\ D_{1D} &= \left(\frac{g}{4\pi^2}\right)\Gamma_1 \frac{S_{D1} T_{1D}}{B_{1D}} \geq \left(\frac{g}{4\pi^2}\right)\Gamma_1 \frac{S_{DS} T_1}{B_{1E}} \quad \text{for} \quad T_{1D} \geq T_S \end{aligned} \tag{13.13}$$

$$D_{mD} = \left(\frac{g}{4\pi^2}\right)\Gamma_m \frac{S_{D1} T_m}{B_{mD}} \leq \left(\frac{g}{4\pi^2}\right)\Gamma_m \frac{S_{DS} T_m^2}{B_{mD}} \tag{13.14}$$

where B_{1E} is the numerical coefficient for effective damping equal to β_I and β_{VI} and period of structure equal to T_1. It is recalled that β_I is the effective damping of the structure caused by the inherent energy dissipation of its elements and β_{VI} is the effective damping of the ith mode of vibration of the structure caused by viscous energy dissipation of the damping system. Both β_I and β_{VI} are calculated at or just below the effective yield displacement of the seismic force resisting system. The design storey drift in the fundamental mode, Δ_{1D}, and higher modes, Δ_{mD}, of the structure can be calculated using the modal roof displacements provided by Eqs. (13.13) and (13.14) in conjunction with the story drift, i.e., the difference of

Table 13.1 Damping coefficients

Effective damping β (%)	B_{V+I}, B_{1D}, B_{1E}, B_R, B_{1M}, B_{mD}, B_{mM}
≤2	0.8
5	1.0
10	1.2
20	1.5
30	1.8
40	2.1
50	2.4
60	2.7
70	3.0
80	3.3
90	3.6
100	4.0

deflections at the top and bottom of the story under consideration. The total storey drift Δ_D is determined by a CQC or SRSS combination of modal design drifts.

Design storey velocity in the fundamental mode (m = 1), ∇_{1D}, and higher modes, ∇_{mD} (m > 1) of the structure are determined as

$$\nabla_{1D} = 2\pi \frac{\Delta_{1D}}{T_{1D}} \quad (13.15)$$

$$\nabla_{mD} = 2\pi \frac{\Delta_{mD}}{T_m} \quad (13.16)$$

Total storey velocity ∇_D is determined by a CQC or SRSS combination of modal design velocities. The expressions for design storey drifts and velocities corresponding to the DBE and provided by Eqs. (13.13)–(13.16) can also be used for drifts and velocities corresponding to the MCE by simply replacing the subscript D by subscript M.

Values of the damping coefficients B_{1D}, B_{1M} etc. as functions of the effective damping β (%) needed for the calculation of C_{S1}, C_{Sm}, D_{1D}, D_{mD}, D_{1M} and D_{mM} are provided in Table 13.1. These damping coefficients are used to modify the response of the structure due to its total (effective) damping consisting of the inherent damping, the hysteretic after yield damping and the damping supplied by the damper devices.

One last thing to note regarding the response spectrum analysis of the structure with dampers is that the displacements and velocities used to determine design forces in damping devices should take into account the angle of orientation of the damping device with respect to the horizontal and also consider the effects of increased floor response caused by torsional motions.

When the equivalent lateral force procedure is employed, one considers not only the effect of the first mode in computing the design base shear or the floor and roof displacements and velocities but in addition the effect of the residual mode. The total response quantities (forces, displacements and velocities) are determined by employing the SRRS rule to the corresponding two modal components. For example, the seismic base shear V is determined from the equation

$$V = \sqrt{V_1^2 + V_R^2} \geq V_{min} \tag{13.17}$$

where V_1 and V_R are the design values of the seismic base shear of the fundamental and residual mode, respectively and V_{min} is given by Eqs. (13.1) and (13.2). Base shears V_1 and V_R are computed from

$$V_1 = C_{S1}\overline{W}_1 \tag{13.18}$$

$$V_R = C_{SR}\overline{W}_R \tag{13.19}$$

where $\overline{W}_1$ is obtained from Eq. (13.5) for $m = 1$ and $\overline{W}_R = W - \overline{W}_1$ with W being the total effective seismic load of the structure, while C_{S1} and C_{SR} are the seismic response coefficients for the first and the residual mode, respectively. Coefficient C_{S1} is given by Eq. (13.6), while coefficient C_{SR} is given by

$$C_{SR} = \left(\frac{R}{C_d}\right)\frac{S_{DS}}{\Omega_0 B_R} \tag{13.20}$$

where B_R is a numerical coefficient for effective damping β_R and period $T_R = 0.4T_1$, which is obtained from Table 13.1. For explicit expressions pertaining to displacements and velocities for the equivalent lateral force method, the reader may consult ASCE/SEI 7-16 (2017).

The rest of this section is devoted to the calculation of the effective damping needed for both response spectrum and equivalent lateral force procedures. In particular, the effective damping at the design displacement, β_{mD}, is

$$\beta_{mD} = \beta_I + \beta_{Vm}\sqrt{\mu_D} + \beta_{HD} \tag{13.21}$$

where β_{HD} is the component of effective damping caused by post-yield hysteretic behavior of the seismic force-resisting systems and elements of the damping system at effective ductility demand μ_D, β_I is the component of effective damping caused by the inherent damping capacity of the structure at or just below the effective yield displacement of the seismic force-resisting system and β_{Vm} is the component of viscous damping of the mth mode of vibration of the structure caused by the damping system at or just below the effective yield displacement of the seismic force-resisting system. The value of β_{Vm} is calculated from the equation

$$\beta_{Vm} = \frac{\sum_j W_{mj}}{4\pi W_m} \tag{13.22}$$

where W_{mj} is the work done by the jth damping device in one complete cycle of dynamic response corresponding to the mth mode of vibration of the structure at modal displacements δ_{im}, $W_m = 0.5 \cdot \sum_j F_{im}\delta_{im}$, is the maximum strain energy of the mth mode of vibration of the structure at modal displacements δ_{im}, F_{im} is the inertia force at level i of the mth mode and δ_{im} is the deflection of level i in the mth mode of vibration at the center of rigidity of the structure. It should be noted that the aforementioned components β_I, β_{Vm}, β_{HD} of effective damping are calculated in the structural direction of interest and, unless supported by other analysis or data, for higher modes of vibration $\mu_D = 1.0$.

An upper value of 3% is considered for the inherent damping β_I of the structure, whereas the hysteretic damping β_{HD} is given by

$$\beta_{HD} = q_H(0.64 - \beta_I)\left(1 - \frac{1}{\mu_D}\right) \tag{13.23}$$

where q_H is the hysteresis loop adjustment factor reading

$$q_H = 0.67\frac{T_S}{T_1} \tag{13.24}$$

and satisfying $0.5 \leq q_H \leq 1.0$. The effective ductility demand for the DBE is

$$\mu_D = \frac{D_{1D}}{D_Y} \geq 1.0 \tag{13.25}$$

with

$$D_Y = \left(\frac{g}{4\pi^2}\right)\left(\frac{\Omega_0 C_d}{R}\right)\Gamma_1 C_{S1} T_1^2 \tag{13.26}$$

where D_{1D} and D_Y are the fundamental mode design displacement and the displacement at the effective yield point, respectively, both considered at the center of rigidity of the top storey (roof) level and in the direction under consideration. The design ductility demand μ_D should not exceed the maximum value of the effective ductility demand μ_{max} given as

$$\begin{aligned} \mu_{max} &= 0.5[(R/(\Omega_0 I_e)^2) + 1] \quad for \quad T_{1D} \leq T_S \\ \mu_{max} &= R/(\Omega_0 I_e) \; for \quad T_1 \geq T_S \end{aligned} \tag{13.27}$$

where I_e is the importance factor determined in the relevant section of ASCE/SEI 7-16 (2017). For $T_1 < T_S < T_{1D}$, μ_{max} can be found by linear interpolation of the values provided by Eqs. (13.27).

The expressions provided by Eqs. (13.21), (13.23) and (13.25) can also be used for the MCE by simply replacing the subscript D by M.

The elements of the damping system, either in the context of response spectrum procedure or equivalent lateral force procedure, should be designed on the basis of the maximum force of the following three loading conditions:

(i) stage of maximum displacement for which the seismic design force Q_E is calculated as

$$Q_E = \Omega_0 \sqrt{\sum_m (Q_{mSFRS})^2} \pm Q_{DSD} \tag{13.28}$$

where Q_{mSFRS} is the force in an element of the damping system equal to the design seismic force of the mth mode of vibration of the structure and Q_{DSD} is the force in an element of the damping system required to resist design seismic forces of displacement-dependent damping devices.

(ii) stage of maximum velocity for which the seismic design force Q_E is calculated as

$$Q_E = \sqrt{\sum_m (Q_{mDSV})^2} \tag{13.29}$$

where Q_{mDSV} is the force in an element of the damping system required to resist design seismic forces of velocity-dependent damping devices caused by mth mode of vibration of the structure.

(iii) stage of maximum acceleration for which the seismic design force Q_E is calculated as

$$Q_E = \sqrt{\sum_m (C_{mFD}\Omega_0 Q_{mSFRS} + C_{mFV} Q_{mDSV})^2 \pm Q_{DSD}} \tag{13.30}$$

The coefficients C_{mFD} and C_{mFV} are determined from Tables 13.2 and 13.3 on the basis of the following requirements: a) for the fundamental mode response (m = 1), the coefficients C_{1FD} and C_{1FV} are calculated using the velocity exponent α, which relates the force of the device with its velocity as described by Eq. (13.3). The effective modal damping should be taken as equal to the total effective damping of the fundamental mode less the hysteretic component of damping at the response level of interest. For higher mode (m > 1) or residual-mode response (equivalent lateral force procedure), the coefficients C_{mFD} and C_{mFV} should be determined for $\alpha = 1.0$. The effective modal damping should be taken as equal to the total effective damping of the mode of interest. For the determination of the coefficient C_{mFD}, the ductility demand should be taken as equal to that of the fundamental mode. In

Table 13.2 Force coefficient C_{mFD}

	$\mu \leq 1.0$				
Effective damping	$a \leq 0.25$	$a \leq 0.5$	$a \leq 0.75$	$a \geq 1.0$	$C_{mFD} = 1.0$
≤0.05	1.0	1.0	1.0	1.0	$\mu \geq 1.0$
0.1	1.0	1.0	1.0	1.0	$\mu \geq 1.0$
0.2	1.0	0.95	0.94	0.93	$\mu \geq 1.1$
0.3	1.0	0.92	0.88	0.86	$\mu \geq 1.2$
0.4	1.0	0.88	0.81	0.78	$\mu \geq 1.3$
0.5	1.0	0.84	0.73	0.71	$\mu \geq 1.4$
0.6	1.0	0.79	0.64	0.64	$\mu \geq 1.6$
0.7	1.0	0.75	0.55	0.58	$\mu \geq 1.7$
0.8	1.0	0.70	0.50	0.53	$\mu \geq 1.9$
0.9	1.0	0.66	0.50	0.50	$\mu \geq 2.1$
≥1.0	1.0	0.62	0.50	0.50	$\mu \geq 2.2$

Table 13.3 Force coefficient C_{mFV}

Effective damping	$a \leq 0.25$	$a \leq 0.5$	$a \leq 0.75$	$a \geq 1.0$
≤0.05	1.0	0.35	0.21	0.10
0.1	1.0	0.44	0.31	0.20
0.2	1.0	0.56	0.46	0.37
0.3	1.0	0.64	0.58	0.51
0.4	1.0	0.70	0.69	0.62
0.5	1.0	0.75	0.77	0.71
0.6	1.0	0.80	0.84	0.77
0.7	1.0	0.83	0.90	0.81
0.8	1.0	0.90	0.94	0.90
0.9	1.0	1.00	1.00	1.00
≥1.0	1.0	1.00	1.00	1.00

Tables 13.2 and 13.3, interpolation can be used for intermediate values of a. Moreover, in Table 13.2, $C_{mFD} = 1.0$ for values of ductility demand μ greater than or equal to the values shown.

It should be noted that all three stages need to be checked for structures with velocity-dependent damping systems, whereas for displacement-dependent damping systems, the first and third stages are identical and the second stage is redundant.

The design procedure of ASCE/SEI 7-16 (2017) has been thoroughly investigated by Kitayama and Constantinou (2018). This investigation determined the design base shear force, the brace-damper-connection system and the amount of damping as the most critical parameters in obtaining an acceptable seismic performance in terms of collapse potential and residual storey drift ratio exhibition.

ASCE/SEI 41-17 (2017) sets the requirements for the evaluation and retrofit of structures using damping systems (devices). Linear and nonlinear analysis procedures are permitted on the basis of maximum and minimum values of the properties of the damping devices to account for the variation of the nominal values of those properties. Linear analysis procedures, i.e., the equivalent lateral force and the response spectrum analyses, are permitted if all the following criteria are met: (i) the framing system remains elastic for the selected seismic hazard level after the consideration of added damping of the damping system; (ii) the effective damping of the damping system does not exceed 30% of critical in the first (fundamental) mode; (iii) the secant stiffness of each damping device, calculated at the maximum displacement in the device, is included in the mathematical model of the retrofitted structure and (iv) damping devices should be included in the model when the regularity of the structure is evaluated, i.e., dampers should be provided in all storeys of the structure.

Focusing interest only in the response spectrum method and considering the case of velocity-dependent devices (solid viscoelastic, fluid viscoelastic and fluid viscous devices), modification of the 5%-damped response spectrum should be performed to account for the additional damping offered by them. The effective damping in the *m*-th mode of vibration is calculated as

$$\xi_{eff,m} = \xi_m + \frac{\sum_j W_{mj}}{4\pi W_{mk}} \tag{13.31}$$

where ξ_m is the *m*th mode damping in the structure, W_{mj} is the work done by device *j* in one complete cycle corresponding to modal floor displacements δ_{mi} and W_{mk} is the maximum strain energy in the structure in the *m*th mode determined by $W_{mk} = 0.5 \sum F_{mi}\delta_{mi}$, where F_{mi} is the *m*th mode horizontal inertia force at floor level *i* and δ_{mi} is the *m*th mode horizontal displacement at floor level *i*. The work done by a linear viscous device j for one complete cycle of loading in the *m*th mode is $W_{mk} = (2\pi^2/T_m)C_j\delta^2_{mrj}$, where T_m is the mth mode period of the structure (including the stiffness of the velocity-dependent devices), C_j is the damping coefficient of the device *j* and δ_{mrj} is the *m*th mode relative displacement between the ends of device *j* along the axis of that device. The design actions are then determined for three stages, i.e., that of maximum drift, maximum velocity and zero drift and maximum floor acceleration, in each significant mode. For more details on these three stages, one can consult ASCE/SEI 41-17 (2017).

It should be noted that the ratio $\sum_j W_{mj}/4\pi W_{mk}$ in Eq. (13.31) is also used in ASCE/SEI 7-16 (2017) to calculate the viscous damping ratio offered by the devices. This ratio of Eq. (13.31) to calculate damping strictly holds for linear viscous dampers distributed in a stiffness-proportional form. With some modifications this ratio can also be used to define the equivalent modal damping ratio coming from nonlinear dampers or even to account for both horizontal and vertical deformation between the ends of the damper (Hwang et al. 2008). However, neither ASCE/SEI 41-17 (2017) nor ASCE/SEI 7-16 (2017) make explicit use of these modifications.

13.3 A Force-Based Design of Steel MRFs with Supplemental Dampers

The seismic performance of steel MRFs can be improved by the addition of elastomeric dampers that are not sensitive to ambient temperature and loading frequency as viscoelastic dampers. On the other hand, as a result of their modest energy dissipation capacity, elastomeric dampers may need to be relatively large in size and as a result of that a change to the properties of the structure when equipped with elastomeric dampers has to be considered (Lee et al. 2009). Compressed elastomeric dampers (constructed by pre-compressing a high-damping elastomeric material into steel tubes), essentially smaller in size than conventional elastomeric dampers, may offer significant reduction in the weight of steel MRFs (Karavasilis et al. 2012).

The steps included in the performance-based seismic design of steel MRFs with elastomeric dampers are presented in detail in the work of Lee et al. (2009). A summary of these steps is presented in the following:

1. The seismic performance objectives and associated design criteria are established. In particular, all members must remain within the design capacities under the design basis earthquake (DBE) and the story drift is limited to a specific value under the DBE.
2. The elastomeric material is idealized as a linear viscoelastic material.
3. The design temperature range (upper and lower bound temperatures) is determined.
4. An appropriate ratio of the brace stiffness per storey in the global direction to the storey stiffness of the MRF without dampers and braces is selected (the usual value for this ratio is 30). Similarly, an appropriate ratio of the damper stiffness per storey in the global direction to the storey stiffness of the MRF without dampers and braces is also selected. The usual values for this ratio range from 0 to 20 even though values above 5 are considered to be impractical.
5. An elastic static analysis is performed and member forces and storey drift are obtained. More specifically, the first-mode deflected shape of the MRF with dampers is estimated by means of a pattern of lateral forces. Then, the first-mode period and equivalent damping ratio are estimated. The latter is added to the inherent viscous damping of the MRF in order to obtain the total damping of the MRF with dampers. Then, the design base shear is computed from a design spectrum and the equivalent lateral forces to be used in the static analysis are derived. Iterations are performed if needed. Moreover, amplification of damper and brace forces are implemented in order to consider the effects of out-of-phase stiffness of the dampers, something which is not included in the elastic static analysis model.
6. The seismic response obtained in step 5 is compared against the seismic performance objectives and design criteria of step 1.

7. A minimum value that satisfies the design criteria and the performance objectives is selected for the ratio of the damper stiffness per storey in the global direction to the storey stiffness of the MRF without dampers and braces.
8. The structural response is assessed at the low-temperature and the design temperature range.
9. Based on the selected value of the ratio mentioned in step 7, the corresponding required area and thickness of the damper are calculated. The values of the ratios mentioned in step 4 are treated as constants for all storeys.

A similar performance-based seismic design procedure has been proposed by Seo et al. (2014) for steel MRFs with fluid viscous dampers. In this procedure, steps 1 and 4 are the same as above and the rest of the steps include: (i) selection of an appropriate damping coefficient for each storey. This coefficient should provide a reasonable number of dampers; (ii) calculation of the equivalent damping ratio and derivation of a damping reduction factor for the total damping ratio which includes both the inherent damping of the steel MRF (usually 2%) and the equivalent damping ratio; (iii) the elastic response spectrum is reduced by means of the damping reduction factor and the peak inelastic storey drifts are calculated on the basis of equivalent lateral forces applied to the frame without dampers and the equal displacement rule. If the storey drifts do not satisfy the already established performance criteria, all steps up to this point are repeated by using a new value of the damping coefficient for each storey; (iv) the final design of the dampers is based on the required damping coefficient and on the expected maximum damper force and stroke values, which are provided to the damper manufacturer. The peak damper force is estimated and used for the capacity design of braces, beams and columns in the force path of the dampers.

The main conclusion drawn from the application of the above performance-based design method to steel MRFs equipped either with elastomeric or with viscous dampers is that (i) the seismic response of the steel MRF with dampers can be reduced in terms of residual storey drift, plastic hinge rotations and peak floor acceleration and (ii) the steel MRF with viscous dampers can exhibit better performance than the conventional steel MRFs under the DBE and MCE.

13.4 A Direct Displacement-Based Design of Steel MRFs with Supplemental Dampers

A direct displacement-based design procedure for the seismic retrofit of existing buildings using nonlinear viscous dampers has been developed by Lin et al. (2008) as an extension of their previous work (Lin et al. 2003) dealing with linear viscous dampers. In this procedure, use of the concept of equivalent viscous damping ratio and of the first mode of the structure is made. This damping ratio is derived from the assumption that the average energy dissipated by linear and nonlinear viscous dampers is equal. The equivalent period needed for the definition of the equivalent linear system is derived from the concept of the average storage energy. The design procedure is implemented on the basis of the following steps:

1. The design maximum roof displacement u_r is determined upon consideration of the importance and the function of the building or of an allowable interstorey drift ratio (*IDR*) as specified in seismic codes.
2. The location for installation of nonlinear dampers is chosen and for each damper the damper coefficient C and the velocity exponent α as appearing in Eq. (13.3) are assumed.
3. Pushover analysis is conducted under lateral forces F_i analogous to $m_i\varphi_{ni}$ with m_i and φ_{ni} being the mass and the *n*th mode shape at storey *i*, respectively and a bi-linearization of the pushover curve is performed. Form this curve, the ductility ratio μ and post-yield stiffness ratio a of the structure are obtained.
4. The equivalent period T_e and damping ξ_e are computed as

$$\frac{T_e}{T_m} = \left[\mathrm{a} + (1 - \mathrm{a})\left(\frac{1}{\mu^3} + \frac{3}{2\mu}\left(1 - \frac{1}{\mu^2}\right)\right)\right]^{-0.5} \tag{13.32}$$

$$\xi_e = \frac{1}{\pi}\frac{(1-\mathrm{a})\left[\frac{1}{2\mu}\left(1+\frac{1}{\mu^2}\right)-\frac{1}{\mu^2}\right]}{(1-\mathrm{a})\left[\frac{1}{4\mu}\left(1-\frac{1}{\mu^2}\right)+\frac{1}{6\mu^3}\right]+\frac{\mathrm{a}}{6}} + \xi_i + \xi_m\frac{T_e}{T_m} \tag{13.33}$$

where T_m is the period of vibration of the primary (first) mode, the first part of ξ_e is the equivalent damping to the inelastic behavior of the structure, ξ_i is the inherent viscous damping of the structure (usually 0.02 for steel structures) and ξ_m is the equivalent viscous damping to the nonlinear viscous damping of dampers having the form

$$\xi_m = \left[3\sum_j\left[(\lambda_j C_j)/(2+a_j)\right](2\pi/T_m)^{a_j-2}u_r^{a_j-1}\varphi_{mr,j}^{\alpha_j+1}f_j^{\alpha_j+1}\right] \Big/ \left(2\pi\sum_i m_i\varphi_{mi}^2\right) \tag{13.34}$$

where subscripts j and i refer to damper and storey numbers, respectively, $\lambda_j = 2^{\alpha_j+2}\Gamma^2(1+\alpha_j/2)/\Gamma(2+\alpha_j)$ with Γ being the gamma function and $f_y = \cos\theta_j$ with θ_j being the angle of the brace/damper j with the horizontal direction.

5. The displacement spectrum reduced for a damping ratio equal to the aforementioned equivalent damping ξ_e is constructed and the roof displacement corresponding to the equivalent period T_e is read.
6. If the ratio of the roof displacement of step 1 to the participation factor of the first mode is nearly equal to the roof displacement of step 5, the procedure has converged. Otherwise, iterations have to be performed by adjusting the damper coefficient and velocity exponent of each damper, i.e., steps 2–6 are repeated.

Similar displacement-based seismic design methods for retrofitting plane steel frames using viscous or viscoelastic dampers have been developed by Kim and Choi (2006) and Sullivan and Lago (2012).

13.5 Additional Design Methods for Steel Frames with Dampers

13.5.1 A Five-Step Design of Steel MRFs with Viscous Dampers

A simplified procedure of the force-based type to dimension viscous dampers and structural elements of framed steel structures has been developed by Palermo et al. (2018). This procedure consists of 5 steps, which can be summarized as follows:

1. The seismic performance objectives are identified by imposing a target reduction factor to the elastic design spectrum. This factor expresses the energy dissipated by the viscous dampers and the hysteretic energy dissipated by the structure and thus is a function of the damping reduction factor and the behavior factor.
2. Assuming the along-the-height relative distribution of the dampers, linear viscous dampers are dimensioned using simple analytical formulas.
3. Peak inter-storey drifts, peak inter-storey velocities and peak damper velocities, strokes and forces expected by the design earthquake are estimated.
4. Non-linear viscous dampers are dimensioned. In particular, for low values of the damping exponent ($a = 0.15 - 0.30$) the damping coefficient can be obtained in terms of the damping coefficient of the corresponding linear damper using a design velocity that equals to 0.8 times the peak damper velocity.
5. Two equivalent static analyses are performed and from the envelope of their results, one dimensions the structural elements. These analyses are needed because forces in structural elements become maximum for maximum displacements, whereas forces in viscous dampers for maximum inter-story velocities.

However, final verification of this five-step procedure can only be done by means of NLTH analyses. From the results of these analyses, adjustments to viscous dampers and to structural elements may be needed in order to fully satisfy the required seismic performance objectives. Alternatively, an optimal damper design may be sought.

13.5.2 Modified Capacity Design in Tall Steel MRFs with Viscous Dampers

Taking into account that tall steel MRFs with linear viscous dampers are prone to column plastic hinging (Karavasilis 2016) and that the formation of a sway plastic mechanism that involves plastic hinges in beams and column bases is a fundamental requirement of current seismic codes (e.g., EC8 2004), Kariniotakis and Karavasilis (2018) make use of a modified capacity design rule for the columns of steel MRFs with viscous dampers in an effort to achieve the desired global plastic mechanism. This modified capacity design rule results in plastic mechanisms similar to those of steel MRFs without dampers.

More specifically, according to EC8 (2004), columns are designed against axial forces, shear forces and bending moments employing Eqs. (3.29)–(3.31) of Chap. 3. The modified capacity design rule proposed for the interior columns in the force path of viscous dampers reads

$$N_{Ed} = N_{Ed,G} + 1.1\gamma_{ov}\Omega(N_{Ed,E}, SF \cdot N_{Ed,E,V}) \tag{13.35}$$

where $N_{Ed,\,E,\,V}$ is the column axial force at the state of peak velocity under the design basis earthquake (DBE) and SF is a scale factor. The rest of the symbols in Eq. (13.35) have been explained in Chap. 3. Thus, Eq. (13.35) is complementary to Eqs. (3.29)–(3.31) for the capacity design of the columns in the force path of viscous dampers. Moreover, N_{Ed} from Eq. (13.35) is used to calculate M_{Rc} in Eq. (3.22). Axial force $N_{Ed,\,E,\,V}$ is obtained from the usual response spectrum in conjunction with the modal synthesis procedure of EC8 (2004).

It should be recalled that for the case of elastic or mildly inelastic frames, $N_{Ed,\,E,\,V}$ is out-of-phase with N_{Ed} and, thus, the modified capacity rule is conservative. However, for seismic intensities higher than the DBE, peak damper forces increase beyond their design values under the DBE, while inelasticity of the steel MRF may result in unfavorable combinations of axial forces, shear forces and bending moments in columns. Therefore, the proposed modified capacity design rule is justified in promoting a global sway plastic mechanism in tall steel MRFs with viscous dampers.

The proposed modified capacity design rule becomes stricter for buildings with more than 10 storeys because conventional analysis methods for structures with dampers underestimate the peak damper forces in the lower storeys of yielding tall MRFs. Thus, the aforementioned scale factor SF equals to 1.0 for steel MRFs with less than 10 storeys and to 3.5 for steel MRFs with 20 storeys. A linear interpolation can be used to estimate SF for steel MRFs with 11–19 storeys. Nevertheless, the values of SF are expected to be different if non-linear viscous dampers are employed not to mention the case of steel structures with different configurations.

13.5.3 Seismic Retrofit of Steel MRFs with Viscous Dampers Using Interstorey Velocity

The seismic retrofit method for steel MRFs equipped with viscous dampers proposed by Papagiannopoulos et al. (2018) starts by using equivalent modal damping ratios given for specific deformation levels (in terms of *IDR*) of steel MRFs (Papagiannopoulos and Beskos 2010 and Chap. 8). Then, these modal damping ratios are employed in the context of linear time-history analyses of a steel MRF subjected to a number of recorded accelerograms. From these analyses, the along-the-height maximum values for elastic interstorey velocity (*IV*) and storey shears are computed. Damping coefficients *C* of the viscous dampers (linear or non-linear) are determined per storey utilizing the aforementioned storey shears and *IV*s in the form

$$C = kV_{storey} / \left(IV^{a} \cos^{2}\theta \right) \tag{13.36}$$

where V_{storey} is the storey shear, *k* is the percentage of storey shear that should be resisted by the viscous damper, *a* is the velocity exponent (equals to 1 for a linear damper and to less than 1 for a non-linear damper), θ is the angle of inclination of the damper/brace diagonal and *IV* is the interstorey velocity given for a specific *IDR* range. At this point, two correction factors have to be employed in order to take into account inelastic effects. The first relates the expected inelastic *IVs* for a specific *IDR* range with the elastic ones and the second the expected inelastic storey shears with the elastic ones. These correction factors are expressed in terms of polynomial functions of height and can be found in Papagiannopoulos et al. (2018). After the application of these factors to the parameters *IV* and V_{storey} of Eq. (13.36), one calculates the damping coefficients of the linear or non-linear viscous dampers on the basis of a preselected value of *k*.

The seismic retrofit method of Papagiannopoulos et al. (2018) seems to work better in the case of steel MRFs equipped with linear viscous dampers, producing good results for *IDR*, *IV* and damper forces. On the other hand, for the case of non-linear dampers, the proposed retrofit method fails to satisfy *IDR* and *IV* and needs improvement. However, the effectiveness of the proposed retrofitting method using either type of dampers is obvious as it reduces the number of plastic hinges in structural elements of the structure without dampers.

Logotheti et al. (2020) also proposed dimensioning of viscous dampers for steel frame retrofitting by using Eq. (13.36). If linear viscous dampers are selected, the damper coefficient *C* of Eq. (13.36) is denoted as C^{lin}. If nonlinear viscous dampers are selected, one can derive the damper coefficient of a nonlinear viscous damper to be $C^{nlin} = C^{lin} IV^{1-a}$. With the aid of extensive parametric studies on many steel frames under many earthquakes performed on the basis of NLTH analyses, Logotheti et al. (2020) were able to derive explicit empirical expressions for *IV* to be used directly in Eq. (13.36) for an easy computation of the damper coefficient. In particular, the maximum and minimum *IV* (in m/s) values, given as functions of the number of stories N ($1 \leq N \leq 20$), for $IDR \leq 0.7\%$ and $0.7\ \% < IDR \leq 1.5\%$, are provided by Eqs. (13.37) and (13.38), respectively, and read

$$\begin{aligned} minIV &= 0.26 \\ maxIV &= 0.152 - 0.006N + 7.23 \cdot 10^{-11} \cdot e^{H} - 0.10/N \end{aligned} \tag{13.37}$$

$$\begin{aligned} minIV &= 0.093 - 0.0007N + 1.81 \cdot 10^{-10} e^{N} - 0.027/N \\ maxIV &= 0.703 - 0.018N + 0.274/N - 0.377/N^2 \end{aligned} \tag{13.38}$$

According to Logotheti et al. (2020), the retrofit of steel MRFs with linear or nonlinear viscous dampers is performed for the minimum values of *IV* that correspond to a specific *IDR* range in order to obtain the maximum values of the damping coefficient of the dampers. Moreover, the value of the parameter k selected in Eq. (13.36) is significant because only certain combinations among k, *IV* and a lead to a good overall performance of the retrofitted steel MRF, especially when nonlinear dampers are utilized. Verification of the method proposed by Logotheti et al. (2020) has been performed by means of NLTH analyses on the basis of response results involving *IDRs*, *IVs* and maximum damper forces. Depending on the level of seismic demand, e.g., DBE or MCE, the aforementioned response results revealed that *IV* can be successfully used for the dimensioning of linear or nonlinear viscous dampers in steel MRFs. Nevertheless, the range of *IVs* where the forces induced by the dampers induce undesirable plastic hinge formation in columns should be further investigated.

13.6 Optimal Design of Steel MRFs with Dampers

A large number of studies regarding the optimization of damping devices in structures has been carried out. The optimization schemes can be divided into heuristic, analytical and evolutionary approaches as described in the review paper of De Domenico et al. (2019). The most popular (standard) of these schemes are those in which the distribution of dampers is uniform or storey-shear-proportional or stiffness-proportional. The first scheme assumes that damping coefficients are identical at every storey, whereas the next two ones assume a distribution of damping devices in proportion to design storey shears and to storey stiffness, respectively. All three schemes have been applied and evaluated by Wang and Mahin (2017) as possible damper distribution options for the seismic retrofit of an existing high-rise steel MRF using fluid viscous dampers. It should be stressed that very few studies in literature make use of realistic steel structures where a proposed damper optimization scheme is applied and evaluated. Among these few studies, one can mention the works of Attard (2007), Cimellaro and Retamales (2007), Apostolakis and Dargush (2010) and Halperin et al. (2016). In the work of Whittle et al. (2012), a comparison of five damper optimization schemes as applied to two steel MRF buildings, one regular and another irregular in elevation, is performed. The methods considered are two standard methods and three advanced methods. The first two methods are the uniform damping and the stiffness proportional damping methods. The three advanced methods are the simplified sequential search algorithm of Lopez-Garcia (2001), the damper placement for minimum transfer functions of Takewaki (1997) and the fully stressed analysis/redesign method of Levy and Lavan (2006). According to Whittle et al. (2012), all five selected schemes achieved the desired drift objective under the DBE and MCE. Moreover, absolute (floor) accelerations and residual drifts were reduced as compared to the steel MRFs without dampers.

As a final note in this section, one should have in mind that current building codes do not prescribe a particular method for the optimal placement of dampers but they only provide an effective damping ratio to be used for the dimensioning of the dampers. Nevertheless, the issue of placement of dampers is very important because the distribution of damping affects the seismic response of the structure as well as the cost associated with the testing and manufacturing of dampers to be used in the final design or retrofit study of the structure.

13.7 Numerical Examples

In this section three numerical examples are presented in some detail in order to illustrate design methods for steel structures with supplemental dampers and demonstrate their effectiveness. What should be kept in mind with respect to these methods is that the number of different dampers finally selected (regarding their type and size) should be the minimum needed in order to avoid the cost of testing different dampers for a single design or retrofit project.

13.7.1 3D Steel Building with MRFs Equipped by Linear Viscous Dampers

The six-storey steel office building of example 3.6.2 consisting of MRFs is assumed to be equipped with linear viscous dampers in a single diagonal configuration, as shown in Figs. 13.3, 13.4 and 13.5. The orientation of the columns and the position of these dampers is shown by dashed lines in Fig. 13.3, whereas their along height distribution is shown in Figs. 13.4 and 13.5 for the horizontal directions x and y, respectively. More specifically, with respect to the elevation views in Figs. 13.4 and 13.5, linear viscous dampers are placed at all stories of the middle bay of the frames along the x_1 and x_5 directions and at all stories of the first and last bays of the frames along the y_1 and y_5 directions. Each viscous damper along the x_1 and x_5 directions has a damping coefficient of 5kNs/mm, whereas along the y_1 and y_5 directions has a

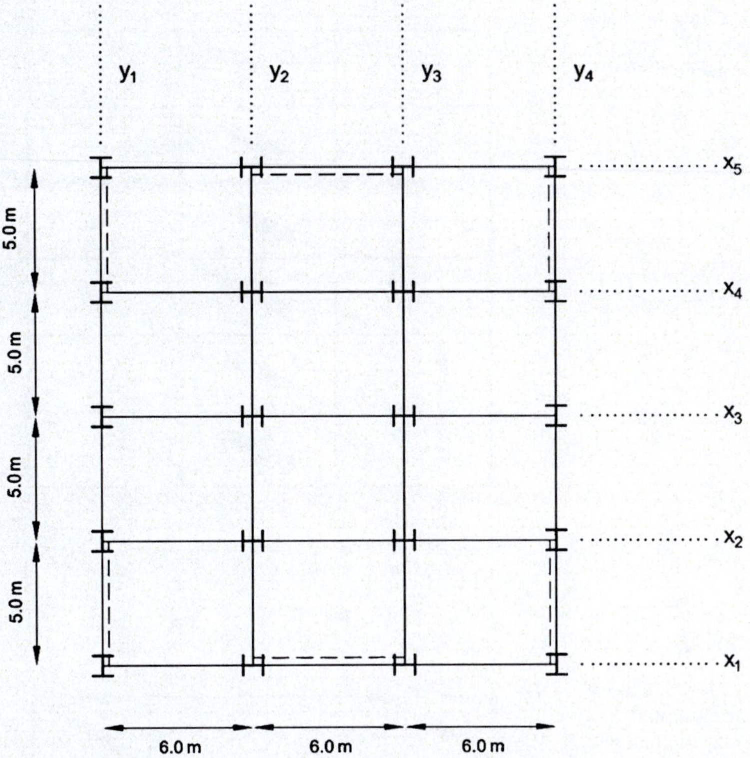

Fig. 13.3 Column orientation and position of dampers along horizontal directions *x* and *y* of six-storey steel building

damping coefficient of 2.5 kNs/mm. The total number of viscous dampers along each one of the directions x_1 and x_5 is $1 \cdot 6 = 6$ and along each one of the directions y_1 and y_5 is $2 \cdot 6 = 12$. Thus, $2 \cdot 6 + 2 \cdot 12 = 36$ viscous dampers are placed in total at the perimeter of the steel building under study. The inherent damping capacity of the structure is assumed to be $\beta_I = 0.05$.

This building with dampers is seismically designed by employing the response spectrum analysis procedure of ASCE/SEI 41-17 (2017). The beam and column sections are the same with those of the example 3.6.2, i.e., IPE400 and HEM340, respectively. Secondary IPE330 beams pinned to the main IPE400 beams are considered per 2.0 m along the *x* direction of the structure of Fig. 13.3. The grade of steel is S355 for both beams and columns. Total weight, including dead and live weight, at each floor is 4168.95 kN. The occupancy category of the building is assumed to be II, implying $I_e = 1.0$ and the MRFs are of the ordinary type. Thus, the response modification factor $R = 3.5$, the overstrength factor $\Omega_0 = 3.0$ and the deflection amplification factor $C_d = 3.0$. The building does not possess vertical or horizontal structural irregularities. It is assumed that for the site class C of the

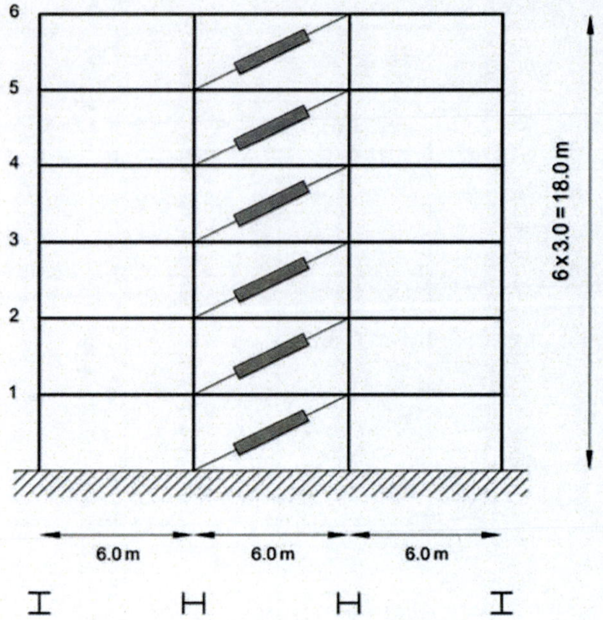

Fig. 13.4 Elevation along the *x* direction of the six-storey steel building

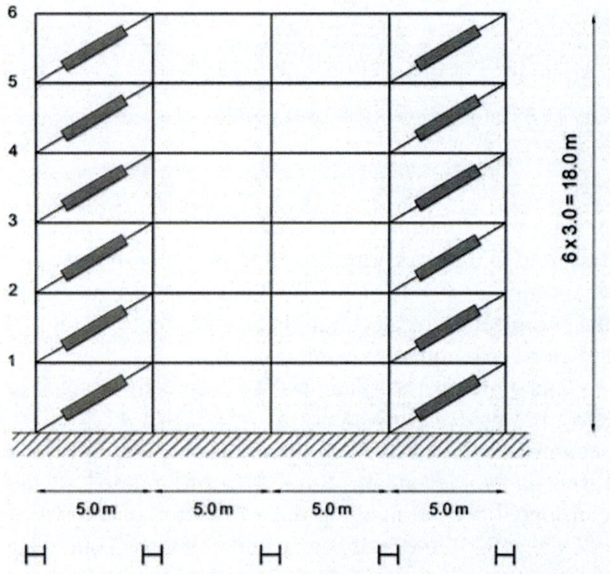

Fig. 13.5 Elevation along the *y* direction of the six-storey steel building

building, the parameters of the seismic design spectrum according to ASCE/SEI 7-16 (2017) are $S_{D1} = 0.5$, $S_{DS} = 0.8$ and $T_S = S_{D1}/S_{DS} = 0.5/0.8 = 0.625$.

A modal analysis of the steel building, performed in SAP 2000 (2020), recovers the same modal periods shown in Table 3.12, taking into account that linear viscous dampers do not add stiffness, while their mass is also small with respect to the total mass of the building. Therefore, one has the following 5 significant structural modes in a period T_i-mode vector φ_i^T-direction format from the top to the bottom storey, where i is the mode number:

$$T_1 = 1.139s - \varphi_1^T = \{1.0\ \ 0.922\ \ 0.786\quad 0.601\quad 0.377\quad 0.146\ \} - \textit{direction } x,$$

$$T_2 = 1.061s - \varphi_2^T = \{\ 1.0\quad 0.925\quad 0.793\quad 0.611\quad 0.389\quad 0.157\ \} - \textit{direction } y$$

$$T_3 = 0.579s - \varphi_{3x}^T = \{\ 1.0\quad 0.937\quad 0.818\quad 0.651\quad 0.443\quad 0.211\ \} - \textit{direction } x$$

$$T_3 = 0.579s - \varphi_{3y}^T = \{\ 1.0\quad 0.937\quad 0.822\quad 0.655\quad 0.449\quad 0.216\ \} - \textit{direction } y$$

$$T_4 = 0.360s - \varphi_4^T = \{\ 1.0\quad 0.376\quad -0.414\quad -0.962\quad -0.970\quad -0.479\ \} - \textit{direction } x$$

$$T_5 = 0.341s - \varphi_5^T = \{\ 1.0\quad 0.412\quad -0.385\quad -0.946\quad -0.977\quad -0.500\ \} - \textit{direction } y$$

Taking into account that the required number of modes to include in the response spectrum procedure of structures with damping systems is not explicitly mentioned in ASCE/SEI 7-16 (2017), it has been decided to use only the above first 5 modes, which contribute significantly to the response (Table 3.12). Moreover, since the third mode is torsional and no torsional excitation is used, the mode shapes along both x and y directions have been considered. Thus, the use of subscripts x and y for the third mode is used in the following with respect to the effective seismic weight $\overline{W}_m$, participation factor Γ_m, response modification factor C_{Sm}, modal base shear V_m, design displacement at the roof level D_{mD}, floor deflection δ_{mD} and storey velocity ∇_{mD}. The design of the steel building is performed for the DBE and for reasons of brevity the variation of the nominal design properties of the dampers is omitted.

One first calculates the effective seismic weight $\overline{W}_m$ from Eq. (13.5). Therefore, in view of the above normalized mode shapes of the first 5 modes and taking into account that the weight of each storey is $w = 4168.95$ kN, one obtains

$$\overline{W}_1 = 4168.95 \cdot \frac{(1 + 0.922 + 0.786 + 0.601 + 0.377 + 0.146)^2}{(1^2 + 0.922^2 + 0.786^2 + 0.601^2 + 0.377^2 + 0.146^2)} = 20456.90kN$$

and similarly $\overline{W}_2 = 20634.23$ kN, $\overline{W}_{3x} = 21396.79$ kN, $\overline{W}_{3y} = 21468.71$ kN, $\overline{W}_4 = 2567.99$ kN and $\overline{W}_5 = 2377.39$ kN. The corresponding values for the participation factor using Eq. (13.11) have as follows: $\Gamma_1 = (4168.95 \cdot 4.907)/$

(4168.95 · 3.832) = 1.281 and in a similar way Γ_2 = 1.277, Γ_{3x} = 1.264, Γ_{3y} = 1.262, Γ_4 = − 0.425, Γ_5 = − 0.408.

The hysteresis loop adjustment factor q_H from Eq. (13.24) is equal to 0.67 · T_S/T_1 = 0.67 · 0.625/1.139 = 0.367 for the first mode (direction *x*) and 0.67 · T_S/T_1 = 0.67 · 0.625/1.061 = 0.395 for the second mode (direction *y*). Both of these q_H values are lower than 0.5, thus q_H = 0.5. Assuming μ_D = 1.1 for the first and second modes (dominant modes in directions *x* and *y*, respectively), using Eq. (13.8), one has $T_{1D} = 1.139\sqrt{1.1} = 1.194$ s $> T_S = 0.625$ s and $T_{2D} = 1.061\sqrt{1.1} = 1.113$ s $> T_S = 0.625$ s. Therefore, considering β_I = 0.05, one determines β_{HD} from Eq. (13.23) as β_{HD} = 0.5 · (0.64 − 0.05)(1 − 1/1.1) = 0.027 for the first and second modes. With respect to higher modes 3–5, μ_D = 1.0, thus from Eq. (13.23) β_{HD} = 0. Furthermore, for the first and second modes, T_1 = 1.139s > T_S = 0.625 s and T_2 = 1.061 s > T_S = 0.625 s, thus, from Eq. (13.27) μ_{max} = 3.5/(3 · 1) = 1.17 > μ_D = 1.1.

Considering the first mode (dominant in the *x* direction), the modal drift is calculated as $\{1.0-0.922, 0.922-0.786, 0.786-0.601,\ 0.601-0.377,\ 0.377-0.146,\ 0.146\}$ and thus reads $\{0.078, 0.136, 0.185, 0.224, 0.231, 0.146\}$. The angle between the linear viscous damper and the horizontal direction is $\theta = \tan^{-1}(3/6) = 26.56°$. Then, one transforms Eq. (13.22) into (Cheng et al. 2008)

$$\beta_{Vm} = \frac{T_m}{4\pi} \frac{\sum_{i=1}^{n}\sum_{j=1}^{k} c_{di,j}\left(\varphi_{i,m} - \varphi_{i-1,m}\right)^2 cos^2\theta_{i,j}}{\sum_{i=1}^{n} m_i \varphi_{i,m}^2} \qquad (13.39)$$

where *i*, *j*, *m* are the number of storey, damper and mode, respectively, and $\varphi_{i,\,m} - \varphi_{i\,-\,1,\,m}$ is the storey drift at mode *m*. Thus, using Eq. (13.39) for the first mode, i.e., *m* = 1 (T_1 = 1.139 s) one obtains

$$\beta_{V1} = \frac{1.139}{4\pi} \cdot \frac{2\cdot 5000\cdot(0.078^2+0.136^2+0.185^2+0.224^2+0.231^2+0.146^2)\cdot\cos^2 26.56^\circ}{(4168.95/9.81)(1^2+0.922^2+0.786^2+0.601^2+0.377^2+0.146^2)} = 0.105$$

and from Eq. (13.21) $\beta_{1D} = 0.05 + 0.105\sqrt{1.1} + 0.027 = 0.187$ and by linear interpolation from Table 13.1 one has B_{1D} = 1.461. For the higher modes along the *x* direction, i.e., *m* = 3 (T_3 = 0.579 s) and *m* = 4 (T_4 = 0.360 s) one finds

$$\beta_{V3} = \frac{0.579}{4\pi} \cdot \frac{2\cdot 5000\cdot\left(0.063^2+0.119^2+0.167^2+0.208^2+0.232^2+0.211^2\right)\cdot cos^2 26.56^\circ}{(4168.95/9.81)\left(1^2+0.937^2+0.818^2+0.651^2+0.443^2+0.211^2\right)} = 0.051$$

$$\beta_{V4} = \frac{0.360}{4\pi} \cdot \frac{2 \cdot 5000 \cdot (0.624^2 + 0.790^2 + 0.548^2 + 0.008^2 + 0.491^2 + 0.479^2) \cdot \cos^2 26.56^{\circ}}{(4168.95/9.81)(1^2 + 0.376^2 + 0.414^2 + 0.960^2 + 0.970^2 + 0.479^2)} = 0.283$$

and from Eq. (13.21) $\beta_{2D} = 0.05 + 0.051\sqrt{1.0} + 0.0 = 0.101$, $\beta_{3D} = 0.05 + 0.283\sqrt{1.0} + 0.0 = 0.333$ and by linear interpolation from Table 13.1 one has $B_{2D} = 1.2$ and $B_{3D} = 1.9$.

Employing the same procedure for the second mode (dominant in the y direction), i.e., $m = 1$ ($T_2 = 1.061$s) the modal drift is $\{0.075 \quad 0.132 \quad 0.182 \quad 0.222 \quad 0.232 \quad 0.157\}$ and

$$\beta_{V2} = \frac{1.061}{4\pi} \cdot \frac{4 \cdot 2500 \cdot \left(0.075^2 + 0.132^2 + 0.182^2 + 0.222^2 + 0.232^2 + 0.157^2\right) \cdot cos^2 26.56^{\circ}}{(4168.95/9.81)\left(1^2 + 0.925^2 + 0.793^2 + 0.611^2 + 0.389^2 + 0.157^2\right)} = 0.096$$

and from Eq. (13.21) $\beta_{1D} = 0.05 + 0.096\sqrt{1.1} + 0.027 = 0.178$ and by linear interpolation from Table 13.1 one finds $B_{1D} = 1.434$. For the higher modes along the y direction, i.e., $m = 3$ ($T_3 = 0.579$ s) and $m = 5$ ($T_5 = 0.341$ s) one finds

$$\beta_{V3} = \frac{0.579}{4\pi} \cdot \frac{4 \cdot 2500 \cdot (0.063^2 + 0.115^2 + 0.167^2 + 0.206^2 + 0.233^2 + 0.216^2) \cdot \cos^2 26.56^{\circ}}{(4168.95/9.81)(1^2 + 0.937^2 + 0.822^2 + 0.655^2 + 0.449^2 + 0.216^2)} = 0.051$$

$$\beta_{V5} = \frac{0.360}{4\pi} \cdot \frac{4 \cdot 2500 \cdot (0.588^2 + 0.797^2 + 0.561^2 + 0.031^2 + 0.477^2 + 0.5^2) \cdot \cos^2 26.56^{\circ}}{(4168.95/9.81)(1^2 + 0.412^2 + 0.385^2 + 0.946^2 + 0.977^2 + 0.5^2)} = 0.261$$

and from Eq. (13.21) one has $\beta_{2D} = 0.05 + 0.051\sqrt{1.0} + 0.0 = 0.101$, $\beta_{3D} = 0.05 + 0.261\sqrt{1.0} + 0.0 = 0.311$ and by linear interpolation from Table 13.1 one obtains $B_{2D} = 1.20$ and $B_{3D} = 1.83$.

The response modification for the first and higher modes along the x direction is calculated using Eqs. (13.6) and (13.7). More specifically, since $T_{1D} = 1.194$ s $> T_S = 0.625$ s, $T_3 = 0.579$ s $< T_S = 0.625$ s and $T_4 = 0.36$ s $< T_S = 0.625$ s, one has $C_{S1} = (3.5/3)(0.5/(1.194 \cdot 3 \cdot 1.461)) = 0.111$, $C_{S3x} = (3.5/3)(0.8/(3 \cdot 1.2)) = 0.259$ and $C_{S4} = (3.5/3)(0.8/(3 \cdot 1.9)) = 0.164$. Thus, the modal base shear of the mth mode of vibration V_m along the x direction is found by using Eq. (13.4) and reads $V_1 = C_{S1}\overline{W}_1 = 0.111 \cdot 20456.90 = 2270.72$ kN, $V_{3x} = C_{S3x}\overline{W}_{3x} = 0.259 \cdot 21396.79 = 5541.77$ kN, $V_4 = C_{S4}\overline{W}_4 = 0.164 \cdot 2567.99 = 421.15$ kN.

Similarly, the response modification for the first and higher modes along the y direction is calculated using again Eqs. (13.6) and (13.7). More specifically, since $T_{2D} = 1.113$ s $> T_S = 0.625$ s, $T_3 = 0.579$ s $< T_S = 0.625$ s and $T_5 = 0.341$ s $< T_S = 0.625$ s, one has $C_{S2} = (3.5/3)(0.5/(1.113 \cdot 3 \cdot 1.434)) = 0.122$, $C_{S3y} = (3.5/3)(0.8/(3 \cdot 1.20)) = 0.259$ and $C_{S5} = (3.5/3)(0.8/(3 \cdot 1.83)) = 0.17$. Thus, the modal base shear of the mth mode of vibration V_m along the y direction is found by using Eq. (13.4) and reads $V_2 = C_{S2}\overline{W}_2 = 0.122 \cdot 20634.23 = 2517.38$ kN, $V_{3y} = C_{S3y}\overline{W}_{3y} = 0.259 \cdot 21468.71 = 5560.40$ kN, $V_5 = C_{S5}\overline{W}_5 = 0.17 \cdot 2377.39 = 404.16$ kN.

The design displacement at the roof level for the first and higher modes along the x direction is then calculated using Eqs. (13.13) and (13.14). Considering $T_{1D} = 1.194$ s $> T_S = 0.625$ s, one obtains $D_{1D} = (9.81/4 \cdot 3.14^2) \cdot (1.281 \cdot 0.5 \cdot 1.194)/1.461) = 0.130$ m, $D_{3Dx} = (9.81/(4 \cdot 3.14^2)) \cdot (1.264 \cdot 0.5 \cdot 0.579)/1.2) = 0.076$ m and $D_{4D} = (9.81/(4 \cdot 3.14^2)) \cdot (-0.425 \cdot 0.5 \cdot 0.3)/1.9) = -0.008$ m. Taking into account that for $\beta_I + \beta_{V1} = 0.05 + 0.105 = 0.155$, the parameter B_{1E} from Table 13.1 is 1.365, one has $D_{1D} = 0.130$ m $< (9.81/(4 \cdot 3.14^2)) \cdot (1.281 \cdot 0.5 \cdot 1.139)/1.365) = 0.133$ m, $D_{3Dx} = 0.076$ m $> (9.81/(4 \cdot 3.14^2)) \cdot (1.264 \cdot 0.8 \cdot 0.579^2)/1.2) = 0.070$ m, $D_{4D} = -0.008$ m $> (9.81/(4 \cdot 3.14^2)) \cdot (-0.425 \cdot 0.8 \cdot 0.36^2)/1.9) = -0.006$ m and, thus, one finally has $D_{1D} = 0.133$ m, $D_{3Dx} = 0.070$ m and $D_{4D} = -0.006$ m.

Similar considerations are employed for the design displacement at the roof level for the first and higher modes along the y direction. Therefore, using Eqs. (13.13) and (13.14) with $T_{2D} = 1.113$ s $> T_S = 0.625$ s, one obtains $D_{2D} = (9.81/(4 \cdot 3.14^2)) \cdot (1.277 \cdot 0.5 \cdot 1.113)/1.434) = 0.123$ m, $D_{3Dy} = (9.81/(4 \cdot 3.14^2)) \cdot (1.262 \cdot 0.5 \cdot 0.579)/1.2) = 0.076$ m and $D_{5D} = (9.81/(4 \cdot 3.14^2)) \cdot (-0.408 \cdot 0.5 \cdot 0.341)/1.83) = -0.009$ m. Taking into account that for $\beta_I + \beta_{V1} = 0.05 + 0.096 = 0.146$, the parameter B_{1E} from Table 13.1 is 1.338, one has $D_{2D} = 0.130$ m $< (9.81/(4 \cdot 3.14^2)) \cdot (1.277 \cdot 0.5 \cdot 1.061)/1.338) = 0.126$ m, $D_{3Dy} = 0.076$ m $> (9.81/(4 \cdot 3.14^2)) \cdot (1.262 \cdot 0.8 \cdot 0.579^2)/1.2) = 0.068$ m, $D_{5D} = -0.009$ m $> (9.81/(4 \cdot 3.14^2)) \cdot (-0.408 \cdot 0.8 \cdot 0.341^2)/1.83) = -0.005$ m and, thus, one finally has $D_{2D} = 0.126$ m, $D_{3Dy} = 0.068$ m and $D_{5D} = -0.005$ m.

The displacement at the effective yield point is obtained using Eq. (13.26) and reads $D_Y = (9.81/(4 \cdot 3.14^2)) \cdot (3 \cdot 3)/3.5) \cdot 1.281 \cdot 0.111 \cdot 1.139^2 = 0.118$ m for the x direction and $D_Y = (9.81/(4 \cdot 3.14^2)) \cdot (3 \cdot 3)/3.5) \cdot 1.277 \cdot 0.122 \cdot 1.061^2 = 0.112$ m for the y direction. Thus, the effective ductility μ_D is found using Eq. (13.25) and reads $\mu_D = 0.133/0.118 = 1.13$ and $\mu_D = 0.126/0.112 = 1.13$ for the directions x and y, respectively. These μ_D values are very close to the initially assumed value $\mu_D = 1.1$, and thus no iterations are required to estimate μ_D.

The seismic base shear V_D of the structure in a given direction is determined employing the SRSS modal combination rule and should satisfy the relation $V_D \geq V_{min}$ with V_{min} given by Eqs. (13.1) and (13.2). In particular, one has $V_D = \sqrt{2270.72^2 + 5541.77^2 + 421.15^2} = 6003.73$ kN and $V_D = \sqrt{2517.38^2 + 5560.40^2 + 404.16^2} = 6117.07$ kN for the x and y directions, respectively. The approximate fundamental period is determined by Eq. (2.45) as $T = 0.0724 \cdot 18^{0.8} = 0.73$ s. Taking into account the upper limit coefficient of 1.4, one has $1.4 \cdot 0.73 = 1.022$ s < 1.139 s and $1.4 \cdot 0.73 = 1.022$ s < 1.061 s and thus the period value of 1.022 s is used to determine C_s (ASCE/SEI 7-16 2017). Therefore, after ASCE/SEI 7-16 (2017), since $C_S = S_{DS}/(R/I_e) = 0.8/(3.5/1) = 0.229 > S_{D1}/(TR/I) = 0.5/(1.022 \cdot 3.5/1) = 0.14$, one has that $C_S = 0.14$, which results in the design base shear $V = C_S W = 0.14 \cdot 6 \cdot 4168.95 = 3501.92$ kN. Using Eq. (13.2), $V_{min} = 0.75 \cdot 3501.92 = 2626.44$ kN. For $\beta_I + \beta_{V1} = 0.05 + 0.105 = 0.155$ and $\beta_I + \beta_{V1} = 0.05 + 0.096 = 0.146$, B_{V+I} has been previously determined (using Table 13.1) and reads 1.365 and 1.338 for the x and y directions, respectively. Thus, from Eq. (13.1) one obtains $V_{min} = 3501.92/1.365 = 2565.51$ kN and $V_{min} = 3501.92/1.338 = 2617.28$ kN for the x and y directions, respectively. Considering the maximum value of V_{min} in each direction, one can easily observe that $V_D = 6003.73 > 2626.44$ kN and $V_D = 6117.07 > 2626.44$ kN in the x and y directions, respectively.

For reasons of easy reference, the conditions under which the response spectrum procedure is permitted to be used for analysis and design of structures with damping systems are: (i) in each principal direction, the damping system has at least two damping devices in each storey, configured to resist torsion; (ii) the total effective damping of the fundamental mode β_{mD} ($m = 1$) of the structure in the direction of interest is $\leq 35\%$ and (iii) the S_1 value for the site is less than 0.6. Obviously, all these conditions hold true for the building studied herein.

The floor deflection due to the fundamental and higher mode contributions at each level along the x direction is determined using Eq. (13.12) as

$$\begin{aligned}\delta_{1D} = D_{1D}\Phi_1^T &= 0.133 \cdot \{1.0 \quad 0.922 \quad 0.786 \quad 0.601 \quad 0.377 \quad 0.146\} \\ &= \{0.133m \quad 0.123m \quad 0.104m \quad 0.080m \quad 0.050m \quad 0.019m\}\end{aligned}$$

$$\begin{aligned}\delta_{3Dx} = D_{3D}\Phi_{3x}^T &= 0.070 \cdot \{1.0 \quad 0.937 \quad 0.818 \quad 0.651 \quad 0.443 \quad 0.211\} \\ &= \{0.070m \quad 0.066m \quad 0.057m \quad 0.046m \quad 0.031m \quad 0.015m\}\end{aligned}$$

$$\begin{aligned}\delta_{4D} = D_{4D}\Phi_4^T &= -0.006 \cdot \{1.0 \quad 0.376 \quad -0.414 \quad -0.962 \quad -0.970 \quad -0.479\} \\ &= \{-0.006m \quad -0.002m \quad 0.002m \quad 0.006m \quad 0.006m \quad 0.003m\}\end{aligned}$$

The storey drift due to the fundamental and higher mode contributions along the x direction is then determined as $\Delta_{1D} = \delta_{1D,\,i} - \delta_{1D,\,i-1}$, where i is the storey number. Therefore, one has

$$\Delta_{1D} = \{0.010\text{ m}\ \ 0.018\text{ m}\ \ 0.025\text{ m}\ \ \ 0.030\text{ m}\ \ \ 0.031\text{ m}\ \ \ 0.019\text{ m}\}$$

$$\Delta_{3Dx} = \{0.004\text{ m}\ \ 0.008\text{ m}\ \ 0.012\text{ m}\ \ \ 0.015\text{ m}\ \ \ 0.016\text{ m}\ \ \ 0.015\text{ m}\}$$

$$\Delta_{4D} = \{-0.004\text{ m}\ \ -0.005\text{ m}\ \ -0.004\text{ m}\ \ \ 0.0\text{ m}\ \ \ 0.003\text{ m}\ \ \ 0.003\text{ m}\}$$

The design storey drift $\Delta_{D,\,i}$ along the x direction is calculated employing the SRSS modal combination rule and reads for storey i (i=1 to 6) as

$$\Delta_{D,6} = \sqrt{0.010^2 + 0.004^2 + (-0.004)^2} = 0.0119\ \text{m}$$

$$\Delta_{D,5} = \sqrt{0.018^2 + 0.008^2 + (-0.005)^2} = 0.0205\ \text{m}$$

$$\Delta_{D,4} = \sqrt{0.025^2 + 0.012^2 + (-0.004)^2} = 0.0274\ \text{m}$$

$$\Delta_{D,3} = \sqrt{0.030^2 + 0.015^2 + 0.0^2} = 0.0332\ \text{m}$$

$$\Delta_{D,2} = \sqrt{0.031^2 + 0.016^2 + 0.003^2} = 0.0349\ \text{m}$$

$$\Delta_{D,1} = \sqrt{0.019^2 + 0.015^2 + 0.003^2} = 0.0246\ \text{m}$$

The storey velocity due to the fundamental and higher mode contributions along the x direction is then found using Eqs. (13.15) and (13.16) and read as

$$\nabla_{1D} = 2\pi\frac{\Delta_{1D}}{T_{1D}} = \frac{2\cdot 3.14}{1.194}\{0.010\ 0.018\ 0.025\ \ \ 0.030\ \ \ 0.031\ \ \ 0.019\}$$
$$= \{0.054\text{ m/s}\ \ 0.095\text{ m/s}\ \ 0.129\text{ m/s}\ \ 0.157\text{ m/s}\ \ 0.162\text{ m/s}\ \ 0.102\text{ m/s}\}$$

$$\nabla_{3Dx} = 2\pi\frac{\Delta_{3Dx}}{T_3} = \frac{2\cdot 3.14}{0.576}\{0.004\ 0.008\ 0.012\ \ \ 0.015\ \ \ 0.016\ \ \ 0.015\}$$
$$= \{0.048\text{ m/s}\ \ 0.091\text{ m/s}\ \ 0.127\text{ m/s}\ \ 0.159\text{m/s}\ \ 0.178\text{ m/s}\ \ 0.161\text{ m/s}\}$$

$$\nabla_{4D} = 2\pi\frac{\Delta_{4D}}{T_4} = \frac{2\cdot 3.14}{0.36}\{-0.004\ \ -0.005\ \ -0.004\ \ \ 0.0\ \ \ 0.003\ \ \ 0.003\}$$
$$= \{-0.065\text{ m/s}\ \ -0.082\text{ m/s}\ \ -0.057\text{ m/s}\ \ 0.0\text{ m/s}\ \ 0.051\text{ m/s}\ \ 0.050\text{ m/s}\}$$

The design storey velocity $\nabla_{D,\,i}$ along the x direction is determined employing the SRSS modal combination rule and reads at each storey i (i=1 to 6) as

$$\nabla_{D,6} = \sqrt{0.054^2 + 0.048^2 + (-0.065)^2} = 0.098\ \text{m/s}$$

$$\nabla_{D,5} = \sqrt{0.095^2 + 0.091^2 + (-0.082)^2} = 0.156 \ \text{m/s}$$

$$\nabla_{D,4} = \sqrt{0.129^2 + 0.127^2 + (-0.057)^2} = 0.190 \ \text{m/s}$$

$$\nabla_{D,3} = \sqrt{0.157^2 + 0.159^2 + 0.0^2} = 0.223 \ \text{m/s}$$

$$\nabla_{D,2} = \sqrt{0.162^2 + 0.178^2 + 0.051^2} = 0.245 \ \text{m/s}$$

$$\nabla_{D,1} = \sqrt{0.102^2 + 0.161^2 + 0.05^2} = 0.197 \ \text{m/s}$$

The lateral force F_i at each storey i (i=1 to 6) along the x direction due to the fundamental and higher mode contributions is calculated using Eq. (13.10) as

$$\begin{aligned}F_{i1} &= \frac{\Gamma_1}{W_1} V_1 w_i \varphi_1^T \\ &= \frac{1.281}{20456.90} 2270.72 \cdot 4168.95 \cdot \{1.0 \ 0.922 \ 0.786 \ \ 0.601 \ \ 0.377 \ \ 0.146\} \\ &= \{592.57 \text{ kN} \ 546.35 \text{ kN} \ 465.76 \text{ kN} \ \ 356.13 \text{ kN} \ \ 223.40 \text{ kN} \ 86.51 \text{ kN}\}\end{aligned}$$

$$\begin{aligned}F_{i3} &= \frac{\Gamma_{3x}}{W_{3x}} V_{3x} w_i \varphi_{3x}^T \\ &= \frac{1.264}{21396.79} 5541.77 \cdot 4168.95 \cdot \{1.0 \ 0.937 \ 0.818 \ \ 0.651 \ \ 0.443 \ \ 0.211\} \\ &= \{1364.97 \text{ kN} \ 1278.98 \text{ kN} \ 1116.55 \text{ kN} \ \ 888.59 \text{ kN} \ \ 604.68 \text{ kN} \ 288.01 \text{ kN}\}\end{aligned}$$

$$\begin{aligned}F_{i4} &= \frac{\Gamma_4}{W_4} V_4 w_i \varphi_4^T \\ &= \frac{-0.425}{2567.99} 421.15 \cdot 4168.95 \cdot \{1.0 \ 0.376 \ -0.414 \ \ -0.962 \ \ -0.970 \ \ -0.479\} \\ &= \{-290.65 \text{ kN} \ -109.29 \text{ kN} \ 120.33 \text{ kN} \ \ 279.61 \text{ kN} \ \ 281.93 \text{ kN} \ 139.22 \text{ kN}\}\end{aligned}$$

The design lateral force $F_{D,\,i}$ at each storey i (i=1 to 6) along the x direction is determined employing the SRSS modal combination rule and reads as

$$F_{D,6} = \sqrt{592.57^2 + 1364.97^2 + (-290.65)^2} = 1516.02 \ \text{kN}$$

$$F_{D,5} = \sqrt{546.35^2 + 1278.98^2 + (-109.29)^2} = 1394.93 \ \text{kN}$$

$$F_{D,4} = \sqrt{465.76^2 + 1116.55^2 + 120.33^2} = 1215.64 \ \text{kN}$$

$$F_{D,3} = \sqrt{356.13^2 + 888.59^2 + 279.61^2} = 997.21 \ \text{kN}$$

$$F_{D,2} = \sqrt{223.40^2 + 604.68^2 + 281.93^2} = 703.52 \text{ kN}$$

$$F_{D,1} = \sqrt{86.51^2 + 288.01^2 + 139.22^2} = 331.36 \text{ kN}$$

To find the maximum design lateral force at each storey level for the design of the dampers along the x direction, the three conditions of Eqs. (13.28)–(13.30) are checked. In particular, at the stage of maximum displacement, one employs Eq. (13.28) in which $Q_{DSD} = 0$ and the quantity under the square root is by definition the design lateral force at the level under consideration. Thus, one has

$$Q_{6,E} = \Omega_0 F_{D,6} = 3 \cdot 1516.02 = 4548.06 \text{ kN}$$

$$Q_{5,E} = \Omega_0 F_{D,5} = 3 \cdot 1394.93 = 4184.80 \text{ kN}$$

$$Q_{4,E} = \Omega_0 F_{D,4} = 3 \cdot 1215.64 = 3646.93 \text{ kN}$$

$$Q_{3,E} = \Omega_0 F_{D,3} = 3 \cdot 997.21 = 2991.63 \text{ kN}$$

$$Q_{2,E} = \Omega_0 F_{D,2} = 3 \cdot 703.52 = 2110.57 \text{ kN}$$

$$Q_{1,E} = \Omega_0 F_{D,1} = 3 \cdot 331.36 = 994.07 \text{ kN}$$

At the stage of maximum velocity, one employs Eq. (13.29), where the square root equals to the sum of $c_{di,\,j} \cdot cos^2\theta_{i,\,j} \cdot \nabla_{D,\,i}$ with i, j being the number of storey and damper, respectively. Therefore, one has

$$Q_{6,E} = \sum c_{dj} \cdot \cos^2\theta \cdot \nabla_{D,6} = 2 \cdot 5000 \cdot 0.8 \cdot 0.098 = 781.98 \text{ kN}$$

$$Q_{5,E} = \sum c_{dj} \cdot \cos^2\theta \cdot \nabla_{D,5} = 2 \cdot 5000 \cdot 0.8 \cdot 0.156 = 1242.87 \text{ kN}$$

$$Q_{4,E} = \sum c_{dj} \cdot \cos^2\theta \cdot \nabla_{D,4} = 2 \cdot 5000 \cdot 0.8 \cdot 0.190 = 1523.83 \text{ kN}$$

$$Q_{3,E} = \sum c_{dj} \cdot \cos^2\theta \cdot \nabla_{D,3} = 2 \cdot 5000 \cdot 0.8 \cdot 0.223 = 1784.44 \text{ kN}$$

$$Q_{2,E} = \sum c_{dj} \cdot \cos^2\theta \cdot \nabla_{D,2} = 2 \cdot 5000 \cdot 0.8 \cdot 0.245 = 1961.28 \text{ kN}$$

$$Q_{1,E} = \sum c_{dj} \cdot \cos^2\theta \cdot \nabla_{D,1} = 2 \cdot 5000 \cdot 0.8 \cdot 0.197 = 1577.37 \text{ kN}$$

Finally, at the stage of maximum acceleration, one makes use of Eq. (13.30) in which for the fundamental mode the total effective damping is equal to $\beta_{1D} - \beta_{HD} = 0.187 - 0.027 = 0.16$ and, thus, from Tables 13.2 and 13.3, for $\mu \geq 1.1$ and $a = 1.0$, $C_{mFD} = 1.0$ and $C_{mFV} = 0.302$. For the higher modes along the x direction, the effective modal damping is taken as equal to $\beta_{3D} = 0.101$ and $\beta_{4D} = 0.311$. Thus, for the third mode $C_{mFD} = 1.0$ and $C_{mFV} = 0.20$, whereas for the fourth mode $C_{mFD} = 0.85$ and $C_{mFV} = 0.52$. Finally, the design lateral force of the damping system at each floor is determined as

$$Q_{6,E} = \sqrt{\begin{matrix}(1.0 \cdot 3 \cdot 592.57 + 0.302 \cdot 2 \cdot 5000 \cdot 0.8 \cdot 0.054)^2 + \\ (1.0 \cdot 3 \cdot 1364.97 + 0.20 \cdot 2 \cdot 5000 \cdot 0.8 \cdot 0.048)^2 + \\ (0.85 \cdot 3 \cdot (-290.65) + 0.52 \cdot 2 \cdot 5000 \cdot 0.8 \cdot (-0.065))^2\end{matrix}} = 4771.62 \text{ kN}$$

$$Q_{5,E} = \sqrt{\begin{matrix}(1.0 \cdot 3 \cdot 546.35 + 0.302 \cdot 2 \cdot 5000 \cdot 0.8 \cdot 0.095)^2 + \\ (1.0 \cdot 3 \cdot 1278.98 + 0.20 \cdot 2 \cdot 5000 \cdot 0.8 \cdot 0.091)^2 + \\ (0.85 \cdot 3 \cdot (-109.29) + 0.52 \cdot 2 \cdot 5000 \cdot 0.8 \cdot (-0.082)^2\end{matrix}} = 4466.90 \text{ kN}$$

$$Q_{4,E} = \sqrt{\begin{matrix}(1.0 \cdot 3 \cdot 465.76 + 0.302 \cdot 2 \cdot 5000 \cdot 0.8 \cdot 0.129)^2 + \\ (1.0 \cdot 3 \cdot 1116.55 + 0.20 \cdot 2 \cdot 5000 \cdot 0.8 \cdot 0.127)^2 + \\ (0.85 \cdot 3 \cdot 120.33 + 0.52 \cdot 2 \cdot 5000 \cdot 0.8 \cdot (-0.057)^2\end{matrix}} = 3978.31 \text{ kN}$$

$$Q_{3,E} = \sqrt{\begin{matrix}(1.0 \cdot 3 \cdot 356.13 + 0.302 \cdot 2 \cdot 5000 \cdot 0.8 \cdot 0.157)^2 + \\ (1.0 \cdot 3 \cdot 888.59 + 0.20 \cdot 2 \cdot 5000 \cdot 0.8 \cdot 0.159)^2 + \\ (0.85 \cdot 3 \cdot 279.61 + 0.52 \cdot 2 \cdot 5000 \cdot 0.8 \cdot 0)^2\end{matrix}} = 3376.87 \text{ kN}$$

$$Q_{2,E} = \sqrt{\begin{matrix}(1.0 \cdot 3 \cdot 223.40 + 0.302 \cdot 2 \cdot 5000 \cdot 0.8 \cdot 0.162)^2 + \\ (1.0 \cdot 3 \cdot 604.68 + 0.20 \cdot 2 \cdot 5000 \cdot 0.8 \cdot 0.178)^2 + \\ (0.85 \cdot 3 \cdot 281.93 + 0.52 \cdot 2 \cdot 5000 \cdot 0.8 \cdot 0.051)^2\end{matrix}} = 2570.86 \text{ kN}$$

$$Q_{1,E} = \sqrt{\begin{matrix}(1.0 \cdot 3 \cdot 86.51 + 0.302 \cdot 2 \cdot 5000 \cdot 0.8 \cdot 0.102)^2 + \\ (1.0 \cdot 3 \cdot 288.01 + 0.20 \cdot 2 \cdot 5000 \cdot 0.8 \cdot 0.161)^2 + \\ (0.85 \cdot 3 \cdot 139.22 + 0.52 \cdot 2 \cdot 5000 \cdot 0.8 \cdot 0.050)^2\end{matrix}} = 1377.74 \text{ kN}$$

From the results provided by Eqs. (13.28)–(13.30), it is concluded that the design lateral force for the dampers is controlled by the stage of maximum acceleration. It should be recalled however, that the final design of the dampers should be done for the MCE, something which is not performed herein due to space limitations. Furthermore, similar calculations regarding floor deflections, storey drifts, storey velocities, design storey velocities, design lateral forces and damper design forces have to be performed for the y direction but they are not presented here simply for reasons of space limitations.

At this point, one should also stress that the analysis of a structure equipped with damping devices, either for the DBE or the MCE, should be performed employing separately the maximum and the minimum properties of these devices. What is typically anticipated is that the analysis in which the maximum properties are considered will produce the largest damping forces, whereas the analysis in which the minimum properties are used will provide the largest storey drifts and velocities as well as the lowest dissipation energy.

It should be finally noted that more rigorous expressions for the computation of maximum damping forces and inelastic (design) storey velocities have been proposed by Hatzigeorgiou and Pnevmatikos (2012) and Hatzigeorgiou and Papagiannopoulos (2012) in an effort to provide more exact expressions to those provided by Eqs. (13.28)–(13.30) and Eqs. (13.15) and (13.16), respectively. Nevertheless, the expressions presented in the aforementioned works can be used only for low-rise steel structures.

13.7.2 *Force-Based Design of a Plane Steel MRF with Viscous Dampers*

Consider a four-storey two-bay plane steel MRF as part of a steel building of square plan view (with sides 7*x*9.144 m) with two such MRFs at each perimeter side for resisting lateral forces. This MRF is seismically designed either as a conventional special moment resisting frame (SMRF) by IBC (2009), or as a MRF equipped with linear fluid viscous dampers, as shown in Fig. 13.6. This design example is taken from the work of Seo et al. (2014). For each of the two types of frames of Fig. 13.6 a tributary seismic mass of 1/4 of the total building mass is considered. The steel yield strength is taken as 345 MPa. The dead and live gravity loads with values 4.0 kN/m^2 and 2.0 kN/m^2, respectively, are used in accordance with IBC (2009) resulting in a

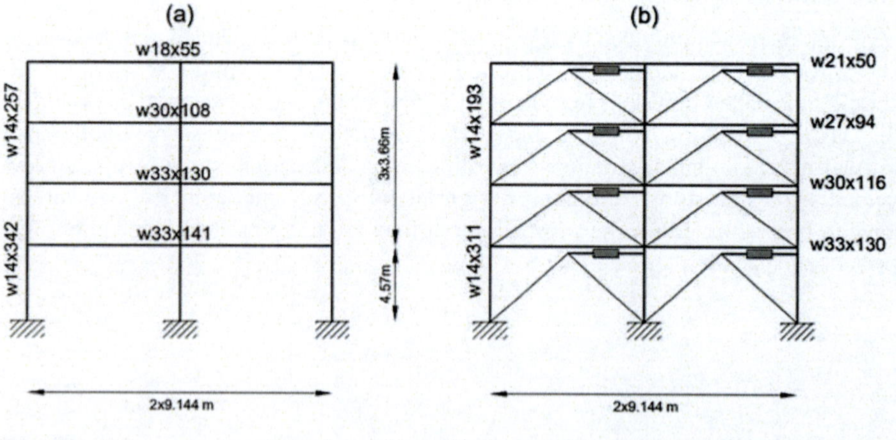

Fig. 13.6 Geometry and sections of steel frames in elevation: (**a**) SMRF; (**b**) MRF with dampers

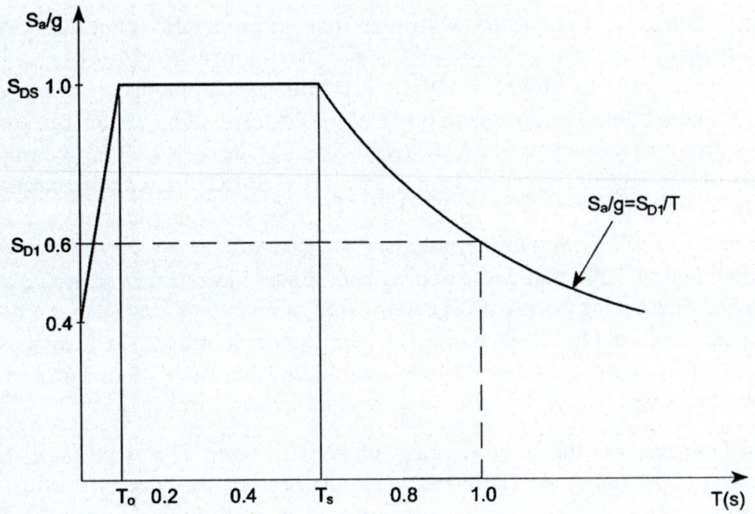

Fig. 13.7 Elastic design spectrum for the DBE in accordance with IBC (2009) for $S_{DS} = 1.0$, $S_{D1} = 0.6$, $T_0 = 0.12$ s and $T_s = 0.6$ s

gravity load for the whole building equal to 75366 kN. The elastic (with 5% damping) spectrum for the MCE, which has a 2% probability of exceedance in 50 years, is associated with values of $S_S = 1.5$ and $S_1 = 0.6$ and site coefficients $F_a = 1.0$ and $F_v = 1.5$ based on a stiff soil site. Thus, one has $S_{MS} = F_\alpha \cdot S_S = 1.0 \cdot 1.5 = 1.5$ and $S_{M1} = F_v \cdot S_1 = 1.5 \cdot 0.6 = 0.9$. Because the intensity of the DBE is 2/3 of that of the MCE, the elastic (with 5% damping) design spectrum is associated with $S_{DS} = (2/3) \cdot S_{MS} = (2/3) \cdot 1.5 = 1.0$, $S_{D1} = (2/3) \cdot S_{M1} = (2/3) \cdot 0.9 = 0.6$, $T_0 = 0.2 \cdot (S_{D1}/S_{DS}) = 0.12$ s and $T_S = S_{D1}/S_{DS} = 0.6/1.0 = 0.6$ s. Figure 13.7 shows the elastic design spectrum for the DBE used in this example.

The conventional SMRF without dampers is designed by the equivalent lateral force method of IBC (2009) using a strength reduction factor $R = 8$, a deflection amplification factor $C_d = 5.5$ and a 2% *IDR* limit. The base shear $V = C_s W$, where W is the seismic weight of 75366 kN and C_s the seismic base shear coefficient, which is given in IBC (2009) as a function of the spectral values S_{DS}, S_{D1}, the factor R, the importance factor I_e taken here as 1.0 and the first natural period T of the frame obtained initially from Eq. (2.45) as 0.65 s with an upper bound value $T = 1.4 \cdot 0.65 = 0.91$ s. Thus, $C_s = S_{DS}/(R/I_e) = 0.125$ with $C_s \leq S_{D1}/T(R/I_e) = 0.082$ result in $V = 0.082 \cdot 75366 = 6180$ kN and for one frame in $V = 6180/4 = 1545$ kN.

The design of the frames was carried out in SAP 2000 (2010), on the basis of center-line modeling in conjunction with the inclusion of a lean-on column in all modes to simulate the effect of the interior gravity columns. The inelastic maximum *IDR* of the frame is obtained by multiplying the elastic one by the coefficient $C_d = 5.5$. The drift controls the design resulting after iterations in the section selection shown in Fig. 13.6a with the first three natural periods reading as $T_1 = 1.70$ *s*, $T_2 = 0.57$ *s* and $T_3 = 0.28$ s and maximum $IDR = 1.93\ \% < 2.0\%$.

The goal is now to design a MRF with fluid viscous dampers arranged as in Fig. 13.6b with lighter sections than those of the corresponding SMRF but similar or better seismic performance. Thus, a MRF is initially designed without dampers on the basis of the same design base shear as that of the SMRF but without enforcing the 2% drift requirements. The section selection based on strength checking alone is shown in Fig. 13.6b with natural periods $T_1 = 1.87$ s, $T_2 = 0.62$ s and $T_3 = 0.30$ s. This MRF is now equipped with supplemental linear viscous dampers, as shown in Fig. 13.6b. These dampers are designed by determining their characteristic damping coefficients C of Eq. (13.23) with $a = 1$ in every storey in order to achieve maximum *IDR* values less than the 2% limit. More specifically, the whole procedure consists of the following steps:

1. Establishment of the performance objective. Here the maximum $IDR = 1.8\ \% < 2.0\%$ under the DBE was set as the performance objective.
2. Selection of an appropriate value for the ratio c of the total brace horizontal stiffness per storey to the MRF storey stiffness. This ratio should provide braces that (i) are stiff enough so that the storey drift results in damper deformation rather than brace deformation; (ii) do not fail under the maximum forces coming from the dampers; and (iii) result in only a small increase of the steel weight of the structure. Typical values of c are in between 5 and 10. Here $c = 10$ is selected following the recommendation of Fu and Kasai (1998). The storey stiffness of the MRF is calculated by dividing the sum of shear forces in each column (including lean-on column) by the interstorey drift on the basis of a static analysis of the frame under lateral loads following the inverted triangle distribution. The storey stiffnesses were found to be $K_1 = 69076$ kN/m, $K_2 = 56538$ kN/m, $K_3 = 42946$ kN/m and $K_4 = 23146$ kN/m for the first to the fourth storey, respectively.
3. Selection of appropriate damping coefficients C_i for every storey *i*. This coefficient is assumed to be here storey stiffness proportional. It should also result in a reasonable number of dampers and satisfy the conditions $\tau/T_n < 0.02$ (Lin and Chopra 2003), where T_n is the natural structural period and τ is the relaxation time equal to C/K_b with K_b being the brace stiffness already selected with the aid of the ratio c. Thus, one can select C_i from the relation $C_i = \lambda(K_i / \sum K_i)$, where λ is a factor of proportionality. Using the storey stiffness values computed in step (2), one has $C_1 = 0.36\lambda$, $C_2 = 0.29\lambda$, $C_3 = 0.22\lambda$ and $C_4 = 0.12\lambda$ and assuming initially $\lambda = 5$, $C_1 = 1.8$, $C_2 = 1.45$, $C_3 = 1.10$ and $C_4 = 0.6$ in units of kNs/mm for the first to the fourth storey, respectively.
4. Calculation of the equivalent damping ratio ξ_{eq}. This ratio is calculated with the aid of the expression (Ramirez et al. 2000)

$$\xi_{eq} = \frac{T_1}{4\pi}\left[\sum_i C_i(\varphi_i - \varphi_{i-1})^2 / \sum_i m_i \varphi_i^2\right] \tag{13.40}$$

where φ_i and φ_{i-1} are the first modal displacement of stories i and $i-1$, respectively and m_i is the mass of storey i. Thus, using the total damping ratio $\xi_t = \xi_{eq} + \xi_i$ with $\xi_i = 0.02$ for steel structures being the inherent damping ratio, one can determine the damping reduction factor B, e.g., from Table 3.1 in Ramirez et al. (2000).

5. Elastic analysis. The elastic response spectrum is reduced by the B factor and the peak inelastic storey drifts are calculated based on response spectrum analysis of the frame without dampers and on the basis of the equal displacement rule. If the storey drifts do not satisfy the already established performance criteria, steps 2 to 4 are repeated starting by selecting a new λ value and hence new C_i values for every storey i. The final values of C_i were found to be 2.11, 1.70, 1.06 and 0.70 in units of kNs/mm from the first to the fourth storey, respectively. For these values of C_i, $\xi_{eq} = 12.5\%$ and the maximum $IDR = 1.8\% < 2.0\%$.
6. Final design of dampers and capacity design rules for structural members. The final design of the dampers is based on the required damping coefficients C_i and on the expected maximum damper force and stroke values which are provided to the damper manufacturer. The peak damper force is estimated from Eq. (13.23) with $\alpha = 1$ and used for the capacity design of braces, beams and columns in the force path of the dampers.

The designed frames of Fig. 13.6a, b were assessed by NLTH analyses involving a set of 22 pairs of recorded far-fault ground motions applied along only one principal axis of the floor plan of the building, as described in Seo et al. (2014). The magnitudes of those motions recorded on stiff soil range from 6.5 to 7.6. The OpenSees (2013) program was used for modeling and analysis purposes. The modeling takes into account cyclic deterioration of beam stiffness and strength but not diagonal brace buckling due to excessive damper forces. Global frame collapse is mainly due to the P-Δ effect. The analysis shows that with respect to the *IDR*s, the MRF with dampers, even though they are designed with a reduced strength (lighter sections), has a performance similar to that of the SMRF with an *IDR* distribution more uniform over the height of the structure and a maximum *IDR* less than or equal to the one developed in the SMRF.

13.7.3 Displacement-Based Design of a Plane Steel MRF with Nonlinear Viscous Dampers

Consider a three-storey one-bay plane steel MRF with storey heights equal to 3.0 m and a bay length equal to 4.5 m, as shown in Fig. 13.8. The floor masses are $12.31 \cdot 10^3$ kg in the first two floors and $11.68 \cdot 10^3$ kg in the third floor. The

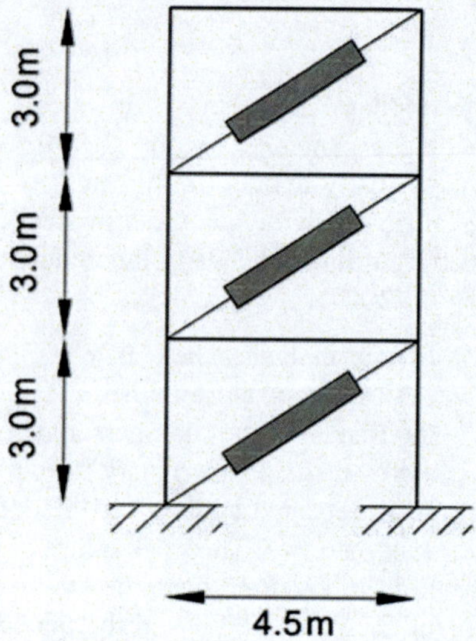

Fig. 13.8 Three-storey one-bay plane steel frame with nonlinear viscous dampers

columns and beams of the frame have sections H200 × 204 × 12 × 12 and H200 × 150 × 6 × 9, respectively. The numbers of those H sections indicate in *mm* height, width and web and flange thicknesses. The fundamental period of the frame is $T_1 = 0.93$ s, the participation factor $\Gamma_1 = 1.25$ and its first mode shape from the first to the third storey is $\varphi_{11} = 0.33$, $\varphi_{12} = 0.74$ and $\varphi_{13} = 1.00$, respectively. The above frame is retrofitted by nonlinear viscous dampers at its diagonal braces, as shown in Fig. 13.8, with a constitutive equation of the form of Eq. (13.3). Thus, the necessary parameter pairs C_j and α_j (j=1, 2, 3) of the dampers are to be determined in order the frame not to exceed a maximum $IDR = 2.78\%$ under an artificial ground motion of $PGA = 0.90$ g. Assuming an inverted triangle type of first mode shape one can determine the maximum (target) roof displacement to be $u_r = 0.00278 \cdot 9.0 = 0.25$ m.

The above example, taken from Lin et al. (2008), proceeds on the basis of the six steps of the design method described in Sect. 13.4. Thus, one has

1. The design maximum roof displacement is $u_r = 0.25$ m.
2. The location of the dampers are shown in Fig. 13.8. Initially all three nonlinear viscous dampers are assumed to have $C_j = 1500\ \mathrm{N(s/m)}^{0.5}$ and $\alpha_j = 0.50$ (j=1, 2, 3) and after iterations these values at the last iteration read

$$C_1 = 2081\ \mathrm{N(s/m)}^{0.47}, \quad \alpha_1 = 0.47$$

$$C_2 = 1500\ \mathrm{N(s/m)}^{0.50}, \quad \alpha_2 = 0.50$$

$$C_3 = 888\ \mathrm{N(s/m)}^{0.55}, \quad \alpha_3 = 0.55$$

In the following, all computations refer to the last iteration.

3. A pushover analysis of the frame is performed and produces a base shear-roof displacement curve, which after bi-linearization, enables one to calculate the ductility and the post-yield stiffness ratio as $\mu = 2.06$ and a = 0.10, respectively.
4. With known values for u_r, T_1, φ_{mi}, C_j, α_j, μ, a, $\theta_j = 33.7^{\circ}$ and assuming $\xi_i = 0.02$, one can use Eqs. (13.25)–(13.27) to compute $T_e = 1.107$ s, $\xi_m = 0.162$ and $\xi_e = 0.37$.
5. From the displacement spectrum corresponding to $PGA = 0.90$ g with high amount of damping, one can determine the roof displacement corresponding to $T_e = 1.107$ s and $\xi_e = 0.37$ equal to $u_r = 0.199$ m.
6. It is observed that $u_r/\Gamma_1 = 0.25/1.25 = 0.20$ m is almost equal to 0.199 m and this implies that iterations have to be terminated. The damper properties are the ones mentioned at the end of step 2.

The above procedure and results are verified by conducting NLTH analysis with the aid of SAP 2000 (2007). The frame properties are those before retrofitting, while the dampers are those obtained at the last iteration cycle through the proposed approach. The input motion is the artificial earthquake with $PGA = 0.90$ g. The maximum roof displacement from NLTH analysis was found to be 0.243m, i.e., very close and from the safe side to the target one of 0.25 m.

13.7.4 Retrofitting of a Steel MRF with Viscous Dampers

Consider the 12-storey 4-bay plane steel MRF with diagonal linear or nonlinear viscous dampers of Fig. 13.9. Storey heights are equal to 3.0 m, while bay lengths are equal to 4.0 m. The vertical load (dead plus 0.3 of live loads) assigned to beams is assumed to be 27.5 kN/m. The frame has been seismically designed according to EC8 (2004) for $PGA = 0.24$ g, behavior factor equal to 3 and soil type B with the aid of SAP 2000 (2010). The resulting HEB (for columns) and IPE (for beams) sections are shown in Table 13.4. In this table expressions of the form 340/360/400/360/340–450 (1–4) mean that for stories 1–4 the section of all beams is IPE450 and the HEB section for all columns, moving from left to right with respect to Fig. 13.9, is 340, 360, 400, 360, 340. Material steel grade for all sections is S275.

Retrofitting of this frame by linear viscous ($a = 1$) or nonlinear ($a = 0.6$) viscous dampers is considered on the assumption that the dampers should resist 80% of the total storey force V_{storey} ($k = 0.8$ in Eq. (13.36)) and the retrofit target value is $IDR \leq 0.7\%$ for the DBE and $0.7\ \% < IDR \leq 1.5\%$ for the MCE (Logotheti 2018; Logotheti et al. 2020). For the computation of the needed damper coefficients C at every storey, use will be made of Eq. (13.36) with the minimum values of IV in Eqs. (13.37) and (13.38) valid for the IDR target range. Minimum IV values are used because these result in maximum C values. The angle θ in Eq. (13.36) is equal to 36.87°, while the storey shears are computed from response spectrum analysis of the frame under study. The resulting values of the damper coefficients C for the target value $IDR \leq 0.7\%$ are listed in Table 13.5. The values of C shown in Table 13.5 are

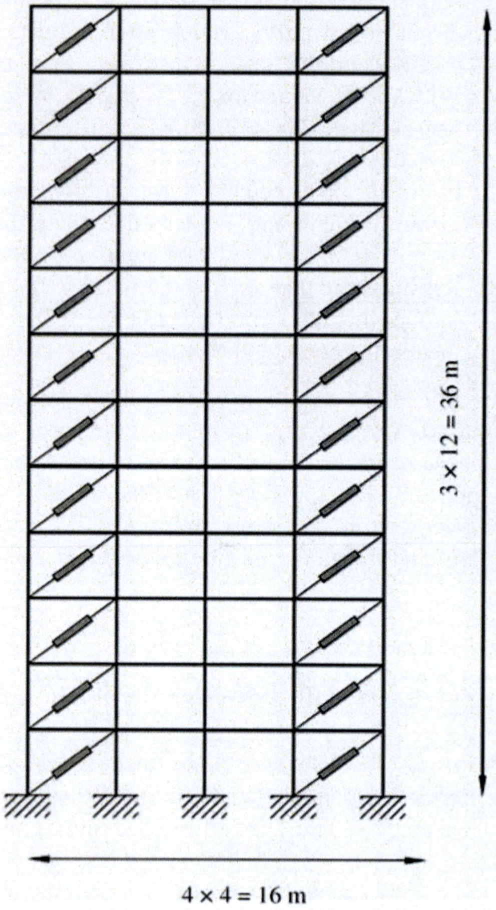

Fig. 13.9 12-storey 4-bay plane steel MRF with diagonal viscous dampers

Table 13.4 Sections of the 12-storey 4-bay frame of Fig. 13.9

Columns (HEB)—Beams (IPE)
340/360/400/360/340–450 (1–4) and 320/340/360/340/320–450 (5–8) and 300/320/340/320/300–450 (9–12)

equally distributed to the number of dampers used in each storey, thus, for the steel frame of Fig. 13.9, each value of C is divided by 2.

The steel MRFs retrofitted by two kinds of dampers ($a = 1$, $a = 0.6$) are subjected to 10 ground motions (No. 3–6, 9, 14, 17, 19–21) assumed to represent the DBE (Logotheti 2018) and seismic response results in terms of *IDR*, *IV* and forces on dampers are obtained by NLTH analyses performed with the aid of SAP 2000 (2010). Figure 13.10 provides *IDR* and *IV* values and maximum damper forces along the height of the frame for the cases of $a = 1$ and $a = 0.6$. Results are shown in the form of response values to the 10 ground motions separately and of mean values.

Table 13.5 Damper coefficients of linear (C^{lin}) and nonlinear (C^{nlin}) viscous dampers for $k = 0.8$ using Eq. (13.29)

Storey	IV_{min} (m/s)	V_{storey} (kN)	C^{lin} (kNs/m)	C^{nlin} (kNs/m)
1	0.0653	2070.6	39,635.8	13,306.2
2	0.0781	1914.1	30,634.8	11,047.8
3	0.0819	1806.6	27,572.9	10,134.4
4	0.0835	1736.1	25,989.8	9626.7
5	0.0841	1664.9	24,746.0	9192.3
6	0.0843	1581.5	23,451.1	8719.5
7	0.0842	1498.8	22,251.0	8269.4
8	0.0840	1409.1	20,969.3	7785.7
9	0.0837	1293.6	19,318.7	7162.6
10	0.0833	1147.0	17,212.0	6369.3
11	0.0829	866.7	13,067.8	4826.4
12	0.0824	454.4	6892.9	2539.7

Figure 13.11 depicts plastic hinge patterns of the frame before and after the addition of linear dampers ($a = 1$) for seismic motions No. 4 and 20 (worst case), while Fig. 13.12 shows plastic hinge patterns of the frame after the addition of nonlinear dampers ($a = 0.6$) for seismic motions No. 1 and 13.

One can observe from all the above results in Figs. 13.10, 13.11, and 13.12 that for the DBE the proposed *IV* expressions provide good estimates of the mean *IV* values of the frame retrofitted with linear ($a = 1$) or nonlinear ($a = 0.6$) viscous dampers. However, for stronger nonlinear dampers ($\alpha = 0.2$) and/or motions of higher seismic demand, Logotheti et al. (2020) have found unsatisfactory results.

Similar results regarding *IV* estimates, with respect to use of linear ($a = 1$) or nonlinear ($a = 0.6$) viscous dampers, can be reached for the MCE where $0.7\% < IDR \leq 1.5\%$ is targeted. Nevertheless, due to space limitations, these results are not presented herein but the reader may consult Logotheti et al. (2020).

13.8 Conclusions

On the basis of the preceding developments and discussion, the following conclusions can be drawn:

1. Use of supplemental dampers in steel frames is an effective way to increase dissipation of seismic energy and thus to seismically design those frames using lighter sections than those required for a design without dampers. Dampers can be also successfully used to retrofit already existing steel frames to enable them to resist ground motions of higher intensity than those for which had been designed.
2. The most widely used type of dampers are the fluid viscous dampers with damping forces depending on velocity in a linear or nonlinear manner. The most usual arrangement of these dampers is in diagonal brace type form. Damper

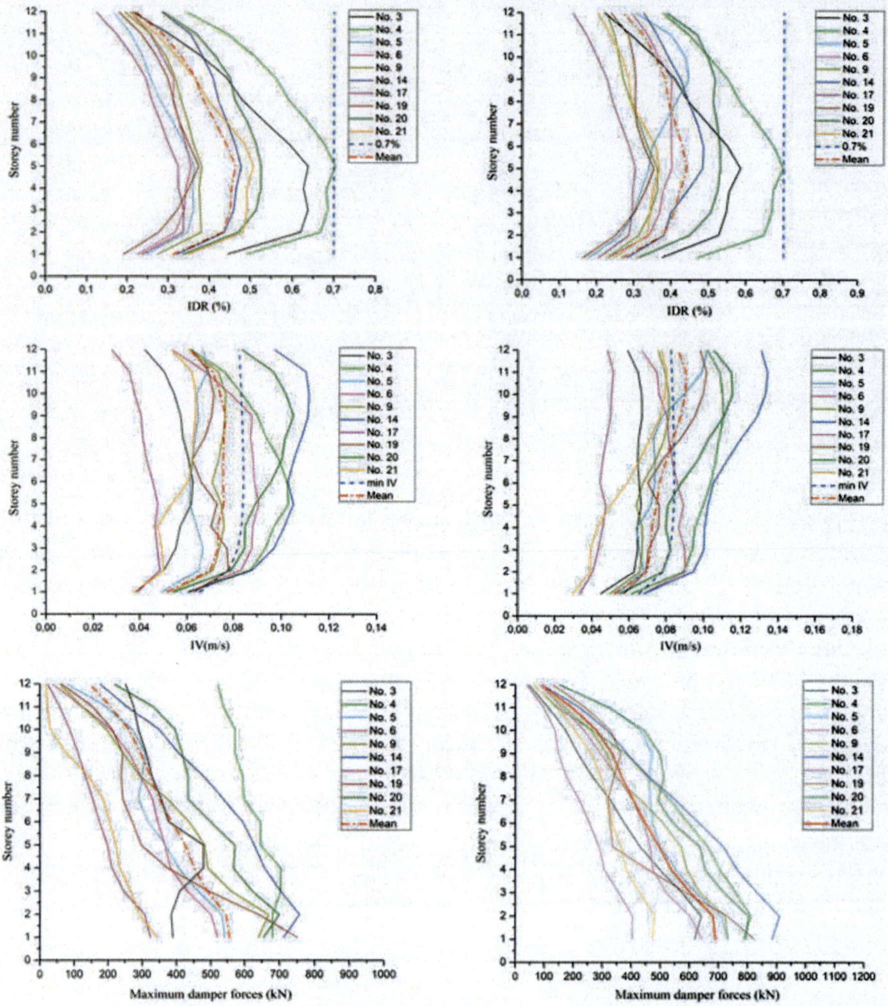

Fig. 13.10 *IDR*, *IV* and maximum damper forces along the height of the frame for $a = 1$ (left) and $a = 0.6$ (right), assuming $k = 0.8$ (after Logotheti 2018, reprinted with permission from UPCE)

arrangement can be done empirically or with the aid of special optimization algorithms.

3. The usual design methods for steel structures with dampers are the force-based simplified methods of the equivalent lateral force and the response spectrum analysis. These methods are approximate and have various limitations. ASCE code provisions present these two simplified methods as well as the method based on nonlinear time-history analyses. Some displacement-based simplified design methods have been also proposed. The suggestion is to use simplified methods for preliminary design and employ nonlinear time-history analyses for the final design.

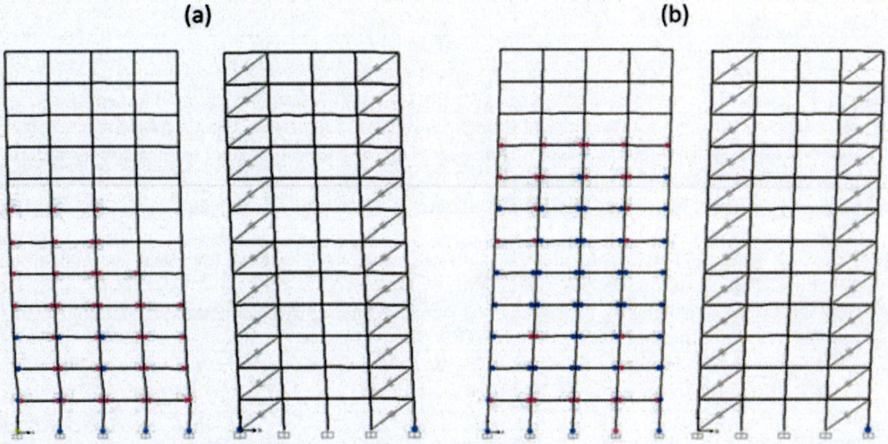

Fig. 13.11 Plastic hinge pattern of the frame before and after the addition of linear dampers ($a = 1$), assuming $k = 0.8$, for (**a**) seismic motion No. 16; (**b**) seismic motion No. 22 (after Logotheti 2018, reprinted with permission from UPCE)

Fig. 13.12 Plastic hinge pattern of the frame before and after the addition of nonlinear dampers ($a = 0.6$), assuming $k = 0.8$, for (**a**) seismic motion No. 4; (**b**) seismic motion No. 20 (after Logotheti 2018, reprinted with permission from UPCE)

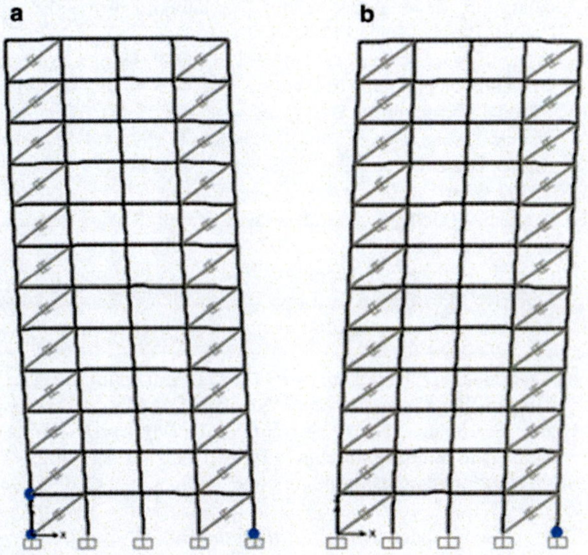

4. Seismic structural design and retrofitting by means of supplemental dampers has been proven to be successful even by using the abovementioned simplified methods. More research is needed in improving those methods with respect to their limitations and/or introducing other more effective and efficient methods, like those using interstorey velocity empirical relations or behavior (strength reduction) factors for the structure-dampers system.

References

Adachi F, Yoshitomi S, Tsuji M, Takewaki I (2013) Nonlinear optimal oil damper design in seismically controlled multi-storey buildings frame. Soil Dyn Earthq Eng 44:1–13

Apostolakis G, Dargush GF (2010) Optimal design of moment-resisting steel frames with hysteretic passive devices. Earthq Eng Struct Dyn 39:355–376

ASCE/SEI 41-17 (2017) Seismic evaluation and retrofit of existing buildings. American Society of Civil Engineers, Reston, VI

ASCE/SEI 7-16 (2017) Minimum design loads and associated criteria for buildings and other structures. American Society of Civil Engineers, Reston, VI

Attard TL (2007) Controlling all interstorey displacements in highly nonlinear steel buildings using optimal viscous damping. J Struct Eng ASCE 133:1331–1340

Cha YJ, Agrawal AK, Phillips BM, Spencer BF Jr (2014) Direct performance-based design with 200kN MR dampers using multi-objective cost effective optimization for steel MRFs. Eng Struct 71:60–72

Chang S-P, Makris N, Whittaker AS, Thompson AST (2002) Experimental and analytical studies on the performance of hybrid isolation systems. Earthq Eng Struct Dyn 31:421–443

Cheng FY, Jiang H, Lou K (2008) Smart structures: innovative systems for seismic response control. CRC Press, Boca Raton, FL

Christopoulos C, Filiatrault A (2006) Principles of passive supplemental damping and seismic isolation. IUSS Press, Pavia

Cimellaro GP, Retamales R (2007) Optimal softening and damping design for buildings. Struct Control Health Monit 14:831–857

De Domenico D, Ricciardi G, Takewaki I (2019) Design strategies of viscous dampers for seismic protection of building structures: a review. Soil Dyn Earthq Eng 118:144–165

EC8 (2004) Eurocode 8, Design of structures for earthquake resistance, Part 1: general rules, seismic actions and rules for buildings, EN 1998-1-1. European Committee for Standardization (CEN), Brussels

EN 15129 (2018) Anti-seismic devices. European Committee for Standardization (CEN), Brussels

Fu Y, Kasai K (1998) Comparative study of frames using viscoelastic and viscous dampers. J Struct Eng ASCE 124:513–522

Halperin I, Ribakov Y, Agranovich G (2016) Optimal viscous dampers gains for structures subjected to earthquakes. Struct Control Health Monit 23:458–469

Hatzigeorgiou GD, Papagiannopoulos GA (2012) Inelastic velocity ratio. Earthq Eng Struct Dyn 41:2025–2041

Hatzigeorgiou GD, Pnevmatikos NG (2012) Maximum damping forces for structures with viscous dampers under near-source earthquakes. Eng Struct 68:1–13

Hwang JS, Huang YN, Yi SL, Ho SY (2008) Design formulations for supplemental viscous dampers to building structures. J Struct Eng ASCE 134:22–31

IBC (2009) International building code. International Code Council, Washington, DC

Karavasilis TL (2016) Assessment of capacity design of columns in steel moment resisting frames with viscous dampers. Soil Dyn Earthq Eng 88:215–222

Karavasilis TL, Sause R, Ricles JM (2012) Seismic design and evaluation of steel moment-resisting frames with compressed elastomer dampers. Earthq Eng Struct Dyn 41:411–429

Kariniotakis K, Karavasilis TL (2018) Modified capacity design rule for columns in tall steel MRFs with linear viscous dampers within the framework of Eurocode 8. Bull Earthq Eng 16:917–932

Kasai K, Fu Y, Watanabe A (1998) Passive control systems for seismic damage mitigation. J Struct Eng ASCE 124:501–512

Kim J, Choi H (2006) Displacement-based design of supplemental dampers for seismic retrofit of a framed structure. J Struct Eng ASCE 132:873–883

Kitayama S, Constantinou MC (2018) Seismic performance of buildings with viscous damping systems designed by the procedures of ASCE/SEI 7-16. J Struct Eng 144:04018050

Lee KS, Fan CP, Sause R, Ricles JM (2005) Simplified design procedure for frame buildings with viscoelastic or elastomeric structural dampers. Earthq Eng Struct Dyn 34:1271–1284

Lee KS, Ricles J, Sause R (2009) Performance-based seismic design of steel MRFs with elastomeric dampers. J Struct Eng ASCE 135:489–498

Levy R, Lavan O (2006) Fully stressed design of passive controllers in framed structures for seismic loadings. Struct Multidiscip Optim 32:485–498

Lin WH, Chopra AK (2003) Earthquake response of elastic single-degree-of-freedom systems with nonlinear viscoelastic dampers. J Struct Eng ASCE 129:597–606

Lin YY, Tsai MH, Hwang JS, Chang KC (2003) Direct displacement-based design for buildings with passive energy dissipation systems. Eng Struct 25:25–37

Lin YY, Chang KC, Chen CY (2008) Direct displacement-based design for seismic retrofit of existing buildings using nonlinear viscous dampers. Bull Earthq Eng 6:535–552

Logotheti VE (2018) Seismic retrofit of steel frames with non-linear viscous dampers using interstorey velocity. M.S. Thesis, Department of Civil Engineering, University of Patras, Patras, Greece (in Greek)

Logotheti VE, Kafetzi TC, Papagiannopoulos GA, Karabalis DL (2020) On the use of interstorey velocity for the seismic retrofit of steel frames with viscous dampers. Soil Dyn Earthq Eng 129:105312

Lopez-Garcia D (2001) A simple method for the design of optimal damper configurations in MDOF structures. Earthquake Spectra 17:387–398

Lu Z, Wang Z, Zhou Y, Lu X (2018) Nonlinear dissipative devices in structural vibration control: a review. J Sound Vib 423:18–49

Makris N, Chang S-P (2000) Effect of viscous, viscoplastic and friction damping on the response of seismic isolated structures. Earthq Eng Struct Dyn 29:85–107

Mazza F, Vulcano A (2014) Equivalent viscous damping for displacement-based seismic design of hysteretic damped braces for retrofiring framed buildings. Bull Earthq Eng 12:2797–2819

McVitty W, Taylor A (2016) Structures with supplemental energy dissipation devices, pp. 16–5 to 16–42 in 2015 NEHRP recommended seismic provisions: design examples, FEMA P-1051, Federal Emergency Management Agency, Washington, DC

OpenSees (2013) Open system for earthquake engineering simulation. Pacific Earthquake Engineering Research Center, University of California at Berkeley, CA. http://opensees.berkeley.edu/

Palermo M, Silvestri S, Landi L, Gasparini G, Trombetti T (2018) A "direct five-step" procedure for the preliminary seismic design of buildings with added viscous dampers. Eng Struct 173:933–950

Papagiannopoulos GA, Beskos DE (2010) Towards a seismic design method for plane steel frames using equivalent modal damping ratios. Soil Dyn Earthq Eng 30:1106–1118

Papagiannopoulos GA, Hatzigeorgiou GD, Pnevmatikos NG (2018) A seismic retrofit method for steel frames with viscous dampers. In: Proceedings of 16th European Conference on Earthquake Engineering, 18–21 June, Thessaloniki, Greece

Providakis CP (2008) Effect of LRB isolators and supplemental viscous dampers on seismic isolated buildings under near-fault excitations. Eng Struct 30:1187–1198

Providakis CP (2009) Effect of supplemental damping on LRB and FPS seismic isolators under near-fault ground motions. Soil Dyn Earthq Eng 29:80–90

Ramirez OM, Constantinou MC, Kircher CA, Whittaker AS, Johnson MW, Gomez JD, Chrysostomou CZ (2000) Development and evaluation of simplified procedures and analysis and design of buildings with passive energy dissipation systems, Technical Report MCEER-00-0010, Revision 1. Multidisciplinary Center for Earthquake Engineering Research, State University of New York at Buffalo, New York

SAP 2000 (2007) Structural analysis program 2000, static and dynamic finite element analysis of structures, Version 11. Computers and Structures Inc, Berkeley, CA

SAP 2000 (2020) Structural analysis program 2000, integrated software for structural analysis and design, Version 22. Computers and Structures Inc., Walnut Creek, CA

SAP 2000 (2010) Structural analysis program 2000, static and dynamic finite element analysis of structures, Version 14. Computers and Structures Inc., Berkeley, CA

Seo CY, Karavasilis TL, Ricles JM, Sause R (2014) Seismic performance and probabilistic collapse resistance assessment of steel moment resisting frames with fluid viscous dampers. Earthq Eng Struct Dyn 43:2135–2154

Soong TT, Dargush GF (1997) Passive energy dissipation systems in structural engineering. Wiley, Chichester

Sullivan TJ, Lago A (2012) Towards a simplified direct DBD procedure for the seismic design of moment resisting frames with viscous dampers. Eng Struct 35:140–148

Symans MD, Charney FA, Whittaker AS, Constantinou MC, Kircher CA, Johnson MW, McNamara RJ (2008) Energy dissipation systems for seismic applications: current practice and recent developments. J Struct Eng ASCE 134:3–21

Takewaki I (1997) Optimal damper placement for minimum transfer functions. Earthq Eng Struct Dyn 26:1113–1124

Takewaki I (2009) Building control with passive dampers: optimal performance-based design for earthquakes. Wiley, Singapore

Tchamo JM, Zhou Y (2018) An alternative practical design method for structures with viscoelastic dampers. Earthq Eng Eng Vib 17:459–473

Wang S, Mahin SA (2017) Seismic retrofit of a high-rise steel moment-resisting frame using fluid viscous dampers. Struct Des Tall Spec Build 26:e1367

Wang SJ, Lee BH, Chuang WC, Chang KC (2018) Optimum dynamic characteristic control approach for building mass damper design. Earthq Eng Struct Dyn 47:872–888

Whittle JK, Williams MS, Karavasilis TL, Blakeborough A (2012) A comparison of viscous damper placement methods for improving seismic building design. J Earthq Eng 16:540–560